Joseph Sauveur and His Time: Music and Mathematicians

Danilo Capecchi · Giulia Capecchi

Joseph Sauveur and His Time: Music and Mathematicians

Decline of the Greek Paradigm of Music

Springer

Danilo Capecchi
Sapienza University of Rome
Rome, Italy

Giulia Capecchi
FachHochschule NordwestSchweiz
Schola Cantorum Basiliensis
Basel, Switzerland

ISBN 978-3-032-04706-9 ISBN 978-3-032-04707-6 (eBook)
https://doi.org/10.1007/978-3-032-04707-6

This Springer imprint is published by the registered company Springer Nature Switzerland AG
The registered company address is: Gewerbestrasse 11, 6330 Cham, Switzerland

To my wife

To my mother

Preface

The history of Western music, especially in its more theoretical aspects, was inextricably linked to the history of science and philosophy for more than two millennia before the 19th century. Since then, this view has been, and continues to be, met with skepticism, largely owing to the prevailing perception of music as a purely artistic pursuit on par with painting, literature, and sculpture. If it is true that a basic understanding of the technical concepts of acoustics and mathematics is necessary, the situation is similar in other artistic disciplines. For instance, the practice of painting requires knowledge of mathematical principles such as perspective, chemical processes related to color composition, and the optical properties of colors.

The status of music as a discipline was somewhat different from that of other arts and sciences. There was a great dichotomy in music; on the one hand it was *musica speculativa*,[1] pursued by scientists and philosophers. On the other hand, it was *musica practica* pursued by those who entertained people by producing pleasant and/or emotional sounds of various kinds (either composers or performers). Musica practica affected the Greek people probably more than any other people. However, unlike today, making music was not an isolated activity; it always accompanied public or private events, symposia included. The educated classes participated in musical performances as well, but they held speculative music in the highest esteem, considering it to be true music.

The classification of speculative music included three categories: harmonics, rhythmics and metrics. The field of metrics focused on the duration of syllables within verses, while rhythmics examined the patterns of short and long syllable groupings. The field of harmonics, on the other hand, dealt with the horizontal organization of pitches, the unique aspect of music inherent in its nature. After all, only sound has high and low pitches, while rhythmics and metrics are also required in poetry. It is harmonics that has had the greatest external influence and has interacted

[1] In the following we will use the term *speculative music* to define the scientific and philosophical part of music, instead of the term theoretical music or music theory, because by music theory today is meant the study of the organization of music, interesting for composition and analysis, activities that would once have been classified as practical.

most closely with patterns of thought characteristic of other intellectual domains. In both Greek and modern contexts, there is a common perception that harmonics represents the primary and most essential aspect of speculative music. Conversely, the disciplines of rhythmics and metrics have emerged as prominent fields of study only in the works of a limited number of authors.

Although the thesis that the history of music and the history of science are inseparable is now explicitly accepted by all historians, only a few have drawn the appropriate conclusion: that in order to study the history of music, it is necessary to use also the tools of the history of science. Many historians have a musical background but only a few a scientific one. In any case, their writings betray their respective training, which gives a biased view of the historical process. This book, authored by scholars with expertise in both science and music, is intended for scientists and historians of science who seek to reacquaint themselves with the history of music and rediscover a tradition that spans millennia. To this end, musical concepts are introduced as the text progresses, when necessary, in a simplified manner without compromising rigor. This approach has been similarly adopted for both mathematical and physical concepts. Therefore, the book can also be read with benefit by musicists and musicologists who, in the spirit of ancient speculative music, wish to delve into the more scientific aspects of their discipline.

The book focuses on the change in the paradigm of music as perceived by the elite, from the Greek conception of music as a predominantly rational enterprise to the modern conception of music as a predominantly artistic enterprise. Rather than studying the interactions between science and music, or the development of music as a scientific activity, as is usually done, we intend to examine how individual scientists, or rather—since the term scientist is of modern coinage—individual mathematicians and natural philosophers of the past addressed musical issues. In this way, the connections between mathematics, physics, and music can be better highlighted.

For reasons of space and to avoid overly fragmented subject matter, we have decided to focus on a single mathematician, Joseph Sauveur, who was active in the early 18th century and who, with his contribution to the mathematical-physical approach, was one of the main parties responsible for the decline of the Greek paradigm.

While historians universally acknowledge Sauveur's foundational contributions to the evolution of music and his instrumental role in the emergence of acoustics as an autonomous field of study, there is a paucity of comprehensive scholarship on his life and work. A notable exception is Robert H. Maxham's Ph.D. dissertation, which was completed in 1976. This dissertation is of particular interest due to its inclusion of all Sauveur's memoirs on acoustics, translated into readable English. This translation renders the texts more accessible to a broader audience than they might otherwise be, given the somewhat archaic nature of Sauveur's French. Another significant contribution was made by Richard Semmens in 1987, who edited, translated and provided commentary on an unpublished early writing by Sauveur, the *Traité de la théorie de musique*.

Joseph Sauveur, a contemporary of Newton and Leibniz, was born in 1653 and died in 1716. He became a highly esteemed mathematician and was admitted to the

Académie des sciences of Paris in 1696, commencing his studies on the science of sound, which he sought to establish on an equal footing with optics. To this end, he published four memoirs in the journal of the Académie, in 1701, 1702, 1707, and 1711. A fifth memoir, completed in 1713, was published posthumously in 1716, the year of his death.

Sauveur was situated at the conclusion of a protracted process of mathematization of the ear and sound. He was neither a musician—he is even said to have been deaf—nor a music theorist; he was a mathematician who viewed music from an external perspective. This enabled him to achieve outcomes that were inaccessible to theoretical musicians of his era, such as Rameau and Tartini, for example.

His first memoir of 1701, *Systême general des intervalles des sons, et son application à tous les systêmes et à tous les instrumens de musique* encompassed most of his discoveries. Here he claimed that the science of sound could no longer be called music. The name he proposed, perhaps not entirely new, was *acoustics*. In the memoir he presented his *Systême general*, a special tempered system, and a means of measuring the duration of notes and rests with a *chronometer*, a closest ancestor of the modern metronome. A certain space in the memoirs was devoted to the *son fixe* (or standard sound), that is, to the way to produce sounds with given pitches, a problem not easy to solve at that time. First he assumed 100 Hz as a standard sound to which all the others of his system could be related, but later adopted 256 Hz, which, being a power of two, allowed the identification of an octave by the exponent of the power of two, simplifying the calculation. In the last part of his memoir, Sauveur introduced the concept and the term *sons harmoniques*, today known in dynamics as higher harmonics, one of his most important contribution to acoustics.

We assumed that to better understand the role played by Sauveur in the abandonment of the Greek paradigm, some information about the history of music is needed; first in ancient Greece, when the paradigm originated, and second in the age of early science, when the paradigm shift began.

In ancient Greece, speculative music mainly concerned *harmonics*, understood as the science of organizing the notes that follow one another in a melody, and in this sense differs from modern harmony, which considers notes played at the same time. Harmonics originated with the goal of studying the structures of melodies played in different places and times of Greece with a prescriptive purpose. At the beginning, harmonics was thus closely linked to the practice of music, but later it was largely separated from it to become a mathematical (and philosophical) science. While arithmetic was the science of discrete quantities, geometry of magnitudes, harmonics was the science of ratios and proportions. Despite its strong mathematical imprint, it also contained themes of natural philosophy, such as the nature of sound, the harmony of the cosmos and the human soul. The importance of the various components, mathematical or philosophical, varied from author to author; writers of more mathematical extraction placed the accent on the mathematical part, those of more philosophical extraction, generally of Platonic mold, also on the more 'metaphysical' part.

The origin of harmonics as a physical-mathematical discipline is usually placed in the 6th century BCE and associated with the name of Pythagoras, although most likely we could find earlier sources. This association is frustrating since we only know the name Pythagoras for certain because most of the late references to him are anecdotal. Pythagoras, or whoever spoke for him, would have related some simple numerical ratios to consonances, that is, to pairs of notes that, when played together, are pleasing to the ear. Given the lack of written evidence from the period, the relationship between sounds and numbers is of uncertain origin; it is a fact that at some time it became precise: the numerical ratios expressing the consonances between two notes were provided by the ratios of the lengths of a vibrating taut string, the monochord, producing the two notes. The Pythagoreans thought that the ratios found with the monochord were also valid for other instruments, although their reasoning was inconsistent; for example, they assumed that hammers of different weights produced consonances when their ratios were in the same simple ratios as the length of the strings of the monochord, which is false. For them, the harmony of consonances, even if perceived by the senses, was justified by arithmetic: it was the magic ratio of certain numbers that gave harmony, regardless of the musical instrument producing them. A similar view was taken up and extended by Plato and his followers. Consonances concerned not only the sounds produced by musical instruments, but also those produced by the planets in their orbits—sounds that could not be perceived by the human ear—and also the organization of the human soul and body. This conception of harmonics will be referred to in the text as *harmony of the spheres*.

The assumption that the origin of harmonics is to be found in the Pythagorean school alone is misleading; it can only be considered correct if one considers as harmonics the discipline of that name which was known between the 4th century BCE and our time. Before then there are no written documents, but witnesses, such as Plato and Aristotle, testify to the existence of musicians who studied harmonics in a more direct contact with experience, without relying on ratios or numerology. A late representative of this tradition was Aristoxenus, who did not have the same fortune as the Pythagoreans: he was partially ignored in the Middle Ages and only revived in the fifteenth century with the rediscovery of Greek musical texts.

Apart from some fragments attributed to Philolaus and Archytas of Tarentum (4–5th century BCE), the oldest texts of harmonics that have come down to us in a fairly complete form are the *Elementa harmonica* of Aristoxenus (4th century BCE) and the *Sectio canonis* of Euclid (3rd century BCE), separated by a few generations. A hundred years later the *Harmonica* of Ptolemy, active in the 2nd century CE, appeared. The three treatises are very different from each other, but also have close similarities. Euclid, like the Pythagoreans, considered harmony to be justified by numerical ratios (but not only), Aristoxenus relied essentially on the judgment of the ear (but not only). Ptolemy took an intermediate approach. In modern language, which must be used with some caution, one could say that Euclid followed a deductive approach, Aristoxenus an inductive approach, while Ptolemy followed a hypothetical deductive approach.

Euclid and Aristoxenus, at least in their surviving writings, avoided metaphysical considerations, such as those concerning the harmonies of the spheres, while Ptolemy devoted a rather long section to these aspects. A major difference between the three lies in the different measurement of distances or intervals between pitches. Euclid and

Ptolemy moved in the Pythagorean tradition and considered intervals to be measured by ratios. Theirs was an *arithmetical approach.* Aristoxenus did not refer to ratios; his interval was defined as the geometric distance between two notes, and pitches were imagined to move continuously along a straight line. His was a *geometrical approach.*

Few written sources of speculative music have survived from the early Latin and Byzantine Middle Ages; in the Byzantine world, the Greek tradition of harmonics continued. In the Latin world, knowledge of Greek music was less documented, at least until the 6th century, when Boethius (c. 480–524 CE) entered the scene—note that the Latin word *music* replaced the Greek *harmonics*. With the university, beginning in the 12th century, speculative music became an integral part of the quadrivium, along with arithmetic, geometry, and astronomy, in the form proposed by Boethius in his *De institutione musica*; here music was divided into three parts: *musica mondana, musica humana,* and *musica instrumentalis*. Musica mondana concerns the macrocosmos, musica humana the microcosmos (soul and body of man) and musica instrumentalis the perception of sound. All were subject to the rules of harmony, which could be better understood by paying attention to musica instrumentalis, as a bridge between sense and reason. According to Boethius, there are three classes of those engaged in musical activity: those who perform on instruments (including the voice), those who compose songs, and those who judge performers and composers. Only the latter is a true musician, a *musicus* indeed.

The majority of the most prominent theorists of the Middle Ages were mathematicians (and/or astronomers). However, theorists also contributed to the development of practical music. This occurred along a distinct trajectory, separate from the traditional university setting. Their work gained significant recognition due to the influence of the Christian Church, which required the continuous production of valuable musical pieces for its rituals. This was particularly evident from the 13th century onwards, with the advent of polyphonic and instrumental music, and especially during the Renaissance period.

As a result of its growing significance, practical music became the subject of comprehensive studies and writings by scholars of the period. This contributed to a change in the perspective of speculative music. The situation is well illustrated by Guido di Arezzo (fl. 1000 CE), unanimously recognized as a leading musician. According to descriptions of Boethius, he was not a musician due to his lack of training in mathematics and philosophy. Nevertheless in his *Regule rithmice* he claimed the title *musicus* for himself and referred to performers as beasts.

During the Renaissance, the Greek paradigm of speculative music experienced a revival, due to the recovery of ancient Greek texts. The mystical philosophical perspectives on the microcosm and macrocosm, which had begun to disappear in the Middle Ages music, also resurfaced. However, speculative music was no longer an end in itself; rather, it was connected to practical music, which by this point was so rich and important that all the most prominent theorists of the era also produced texts on practical music. Despite the absence of a specific designation, these texts were, in essence, theoretical treatises in the modern sense of the term. They addressed, in particular, the rules of counterpoint. For example, one of the first important musicians of the time, Franchino Gaffurio (1451–1522), published a speculative treatise on music entitled *Theorica musice* in 1492 and another treatise entitled *Practica musice* in 1496. The latter could be seen as a continuation of the former.

While practical music underwent significant developments, speculative music remained largely aligned with the Greek harmonics. The most notable shifts occurred in the transition from the Pythagorean intonation of consonances, based on fifths, fourths, and octaves, to the just intonation, which incorporated thirds and sixths. Subsequently, the concept of temperament emerged as a crucial element for fretted instruments and keyboards.

It is only with the advent of the 16th century that we encounter a profound innovation in speculative music. During this period, mathematics had already undergone remarkable developments. These included the rediscovery of the works of the Greeks, particularly Archimedes, and of those of the Arabs. Additionally, internal developments in the abacus schools led to the birth of modern algebra. The technical development of mathematics was accompanied by a gradual change in its role in natural philosophy. This resulted in the appropriation of an increasing proportion of natural philosophy by mathematicians. The initial fields to be affected were mechanics and astronomy, followed by optics. However, this process of reappropriation also occurred in music. Concurrently, music became more intricate with the advent of counterpoint, necessitating a comprehensive examination of compositional rules.

Giovanni Battista Benedetti was among the first professional mathematicians to show an interest in speculative music, more than a millennium after the mathematician Ptolemy. In his 1585 work, *Diversarum speculationum mathematicarum and physicarum liber*, he advanced two distinct fronts of inquiry. The mathematical front concerned the division of the octave and the construction of temperaments, whereas the physical front focused on the nature of sound and consonances. Many mathematicians, beginning with Beeckman and continuing with Galileo, Stevin, Kepler, Salomon De Caus, Descartes, Mersenne, Wallis, and Newton, engaged with these two distinct arguments.

Perhaps the most important contribution of Benedetti and his followers was to spread the idea that sound is due to periodic vibrations of the air and that consonances are related to simple ratios of frequencies rather than abstract string lengths, which could at best explain the pitches of string instruments but not of other instruments. This allowed for a mechanistic explanation of consonances and removed the mystical aura associated with Pythagorean and Platonic numerological concepts.

Since the 17th century, mechanistic theories have also been formulated in astronomy, explaining the movements of the planets in terms of efficient causes. The existence of specific causal explanations that differ from one another for the various actual or presumed harmonies, challenged the very concept of a single universal harmony, or the harmony of the spheres. The only rationale that could be offered in support of this concept was associated with the existence of a demiurge who created humanity and the universe by orchestrating a multitude of independent causal mechanisms. With the advent of rationalism in the 17th century and the Enlightenment in the 18th century, explicit references to divine intervention to explain natural phenomena became increasingly scarce. As a result, the ontology that had previously guided traditional speculative music no longer found a following among musicians and scientists, with a few exceptions. A further factor that contributed to the dissatisfaction with the concept of the harmony of spheres was the discovery of music from other cultures, which in some cases was markedly different from Western music.

With the fundamental role of Sauveur in the 18th century, the process begun in the 16th century brought about a dramatic shift in the Greek paradigm. The various components of speculative music, through an inevitable process of ongoing specialization in all areas of knowledge, gave rise to a number of distinct disciplines:

1. The harmony of the spheres, which played a fundamental role in the Greek paradigm, as reported by Boethius, for example, found a place near the occult sciences.
2. The study of pitches and consonances, which in speculative music was pursued by means of ratios and referred to the monochord, gave rise to musical acoustics, which dealt with sounds in general (transmission, the concept of frequency, echo, beats) and was continued by mathematicians, physicists and physiologists.
3. The problems with the organization of notes and temperament were continued by the mathematicians, giving rise to a discipline that could be called musical mathematics. Especially since the beginning of the 20th century, after the introduction of the non-tonal music, it makes use of non-trivial mathematical tools such as group theory, set theory, graph theory, and so on.

If we maintain the distinction between theoretical and practical aspects of music, it can be argued that the role previously occupied by musica speculativa is now fulfilled by the theory of music, which also addresses issues that were previously the domain of musica practica. The theory of music is a discipline that encompasses a multitude of subjects, including the study of harmony and counterpoint, ear training and musical analysis, both for tonal music (as exemplified by Schenker) and non-tonal music (as evidenced by Schoenberg). It certainly does not include the harmony of the spheres, and this is probably the main difference between the ontologies of modern music theory and ancient speculative music. Music theory, of course, makes use of the results of musical acoustics and mathematical music, but it is hardly interested in their development.

Editorial Considerations

Illustrations, when taken from original sources, may have been redrawn to make them easier to read. However, they are as close to the original as much as possible. Symbols of formulas are those of the authors, except in a few cases that are easy to identify. Translations of texts into English from various languages, Italian, German, French, Latin, are ours, unless otherwise indicated. For the names of the various characters and the titles of the written documents, the native language is generally preferred. Exceptions are made for the ancient Greeks, for whom English is used, and some medieval people, for whom Latin is used. In particular, we have preferred the original spelling for the titles of French works, where the accents do not correspond to modern usage.

For the names of the notes, we followed the Anglo-Saxon notation (C, D, E, F, G, A, B), except for the chapter dedicated to Sauveur, where we used the Latin notation (Ut, Re, Mi, Fa, Sol, La, Si), according to French usage.

Throughout the text, we have tried to avoid modern terms and expressions as much as possible when referring to past theories. In a few clearly identifiable cases, however, we have transgressed this resolution for the sake of simplicity.

Rome, Italy
July 2025

Danilo Capecchi
Giulia Capecchi

Competing Interests The authors have no competing interests to declare that are relevant to the content of this manuscript.

Fundamental Musical Notions

Note

A note is a sound that remains unchanged over time (it is also the symbol denoting this sound [4]).[2] Each note can be viewed into two different ways, either as a physical magnitude or as a feeling of sharpness. In the first view a note is a periodic sound characterized by a constant fundamental frequency f_0 and its higher harmonics. Any sound of frequency $|f_0 \pm \varepsilon|$, with ε sufficiently small can be identified with f_0.[3] In the second view, we speak of pitch, defined by the American National Standard Institute as "that attribute of auditory sensation in terms of which sound may be ordered on a scale extending from high to low"; pitch is somewhat dependent on frequency. A scale has been proposed, called the *mel scale*, in which the intensity of the sensation of pitches is given as a function of frequencies; a more simple and natural scale is given by the logarithm of frequencies. However, pitches do not depend on frequency alone, but other factors intervene, such as intensity, duration, temporal envelope, and other factors of a psychological nature [2].

The notes allowed in Western music are a numerable infinity. The frequencies within which they can vary range from a minimum f_m to a maximum f_M.

Tone

Different views:

1. Timbre or color of a sound.
2. Note.
3. Either the sensation (pitch) or the frequency of a sine wave form. A component of a note.

---

[2] Entry: *Note*.

[3] For a formal definition of a note as an interval of pitches not very different from each other see [3], p. 20.

Octave

An octave is a part of the space of frequencies $f_M - f_m$ such that the ratio between the final and initial values of the frequencies defining it is equal to 2; it is currently assumed that there are 10 contiguous octaves. The first denoted by C_0 conventionally starts at frequency 16.35160 Hz. Subsequent octaves are denoted by C_i, with $i = 1, 2, \ldots, 9$. To give an idea of the range of frequencies we are dealing with, consider the keyboard of a piano. The octaves that it can play are those from C_1 to C_8, with frequencies ranging from 32.7 to 8351.54 Hz (there are only a few notes in C_0 and C_9). Consider that the threshold of audibility varies from about 20 to 20,000 Hz.

Within each octave, Western music, as it has developed from ancient Greece to the 20th century, assumes only seven fundamental notes, plus other notes called *altered notes*, the number of which varies according to historical period.

Scale

A scale is a set of notes ordered according to their pitches. The fundamental notes are denoted by the *seven* symbols C D E F G A B, which are derived from the notation established in the 10th century CE; alternatively, in neo-Latin speaking countries the symbols Ut (Do) Re Mi Fa Sol La Si are used. A scale is thus the sequence of the fundamental notes C, D, E, F, G, A, B and the altered notes delimited by them.

In the international system of pitch notation (IPN) [5], notes of the same octave are denoted by the same subscript, so it is C_0 D_0 E_0 F_0 G_0 A_0 B_0 for the octave beginning with C_0. Particularly interesting is the middle octave beginning with C_4: C_4 D_4 E_4 F_4 G_4 A_4 B_4, where $A_4 = 440$ Hz.

The notes are represented by dots or circles arranged on a series of parallel lines from bottom to top to give a more or less sharp image. However, there are too many notes in the interval C_0–C_9, and too many lines (70) would be needed to represent them all. The problem is solved by dividing the entire C_0–C_9 space into windows of five lines, called staffs, whose position within the entire frequency space is defined by symbols called clefs. The most important of these today are the violin or G clef, symbol 𝄞, and the bass or F clef, symbol 𝄢, as illustrated in Fig. 1.

As clear from the figure, each key by extending the staff rows above and below can represent up to a maximum of two octaves. Figure 2 reports all octaves to give the whole picture.

Interval

An interval is the (vertical) distance between (two notes or) two pitches p_1 and p_2. These distances can be given by ordinal numbers: second, third, fourth, fifth, etc., which indicate the distance between two notes in the staff, or the number of steps

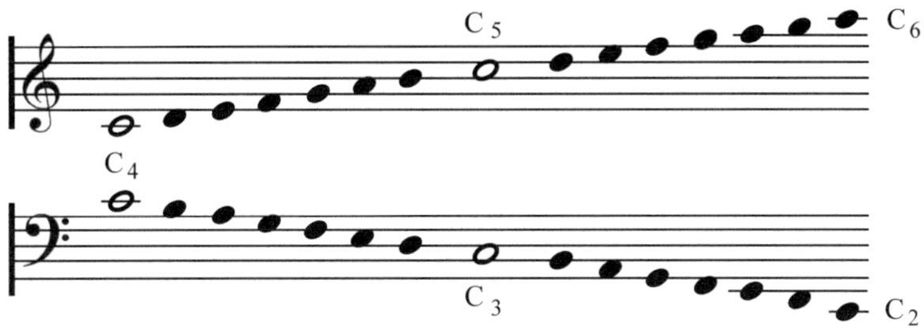

Fig. 1 Clefs of G and F

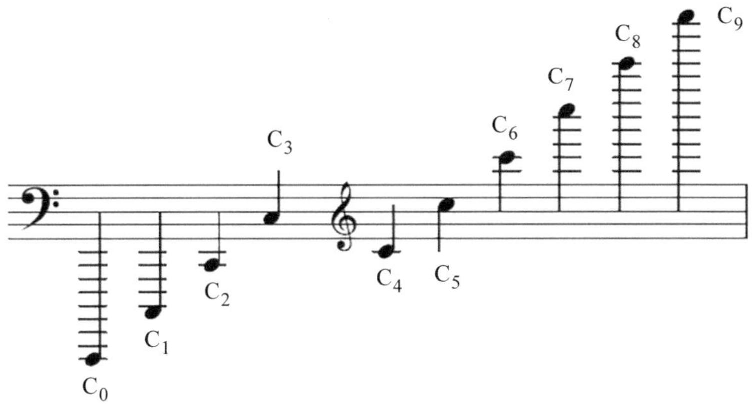

Fig. 2 The whole system of octaves

they embrace in the diatonic scale [4].[4] Two consecutive notes are separated by an interval of second (or more simply by a second), such as C and D; they form an interval of fifth (or more simply a fifth) if they are separated by five degrees in the staff, such as C and G.

The distance can also be measured basing on the numerical value of frequencies of the notes, and this a quite common interpretation in modern theory of music. Depending on the measure of pitches the intervals can be represented by a ratio $(p_2 : p_1)$ between the extreme pitches—if they are measured with frequencies—or by the difference $p_2 - p_1$—if they are measured with logarithms of frequencies.

The most common intervals are the octave, the major sixth, the minor sixth, the fifth, the fourth, the major third, the minor third, the major whole tone, the minor whole tone. They correspond to ratios of frequencies that have varied from time to time. Those that have gained the most acceptance in the recent past are (2 : 1), (5 : 3), (8 : 5), (3 : 2), (4 : 3), (5 : 4), (6 : 5), (9 : 8) and (10 : 9). Notes separated by the above intervals form what is called a just intonation scale.

[4] Entry: *Interval*. Lindley M.

Today, at least in musical acoustics, intervals are measured with logarithms. The unit of measurement is named cent (c) defined by the relation [1]:

$$c = 1200(p_2 - p_1) = 1200(\log_2 f_2 - \log_2 f_1) = 1200\log_2 \frac{f_2}{f_1}$$

This measure is particularly useful in contemporary music where the interval of an octave is divided into 12 equal parts and each part is measured by 100 c; the set of the 12 sounds makes an equally tempered scale. In this scale the intervals of octave, fifth, fourth and tone are worth 1200, 700, 500, 200 respectively, which are integer numbers. Expressing these intervals as ratios of frequencies would have produced irrational numbers.

Characteristic small intervals:

1. Tone. Major (whole) tone (9 : 8), minor (whole) tone (10 : 9).
2. Semitone. Major or diatonic semitone, (16 : 15); minor or chromatic semitone (25 : 24).
3. Syntonic comma. Difference between major and minor tone, (81 : 80).
4. Limma. A minor semitone in the Pythagorean scale, (256 : 243).
5. Pythagorean (or ditonic) comma. The falling to the closure of the circle of 12 fifths, (531, 441 : 524, 288).
6. Schisma. Pythagorean comma minus syntonic comma, (32, 805 : 32, 768).
7. Diesis. Difference between major and minor semitones, (128 : 125).

Value

The value of a note is its duration, which can be either relative or absolute.

Relative Duration

Relative duration concerns the relative duration of individual notes within a bar; the idea is well illustrated by Fig. 3.

The names of the notes with their values are as follow:

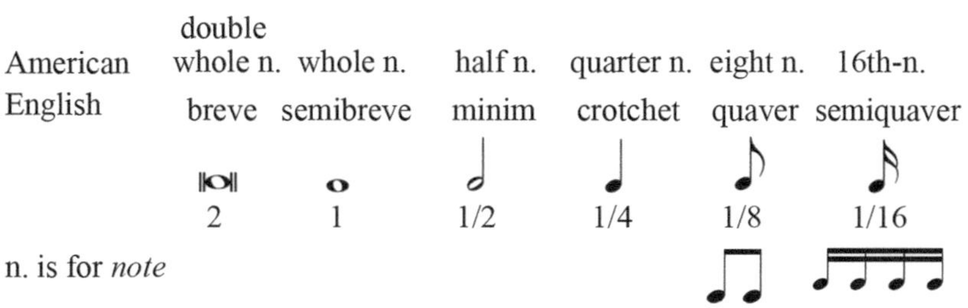

Fig. 3 Values of notes

1. Double whole note (breve). This note (breve) fell out of use as composers invented smaller notes.
2. Whole note (semibreve). The whole note is the longest musical note in common use today. It is identified an open circle without a stem.
3. Half note (minim). Its symbol is distinguished from the symbol representing the whole note by the presence of a stem..
4. Quarter note (crotchet). The quarter note has become the de facto the standard one beat music note. It is identified by a black circle with a stem.
5. Eighth note (quaver). The eighth note is worth 1/2 of a quarter note; it has a flag. Eighths can also be grouped into 3s, 4s, 5s, or even 6s, depending on the time signature.
6. Sixteenth note (semiquaver). The sixteenth note is worth 1/4 of a quarter note, it has two flags. It can be beamed in the same way as the eighth note. It differs from the eighth note in that it has an additional marker.

The unit of measure, the conventionally assigned duration 1, is represented by a white note, the semibreve. Notes whose duration is less than 1/2 of the reference duration are black.

Absolute Duration

The absolute duration is the duration of a note in seconds. It is assigned by specifying the absolute duration of a reference note; today this note is the crotchet. The duration of the crotchet is measured by the metronome, which indicates the number of crotchets contained in one minute. Thus, if the metronome indicates 120, each crochet lasts 0.5 s; thus, for example, the semibreve lasts 2 s and the semiquaver 0.125 s.

References

1. Ellis A (1885) On the musical scales of various nations. Journal of the Society and Arts 33:485–527
2. Houtsma AJ (1995) Pitch perception. In: Moore BC (ed) Hearing, Academic Press, New York, pp 267–296
3. Lindley M, Turner-Smith R (1993) Mathematical models of musical scales. A new approach. Verlag für systematische Musikwissenschaft, Bonn
4. Various (2025) New grove dictionary of music and musicians. https://doi.org/10.1093/gmo/9781561592630.article.27643, accessed 1 January 2025
5. Young RW (1939) Terminology for logarithmic frequency unit. Journal of Acoustical Society 11(1):134–139

Contents

Chapter 1
Harmonics: The Greek Paradigm of Music

Abstract Although there was a rich tradition of Greek music dating back to pre-Homeric times (9th century BCE), written records of its theorization only emerged from the 4th century BCE onwards, with the works of Aristoxenus and Euclid. The two authors presented markedly disparate conceptions of music. The former was firmly anchored in auditory experience, whereas the latter was rooted in the abstract realm of numbers, an approach that is now known as Pythagoreanism. In addition to the treatises of Euclid and Aristoxenus, numerous other ancient works have been preserved. Of particular significance is Claudius Ptolemy's *Harmonica* from the 2nd century CE, which represents a synthesis of the musical traditions prevalent during that period. The theoretical framework of Greek music serves as a foundation for modern Western music, with a similar number of notes and interval sizes. A fundamental characteristic of the ancient Greek music, which distinguishes it from modern music, is the idea that the harmonies observed in it are only the tip of the iceberg of a larger, more comprehensive harmony that encompasses the whole of creation.

1.1 Harmonics as a Mathematical Physical Science

The Greeks called harmonics the theory of combining or juxtaposing of a higher and a lower note [72].[1] Modern harmonics, today called harmony, deals with groups of sounds played simultaneously; Greek harmonics instead dealt with sounds played in sequence; i.e., it was melodic [31]. In modern terms: "The theory of harmony is that part of the logically rational and technically correct connection of chords (the simultaneous sounding of several sounds of different pitch) [...]. Melody is the logically rational and aesthetically satisfactory motion of a part through sounds of different pitch" [68].[2]

In ancient Greece, harmonics was dealt with as a physical mathematical science, and was paired with astronomy, optics, and mechanics. Astronomy was the less

---

[1] Entry: *Harmony*. Dahlhaus.

[2] p. 1.

D. Capecchi and G. Capecchi, *Joseph Sauveur and His Time: Music and Mathematicians*,
https://doi.org/10.1007/978-3-032-04707-6_1

Table 1.1 Ancient Greeks fonts that have reached us. From [52], p. 113

Aristoxenus	c 360-c 320 BCE	Harmonic elements and rhythmic elements
Euclid	4th-3rd century BCE	Division of the canon
Cleonides	2nd century common era (CE)	Introduction to harmonics
Nicomachus of Gerasa	fl. 100–150 CE	Manual of harmonics
Theon of Smyrna	Fl. 115–140 CE	On mathematics useful for the understanding of Plato
Claudius Ptolemy	fl. 127–148 CE	Harmonics
Gaudentius	3rd or 4th century CE	Harmonic introduction
Porphyrius	232/3–c. 305 CE	On Ptolemy's harmonics
Aristides Quintilianus	late 3rd–mid 4th century CE	On music
Bacchius Geron	4th century CE or later	Introduction to the art of music
Alypius	4th-5th century CE	Introduction to music

physical, as it was concerned only with motion; extra mathematical components were given by the concept of time and by principles on the organization of the motion of the planets derived from natural philosophy. Optics had as its extra mathematical components theories about the nature of light and its law of propagation. Mechanics had as its extra mathematical component the concept of weight and equilibrium. The extra mathematical principles on which harmonics was based could be of empirical or metaphysical nature, they mainly concerned the nature of sound, terrestrial or celestial, and the way in which it was perceived [24, 33].

Harmonics has remote roots, stemming from the attempt to rationalize musical production as it began to become socially important. Knowledge of ancient music compositions, though fragmentary, is relevant today, extending back to the time of Euripides (c. 480-c. 406 BCE) (for a collection and commentary on Greek musical practice see [45, 55, 64, 75]). The situation is even better for music theory, which is documented in numerous fragments and treatises. Table 1.1 regards the fundamental Greek treatises that has reached us in more or less complete and reliable form. Notice that only the first two of them are actually ancient in the usual meaning of the word. For the sake of simplicity we have used the English title of the works, while for the authors, Latin names are used also.

The recent acquisition of samples of ancient Greek music makes it possible to compare the extent to which theory and practice were correlated. Although the analysis is far from being conclusive, it can be said that theory and practice were not as independent as historians believed until recently [45, 70] and that three phases can be hypothesized for harmonics [7, 9, 15, 45]. A Pythagorean phase, mainly concerned with numerology and the relation of music to the cosmos and the soul, in which actual musical phenomena were not the focus. A phase associated with a group of theorists known as the *harmonicists*, who showed some affinities with the Pythagoreans but were also interested in actual musical phenomena. A third phase of the Aristoxenians, more concerned with the empirical aspects of music and practical performance. Of course, the story doesn't end with the Aristoxenians. Their various traditions were mixed with the Pythagorean approach, which was given a new lease of life.

As a mathematical physical discipline, harmonics was essentially cultivated by mathematicians and philosophers. But every change in theory was accompanied by changes in musical performance, and vice versa. Independently of the actual historical development, we should consider two main schools of harmonics, which we conventionally call Pythagorean and Aristoxenian. The former school was supported by music theorists such as Archytas of Tarentum, Plato and with some distinction Euclid and Ptolemy; the latter school, instead, which denied the fundamentality of arithmetic and developed an empirical though quantitative (geometric) approach, was associated mainly with Aristoxenus.

Although Greek harmonics was a mathematical physical discipline, it has received little attention from historians of science; it was, and still is, of course, the subject of historians of music, who have written extensively on the theme [11–14, 25, 30, 59, 60, 73]. Of some interest are also the accounts of the ancient Greek writers on their musical past [16].

1.2 Fundamentals of Greek Harmonics

Sounds were classified as discrete or continuous [67];[3] music and harmonics deal only with discrete sounds; a discrete sound of a given pitch was called a *note* [67].[4]

The Greeks had no reliable measure for the pitch of a note, nor even a clear idea of its nature; they could only say how much a note was either higher or lower than another; for this reason a note alone had no meaning, it was 'irrational' [67].[5] So a note should be associated with other different notes. Associations of notes are *emmelic* (melodic) if they are acceptable to the ear and consonant if they form a single blend; otherwise they are *ecmelic* (unmelodic). One of the basic principles of Greek harmonics is that two notes played one right after the other may be consonant for a given combination of their pitches, otherwise they are dissonant [10].[6] The words for consonant and dissonant notes are symphōnoi and diaphōnoi, respectively.

According to the Greeks, only four consonances, i.e. couples of consonant sounds, could be clearly heard. The first was the unison, when the two sounds produced the same pitch; the second consonance was called *diatessaron* (literally through the four), the third *diapente* (through the five), the fourth *diapason*. This even before any association of consonances with ratios was claimed. The name of the second and third consonances comes from the position of the string in the lyra—fourth or

[3] p. 15.
[4] p. 15.
[5] p. 15. Barker translates this as "has no relation" [12], p. 284.
[6] p. 117.

fifth, respectively—which, played with the first string, produces the consonances; diapason means literally through all notes.

1.2.1 Musical Intervals and Consonances

It must be said that neither school, rationalist and empiricist or Pythagorean and Aristoxenian, was monolithic and that the work of each did not develop independently of the other. A major difference between the two schools is the way they thought about pitches and intervals.

Two different pitches of musical notes are said to be separated by a (musical) *interval*. This was a difficult concept to grasp given the elusive sensory experiences of pitch. Above all interval means a distance, and the name, not only for a modern reader, points to geometry. Euclid used the same Greek word diastēma both in his geometric treatise, *Elements*, and in his musical treatise, *Sectio canonis*, equating the distance between two points to the distance between two pitches; the same term was used by Aristoxenus in his *Elementa harmonica*.

The Pythagoreans considered the interval between two pitches as a ratio between two integers; for them, harmonics was largely a theory of ratios. Here a role of composition (or summation) of intervals was introduced, as well as a rule of subtraction as an inverse operation. Imagine, for example, two intervals defined by the ratios $(a : b)$ and $(b : c)$; of these the Pythagoreans, as well as any Greek mathematician, could not define a sum, because for them ratios were not numbers and therefore could not be added with the ordinary rules of arithmetic.[7] If the Pythagoreans could not sum the two intervals $(a : b)$ and $(b : c)$, they could speak of their composition, which gives the ratio $(a : c)$. We will discuss the legitimacy of the rule of composition in another section, and also show that once the intervals are defined as ratios, there is only one way to composing them.

In the following we indicate the composition of two intervals with the sign "+", the subtraction with "−". The rules apply:

$$(a : b) + (b : c) = (a : c)$$
$$(a : c) - (a : b) = (b : c)$$

A modern reader intuitively associates ratios with fractions and identify $(a : b)$ with a/b, so the composition of the two ratios $(a : b)$ and $(b : c)$ is interpreted as:

$$\frac{a}{b} \cdot \frac{b}{c} = \frac{a}{c}$$

[7] For Greek mathematicians a ratio is a comparison of two magnitudes, or integer numbers; for instance assume that the ratio $(3 : 2)$ refers to two lengths l_1 and l_2; it simply means that joining two segments of length l_1 and three segments of length l_2, one obtains segments of equal length. A modern reader would simply say that l_1 is 1.5 times l_2.

That is the composition of two intervals is related to the product of the corresponding fractions. Similarly, subtraction corresponds to division, multiples of ratios correspond to powers of fractions, so $n(a : b) + m(b : c)$ is interpreted as $(a/b)^n \cdot (b/c)^m$.

The ratios that define the consonances diatessaron, diapente, and diapason according to Pythagoreans are given by $(4 : 3)$, $(3 : 2)$, and $(2 : 1)$, respectively. For the given definition of sum, diatessaron added to a diapente gives a diapason; indeed, $(4 : 3) + (3 : 2) = (4 : 2) \equiv (2 : 1)$. The difference between a diapente and a diatessaron is an interval with a ratio of $(9 : 8)$, called a *tone*.[8]

The Aristoxenians did not refer to ratios because they had no musical relevance for them, at least officially. They did, however, assume that pitches could be quantified from high to low, which required a geometric analogy with the points of a straight line, so that each point corresponds to a pitch. An interval, according to the Aristoxenians, was simply a segment of this line; the sum of two intervals was the union of the two segments representing them. This conception of intervals only makes sense if there is a unit of measurement. Aristoxenus solved this difficulty by considering the tone, the interval difference between diapente and diatessaron, as the unit of measurement.

Both approaches require a high degree of abstraction, but Aristoxenus' is closest to sensory experience and was probably developed first. A modern reader who associates peaks with vibrational frequencies realizes that the Pythagorean and Aristoxenian approaches are not as different as they seem. It is as if the Pythagoreans measured peaks directly with frequencies, and the Aristoxenians with logarithms of frequencies (of course, Aristoxenus did not know logarithms, and neither Euclid nor Aristoxenus had the modern concept of vibration frequency).

Consider two pitches p_1 and p_2 associated with frequencies f_1 and f_2, respectively. If the pitches are measured directly in frequencies, $p_1 = f_1$, $p_2 = f_2$; if they are measured in logarithms $p_1 = \log f_1$, $p_2 = \log f_2$, with the basis of the logarithm left undefined for the moment. In the first case, equating as moderns do, ratios and fractions, an interval i is given as $i = p_1/p_2 = f_1/f_2$; in the second case, $i = p_1 - p_2 = \log f_1 - \log f_2 = \log f_1/f_2$.

The choice to define peaks as frequency or as its logarithm is completely arbitrary and free; there is no reason to make one choice over the other; only one choice may be more or less convenient. In fact, when referring to the intensity of sound defined by air pressure, the decibel, a logarithm of pressure, is often used instead of the pressure itself. This is because the perceived intensity of sound is not proportional to the pressure of the air, but rather to its logarithm. It is therefore not surprising that the logarithm of frequency can be used to define pitches, rather than the frequency itself.

The two concepts of the measure of intervals clash to some extent. For example, the Aristoxenians, in their geometric approach, claimed that the diatessaron measured two and a half tones, the diapente three and a half tones (and the diapason six tones). For the Pythagoreans, however, six tones spanned an interval of $6(9 : 8) \equiv (531441 : 262144)$, larger than a diapason $(2 : 1)$ by a Pythagorean comma.[9] Moreover, the

---

[8] In fact, $(3 : 2) - (4 : 3)$, corresponds to $3/2 : 4/3 = 9/8$, or $(9 : 8)$.

[9] $9/8^6 : 2 = 531441/524,288 = 1.0136$.

Aristoxenians could divide the tone into equal parts, at least in principle, geometry allows it; the Pythagoreans could not, arithmetic forbids it, because dividing the ratio (9 : 8) of the tone into n equal parts requires irrational numbers. In modern terms, the n square root of (9 : 8) should be evaluated, which is not rational and thus not an interval.

1.2.2 The Tetrachord

The basis of Greek music is the tetrachord. The name comes from the stringed instruments, such as the lyra, which in a certain historical period had four (tetra) strings (chorda).[10]

The tetrachord of the scholars of harmonics, however, is immaterial; it consists of a set of four notes of different heights. The lowest and highest notes are separated by an interval that is always equal to a diatessaron. The notes are indicated by the names of the strings in a lyra: hypathē, the string furthest from the singer and the lowest note; parhypathē, the string closest to hypathē; lichanos, the string touched by the index finger, and mesē, the middle string, in a system of two tetrachords. The height of the pitches increases from hypathē to mesē. The intervals between the intermediate notes and the fixed notes can vary and define the *genus* of the tetrachord.

There were three genera, categorized at least since Archytas of Tarentum: diatonic, chromatic and enharmonic. The diatonic genus is distinguished from the other two by having two large intervals of comparable size and one small interval; the other genera have two small intervals and one larger interval. To simplify, we could say that the diatonic genus possesses two (macro) tones and one micro tone while the enharmonic and chromatic possess two micro tones and one macro tone. The micro tones are always arranged close to the lowest note; in the case of the enharmonic and chromatic systems, the two micro tones together form an interval of overall amplitude smaller than the macro tone and constitute what is called a *pyknon*. The situation is illustrated in Fig. 1.1.

Tetrachords were joined to form a systēma; more generally, this term refers to the union of contiguous musical intervals which is the Greek counterpart of the modern musical scale. A basic systēma consists of two tetrachords. They may be in disjunction when separated by a tone, as shown in Fig. 1.2a; or in conjunction when not separated, Fig. 1.2b. A systēma of two disjoint tetrachords forms a diapason and has seven intervals, as in the octave of a modern musical scale.

A more complex systēma is the *greater perfect system* (teleion systēma) shown in Fig. 1.2c, which is made up of four tetrachords, both in conjunction and in disjunction, with a note at the bottom that is separated by a tone known as the *proslambanomenos*.

---

[10] The Greek word for lyra, *lura*, was used in two ways in classical times; as a general term for any instrument of the lyre family (i.e. barbito, kitara etc.) and as a common name for one of the two kinds of lyre, made from the shell of the tortoise, namely the chelys lyra [72], Entry: *Lyra*. Maas M.

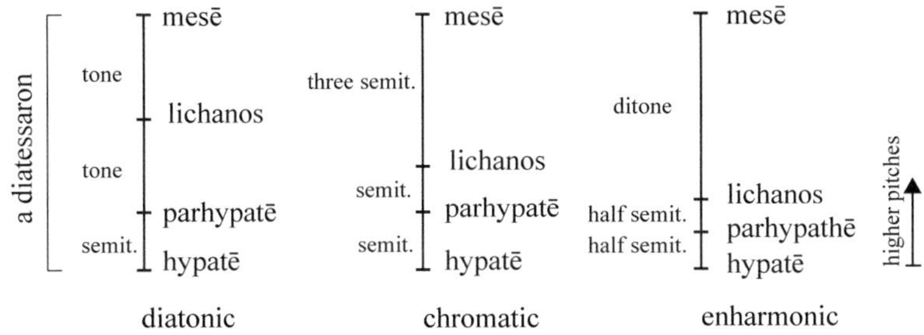

Fig. 1.1 Tetrachord genera

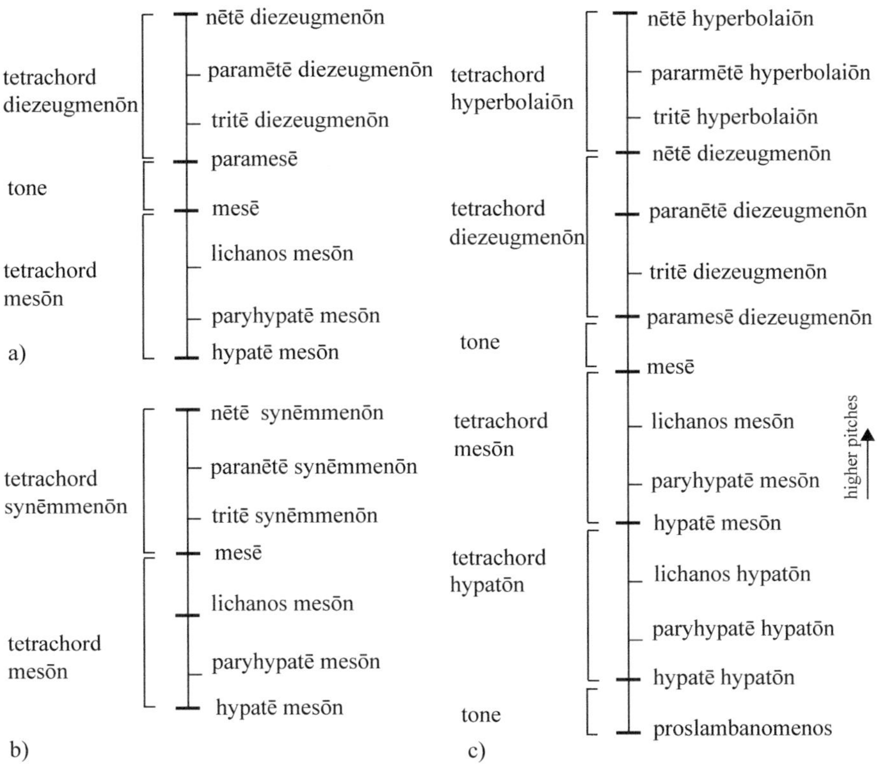

Fig. 1.2 Various systēmata: The central diapason **a**, the synēmmenōn tetrachord **b** plus the greater perfect system **c**. Redrawn from [15], pp. 13, 14

Each tetrachord is named according to its position: from the bottom, i.e., from the tetrachord with the lowest pitches, they are: hypatōn, mesōn, diegzeumenōn, and hyperbolaiōn. The four notes of each tetrachord have names preceded by the name of the tetrachord to which they belong. Another interesting systēma is the *lesser perfect*

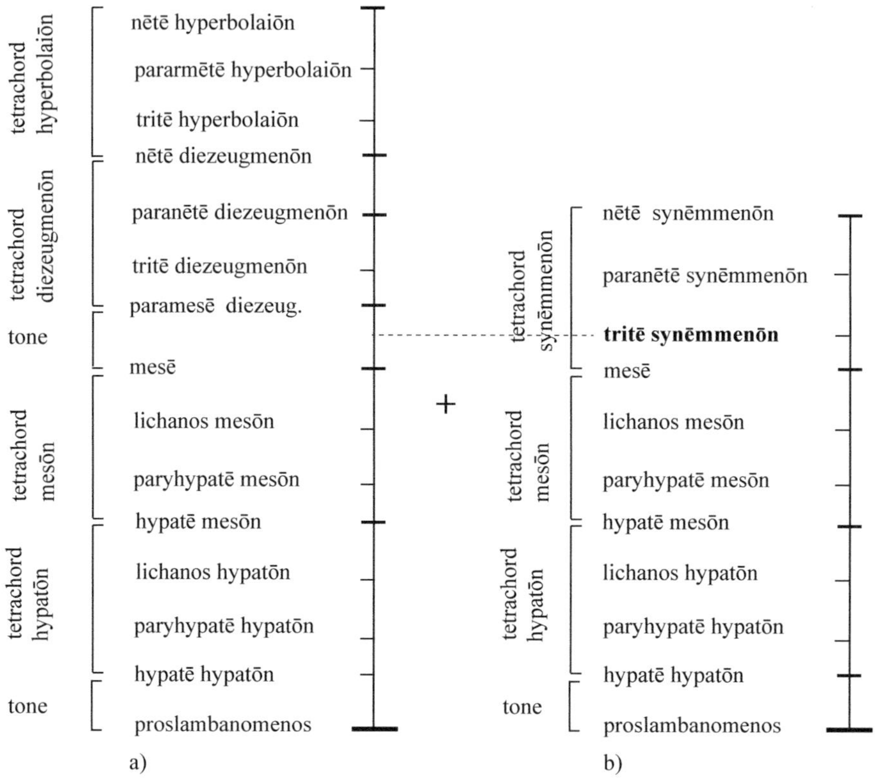

Fig. 1.3 The unchanging system: greater perfect system **a** and lesser perfect system **b**

system, which is made up of three joined tetrachords, hypatōn, mesōn, synēmmenōn, separated by a tone from the *proslambanomenos*.

In any regular systēma, all tetrachords have the same pattern of intervals; this means that each note of a tetrachord forms a diatessaron with the corresponding note of the next tetrachord if they are in conjunction, or a diapente if they are in disjunction.

Figure 1.3 shows the most complete set of tetrachords. The greater perfect system is connected to the *lesser perfect system* to form the *unchanging system*. As can be seen from the figure, the main effect of the conjunction is to add the note tritē synēmmenōn to the greater perfect system, a semitone away from mesē.

A modern reader, accustomed to an organization of notes by octaves, is tempted to see the same organization in Greek systems, particularly in the greater perfect system. Indeed, if one considers only the values of the intervals and not their function, such a system can be seen as consisting of two octaves, and a Greek would see it as the repetition of two diapason. However, this does not imply that there is the same relationship of equivalence between the two diapason as there is in modern music, where two notes separated by an octave are referred to by the same name and considered somehow equivalent (octave equivalence), whereas in the greater perfect system of the Greeks the notes separated by a diapason have different names, indicating that such equivalence does not apply; it is the notes belonging to different tetrachords but similarly situated in them that have the same name and are somehow equivalent.

1.3 The Oldest Surviving Treatises on Harmonics

In the following we will present two typical approaches of the two schools, the Aristoxenian and the Pythagorean, from two of the oldest treatises of musical theory. One of these is the Aristoxenian *Elementa harmonica*, the other the Euclidean *Sectio canonis*.

1.3.1 Aristoxenus' Elementa Harmonica

Aristoxenus (c. 360-c. 320) BCE, if not a professional musician himself, was the son of a musician and knew practical music very well. After an education as a Pythagorean in Tarentum (witnessed in some fragments of a his philosophical writing, known as *Pythagorean precepts* [46]), he came to Athens to become an important student of Aristotle and even aspired to succeed him in Liceum, but without success. He was (one of) the most important figures in the history of Greek music theory [35]. A prolific writer in philosophy, though little interested in metaphysics, he wrote several treatises on music, including the *Elementa harmonica* and *Elementa rhythmica*. The *Elementa harmonica* has come down to us relatively intact. At least two editions exist in English [3, 5]. A large commentary [47] and a good Italian edition [4] are also available. Different is the situation of the *Elementa rhythmica* of which only some fragments are available [51].

The *Elementa harmonica* is the most complete known ancient work on music and a repository of ancient thought and achievement. The extant text is divided into three books: the first two are partially overlapping and report the main elements of theory and epistemology and concern the way systēmata can be correctly constructed. The third book contains the demonstration of some propositions of a musical nature.

Aristoxenus was influenced by Aristotle's epistemology; however, in his writings on harmonics he adopted in practice the more pragmatic epistemology of mathematical physics, which can accept principles as true as a matter of fact, rather than as logically necessary or as part of a coherent system of natural philosophy. For him, the principles of harmonics should be derived directly from harmonics itself, and not from other sciences such as arithmetic.

One important reason why Aristoxenus believed that arithmetic could not exhaust the study of harmonics was that the musical value of a set of notes is given not only by the measure of their intervals, but also by their dynamis in a melody. For example, the intermediate notes of a tetrachord of the same genus could be described by different (infinite) mathematical ratios and thus be considered as different notes. However, if the pitch variation is not very large, the ear judges the notes as belonging to the same genus, and this cannot be explained by arithmetic [5, 11].[11]

The first two books of the *Elementa harmonica* clarified what harmonics is; meanwhile, it does not coincide with music but only affects the organization of musical intervals:

> Harmonics is only a part of the musician's accomplishment, as are the sciences of rhythm, metre and instruments. We must therefore give an account of harmonics and its parts. It is to be understood, speaking generally, as the science which deals with all melody, and enquires how the voice naturally places intervals as it is tensed and relaxed. For we assert that the voice has a natural way of moving, and does not place intervals haphazardly [5].[12]

Harmonics depends on hearing and reason. And hearing, plays a major role.

> The geometer makes no use of the faculty of perception: he does not train his eyesight to assess the straight or the circular or anything else of that kind either well or badly: it is rather the carpenter, the wood-turner, and some of the other crafts that concern themselves with this. But for the student of music accuracy of perception stands just about first in order of importance, since if he perceives badly it is impossible for him to give a good account of the things which he does not perceive at all [5].[13]

Aristoxenus also gave great prominence to memory, for the understanding of music comes from two things, perception and memory [5].[14] He divided harmonics into six parts:

1. Nature of sound and notes.
2. Distinction between intervals.
3. Distinction between genera of tetrachords.
4. Description of tonoi.
5. Modulation.
6. Melodic compositions.

Sound is presented in Book I of the *Elementa harmonica* as a process in which pitches vary continuously from high to low and vice versa, like a point on a line. A note is essentially the stabilization of sound at a single pitch [12].[15] There are physical limits to pitches, both for generation and for the ear. However, Aristoxenus was referring to intervals rather than the absolute values of the pitch of a note.

The geometric image of pitches moving on a straight line only makes sense if one can imagine them as magnitudes. One possibility (though for sure not an idea of Aristoxenus) could be offered by the acoustic theories of the time, for which a measure of a pitch could be given by the speed of the air waves carrying that pitch: the higher the speed, the higher the pitch.

---

[11] p. 12; pp.161–163.

[12] p. 149.

[13] pp. 150–151.

[14] p. 155.

[15] p. 136.

The interval is what is bounded by two notes of different pitch, analogous to the interval of a straight line defined by the two end points. Intervals are classified according to:

1. Their measure.
2. Their consonance or dissonance.
3. Their aggregation.
4. The genus of the tetrachord to which they belong.

Although Aristoxenus said that there are eight kinds of consonance, he referred to only three of them: the diatessaron, the diapente, and the sum of a diatessaron and a diapente, i.e., the diapason. Other consonant intervals are obtained by adding a diapason, in fact experience says that for any consonance the result of adding a diapason is a consonance [5].[16]

The geometric image of intervals allows easy mathematical operations. For example, one can subtract a diatessaron from a diapente; the result is the *tone* which can be used for metrics. The tone can be divided into two equal parts (half-tone), three parts, four parts. Aristoxenus also accepted that, at least from a purely abstract point of view, there is no limit to the division, as occurs in a geometric line, although he qualified as melodic intervals not less than a quarter of a tone, the limit for the ear to distinguish them clearly. Aristoxenus also mentioned intervals as small as a twelfth of a tone, however, there is no procedure for measuring intervals smaller than a half-tone in the *Elementa harmonica*. The division proposed by Aristoxenus is criticized in [5][17] as not being supported by any empirical evidence; we do not share this criticism, since in any mathematical-physical discipline aspects are proposed that cannot be verified experimentally, but this does not prevent the discipline from being classified as empirical.

With his unit of measure, Aristoxenus associated numbers with consonances: the diatessaron is two and a half tones, and the diapente is three and a half tones; the fact that the diapason is six tones is a natural consequence, but one that Aristoxenus did not draw.

Although mathematics had no ontological or epistemological role for Aristoxenus in judging intervals and consonances, he made extensive use of geometry, which makes his harmonics a mathematical physical discipline. Since the discovery of immeasurable quantities in arithmetic by the Pythagoreans, from the 4th century BCE on, mathematicians concentrated their efforts on geometry; the mathematician Aristoxenus moved along this new path. Having introduced a metric, the tone, he was able to measure musical intervals and prove theorems with the aid of geometry.

Aristoxenus only proposed methods for measuring intervals that are multiples of a tone or half a tone. For example, to measure the interval defined by a note E below or above a given note A (see Fig. 1.4), one can move alternately by a diapente and a diatessaron, advancing or retreating by one tone at each step. The following quotation refers to the case when the interval EA is one ditone (two tones) below A:

---

[16] p. 160.
[17] p. 16.

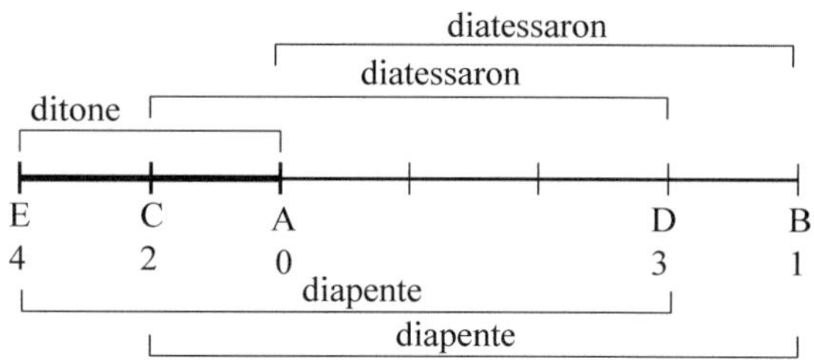

Fig. 1.4 Measure/construction of a ditone interval

> Thus, if we have the task of constructing from a given note a discord such as a ditone (or any of those which can be constructed by means of consonances) we should construct from the given note [A] a diatessaron upwards [B], from there a diapente downwards [C], then another diatessaron upwards [D], and then another diapente downwards [E]. In this way the ditone [AE] downwards from the given note will have been constructed [5].[18]

Using a variant of this approach Aristoxenus was able to measure the interval of a semitone. The procedure is described below, where it is proved that the diatessaron consists of two and half tones:

> Take a diatessaron, and starting from each of its extremes in turn, mark off a ditone by means of consonances. It is clear that the remainders are equal, since equals have been taken from equals. Next, take a diatessaron upwards from the lower bounding note of the upper ditone, and take another diatessaron downwards from the upper bounding note of the lower ditone. It is clear that next to each of the notes which bound the resulting *systēma* there will be two consecutive remainders which must be equal, for the reasons stated before. When this construction has been set up, we must bring to the judgement of perception the outermost of the notes that have been located. If they appear to perception as discordant, it will be evident that the diatessaron is not two and a half tones, but if they sound the consonance of a diapente, it will be evident that the diatessaron is two and a half tones. For the lowest of the notes constructed was tuned to make the consonance of a diatessaron with the upper bounding note of the lower ditone, and the highest of the notes constructed has turned out to form the consonance of a fifth with the lowest; so that since the difference is a tone, and since it is divided into equal parts, each of which is a semitone and is also the excess of a diatessaron over a ditone, it is clear that the diatessaron consists of five semitones [5].[19]

There are no figures in the above quotation (this also applies to the construction of a ditone). But the reasoning is purely geometric, based on the addition and subtraction of segments representing musical intervals. Why then the absence of figures? It is possible that this is a rhetorical device to keep geometry, in which the use of figures is essential, out of a theory that declares mathematics inessential.

In the following, in summarizing the proof of Aristoxenus, we will make its geometric nature explicit, relying on Fig. 1.5. Similarly, Giovanni Maria Artusi (1600) resumed the proof of Aristoxenus and interpreted it with geometry. Figure 1.6 illustrates Artusi's construction, wherein notes and intervals are reported.

Let AB be a diatessaron and D a ditone from B (i.e., the segment DB is a ditone); similarly, let point C be one ditone from A. Obviously AD = CB. Aristoxenus then

[18] p. 167.

[19] p. 169.

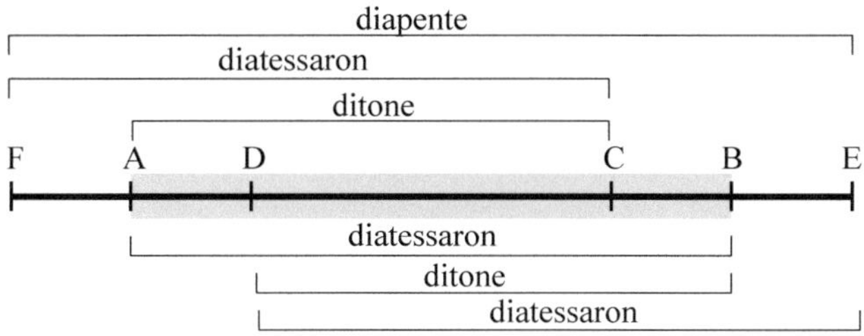

Fig. 1.5 A diatessaron versus two and a half tones

considered the pitches E and F to be one diatessaron away from D and C respectively. At this point he called for a musical experiment to see if the interval EF is consonant. And without being particularly clear, he claimed that EF was consonant. It is then easy to argue that EF is a diapente, since it is a consonance closer to a diatessaron than to a diapason, and there are no other nearby consonances. But if EF is a diapente, then, by the definition of tone, FA + BE is a tone (the difference between a diapente and a diatessaron), BE (= FA) is a semitone (half a tone), and the diatessaron DE, the sum of the ditone AC and the half-tone BE, is two and a half tones, QED.[20]

Aristoxenus' proof, though very interesting, can be questioned. From a formal point of view, because in order to be considered a true proof *more geometrico*, the result of experimental observations should appear in the premise. From a substantive point of view, because the proposition could be considered false (this is the opinion of most historians); this issue is discussed in [49].[21] However, we believe that Aristoxenus's account is essentially correct. The criticisms are all based on the assumption that the diapente and diatessaron named by Aristoxenus are also Pythagorean dia-

Fig. 1.6 Artursi's division of the diatessaron. Redrawn from [6], p. 32

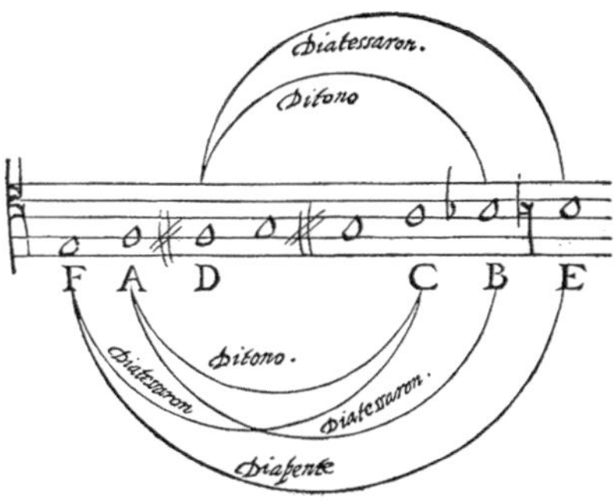

[20] We repeat here, using an algebraic language, the reasoning of Aristoxenus. Let $\omega_A, \omega_B, \omega_C, \omega_D, \omega_E, \omega_F$, be the pitches associated with A, B, C, D, E, F, respectively (measured, for example, by the logarithm of their frequencies expressed in Hz), in Fig. 1.5. T is the interval of a tone and F is the interval of a diatessaron. The construction is: $\omega_D = \omega_B - 2T$; $\omega_C = \omega_A + 2T$; $\omega_F = \omega_C - F$; $\omega_E = \omega_D + F$. Simple calculation gives: $\omega_E - \omega_F = 3F - 4T$. Assuming $\omega_E - \omega_F = F + T$, which is a diapente, the previous relation gives: $F + T = 3F - 4T$, or $F = 2.5T$, thus Aristotle's thesis.

[21] p. 64.

pente (3 : 2) and diatessaron (4 : 3). If this were the case, the interval EF would measure (see footnote 20 below):

$$\mathrm{EF} = 3F - 4T = 3(4:3) - 4(9:8)$$

which, as simple calculations show, is one Pythagorean comma less than the Pythagorean fifth,[22] and the ear of trained musicians, most likely Aristoxenus himself, would find it dissonant.

But Aristoxenus nowhere gave ratios for his diapente and diatessaron. So if there are any flaws in his proof, they are not logical, but experimental: may be his ear was not very good, and he was unable to detect consonances. Therefore, rather than being wrong, Aristoxenus' results should at most be considered approximate. Of course, this would only be true if Pythagoras was correct. Aristoxenus' position will be examined again in Sect. 3.2.3.1, where the partition of a diapason is discussed.

1.3.2 Euclid's Sectio Canonis

The attribution of the *Sectio canonis* to Euclid (3rd century BCE) is controversial. For the text, the authorship and its translation we refer to [10, 37] and to [12, 39, 52]). But luckily this is not our concern, and for the sake of simplicity we will refer to Euclid as the author of the *Sectio canonis*.

A modern reader cannot but be fascinated by the reading of the *Sectio canonis* for its elegance in most parts, which can be compared to the *Elements*, even though the exactness is less evident. Here Euclid introduced intervals only in abstract, without any reference to the way sounds are generated; in fact, apart from the introductory part of his writing, he hardly referred to sound. This is a typical way of exposition of the Greek mathematicians of the 3rd century BCE; and the structure of the *Sectio canonis* resembles that of the other two treatises on mathematical physics attributed to Euclid, the *Book of balance* [28][23] and the *Optics* [38].[24]

Despite the abstract approach, Euclid most likely had a concrete model on which to base his definitions and theorems. This is also common with modern mathematicians and physicists who present their theories in an abstract axiomatic way, while inventing them by resorting to a 'trick' that does not necessarily have very rigorous basis. Euclid may have used the monochord or canon to elaborate the theory of intervals;[25]

---

[22] Using fractions instead of ratios, EF is given by $(4/3)^3 : (9/8)^4 = 1.47981$ to compare with 1.5.

[23] pp. 24–30.

[24] According to Barker [15], p. 27, Euclid would have followed this abstract approach to give greater generality to ratios after the Pythagoreans.

[25] Monochord and canon (*kanōn*) are often considered synonymous. Strictly speaking, the monochord is a taut string, while the canon is a graduated ruler located below the string and used to place the bridge so that it determines the lengths of the string sounding the due notes.

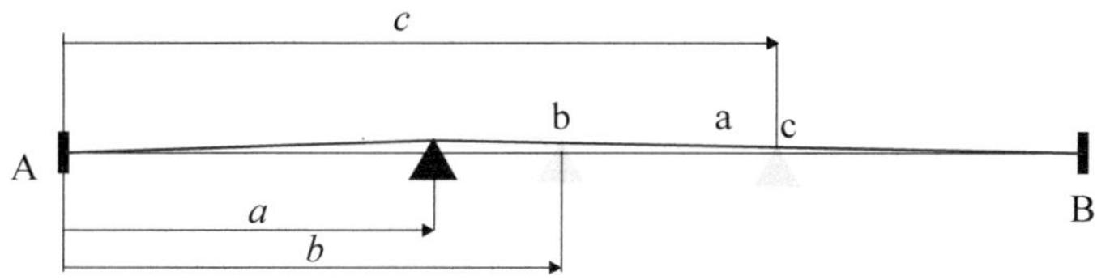

Fig. 1.7 A simple monochord

it consists of a taut string with a bridge that can divide it into two parts, as shown in Fig. 1.7.

For instance, take the left part A of the string as a reference and measure the length of the vibrating string from it. Let a, b, c be the distances of three different positions of the bridge. The intervals associated with the pairs of notes a, b and b, c are given by $(a : b)$ and $(b : c)$; the interval associated with the notes a and c is $(a : c)$, according to the Pythagorean theory, where intervals are defined by the ratio of the lengths of the monochord. As Fig. 1.7 suggests, the ratio $(a : c)$ can be assumed to be the composition, or summation, of $(a : b)$ and $(b : c)$. It is clear from the figure that $a/b \times b/c = a/c$; i.e., the ratio of a summation of two intervals must be given by the product of their ratios. A 'strange' but inescapable conclusion.

Although the monochord was most likely the inspiration for Euclid's exposition of the theory of musical ratios, it must be said that this theory predates the use of the monochord, since there is no evidence of its existence before 300 BCE, while the use of ratios is much earlier. The story of Pythagoras, for example, refers to ratios of hammer weights. Here is how Nicomachus told the story: while he [Pythagoras] was passing by a blacksmith's shop, he heard the hammers beating iron on the anvil and producing in combination sounds that were most harmonious with each other, except for one combination. He recognized in these sounds the consonance of the diapason, the diapente, and the diatessaron. He found by various experiments that the difference in sound was due to the weight of the hammers, but not to the force of the blows, nor to the shapes of the hammers, nor to the alteration of the iron being forged [57].[26] Nicomachus was not precise, but he seems to suggest that the various consonances are assigned to hammers with weights proportional to 4 : 3 : 2.[27]

An alternative to the use of strings to define ratios as witnessed by a Pythagorean of the early 5th century BCE is reported in [32]. The various consonances can be obtained from bronze disks of different sizes:

> For a certain Hippasus made four bronze discs in such a way that while their diameters were equal, the thickness of the first disc was epitritic [4:3] in relation to that of the second,

[26] p. 83.

[27] There is a fundamental flaw in his account; indeed, a single anvil struck by four different hammers would not sound four different pitches; it would sound one and the same pitch. This is because the pitch produced by percussion depends on the object being struck, in this case the anvil, and not on the objects doing the striking, in this case the hammers. For an anvil, when set into vibration by a hammer blow, functions exactly like a bell.

> hemiolic [3:2] in relation to that of the third, and double [2: 1] that of the fourth, and when they were struck they produced a concord [32].[28]

But consonant ratios could also be observed elsewhere, and have been known for centuries, as they could be derived from the distance between the holes of a Panpipe.

1.3.2.1 The Premise

The *Sectio canonis* consists of a short premise and some propositions (theorems). While the propositions are of a mathematical nature, the premise refers to quite interesting general physical and epistemological considerations. Because of its relevance and brevity, the premise is given in full below. For the sake of explanation, it has been divided into six relevant items.

1. If there were stillness and non-motion, there would be silence. If there were silence and nothing moving, nothing would be heard. If therefore anything at all will be heard, it is necessary first for there to be percussion and motion. Thus, since all notes come about from some existing percussion, it is impossible for there to be percussion without there first being motion.
2. Of motions, there are the more dense and the more rare. The more dense produce higher notes, and the more rare, lower. It is necessary for there to be higher notes, since they are put together from more dense and greater motions, and lower notes since really they are put together from more rare and fewer motions. Thus, those notes higher than desired are loosened by subtracting motions and attain the desired pitch; those lower are tightened by adding motion and attain the desired pitch.
3. Wherefore, one must say that notes are put together from parts, since by addition and subtraction, they attain the desired pitch. All things put together from parts are related to one another by numerical ratio; and thus with notes, it is necessary that they be related to one another in numerical ratio.
4. Of numbers, some are related in multiple ratio, some in epimoric, and some in epimeric;[29] and thus it is necessary that notes be related to one another in ratios such as these. Of these, then, the multiples and the epimoric are related to one another by one name.[30]
5. And we know of notes that some are consonant, some dissonant; and the consonant notes make a single blend from both notes, but dissonant notes do not.
6. This being the case, it is reasonable that consonant notes, since they make a single blend of sound from both notes, are related numerically to one another in one name, since they are either multiple or epimoric [10].[31]

Sound is associated with motion (item 1), "If there should be rest and absence of motion, there would be silence". From the nature of motion derives the nature of notes, that are however not explicitly defined (item 2). As there are controversial

---

[28] p. 93.

[29] An epimoric ratio has the form $[(n+1):n)]$, an epimeric ratio has the form $[(n+a):n)]$ $a > 1$. Following Nicomachus' nomenclature they are respectively superparticular and superpartient [56].

[30] It is not clear what Euclid intended; a possible interpretation is similar numbers.

[31] Greek text and translation into English, pp. 114–117.

interpretations of the passage concerning the transmission of sound we present two more translations.

> And since of movements some are closer packed, others more widely spaced, those which are closer packed producing higher notes and those which are more widely spaced lower ones, it follows that some notes must be higher, since they are composed of closer packed and more numerous movements, and others lower, since they are composed of movements more widely spaced and less numerous [12].[32]

> And, of motions, there are the more frequent and the intermittent; and the more frequent make higher notes, the intermittent, lower [notes]—it is necessary that there be on the one hand higher [notes] since they are compounded of more frequent and abundant motions, and on the other hand lower [notes], since they are compounded of intermittent and fewer motions [39].[33]

The first of these two quotations gives respectively "closer packed" and "more widely spaced"; the second "more frequent" and "intermittent", while in the full quotation of the premise the translations give respectively "more dense" and "more rare".

It is thus possible that Euclid asserted that a closer packed motion has a higher frequency of vibration and a wider packed one a lower one; a view of sound not common in ancient Greece, where the association of pitches with air speed was more common. It is possible, however, that Euclid shared the common view and that the closer and wider packing was not the cause of pitches, but simply a concomitant effect, since when the speed of sound is higher, the packing is closer and vice versa. In any case, although it is not possible to say for certain what Euclid's thinking was, it can be said that the readers of the early modern era could see in his writings the suggestion to associate the pitches of sounds with the frequency of air vibrations.

According to Euclid, since packing varies in a discrete way, notes do the same and could therefore be represented by numerical discrete values, i.e. integers (item 3), rather than magnitudes; intervals by ratios. Consonant intervals should be represented by agreeable ratios (first part of item 6).

To a modern reader, Euclid's reasoning does not seem very rigorous; it resembles the reasoning of a natural philosopher rather than that of a mathematician, and its acceptance requires the acceptance of principles of natural philosophy that is difficult to understand today. In any case, a criterion is introduced to decide on consonances (second part of item 6).

> It is reasonable that consonant notes, since they make a single blend of sound from both notes, are related numerically to one another in one name, since they are either multiple or epimoric [10].[34]

This quotation expresses the fundamental principle of Euclid and of all the classical Greek harmonics, whose formulation is usually attributed to Archytas of Tarentum [17]:[35]

---

[32] p. 192.

[33] p. 239.

[34] p. 117. Multiple are ratios in the form $(mn : n)$ or $(m : 1)$. Epimoric ratios are those in the form $(n + 1) : n$.

[35] p. 271.

E_1 Consonant intervals $\longrightarrow$ multiple or epimoric.

This principle is what makes the *Sectio canonis* a treatise on mathematical physics and not simply on the ratios that Euclid had studied in Book V of his *Elements* [54]. Note that the reverse:

E_2 Multiple or epimoric intervals $\longrightarrow$ consonant.

is not true, because not all epimoric or double intervals are consonant. In fact, in ancient Greece the consonances are represented by ratios derived from the quaternary only $(1, 2, 3, 4)$, i.e. $(2:1)$, $(3:1)$, $(4:1)$, $(3:2)$, $(4:3)$, according to Pythagorean numerology (a first removal of this limit can be found in Ptolemy's *Harmonica*, which accepted ratios with higher numbers).

For many Pythagoreans the principle E_1 was metaphysical in nature, deriving from the idea that the phenomena of the physical world are governed by numbers and numerical relations that obey certain rules of 'symmetry'. For other Pythagoreans, the principle was accepted on the basis of musical practice. In both cases, the principle was taken for granted and assumed as a premise of *more geometrico* arguments.

Euclid seems to take an intermediate position, he said that the simplicity of ratios *it is to be expected to give consonances*, which is weaker than to say *it gives*. Anyway, Euclid is a clever mathematician, and one wonders why he could accept as rigorous the deduction of E_1, based on items 4 and 5, where the simplicity and elegance of ratios in mathematics is transferred to the elegance of ratios in harmonics: since consonant notes make a single blend, it is natural to assume that they are represented by numbers that are closest to each other. Perhaps a better move, at least for a modern, would have been to take E_1 as simply empirically true, which could have been accepted for the epistemology of mathematical physics at the time. In any case, Euclid's explanation gave no reason why the consonances should be chosen only in the quaternary $(1, 2, 3, 4)$.

1.3.2.2 Propositions

After the brief preface, Euclid presented the propositions, numbered from 1 to 20 in [12], listed below. The first nine were purely mathematical, covering the theory of ratios. In these propositions, however, musical terminology is evident; the term *interval* is used to indicate a ratio of integers; the union of intervals is an interval given by the composition of the ratios of the constituent intervals. From the tenth to the sixteenth propositions, Euclid entered the realm of harmonics. The seventeenth and eighteenth concern the enharmonic genus and are referred to as the *enharmonic passage* in [10]. The last two propositions refer to the division of the monochord and justify the title of the treatise: *Section canonis* (in English: *Division of the canon*).

1. If a multiple interval put together twice makes some interval, this interval too will be multiple.
2. If an interval put together twice makes a whole that is multiple, then that interval will also be multiple.
3. In the case of an epimoric interval, no mean number, neither one nor more than one, will fall within it proportionally.
4. If an interval which is not multiple is put together twice, the whole will be neither multiple nor epimoric.
5. If an interval put together twice does not make a whole that is multiple, that interval itself will not be multiple either.
6. The duple interval [the diapason] is composed of the two greatest epimoric intervals, the hemiolic [3 : 2] and the epitritic [4 : 3].
7. From the duple interval and the hemiolic [3 : 2], a triple interval is generated.
8. If from a hemiolic [3 : 2] interval an epitritic interval [4 : 3] is subtracted, the remainder left is epogdoic (9 : 8).
9. Six epogdoic (9 : 8) intervals are greater than one duple interval.
10. The diapason interval is multiple.
11. The interval of the diatessaron and that of the diapente are each epimoric.
12. The diapason interval is duple.
13. It remains to consider the interval of a tone, to show that it is epogdoic.
14. The diapason is less than six tones.
15. The diatessaron is less than two and a half tones, and the diapente is less than three and a half tones.
16. The tone will not be divided into two or more equal intervals.
17. The *paranētai* and *lichanoi*[36] will be found by means of consonances.
18. The *parhypatai* and *tritai* [37] do not divide the *pyknon*.
19. To mark out the kanōn according to the so-called immutable systēma.
20. It remains to find the movable notes.

In the following, for reasons of space, we will consider in detail only the proofs of Proposition 11—limited to the diatessaron—which is enough to understand Euclid's way of reasoning.

This proposition differs from the first nine purely mathematical propositions in that it clearly belongs to harmonics, not so much because of the use of musical jargon (by naming the notes of a systēma), which is inessential, but because of the use of an empirical principle.

> *Proposition 11*. The interval of diatessaron and that of the diapente are each epimoric.
> Let A be nētē synēmmenōn, let B be mesē, and let C be hypatē mesōn (Fig. 1.8). The interval AC, being a double diatessaron, is therefore dissonant: it is therefore not multiple.Thus since the two equal intervals AB and BC when put together make a whole which is not multiple, neither is AB multiple. And it is concordant: therefore it is epimoric. The same demonstration applies also to the diapente [12].[38]

---

[36] The paranētai and lichanoi refer to the two movable notes of the tetrachord.

[37] The tritai refers to a movable note of the tetrachord.

[38] p. 199.

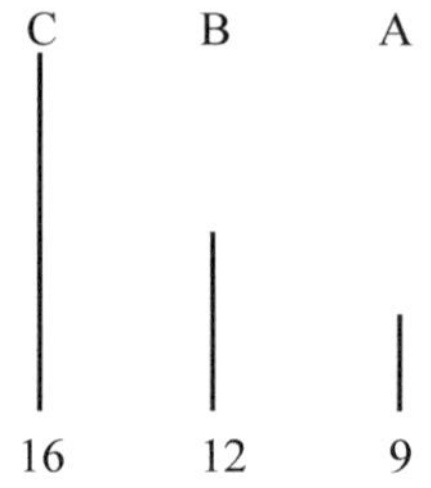

Fig. 1.8 Interval of a double diatessaron. Redrawn from [12], p. 200

Below is the sketch of the proof, (T) stands for theorem, (P) for empirical principle or experiment, (A) for assumption, H for Euclid's proposition:

A The interval AC between a nētē synēmmenōn and a hypatē mesōn is a double diatessaron—see Fig. 1.2b.
P Empirical observation: A double diatessaron is dissonant.
T So AC is dissonant.
P Principle E_2, reversed: Dissonant intervals $\longrightarrow$ neither multiple nor epimoric.
T AC is dissonant, so it cannot be multiple.
H Proposition P_5: If an interval put together twice does not make a whole that is multiple, then the interval itself is not multiple.
T Since AC is not multiple, AB (a diatessaron) is not multiple either.
P Empirical observation: A diatessaron is consonant.
T AB is a diatessaron, so it is consonant.
H Principle E_1: Consonant intervals $\longrightarrow$ are multiple or epimoric.
T AB is consonant and not a multiple, so it should be epimoric.

Euclid said that AC, the union of two diatessaron, could not be a multiple because it was dissonant, but he could not make this assertion, because for the Greeks not all multiples were consonant; for example, the interval of four diapente and a major third (5 : 1) [15].[39] However, Euclid is concerned with the division of the canon in his treatise, and usually the Greeks did not go beyond two diapason in such a construction, and up to that value all multiples were assumed to be consonant [8].[40]

Demonstrations of other propositions proceed in much the same way. Note Proposition 13, which shows that the range of a tone is expressed by the ratio (9 : 8), i.e., it is *epogdoic*.

> *Proposition 13*. It remains to consider the interval of a tone, to show that it is epogdoic. We have learned that if an epitritic [4 : 3] interval is subtracted from a hemiolic [3 : 2] interval, the remainder left is epogdoic [9 : 8]. And if the diatessaron is taken from the diapente, the remainder is the interval of a tone. Therefore, the interval of a tone is epogdoic [12].[41]

The starting point is the definition of tone as the distance between a diapente and a diatessaron. Since Propositions 11 and 12 (not discussed here) show that the intervals

---

[39] p. 386.
[40] p. 160.
[41] p. 201.

of diatessaron and diapente have ratios of (4 : 3) and (3 : 2) respectively, the proof of Proposition 13 is immediate, since the distance of the intervals is obtained as the inverse of the summation: (3 : 2) − (4 : 3) = (9 : 8).

Also noteworthy are propositions 14 and 15, later emphasized by Ptolemy, to counter the thesis he attributed to Aristoxenus, according to which the diapason is six tones, being the sum of a diatessaron of two and a half tones and a diapente of three and a half tones. Proposition 16 shows the limit of the arithmetic approach to harmonics. A tone cannot be divided into two equal intervals, because an epimoric ratio cannot be divided into two equal ratios; this would correspond to the evaluation of the square root (modern term), which is irrational and thus not a ratio.

1.3.3 The Division of Tetrachords. Greek Notes

Each genus of tetrachord is characterized by the position of the two mobile notes, such as lichanos and parhypathē. The first dispositions of the mobile notes were certainly dictated by the feeling of the ancient musicians, then there were attempts at systematization by mathematicians.

The best known scholar from the first phase of the history of harmonics is Archytas of Tarentum (4–5th century BCE). Although there is no original writing by him, his ideas are reported by Ptolemy, active nearly five centuries later, in his *Harmonica* [67],[42] as given in Table 1.2.

Note the presence of the semitone (28 : 27) in all the three genera and of the just major third interval, (5 : 4), between mesē and lichanos. The first written evidence that has reached us, however, is reported in the *Timaeus* of Plato, a younger contemporary of Archytas. Here the author divided the human soul according to the division of the diatonic tetrachords, characterized by the intervals (9 : 8), (9 : 8), (256 : 243) (starting from the mesē), with the last interval (256 : 243) being known as *limma*.

> Where there were intervals of (3 : 2) and of (4 : 3) and of (9 : 8), made by the connecting terms in the former intervals, he filled up all the intervals of (4 : 3) with the interval of (9 : 8), leaving a fraction over; and the interval which this fraction expressed was in the ratio of 256 to 243 [63].[43]

Table 1.2 Archytas' tetrachords; from [7], p. 296

Intervals		Enharmonic	Chromatic	Diatonic
Mesē-lichanos		5:4	32:27	9:8
Lichanos-parhypathē		36:35	243:224	8:7
Parhypathē-hypathē		28:27	28:27	28:27

[42] I 31.6–31.15; pp. 44–45.

[43] 36a–36b.

Table 1.3 Genera and shades of Aristoxenus' tetrachords; from [7], p. 298

Intervals	Enharmonic	Soft Chromatic	Hemiolic Chromatic	Tonic Chromatic	Soft Diatonic	Intense Diatonic
Mesē-lichanos	2	1+5/6	1 + 3/4	1 + 1/2	1 + 1/4	1
Lichanos-parhypathē	1/4	1/3	3/8	1/2	3/4	1
Parhypathē-hypathē	1/4	1/3	3/8	1/2	1/2	1/2

Table 1.4 Genera and shades of Ptolemy's tetrachords; from [7], p. 307

Intervals	Enharmonic	Soft Chromatic	Syntonic Chromatic	Soft Diatonic	Tonic Diatonic	Ditonic Diatonic
Mesē-lichanos	5 : 4	6 : 5	7 : 6	8 : 7	9 : 8	9 : 8
Lichanos-parhypathē	24 : 23	15 : 14	12 : 11	10 : 9	8 : 7	9 : 8
Parhypathē-hypathē	46 : 45	28 : 27	22 : 21	21 : 20	28 : 27	256 : 243
Intervals	Syntonic	Smooth				
	Diatonic	Diatonic				
Mesē-lichanos	10 : 9	10 : 9				
Lichanos-parhypathē	9 : 8	11 : 10				
Parhypathē-hypathē	16 : 15	12 : 11				

Though Plato was in touch with Archytas it is most probable that his division of the tetrachord had an earlier source. It will define the Pythagorean standard of the diatonic genus for over a millennium.

Aristoxenus, a fifty year after Plato, had an easier task to divide the tetrachord. It was considered a geometric segment, and like any geometric segment, it could be divided at will. This allowed a rational and simple representation of the different genera and shades. Table 1.3 illustrates the situation [3].[44]

Here 2 means two tones, 1 one tone, 1/4 a quarter of tone, and so on. According to Aristoxenus an interval of the tetrachord cannot be less that 1/4, otherwise it would be unmelodic [3].[45]

Ptolemy much later presented his own divisions, as shown in Table 1.4. Of particular interest is the syntonic diatonic shade, which since the Renaissance has been considered the model of just intonation.

Note that all of Ptolemy's intervals, apart from those of the ditonic diatonic shade, are epimoric. The intervals between mesē and lichanos follow an arithmetic

[44] II 44–52.

[45] II 46, p. 199.

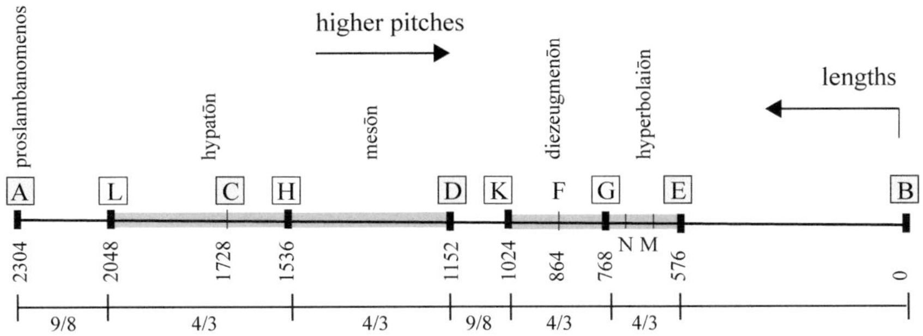

Fig. 1.9 The tetrachords of the greater perfect system

progression; the other intervals are not obtained on the basis of aesthetic considerations, but more simply according to a well-defined mathematical algorithm [7].[46]

The Pythagorean Tuning
Euclid, in propositions 19 and 20 of his *Sectio canonis*, proposed an algorithm for finding the intervals of the fixed and mobile notes of the tetrachord as a by-product of the division of the monochord to obtain the greater perfect system (see Sect. 1.2.2). Euclid's is the oldest surviving text on the subject. His approach, discussed in some detail below, led to tetrachords in the diatonic genus—with the same intervals between the mobile notes as those reported by Plato in *Timaeus*: (256 : 243), (9 : 8), (9 : 8). The resulting sequence of notes in a diapason, for instance in the central diapason, is today called *Pythagorean tuning* [12].[47]

To start with, Euclid's algorithm searches for the points E, G, K, D, H, C, L of a monochord AB that represent the fixed notes of the four tetrachords of the greater perfect system, as shown in Fig. 1.9 (numbers indicating lengths are ours).
EG is for the tetrachord hyperbolaiōn, GK for the diegzeumenōn, DH for the mesōn, and HL for the hypatōn.

The first step of the procedure is to divide AB into four equal parts BE, ED, DC, CA (gray boxes of Fig. 1.9). Segment ED is representative of the highest diapason, between mesē and nētē hyperbolaiōn; DA of the lowest, between proslambanomenos and mesē (see Fig. 1.3a). The next step is to find the point F such that BF = $\frac{1}{2}$ BC. Assuming that intervals are defined by lengths measured from B, we can say that F is a diapason below C. Euclid was not very clear, but F is introduced because it is a fixed point with D of the tetrachord synēmmenōn, which is not part of the greater perfect system, but can be associated with it to give the *unchanging perfect system* [15].[48]

[46] p. 307.

[47] p. 15.

[48] p. 17.

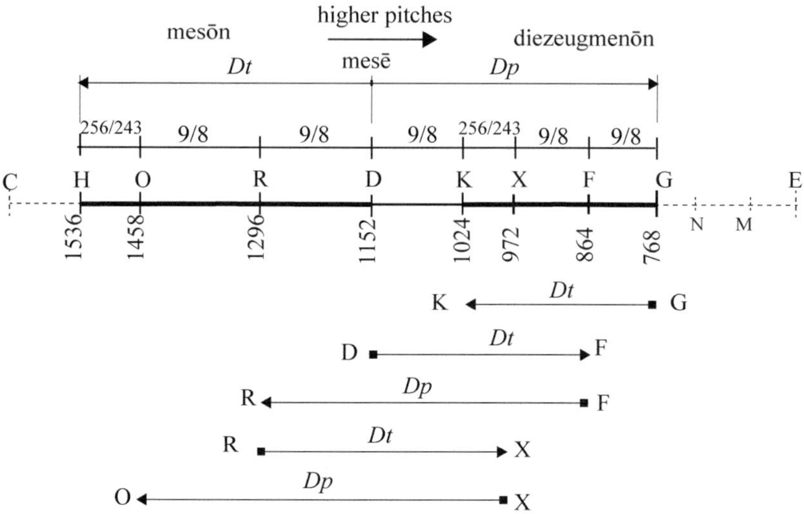

Fig. 1.10 Tuning central diapason

The next point proposed by the algorithm is G, which is located a diapente below D. H is a diapason above G, K is a diapente below H, and finally L is a diapason above K. Resuming in a compact notation, one can write:

$$\begin{aligned} G &= D - Dp \\ H &= G + Do \\ K &= H - Dp \\ L &= K + Do \end{aligned}$$

where *Dp* is for a diapente and *Do* for a diapason.

Euclid presented his division in a purely geometric way, and when he used musical terms such as diapason diapente, diatessaron, he seemed to be referring only to the ratios between the lengths of the segments of a monochord. However, the division of the monochord could (at least theoretically) be carried on by ear, which required adherence to consonances unanimously accepted by Greek musicians.

For the mobile notes the procedure is more articulated. Although Euclid could have proceeded by diapason and diapente, he chose a less elegant but simpler approach by using the major tone (9 : 8) and the diatessaron (4 : 3). In the following, attention is focused on the central diapason shown in Fig. 1.10 (the segment GH of Fig. 1.9). Euclid, however, started with the mobile points M and N of the tetrachord hyperbolaiōn EG shown in Fig. 1.9, which is not part of the central diapason. They are given as:

$$\begin{aligned} M &= E + T \\ N &= M + T \end{aligned}$$

where T is the major tone, assumed as (9 : 8) by Euclid.

Table 1.5 Fixed and mobile notes of the central diapason

Notes	Lengths from G	Pitches from H	Relative intervals
G	1	2	
$F = G + 2Dp - Do$	9 : 8	16:9	(9 : 8)
$X = G + 4Dp - 2Do$	81:64	128:81	(9 : 8)
$K = G + Do - Dp$	4:3	3:2	(256 : 243)
$D = G + Dp$	3:2	4:3	(9 : 8)
$R = G + 3Dp - Do$	27:16	32:27	(9 : 8)
$O = G + 5Dp - 2Do$	243:128	256:243	(9 : 8)
$H = G + 2Do$	2	1	(256 : 243)

The modern reader is left perplexed by Euclid's approach, which involves the re-proposition of the sequence (256:243), (9:8), (9:8), without providing a musical rationale. This aspect of Euclid's methodology fails to align with the principles of consonance, given that the major tone does not constitute a consonance and, consequently, cannot be identified through ear alone. This is probably just a shortcut, since Euclid knew that a tone was given by the difference between a diapente and a diatessaron. So, for example, M could have been found by moving a diapente forward from E to F and a diatessaron back.

The first mobile point F of the tetrachord diegzeumenōn GK was already obtained; points X, R, O are given as:

$$X = N + Dt$$
$$O = X + Dp$$
$$R = C - Dt$$

where Dt is a diatessaron.

Once the mobile points of the tetrachord EG have been obtained, the mobile points the other tetrachords, including those the central diapason, can easily be obtained by moving by diapente and diatessaron.

Using a modern algebraic approach the procedure to locate all the points of the central diapason can be made more elegant by considering only diapason and diapente, according to the Table 1.5.

Here the cycle of fifths in Col. 1 refers to lengths, the ratios of which are given in Col. 2. Col. 3 gives pitches, assuming 1 for the lowest value in H. Col. 4 gives the relative distances in both lengths and pitches. The intervals in this table are those typical of the Pythagorean scale, with only one pitch and one semitone. The process can be repeated for all tetrachords, and the greater perfect system is obtained.

Euclid's approach is fine for tuning a monochord, but not for a musical instrument such as a lyra. To tune the strings of a lyra, one could, in principle, use a monochord built according to Euclid's rules. The individual strings of the lyra could be made to resonate one by one with the monochord, in which the bridge is suitably placed.

However, this approach is cumbersome and requires a culture of speculative music, which not all practical musicians had. They could tune the lyra by respecting the relationships of consonance, diapente and diatessaron, between the different strings. For example, to tune a lute with eight strings, which for convenience were marked with the letters of Fig. 1.10, one can proceed according to the following table:

$$\begin{aligned}
&\mathrm{D}\\
&\mathrm{H} = \mathrm{D} - Dt\\
&\mathrm{G} = \mathrm{D} + Dp\\
&\mathrm{K} = \mathrm{G} - Dt\\
&\mathrm{F} = \mathrm{D} + Dt\\
&\mathrm{R} = \mathrm{F} - Dp\\
&\mathrm{X} = \mathrm{R} + Dt\\
&\mathrm{O} = \mathrm{X} - Dp
\end{aligned}$$

whose steps are illustrated in Fig. 1.10, below. Notice that the tuning starts from the mesē [41, 48].[49]

1.3.3.1 Absolute Pitches

As already noted, the Greeks could not define the pitch of a single note, and the various systems developed by theorists, such as the greater perfect system, were mainly used to define sequences of intervals rather than absolute pitches. Nevertheless, it would be erroneous to assume that Greek melodies were entirely devoid of any underlying rationale (until the 18th century, when the precise measurement of the frequencies of notes became available, the perception of pitches of sound by Western musicians was somewhat similar to that of the Greeks). One limitation to the arbitrary selection of a note's pitch was the range of the human voice. While this range differs between individuals, it remains relatively consistent across genders and age groups. Additionally, it is plausible that the Greeks, like the modern people, had access to pitch pipes and other instruments with a relatively consistent register, which could have been used to maintain a standard pitch. Otherwise there would be no reason to choose one tonos (see below for a precise definition) rather than another, for a given piece of music.

Even though the standard pitch was not precisely defined and could reasonably vary from place to place and from time to time, it had some value in terms of the absoluteness of the notes, which were indicated by symbols referring to pitches rather than dynamis, as the theorists' letters did.

Table 1.6, reports Greek notation. The first line of the table contains numbers assigned to the notes for ease of reference; the second line contains the modern notes conventionally accepted as equivalent to the Greek notes,[50] which here have only the role to indicate the compass of the Greek notes. The third row contains the symbols

[49] p. 93.

[50] The symbol x can be considered as a sharp.

Table 1.6 Greek notation of pitches; from [12], vol. II, p. 426

1	4	7	10	13	18	21	25	28	31	34	39	42	46	49	52	55	60	63	67	70
E♭	F	G	A	B	cx	dx	f	g	a	b	c′x	d′x	f′	g′	a′	b′	c″x	d″x	f″	g″
⊐	⊸	ɜ	ϙ	W	ᖷ	∇	Ω	Φ	C	O	K	H	Γ	Ⴎ	Θ	O′	K′	H′	Γ′	Ⴎ′
⊏	⟜	ε	H	h	∃	⊣	ϒ	F	C	K	Λ	>	N	Z	Ϻ	K′	Λ′	>′	N′	Z′

3	6	9	12	16	19	22	27	30	33	37	40	43	48	51	54	58	61	64	69
E♭x	Fx	Gx	Ax	c	d	e	fx	gx	ax	c′	d′	e′	f′x	g′x	a′x	c″	d″	e″	f″x
✱	⊣	⊔	Ϻ	—	7	ℸ	X	T	Π	M	I	Z	A	✱	⊥	M′	I′	Z′	A′
⋇	T	ɜ	⊓	E	⊢	Γ	Ч	ꟻ	Ↄ	ᒣ	<	⊏	⟍	⋋	⋌	ᒣ′	<′	⊏′	⟍′

for voices and the fourth the symbols for instruments. Notation of Table 1.6 was already firmly established by the middle of the 3rd century BCE [45],[51] and a full account of it is given in the treatise of Alypius of Alexandria (fl 4th century CE) [1, 40].

Figure 1.11 shows a distribution of pitches derived for a number of ancient Greek aulos from museums [71].[52] On the left are the names of the notes often associated with the unmodulating greater perfect system, and on the right are the numerical values of the frequencies. The figure shows that the modern A_4 of 440 Hz is one of the highest pitches found; the same is verified for the pitches of the strings of the lyra [45].[53]

The association of pitches with the notes of the various tonoi (see below) can also be established on the basis of manuscripts. To cite an example—according to Ptolemy's description of modes—the central diapason of the Lydian tonos is situated between F_3 ($F_3\sharp$) and F_4 ($F_4\sharp$) [45],[54] alternatively the mēse is B_3 ($B_3\sharp$) [45][55] and the corresponding proslambanomenos is one diapason below this value, specifically B_2 ($B_2\sharp$). For the Dorian proslambanomenos, two tones below, the pitch is G_2 ($G_2\sharp$) and the central diapason is defined by D_3-D_4, much lower—more or less one diapason—than the modern central diapason (octave) of the violin clef, between C_4 and C_5.

---

[51] p. 434

[52] The aulos (Latin *tibia*) is a reed-blown double pipe [72], Entry: *Aulos*. Bélis A.

[53] p. 90.

[54] p. 332.

[55] p. 71.

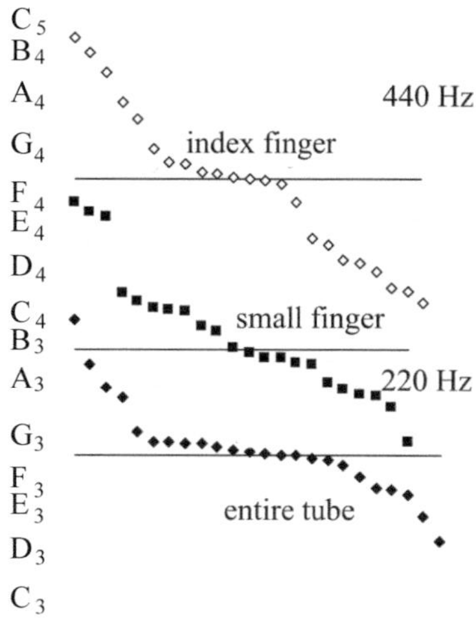

Fig. 1.11 Distribution of pitches in historical auloi. Redrawn from [45], p. 331

1.4 Later Scholars

1.4.1 Ptolemy's Synthesis. The Harmonica

Claudius Ptolemy (fl. 127–148) CE wrote a fundamental treatise on harmonics, the *Harmonica* [66]. Unfortunately, the surviving text is not completely reliable. In particular the last three chapters of the third book are missing and the authenticity of this book as a whole is doubtful [67].[56] Its history and translation in the late Middle Ages and in the Renaissance are documented in [50] and [59], respectively; we simply note that since the invention of printing, *Harmonica* has fared badly. It was printed in Venice, with a Latin translation from Greek by Antonius Gogava (1529–1569) in 1562, preceded by a Latin version of Aristoxenus' *Elementa harmonica* and even by *Aristotelis de objecto auditus* [44].

When the *Harmonica* became available in Europe, it was no more than a historical document, and musical theory had evolved. However, it influenced the history of music largely in an indirect way, as its excerpts were disseminated in the writings of Boethius (see below) and elsewhere. In modern times, the Greek version was edited by Ingemar Düring [34] in 1930.[57] To our knowledge there are two modern English translations, one by Andrew Barker [12], the other by Jon Solomon [67]. Solomon's translation and critique will be largely used here, with some reference to Barker's.

The treatise consists of three books, each of which is divided into sixteen chapters. In Book I Ptolemy dealt with the basics of the subject, defined epistemology, and criticized previous approaches; here he presented his theory of the tetrachord. Book II fully develops his theory of music. Book III describes the harmony of the spheres.

56 p. XXX.

57 p. 319.

From the very first pages, Ptolemy's efforts to show that his epistemology was not only about music, but about all sciences, are quite clear. The epistemological positions are—surprisingly for us today—fully explicit, more so than, for example, in the *Almagestum*, which could appear to be a much more scientific text. Ptolemy sought an agreement between a mathematical theory, made explicit by assumptions about numerical relationships, and experience, concerning the agreement of the human ear to sounds. Ptolemy was convinced that this agreement should exist because nature itself has an intrinsic harmony which should be recognized through the use of reason and senses (hearing) [67].[58]

The roles of reason and hearing are distinct "hearing is the criterion of matter and condition, while reason is the criterion of form and causes" [67].[59] Ptolemy specified that the role of hearing is to "discover the approximate and accept the exact", while the role of reason is to "accept the approximate and discover the exact". That is, given two or more sounds, the role of hearing is to discover the approximate ratio of their pitches for which the sounds appear to be consonant. Reason then examines these ratios and corrects them if they appear to be 'irrational'. Ptolemy stated that sense perception is characterized by uncertainty [67], [60] but that with proper training it can distinguish between similar consonances even when the difference is minimal at first glance. In any case, there is a limit to human perception.

A continuous hearing can help to perceive even very small differences. Indeed, something that is only once quite short of the truth and cannot make the amount of its shortness perceptible by a few comparisons, will be most perceptible and comprehensible to all when a number of comparisons are made [67].[61] This consideration is interesting because it refers to actual experimentation, similar to that carried out in a modern laboratory, where an experience is repeated many times to prove the regularity of a phenomenon.

Ptolemy defined his musical hypotheses/principles in the negative sense, stating how they differed from those of the Pythagoreans and Aristoxenians. Differences regarding the Pythagoreans are of greater interest, and only these will be remarked upon.[62] Ptolemy said that their rational criteria were quite satisfactory, but they failed in some aspects; in particular, they were unable to provide rational criteria for all consonances; for example, they could not justify the concord $(8:3)$.

Ptolemy first solved this particular problem by observing that the interval $(8:3)$ is the conjunction of a diapason and a diatessaron, $(8:3)=(2:1)+(4:3)$. Take three notes, Ptolemy said, A, B, and C. A and B are separated by a diapason, B and C by a diatessaron; consequently, A and C are separated by an interval of $(8:3)$. On the basis of hearing, A and B are nearly equal, so if there is consonance between B and C (and there is), there should also be consonance between A and C [67].[63]

---

[58] p. 8.

[59] p. 3.

[60] p. 4.

[61] p. 5.

[62] Probably Ptolemy was referring to Archytas of Tarentum [13], p. 132.

[63] p. 20.

Then Ptolemy gave his view about epimoric ratios; when they are not consonant they are at least melodic. Ptolemy's interest is centered on intervals smaller than a diatessaron, those within a tetrachord. His position is quite articulate; to fully grapple with his ideas, reference can be made to [14]. In essence, he proposed the following architectonic principle, which can be classified as rational because it can be deduced by reason according to symmetry and simplicity:

> P_1.The homophones would come from both the first multiple ratio [(2 : 1)] and those measured from it. Consonances are the first two epimoric ratios [(3 : 2) and (4 : 3)] and those composed from them plus the homophones, while the emmelic [melodic] are those epimoric ratios after the epitritic [(4 : 3)] [67].[64]

The homophones are notes separated by diapason intervals, the consonances are the diatessaron and the diapente (the first two epimoric ratios); then follow the simple melodic intervals, which are epimoric ratios smaller than a diatessaron. Ptolemy also proposed a further hypothesis, according to which the epimoric ratios $(n + 1) : n$ with the smaller values of n are more melodic [12].[65]

To these fully rational criteria, Ptolemy added another of an empirical nature, which applies to the arrangement of the two moving notes of the tetrachord: they should be arranged so that of the three intervals, the lowest is smaller than the other two.

> Secondly, on the basis of agreed perception, we adopt similarly, as common to all the genera, the thesis that the 'following' [lowest] value of the three are smaller than each of the remaining ones; as peculiar to the genera that have pykna[66] the thesis that the two values next to the lowest note together are smaller than the one next to the highest note; and as peculiar to the apykna[67] the thesis that none of the values is greater than the remaining two together [12].[68]

Ptolemy was not worried about mixing rational and empirical principles; his epistemology allowed this, organized as it was in two distinct phases. The first phase is inductive. Here Ptolemy is completely free in his choice, which could even be made by chance—although this is quite a modern approach. So any source could be assumed for the hypotheses. In the second deductive phase, theorems are derived from the hypotheses, assumed as principles of a *more geometrico* argument. Theorems consist of propositions to be verified by hearing. If consent is obtained, the hypotheses are validated regardless of how they were arrived at.

Ptolemy believed that his criteria of melody had absolute value. For him, it is part of human nature to distinguish which sounds are melodic and which are not. Today it is more commonly accepted that musical taste is largely influenced by culture. Specifically, what Ptolemy and the Greeks considered pleasing melodies, if the reconstruction of Greek sounds is reliable, are not such for us [3].[69]

---

[64] p. 24.

[65] p. 290.

[66] Singular: pyknon.

[67] Peculiar of the diatonic genus when neither one of the three ratios is greater than both the remaining two.

[68] p. 307.

[69] Introduction.

The basic assumption of *Harmonica* is that pitches are represented by ratios of integer numbers. But since numbers cannot be found in sounds, they must be found in their causes. More precisely, they are to be found in the observable and measurable aspects of musical instruments.

After some fairly convincing arguments, Ptolemy came to the conclusion that the most suitable instrument for identifying the relations between notes is the *monochord*; it is simple to construct and is controllable, more than the aulos, for example, where the height of the sound also depends on the power of the player, and this is not observable.

When a string is plucked or drawn with a bow, it produces a sound whose pitch depends on several parameters. Among them Ptolemy specified: the length of the string, the diameter of the section, the density of the matter; the best observable quantity is the length and the number which characterizes it serves as a measure of the height of the pitch. Only later did Ptolemy mention tension (note that here and in the following, tension means force of traction, not to be confused with stress, which is force per unit of area).

> There are three reasons for the difference between high and low sounds; the *density* [emphasis added], the diameter, and the lengths of the strings. The higher sound results from a denser and thinner string and a shorter length; tension here takes the place of density, for it creates greater tension and hardens also what would be similar if the string had a shorter length [67].[70]

Here Ptolemy said that the greater the density—a term with which he probably (?) associated the present meaning—the higher the sound. Actually, the opposite is true for strings. The reason for this 'error' may be sought in the fact that Ptolemy probably compared sounds produced not by strings but by other means, such as bells of bronze and wood; in such a case the denser matter actually produce higher sounds [67].[71] With some imprecision, he added that tension (force) and diameter can compensate each other. That is, strings of different diameters will sound the same if their tension is proportional to their diameter. Actually, the statement is true if the area, not the diameter, of the string's cross section is taken.

Once the correct construction of the monochord is verified, it can be used for musical experiments. Ptolemy expressed no doubt about this, although the modern reader may remain somewhat skeptical about the sense of experimenting to verify mathematical hypotheses that seem to leave no alternatives. Nowadays, but probably even then, the monochord is mostly used to demonstrate to students the goodness of the theory of harmony based on ratios, for certain values of them; for example, (2 : 1), (4 : 3), (3 : 2). If, for some reason, the ear suggests a lack of consonance, we immediately believe that the monochord has serious construction flaws.

Ptolemy argued that the difference between six tones and the diapason could be observed experimentally. To do this, he proposed to build an instrument with eight

---

[70] pp. 37–38.

[71] p. 12. This depends mainly on the greater stiffness of bronze, which has an influence on the sounds due to solicitation in bells, but not in strings.

strings (octachord canon), the first seven separated by an interval of one tone, the last one equal in length to half the first string (then separated from it by a diapason). The purpose of the experiment, we believe, was to empirically verify the difference between the union of six tones—represented by the seventh string—and the diapason. The use of several strings should have the function of being able to achieve a better control of the division into six tones. Ptolemy also discussed, albeit very briefly, the possibility of tuning the six strings separately by adjusting the tension to which they are subjected.

1.4.1.1 Species and Tonoi

At least until the 5th, and probably the 4th, century BCE Greek music knew many styles of melodies, or *harmoniai*, different in character and named after Hellenic or barbarian tribes. The most widespread harmoniai were Dorian, Phrygian, Lydian. The notes required for each type of melody constituted a tuning of the lyra. There is little historical evidence about the nature of these harmoniai [12, 53, 58, 70, 74, 76]; in particular, we know almost nothing about their early history, including the reason for their emotional role—see [69]—but we can say that the use of the word harmoniai in the 7th–5th centuries BCE was not technical. Centuries after the classical age, Ptolemy proposed a fairly clear exposition of these harmoniai with the theory of *tonoi*. They are seven in number, and are essentially the species of the diapason [76].[72]

As tonoi are connected with species, it will be useful to explain the theory of species according to Ptolemy. "A species is a particular positioning of the proportions [intervals] characteristic of each genus, within its proper limits" [67].[73] Ptolemy considered the species of the principal consonances: diatessaron, diapente, and diapason. To illustrate them he referred to the 'two-diapason' system of Fig. 1.12, which he constructed as follows. Consider a diatessaron ΑΒΓΔ; on the right side and with lower pitches construct another ΔΕΖΗ, in conjunction, and then add a tone ΗΘ, which serves by disjunction to another tetrachord ΘΚΛΗ; one more tetrachord ΜΝΞΟ closes the series. Here Ptolemy ordered the letters from the higher note to the lower note (however, he was not always consistent in his treatise on this order).

Note that what Ptolemy considered, is not really a two diapason system, because it lacks a note, the proslambanomenos. Otherwise it is equivalent to the greater perfect system.

Referring to Fig. 1.12, from right to left, the species considered for the diatessaron, three in number, are represented by ΜΟ (TTS), ΛΞ (STT), and ΚΝ (TST), where T and S have the usual meaning of tone and semitone, respectively; all in the lower part of the two-diapason system. Other species reproduce the same interval pattern. For example, the diatessaron ΘΜ (TTS) is the same as ΜΟ (TTS). Of the three species, only the first ΜΟ is bounded by two fixed notes in the system. For the diapente there

[72] p. 81.
[73] p. 69.

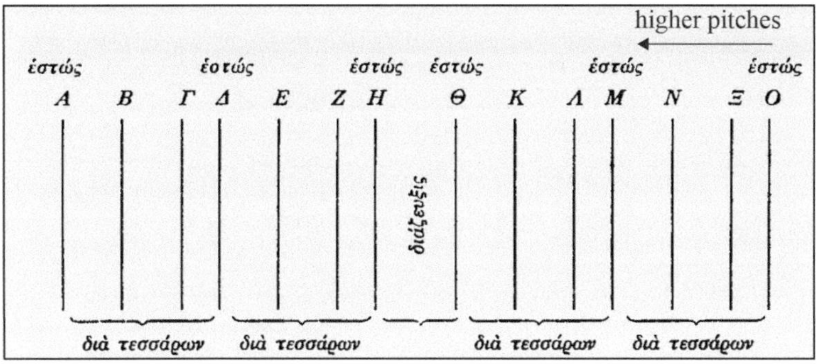

Fig. 1.12 The two diapason system. Redrawn from [66], p. 50

are four species: HM, ZΛ, EK, ΔΘ. Ptolemy started from HM with intervals (TTTS) to take three tones and one semitone. In this case there are two species limited by fixed strings, HM and ΔΘ.

There can only be seven species of diapason, listed in the following Table 1.7, ordered from higher pitches (left) to lower pitches (right).

The connection between tonoi and species is not as direct as Ptolemy seems to suggest. To see this, it is necessary to remember that he named notes in two different ways. In the first, according to their position (pitch), or thesis; in the second, according to their function, or dynamis. Imagine a staff with many rows that can be tuned in different ways. Each row can be given a name; in the first way, according to thesis, using the letters of the Greek alphabet (see Table 1.6), starting from the higher pitches for example. In the second way, according to the dynamis, the name is drawn from the unmodulating system (see Fig. 1.2), with the higher note known as nētē hyperbolaiōn and the lowest as proslambanomenos, independently of their pitches.

> The notes of the real, perfect double diapason system—it consists of fifteen note since one note, the middle note [the mesē] is common to the lower and higher diapason—we name sometimes according to their position, that is simply by criteria of being higher or lower […]. At other times we name the notes according to their function. When using the name by

Table 1.7 The seven species of diapason according to Ptolemy

HO	T	T	T	S	T	T	S
ZΞ	S	T	T	T	S	T	T
EN	T	S	T	T	T	S	T
ΔM	T	T	S	T	T	T	S
Γ Λ	S	T	T	S	T	T	T
BK	T	S	T	T	S	T	T
AΘ	T	T	S	T	T	S	T

Fig. 1.13 Position of the tonoi. Redrawn from [66], p. 64

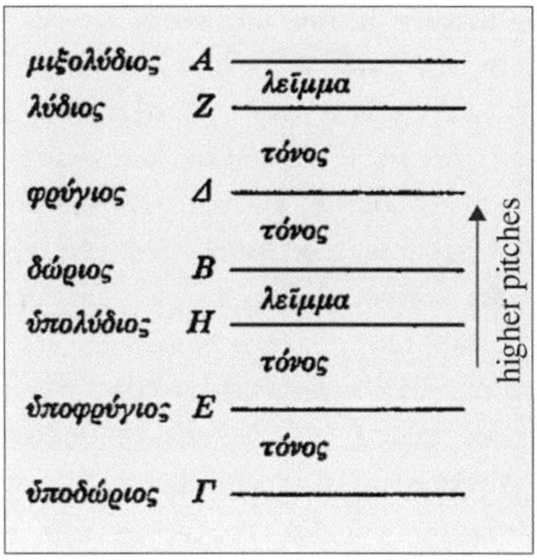

function, a mesē is simply individuated as the string which has a tone above and two tones below, independently of the position that the string occupies in the lyra.

Ptolemy labeled his tonoi according to the traditional names they had acquired over centuries of history: Mixolydian, Lydian, Phrygian, Dorian, Hypolydian, Hypophrygian, Hypodorian; they are characterized by different heights of sound. With reference to Fig. 1.13, start by the highest tonos, which we will call A (Mixolydian); then take one lower by one diatessaron, B (Dorian), and the next one lower than B by one more diatessaron, Γ (Hypodorian), which is still within the compass of a diapason. If you move down from Γ of a diatessaron, you fall out of the diapason. This tonos can be found by still descending a diatessaron from Γ but ascending a diapason, resulting in the tonoi Δ (Phyrigian) a diapente above Γ. From Δ go down one diatessaron and get the position of the tonos E (Hypophrygian). By going down of a diatessaron one goes far from the compass of a diapente, so instead of going down he raises of a diapente, giving Z (the Lydian), from which by going down of a diatessaron one eventually obtains H (Hypolydian).

But the order of the tonoi described above is not sufficient to reveal their nature. It is necessary to refer to their internal structure, which for Ptolemy is known if the dynamic position of the mesē is known.

> The dynamic mesē of the Mixolydian will be attuned to the position of paranētē diezeugmenon, so that the tonos may make the first species of the diapason in the range set out; that of the Lydian will be attuned to the position of tritē diezeugmenon, corresponding to the second species; that of Phrygian to the position of paramesē, corresponding to the third species; that of Dorian to the position of mesē, making the fourth and central species of the diapason; that of Hypolydian to the position of lichanos mesōn, corresponding to the fifth species; that of Hypophrygian to the position of parhypatē mesōn, corresponding to the sixth species; and that of Hypodorian to the position of hypatē mesōn, corresponding to the seventh species [66]. [74]

That is, the internal structure for all tonoi is the same, that of the unmodulating greater perfect system.

[74] p. 65. Translation into English in [12].

Table 1.8 Ptolemy's tonoi

A	nētē hyperbolaiōn
B	pararmētē hyperbolaiōn
Γ	tritē hyperbolaiōn
Δ	nētē diezeugmenōn
E	paranētē diezeugmenōn
Z	tritē diezeugmenōn
H	paramesē
Θ	mesē
K	lichanos mesōn
Λ	paryhypatē mesōn
M	hypatē mesōn
N	lichanos hypatōn
Ξ	paryhypatē hypatōn
O	hypatē hypatōn
P	proslambanomenos

	Hypodorian (Γ)	Hypophrygian (E)	Hypolydian (H)	Dorian (B)	Phrygian (Δ)	Lydian (Z)	Mixolydian (A)
							A
						A	
							B
					A	B	
							Γ
A_4				A	B	Γ	Δ
$G_4\#$			A			Δ	
G_4				B	Γ		E
$F_4\#$		A	B		Δ	E	
F_4				Γ			Z
E_4	A	B	Γ	Δ	E	Z	H
$E_4\flat$			Δ			H	
D_4	B	Γ		E	Z		Θ
$C_4\#$		Δ	E		H	Θ	
C_4	Γ			Z			K
B_3	Δ	E	Z	H	Θ	K	
$B_3\flat$			H				Λ
A_3	E	Z		Θ	K	Λ	M
$G_3\#$		H	Θ			M	
G_3	Z			K	Λ		N
$F_3\#$	H	Θ	K		M	N	
F_3				Λ			Ξ
E_3	Θ	K	Λ	M	N	Ξ	O
$E_3\flat$			M			O	
D_3	K	Λ		N	Ξ		P
$C_3\#$		M	N		O	P	
C_3	Λ			Ξ			
B_2	M	N	Ξ	O	P		
$B_2\flat$			O				
A_2	N	Ξ		P			
$G_2\#$		O	P				
G_2	Ξ						
$F_2\#$	O	P					

Table 1.8 illustrates one possible interpretation of Ptolemy's words. Each row of the table represents a semitone division of notes, ordered with increasing pitch from bottom to top. For the sake of simplicity, only the diatonic mode is assumed, which has two equal tones and one semitone.

On the left of the table are the names of the notes of the unmodulating system. To facilitate analysis, a possible modern names of the notes are also given on the right. A distinction is made between ♯ and ♭ according to modern usage for the 12 note chromatic scale, although this distinction is not necessary here.

The Dorian tonos is listed on Col. 4 of the table. It is defined by Greek letters arranged from top (A) to bottom (P). These letters denote the notes according to their function. As you can see, the mesē, indicated by a darker box, functionally defined as the note with two tones below and one tone above, is in the standard position. The

Hypodorian tonos, the lowest, is that of Col. 1. According to Ptolemy, it has its mesē at the hypatē mesōn. All other tonoi follow according to Ptolemy.

Table 1.8 can be derived in two ways. The first, suggested by Fig. 1.13 and made explicit by the instructions reported in the quotation just mentioned, consists in placing the mesē in its right position; then all the notes are located according to their dynamis. The second way is to fill in the gray box, delimiting the central diapason with the species of diapason defined in Table 1.7. The other notes can easily be located according to their dynamis. Note that since each mesē can be located by moving it by fifths or fourths, the first approach has the advantage of being tunable by ear.

A modern reader will not fail to see an analogy between the tonoi and the modern minor keys, extended to two diapason. There are seven tonoi: the Hypodorian is a key on E_2, the Hypophrygian on $F_2\sharp$, the Hypolydian on $G_2\sharp$, the Dorian on A_2, the Phrygian on B_2, the Lydian on $C_3\sharp$, and the Mixolydian on D_3 Observe that the association of A_2 with the proslambanomenos of the Dorian tonos should be considered conventional. In fact, while commenting Fig. 1.11, it was noted that a most likely pitch for the proslambanomenos of the Dorian tonos should be G_2 (or $G_2\sharp$) and not A_2, i.e. one tone below. This notwithstanding, in the literature the association of A_2 with the proslambanomenos is common [12],[75] because:

1. Only in the Dorian tonos are the names of the notes arranged according to their dynamis the same as those of the notes arranged according to their thesis.
2. The greater perfect system can easily be associated with an A minor scale.

1.4.2 Disseminators

1.4.2.1 The Aristoxenian Cleonides

Nothing is known about the life of the Greek Cleonides. It is a common opinion that he lived around the 2nd century CE and that he was the author of a small treatise known as the *Eisagoge harmonike* (Harmonic introduction), but there is no support for this assumption. Independently of the authorship this treatise is important for at least three reasons:

1. It is so close to Aristoxenus, even in its terminology and wording, that it can be assumed that the *Eisagoge harmonike* is a compensation for the loss of part of Aristoxenus' *Elementa harmonike*.
2. It appeared in a Latin translation by Giorgio Valla in 1497 and became one of the sources from which Renaissance musicians learned about ancient Greek music [59].[76]
3. It is well written.

---

[75] vol. 2, p. 208.

[76] Chapters 5–6.

Cleonides' little treatise begins with definitions of the basic concepts of harmonics, such as note, interval, system, tone. He did this by giving them very succinctly, as if they were familiar concepts. Particularly interesting are the definitions of note and tone, which have a non-quantitative character:

> A note is a harmonious incidence of the voice upon a single pitch. [...] A tone is any region of the voice, apt for the reception of a system; it is without breadth [29].[77]

These definitions are taken up and expanded in the following:

> The causes of the difference between pitches are tension and relaxation, their effects are height and depth. For the result of tension is to lead toward the high, that of relaxation toward the low. And height is the effect resulting from tension, depth that resulting from relaxation. Pitches are also called notes: one calls them *pitches* when they are produced by instruments that are struck, because of their being stretched; one calls them *notes* when they are produced by the voice. For to be stretched is a property of both. Considered as pitches, the number of notes is infinite; considered as functions, there are in each genus eighteen [29].[78]

Table 1.9 lists the eighteen notes he presented for each genus.

Cleonides added to the fifteen notes of the greater perfect system the notes of the synēmmenōn tetrachord, which together give the unchanging system of Fig. 1.3.

Intervals are classified according to various criteria: size, genus, symphonic and diaphonic, composite and incomposite, rational and irrational.

1. They differ in magnitude in that some intervals are major and others minor, for example, the ditone, the tone, the semitone, the diatessaron, the diapente, the diapason, and the like.
2. They differ in genus in that some intervals are diatonic, some are chromatic, and some are enharmonic.
3. They are distinguished as symphonic or diaphonic (i.e., consonant or dissonant) respectively. Symphonic intervals are: diatessaron, diapente, diapason, and the like. Diaphonic intervals are all intervals smaller than a diatessaron and all intervals between symphonic intervals. The diaphonic intervals smaller than a diatessaron are the diesis, semitone, tone, triemitone, and ditone; those between symphonic intervals are the tritone, tetratone, pentatone, and the like.
4. Intervals differ in composition in that some are incomposite and some are composite. Incomposite intervals are those bounded by successive notes, such as those bounded by hypatē and parhypatē and by lichanos and mesē; the same is true of the remaining intervals. Composite intervals are those not bounded by consecutive notes, for example, the intervals bounded by mesē and parhypatē and mesē and nētē and paramesē and hypatē. Some intervals can be classified as either composite or incomposite, namely those from semitone to ditone. In fact, the semitone is composite in the enharmonic genus (composed of two diesis), but incomposite in the chromatic and diatonic; the tone is composite in the chromatic genus, but incomposite in the diatonic; the triemitone is incomposite in the chromatic genus,

[77] p. 34.
[78] p. 34.

Table 1.9 The notes of the three Cleonides' genera; from [29], p. 36

	Diatonic	Chromatic	Enharmonic
1	Proslambanomenos	Proslambanomenos	Proslambanomenos
2	hypatē hypatōn	hypatē hypatōn	hypatē hypatōn
3	parhypatē hypatōn	parhypatē hypatōn	parhypatē hypatōn
4	diatonic lichanos hypatōn	chromatic lichanos hypatōn	enharmonic lichanos hypatōn
5	hypatē mesōn	hypatē mesōn	hypatē mesōn
6	parhypatē mesōn	parhypatē mesōn	parhypatē mesōn
7	diatonic lichanos mesōn	chromatic lichanos mesōn	enharmonic lichanos mesōn
8	mesē	mesē	mesē
9	tritē synēmmenōn	tritē synēmmenōn	tritē synēmmenōn
10	diatonic paranētē synēmmenōn	chromatic paranētē synēmmenōn	enharmonic paranētē synēmmenōn
11	nētē synēmmenōn	nētē synēmmenōn	nētē synēmmenōn
12	paramesē	paramesē	paramesē
13	tritē diezeugmenon	tritē diezeugmenon	tritē diezeugmenon
14	diatonic paranētē diezeugmenon	chromatic paranētē diezeugmenon	enharmonic paranētē diezeugmenon
15	nētē diezeugmenon	nētē diezeugmenon	nētē diezeugmenon
16	tritē hyperbolaiōn	tritē hyperbolaiōn	tritē hyperbolaiōn
17	diatonic paranētē hyperbolaiōn	chromatic paranētē hyperbolaiōn	enharmonic paranētē hyperbolaiōn
18	nētē hyperbolaiōn	nētē hyperbolaiōn	nētē hyperbolaiōn

but composite in the diatonic; the ditone is incomposite in the enharmonic genus, but composite in the chromatic and diatonic. All intervals less than a semitone are incomposite; all intervals greater than a ditone are composite.

5. Intervals can be either rational or irrational. Rational intervals are those whose magnitudes can be defined, such as tone, semitone, ditone, tritone, and the like. Irrational intervals are those that deviate from these magnitudes more or less by some irrational amount [29].[79]

This last definition is somewhat puzzling, as it seems to extend the common concept of ratio. Perhaps Cleonides meant by rational those intervals which have a unit of measure, such as the tone or a simple fraction of it.

The definition of shade is developed by Cleonides in a clearer and more complete way than by Aristoxenus. A shade of a genus is a subspecies of it. The note arrangements of a tetrachord with different shades, but belonging to the same genus, differ by a very small amount. Cleonides, with Aristoxenus, assumed that the smallest

[79] p. 38.

variation to which the notes of a tetrachord can be subjected is an enharmonic diesis equal to a quarter of the tone.

In order to understand the concept of shade, Cleonides clearly defined that of genus. The difference between the diatonic, chromatic and enharmonic genera arises in connection with the movable notes, for the lichanos is moved within the locus of one tone, the parhypatē within that of one diesis. Thus, the highest lichanos is that one tone away from the upper boundary of the tetrachord, the lowest that one ditone away. Likewise, the lowest parhypatē is a diesis away from the lower boundary of the tetrachord, the highest a semitone. The enharmonic genus is when the lichanos is at the maximum distance from the upper note, which is 2 tones. In the chromatic genus the lichanos has a distance less than 2 tones but greater than or equal to 1.5 tones. In the diatonic genus the distance should be less than or equal to one and a quarter of a tone.

There is only one shade of the enharmonic genus; it uses the division of the genus itself, since moving the lichanos above or below an enharmonic diesis—a quarter of a tone—results in a change of genus.

There are three shades of the chromatic genus:

1. The soft chromatic; it is defined by a chromatic diesis of a third of a tone, another diesis equal to it, and the equivalent of a tone plus a semitone plus a third of a tone.
2. The hemiolic chromatic; it is defined by a hemiolic diesis equal to one and a half times the enharmonic diesis, another diesis equal to it, and an incomposite interval equal to seven enharmonic diesis.
3. The tonic chromatic uses the shade of the genus itself, since it is sung by semitone, semitone, and triemitone.

There are two shades for the diatonic genus; the soft and the intense. The shade of the soft diatonic is defined by a semitone, plus an uncomposed interval equal to three enharmonic diesis, plus an interval equivalent to five enharmonic diesis. That of the intense diatonic participates in the division of the genus itself, since it is made up of semitone, tone and tone.

The shades defined by Cleonides are exactly those of Aristoxenus, as shown in Table 1.3. They, according to Cleonides, are more conveniently represented by integer numbers. It is assumed that the tone is divided into 12 minimal parts, each of which is called a twelfth of a tone. It is also assumed that the remaining intervals are divided in the same proportion; the semitone into 6 twelfths, the diesis of a quarter of a tone into 3 twelfths, the diesis equivalent to a third of a tone into 4 twelfths, the whole diatessaron into 30 twelfths. Quantitatively, therefore, the enharmonic has 3, 3, and 24 twelfths; the soft chromatic 4, 4, and 22; the hemiolic chromatic $4^1/_2$, $4^1/_2$, and 21; the tonic chromatic 6, 6, and 18; the soft diatonic 6, 9, and 15; the intense diatonic 6, 12, and 12 [29], [80] as summarized in Table 1.10.

In the definition of species, Cleonides departed from Aristoxenus from a methodological point of view. The latter, at the end of his Book III of *Elementa harmonica*,

---

[80] p. 40.

Table 1.10 Genera and shades according to Cleonides

Intervals	Enharmonic	Soft Chromatic	Hemiolic Chromatic	Tonic Chromatic	Soft Diatonic	Intense Diatonic
Mesē-lichanos	24	22	21	18	15	12
Lichanos-parhypathē	3	4	4 1/2	6	9	12
Parhypathē-hypathē	3	4	4 1/2	6	6	6

Fig. 1.14 The species of the diatessaron

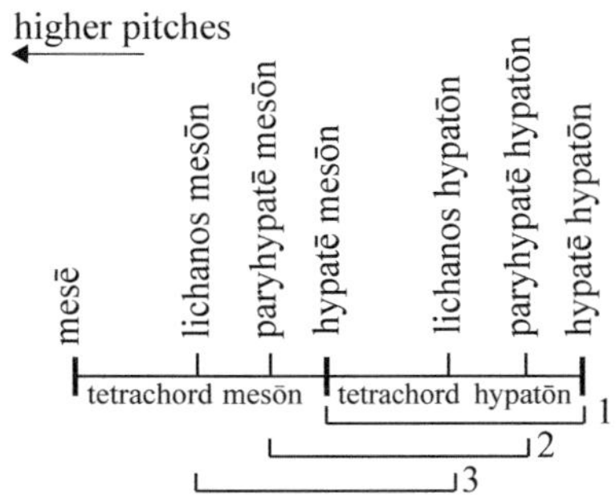

defined the species of a composite interval as a variation of the order of its simple part, while both the number and magnitude of these parts remain the same. Thus, Aristoxenus considered as species all possible permutations of the interval under consideration. Starting from this definition, it can be shown that there are, for example, three kinds of the diatessaron. First, there is that in which the pyknon is at the bottom; second, that in which a diesis is on each side of the ditone; third, that in which the pyknon is above the ditone. It will be readily seen that there are no other possible relative positions of the parts of the diatessaron [3].[81]

Cleonides, although obtaining the same kind of species as Aristoxenus, had an approach closer to that of Ptolemy, for whom the species of the various intervals, diatessaron, diapente, diapason, are found by taking different windows of the greater perfect system.

Below we report Cleonides' description of the species only for the diatessaron; for clarity, the first two tetrachords of the greater perfect systems are redrawn in Fig. 1.14.

Of the three species of the diatessaron of the enharmonic and chromatic genera, the first [1] is that bounded by barypykna (lower pykna), as is that from hypatē hypatōn to hypatē mesōn (the first tetrachord of the greater perfect system); the second [2], bounded by mesopykna, as is that from parhypatē hypatōn to parhypatē mesōn (a

[81] p. 222. In fact, the possible dispositions of the intervals in a diapente are six, but only three of them are musically interesting [9], p. 232.

tetrachord shifted by a semitone); the third [3] bounded by oxypykna (upper pykna), such as that from lichanos hypatōn to lichanos mesōn (a tetrachord shifted by two semitones). In the diatonic genus, however, the species do not occur in conjunction with a pyknon, since this genus has no pykna. Thus, in the diatonic genus, the first species of the diatessaron is that in which the semitone is below the tones, the second is that in which it is above the tones, the third is that in which it is between the tones. And these species begin and end with the same notes as in the other genera [29].[82]

To conclude with Cleonides, we report his comments on the words tone/tonoi.

> The word *tone* is used in four senses: as note, as an interval, as a region of the voice, and as a pitch [...]. We use it [the word tone] of the region of the voice whenever we speak of Dorian, or Phrygian, or Lydian, or any of the other tones. According to Aristoxenus there are 13 tones [tonoi]:[83]
>
> a Hypermixolydian, also called Hyperphrygian;
>
> b Two Mixolydians, a higher and a lower, of which the higher is also called Hyperiastian, the lower Hyperdorian;
>
> c Two Lydians, a higher and a lower, of which the lower is also called Aeolian; Two Phrygians, a higher and a lower, of which the lower is also called Iastian;
>
> d One Dorian;
>
> e Two Hypolydians, a higher and a lower, the latter also called Hypoaeolian;
>
> f Two Hypophrygians, of which the lower is also called Hypoiastian;
>
> g One Hypodorian.
>
> Of these the highest is the Hypermixolydian, the lowest the Hypodorian. From the highest to the lowest, the distance between consecutive tones is a semitone, between two parallel tones a trihemitone; with the distance between the remaining tones the case will be similar. The Hypermixolydian is a diapason above the Hypodorian.[29].[84]

That is, Cleonides attributed 13 tonoi to Aristoxenus, corresponding to the 13 semitones of the first diapason of the greater perfect system. The keys of the tonoi begin at A_2 and move by semitones to A_3 (one diapason higher). After locating the proslambanomenos, the next 14 notes follow according to the organization of the intervals of the greater perfect system.

Table 1.11 shows in Col. 1 the names of notes of the unmodulating system, in Col. 2 the names of the tonoi having as proslambanomenos the notes of Col. 1; in Col. 3 the names of the modern notes; stars indicate tonoi corresponding to Ptolemy's tonoi. With respect to Cleonides, the table changes *iastian* to *ionian* and omits the optional names. As you can see, by comparing Tables 1.11 and 1.8 Cleonides' notes are one fourth higher [36].[85]

---

[82] p. 41.

[83] Actually in Aristoxenus extant writings there is not such precise reference to tonoi.

[84] p. 44.

[85] pp. 8–12.

Table 1.11 Cleonides's tonoi; from [77], vol. 1, p. 41

Note in Greek Scale.		Greek Key.	Modern Equivalent for Greek Key
Mese	A	Hyperphrygian (= *Hypodorian*)	A minor
Semitone		Hyperionian	G♯ minor
Lichanos mesôn	G	*Mixolydian	G minor
Semitone		*Lydian	F♯ minor
Parhypate mesôn	F	Aeolian	F minor
Hypate mesôn	E	*Phrygian	E minor
Semitone		Ionian	D♯ minor (*or E♭ minor*)
Lichanos Hypatôn	D	*Dorian	D minor
Semitone		*Hypolydian	C♯ minor
Parhypate Hypatôn	C	Hypoaeolian	C minor
Hypate Hypatôn	B	*Hypophrygian	B minor
Semitone		Hypoionian	A♯ minor (*or B♭ minor*)
Proslambanomenos	A	*Hypodorian	A minor

1.4.2.2 The Last of the Greeks: Severinus Boethius

The Latin polymath Anicius Manlius Severinus Boethius (480–524), about the early 6th century CE, began to transfuse into Latin the great deal he had learned from Greek. As first he wrote treatises on mathematics and physical mathematics, part of what he called *quadrivium*, a term adopted since then to indicate the four liberal arts: geometry, arithmetic, astronomy and music. Only three of these treatises have survived, *De institutione arithmetica*, *De geometria* and *De institutione musica* [19].[86]

The *De institutione musica*—like the other writings of Boethius—is not an original work, being mainly a collection of free and clever translations from Greek sources; for this reason we consider the Latin Boethius as the last of the Greeks. His treatise is very important from a historical point of view. In fact, it played a key role in the history of music from the Middle Ages until at least the 16th century, and it contains fragments of Greek music theorists who are now lost, such as Nicomachus. Since the rise of the universities, the *De institutione musica* was the fundamental text on music, studied in the quadrivium. It was one of the first musical works to appear in print, in Boethius' *Opera* in 1491–1492 [18]. Its English translation, however, is quite recent [20, 22]. In the following we have based on the edition by Gottfried Friedlein (1828–1875), the first and only critical edition [19].

[86] Introduction.

De institutione musica has reached us in an incomplete form; it breaks off in the middle of book 5, while a whole of eleven books was planned [19].[87] Except for brief sections with philosophical aims, the book is a strictly physical mathematical treatise, quite technical, in the Euclidean tradition. Probably because of the supposed lack of originality of the technical part, the philosophical sections, dealing with both natural philosophy and metaphysics, are given prominence in textbooks on the history of music. These sections concern the classification of the various types of music and musicians, some considerations on the nature of sound, a legend about Pythagoras' discovery of the ratios of consonances. A limited space is also devoted to the harmony of the spheres.

The following is only a brief summary of Boethius' work. To begin with, physical considerations are dealt with, then some relevant points of Greek harmonics, possibly derived from Euclid, Nicomachus and Ptolemy. Finally, the theory of tonoi.

Notations

Boethius introduced in speculative music a new nomenclature for the ratios, based on Nicomachus *Aritmetica*. In particular ratios of the form $(m : n)$ are named superparticular if $m = n + 1$, superpartient if $m > n + 1$. More specific names, still taken from Nicomachus, are associated to particular values of m and n. This nomenclature is simpler and clearer than the traditional Greek one, i.e. epimoric, epgodic, eptrictic, hemiolic.

There are Three Kinds of Music and Musicians According to Boethius, there are three kinds of music. "The first is cosmic; the second is human; the third is that which resides in certain instruments, such as the kithara or the aulos or other instruments that serve melody" [19].[88]

1. Cosmic music concerns those things that are observed in the heavens, or in the combination of the elements, or in the variety of the seasons. Although the sound of the heavens cannot be heard, it is impossible for an extremely fast motion of large bodies such as planets not to make a sound. Similarly, how could the four elements unite without some harmony?
2. Human music is perceived by those who look within themselves. For what is it that unites the immaterial nature of reason with the body, if not a certain harmony, as it were, a careful tuning of low and high pitches, as if to produce a consonance? What but this unites the parts of the soul, which, according to Aristotle, "is composed of the rational and the irrational" [19].[89]
3. Instrumental music is that which is produced by musical instruments or the human voice. "This music is governed either by tension, as in strings, or by breath, as in

[87] p. 177.

[88] p. 187.

[89] p. 189. Translation into English in [20].

the aulos or those instruments activated by water, or by a certain percussion, as in those which are cast in concave brass, and various sounds are produced from these" [19].[90]

Accordingly, there are three classes of those who are engaged in music. The first class consists of those who perform on instruments, the second of those who compose songs, and the third of those who judge instrumental performances and songs [19].[91] Those who are dependent upon instruments and devote all their efforts to them—such as citharists and players of organ and other musical instruments—do not have complete musical knowledge, for they are practically slaves, obeying the composers. The second class of those who practice music is that of composers. This class is also separated from music because they are guided by a certain natural instinct rather than by reason. The third class is made up of those who have enough ability to judge rhythms, melodies, and composition as a whole. Since they are based on reason and thought, they are the only ones who can rightly be called musicians.

Elements of Voices and Music
In the following quotation one can read the hypothesis of a periodic motion of sound. But the pairing of *fast* with *frequent* makes it difficult to understand what Boethius really meant:

> For this reason, *sound is defined as a percussion of air* [emphasis added] remaining undissolved all the way to the hearing. Some motions are faster, others slower; some motions are less frequent, others more frequent. If someone regards an uninterrupted motion, he will necessarily observe in it either speed or slowness; moreover, if someone moves his hand, he will move it in either a frequent or less frequent motion. If motion is slow and less frequent, low sounds are necessarily produced by the very slowness and infrequency of striking. But if motions are fast and more frequent, high sounds are necessarily produced. For this reason, if the same string is made tighter, it sounds high, if loosened, low. For when it is tighter, it produces faster pulsation, recurs more quickly, and strikes the air more frequently and densely. The string that is looser brings about lax and slow pulsations, and, being less frequent because of this very weakness of striking, does not vibrate very long [19].[92]

How we Hear

> At this time we should discuss how we hear. The same thing happens in sounds that happens when a stone, thrown from above, falls into a puddle or into quiet water. First it causes a wave in a very small circle; then it disperses clusters of waves into larger circles, and so on until the motion, exhausted by the spreading out of waves, dies away [...]. In the same way, then, *when air pulses creates sound* [emphasis added], it affects other air nearby and in this way sets in motion a circular wave of air; and so it is diffused and reaches the hearing of all standing around at the same time. The sound is fainter to someone standing at a distance, since the wave of activated air approaches him more weakly [19].[93]

Here is the celebrated analogy between sound waves and the circular propagation of waves in a puddle of still water disturbed by a thrown stone. The analogy is not

[90] p. 189. Translation into English in [20].
[91] pp. 224–225.
[92] pp. 189–190. Translation into English in [20].
[93] p. 200. Translation into English adapted from [20].

original to Boethius however; it is found in an even more expressive form in some fragments attributed to Chrysippus (3rd century BCE) and the Stoics.

> The Stoics teach that the air does not consist of small particles, but is a continuum without any empty space between. If the air receives an impact by compressed air (pneuma), it oscillates spherically until the end, until it has filled the surrounding air. The same happens if a stone is thrown into a pond, where the oscillation is cyclical, while in the air it is spherical [27].[94]

but probably the analogy can be traced back to the fifth century BCE.

In the two passages cited above, Boethius presented two alternative conceptions of the propagation of sound. In the first quotation sound is conceived as a movement (percussion) in the air; in the second as a movement of the air (pulses). Both positions were advocated in antiquity. The first, of a more abstract nature, was supported by Aristotle, among others, for whom the air [the medium] itself does not move but only carries the form impressed to it [65].[95] Only in the 17th century, with air conceived as a collection of elastic particles colliding with each other, a mechanistic convincing (for a modern reader) explanation was given for the motion in the air.

Pythagoras Tale

An ancient written version of how Pythagoras discovered the ratios of consonances are attributed to Nicomachus of Gerasa, see Sect. 1.3.2. Here we consider the version of Boethius, which, although derived from Nicomachus, is slightly different and therefore worth repeating, also because in the Middle Ages Pythagoras' role in music was known through Boethius' narrative.

The tale goes that Pythagoras, passing by a blacksmith's shop, heard the sound of hammers, producing consonances of different sounds. He thought for a moment that the variety of consonances was due to the strength of the men hammering. But after an experiment in which he changed the force of the hammering, he realized that the property of the sounds did not depend on it. Having established this, Pythagoras examined the weight of the hammers. There were five hammers, and those that sounded together the consonance of the diapason were twice as heavy. He then found that by using pairs of different hammers, the consonance of diapente and diatessaron could be obtained. Boethius emphasized that although some musical consonances had been called diapason, diapente, and diatessaron before Pythagoras, he was the first to determine that the weights of the four hammers were contained in the numbers $12 : 9 : 8 : 6$. The weight of the first hammer to the second, which gives a diapason, obeys the ratio $(12 : 6) = (2 : 1)$, that of the second to the fourth, which gives a diapente, obeys the ratio $(12 : 8) = (3 : 2)$, and that of the first to the second, which gives a diatessaron, obeys the ratio $(12 : 9) = (4 : 3)$.[96]

The tale continues. After returning home, Pythagoras made various experimental observations to see whether or not the consonances continued to obey these ratios. First, he attached different weights to strings of the same property and length and

[94] 425, p. 140. Translation into English in [65].

[95] p. 255.

[96] Note that in the version referred to by Nicomachus there is no reference to the numerical values of the weights of the hammers.

determined their consonances by ear; then he considered pipes of different lengths. By means of measurement, he also poured ladles of different weights into glasses. Finally, he turned to the length and thickness of strings. In conclusion, he claimed that the rules he found for hammers were also valid in all cases [19].[97]

Of course, not only the account of the hammers is against the evidence, as already commented in Sect. 1.3.2, but also the actual behavior of strings stretched by tensions in the ratio 12 : 9 : 8 : 6 is contrary to what is claimed in Pythagoras' tale, since pitches are proportional not to the tension but to its square. Despite the lack of foundation, Boethius' account was taken for granted for centuries, at least by theorists. It was not until Vincenzo Galilei in the 16th century that the tale was demistyfied, see Chap. 2.

Elements of Music

Boethius reported the main notions of Greek harmonics. The references to the authors he drew on, and there are many, are not always clear. Certainly Nicomachus and Euclid play a major role in the first four books, while the last book, the fifth, is largely derived from Book I of Ptolemy's *Harmonica*. The possible sources are specified quite precisely in the annotated translation of *De institutione musica* [20].

One of Boethius' earliest definitions is that of sound: "sound [sonus] is a melodic instance of pitches" [19].[98] That is, sound is introduced with a narrow meaning, that of note, also called *phthongos* from Greek. The interval is the distance between a high and a low note. Consonance is a merging of two sounds that fall pleasantly on the ears. Dissonance, on the other hand, is an unpleasant combination of two sounds that come to the ear intermingled, but not blended.

The pleasantness of consonances is associated with Pythagorean mysticism, according to which the greater the pleasantness of a consonance, the greater the elegance of the ratio expressing it, that is, for ratios expressed by small numbers. Thus the following hierarchy is established: diapason (2 : 1), diapente (3 : 2), diatessaron, ratio (4 : 3).

Ratio, Proportion and Mean

> A ratio is a certain comparison of two terms measured against themselves. A proportion is a collection of equal ratios. A proportion consists of at least three terms, for when a first term related to a second holds the same ratio as the second to the third, then this is called a "proportion", and the "mean" among these three terms is that which is second [19].[99]

According to Boethius, there are three different kinds of proportions, and consequently of means.

1. The arithmetic proportion. When the difference between the smallest term and the mean term is equal to the difference between the mean and the largest term. For example, for the numbers 1 : 2 : 3, unity is the difference between 1 and 2 and 2 and 3. The arithmetic mean is 2, the second term.

---

[97] p. 197. Translation into English adapted from [20].

[98] p. 195.

[99] p. 241. Translation into English in [20].

2. The geometric proportion. If there is an equal ratio between between both pairs. For example, for the numbers 1 : 2 : 4, 2 : 1 is a double, as is 4 : 2. The geometric mean is 2.
3. The harmonic proportion. When the largest term is related to the smallest in the same way that the difference between the larger terms is related to the difference between the smaller terms. For example, in the numbers 3 : 4 : 6, 2 is between 6 and 4, while unity is between 4 and 3, but 2 is twice as large as unity. The harmonic mean is 4.

Boethius then gave a more operational definition of the mean by considering the case of the numbers 10 and 40. For the arithmetic mean, the difference of the given terms is divided by two and added to the smaller term (the arithmetic mean is 25). Similarly, for a geometric mean, we multiply the extremes and take their square root (the geometric mean is 20). More complicated is the definition of the harmonic mean, which is given as an algorithm. Add the extremes, 10 plus 40 equals 50. Multiply the difference of these terms, which is 30, by the smaller term, 10, which is 300. Divide that by 50, which is 6. This is added to the smaller term 10, which gives 16. This is the harmonic mean [19].[100]

Ratio of a Semitone

An interesting part of Boethius' work concerns the evaluation of the ratio of a semitone. It is immediately made clear that a semitone does not mean half a tone, but simply an interval smaller than a tone. It is defined as the interval obtained by subtracting two tones from the diatessaron. It is very easy for a modern reader to obtain the semitone ratio by identifying ratios with fractions. The tone is represented by the ratio (9 : 8), the diatessaron by (4 : 3), so the semitone is given by

$$(4:3) - 2(9:8) \equiv 4/3 : (9/8)^2 \equiv (256:243)$$

Boethius, using Nicomachus' arithmetic, had to proceed in a somewhat more complicated way. The diatessaron is defined by the ratio (4 : 3), conveniently rewritten as (256 : 192), by multiplying each term by 64. Dividing 192 by 8 and adding the result is 216 (in modern terms, this means multiplying 192 by 9/8). Repeating the process gives 243. The series of ratios can thus be written as:

$$192 : 216 : 243 : 256$$

It is clear that the semitone is given by the last ratio, since the first two ratios represent two tones.

The semitone (256 : 243), commonly known as the *limma*, is also called by Boethius *diesis*: "the interval which we call a semitone was called by the ancients *limma* or *diesis*" [19].[101] Note, however, that the term diesis is not used consistently

[100] p. 241.
[101] p. 260.

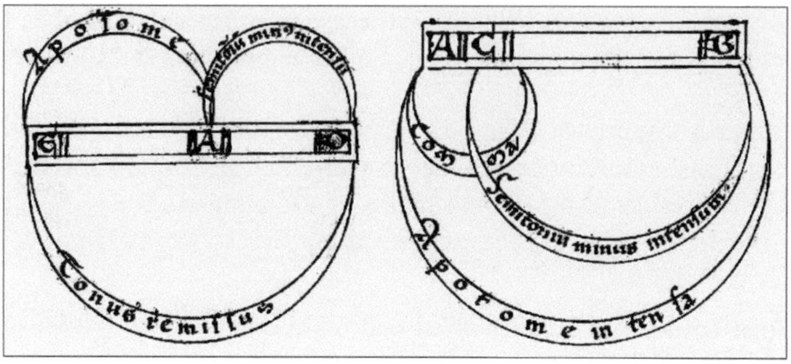

Fig. 1.15 Tone, limma, apotome and comma. Redrawn from [21], p. 52, unnumbered

in the *De institutione musica*. It is generally reserved for very small intervals, such as the quarter tone of Aristoxenus [19].[102]

The remaining part of the tone, after a semitone is subtracted is named *apotome* or major semitone. Its ratio is found by considering the following sequence of ratios:

$$1944 : 2048 : 2187$$

where (2187 : 1944) is the same as (9 : 8), that is a tone; the ratio (2048 : 1994) is the same of the ratio of the semitone (256 : 243). This means that the ratio (2187 : 2048) represents the apotome, and is greater than the ratio of the semitone. The difference between the apotome and the semitone is called a *comma* (Fig. 1.15) and is given by:

$$(2187 : 2048) - (256 : 243) = (531,441 : 524,288)$$

Notice that Boethius' comma is the same as the Pythagorean comma, the failure to the closure of the circle of 12 fifths.

Theory of Modes
Boethius' theory of tonoi, which he called modes, is closely related to that of the species of consonances, diatessaron, diapente, diapason, with a certain kinship to Ptolemy. The species of diapason, the only ones discussed here, are the most interesting; they are derived from the greater perfect system by translating a window that spans a diapason. Their nature can be understood by referring to Table 1.12, which shows a two diapason system, without the proslambanomenos, as in Ptolemy.

Note that Boethius used the (Latin) letters from A to O (and P) in a quite consistent way; that is, each Greek name is assigned the same Latin letter. This was not the case with Ptolemy, who used different Greek letters in different parts of his work. For this reason, Boethius' way of using letters is now called *Boethian notation*.

102 p. 213.

Table 1.12 The two diapason system according to Boethius. Adapted from [19], p. 341

A	Hypatē hypatōn
B	Parhypatē hypatōn
C	Lichanos hypatōn
D	Hypatē mesōn
E	Parbypatē mesōn
F	Lichanos mesōn
G	Mesē
H	Paramesē diezeugmenōn
I	Tritē diezeugmenōn
K	Paranētē diezeugmenōn
L	Nētē diezeugmenōn
M	Tritē hyperboleōn
N	Paranētē hyperbōleon
O	Nētē hyperboleōn

higher

Table 1.13 The seven Boethius's species of the diapason

OG	T	T	S	T	T	S	T
NF	T	S	T	T	S	T	T
ME	S	T	T	S	T	T	T
LD	T	T	S	T	T	T	S)
KC	T	S	T	T	T	S	T
IB	S	T	T	T	S	T	T
HA	T	T	T	S	T	T	S

Unlike Ptolemy, Boethius' windows move from lower to higher pitches. The resulting species of diapason are listed in Table 1.13. This table is the same as Table 1.7, with the rows in reverse order.

Boethius more strongly than Ptolemy declared that modes come from the species of diapason:

> From the species of the consonance of the diapason exist (come into existence) those things called modes. They call them also *tropes*, or *tones* [19].[103]

If, continued Boethius, one were to raise or lower whole systems, according to the species of the diapason, he will bring about the seven modes: Hypodorian, Hypophrygian, Hypolydian, Dorian, Phrygian, Lydian and Mixolydian, according to the rules:

> The arrangement of the modes proceeds in the following manner. Set out the succession of pitches in the diatonic genus from the proslambanomenos to the nētē hyperboleon. Let this be the Hypodorian mode. If one were to raise the proslambanomenos by one tone, and further raise the hypatē hypatōn by the same tone, thereby making the whole disposition higher by

[103] p. 341. Translation into English in [20].

Table 1.14 Boethius' modes. Adapted from [23], p. 261

	Hypodorian	Hypophrygian	Hypolydian	Dorian	Phrygian	Lydian	Mixolydian	Hypermixolydian
								A5
							G5	G5
						F5#		
							F5	F5
					E5	E5		E5
							E5♭	
				D5	D5	D5	D5	D5
			C5#			C5#		
				C5	C5		C5	C5
		B4	B4		B4			B4
				B4♭			B4♭	
nētē hyperbolaiōn	A4	A4	A4	A4	A4	A4	A4	A4
			G4#			G4#		
pararmētē hyperbolaiōn	G4	G4		G4	G4		G4	G4
		F4#	F4#		F4#	F4#		
tritē hyperbolaiōn	F4			F4			F4	F4
nētē diezeugmenōn	E4	E4	E4	E4	E4	E4		E4
			D4#				E3♭	
paranētē diezeugmenōn	D4	D4		D4	D4	D4	D4	D4
		C4#	C4#			C4#		
tritē diezeugmenōn	C4			C4	C4		C4	C4
paramesē	B3	B3	B3		B3	B3		B3
				B3♭			B3♭	
mesē	A3	A3	A3	A3	A3	A3	A3	A3
			G3#			G3#		
lichanos mesōn	G3	G3		G3	G3		G3	
		F3#	F3#		F3#	F3#		
paryhypatē mesōn	F3			F3				
hypatē mesōn	E3	E3	E3	E3	E3			
			D3#					
lichanos hypatōn	D3	D3		D3				
		C3#	C3#					
paryhypatē hypatōn	C3							
hypatē hypatōn	B2	B2						
proslambanomenos	A2							

a tone, then the entire succession would turn out higher than it was before it was raised by a tone. Thus this whole system, having been made higher, forms the Hypophrygian mode. Now if the pitches of the Hypophrygian mode were similarly raised by a tone, the collection of pitches for the Hypolydian mode would be born. If someone raised the Hypolydian by a semitone, he would make the Dorian. The progression to higher pitch is similar in the other modes [19].[104]

The above instructions are slightly different from those of Ptolemy and close to those of Cleonides. In fact, Ptolemy defined the position of the various tonoi in terms of fourths or fifths, using an acoustic criterion. Boethius, referring to the species of diapason, said more simply that each mode is higher or lower than the next by a tone or a semitone, which is a much more abstract approach.

Table 1.14 makes clear the way Boethius' modes are generated; here each row of the table is a semitone away from the next, and can be thought of as a line of a staff.

[104] p. 343. Translation into English in [20].

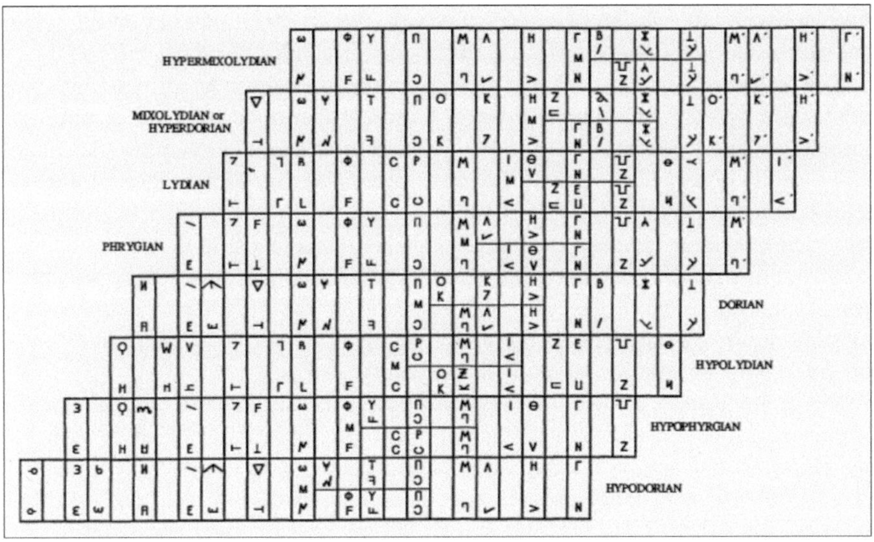

Fig. 1.16 Boethius's modes with ancient notation of pitches; from [19], p. 343 addendum

Unlike Ptolemy, Boethius preferred to name the notes according to their pitches. In the table, the notes are indicated by modern symbols for the sake of uniformity; Boethius used Greek notation instead. Although it is difficult to interpret the meaning of the symbols of Fig. 1.16, it can be seen that the general trend is similar to that of Table 1.14.

The first mode, the Hypodorian, is obtained by placing the proslambanomenos at the lowest pitch, A_2. The second mode begins one tone higher than the first; tones and semitones follow with the proslambanomenos starting from B_2. The third mode starts from $C_3\sharp$, since C_3 is only a semitone higher than B_2, while we need a tone. And so on until we reach the seventh mode, which starts at G_3. Boethius added an eighth mode in the last column (the Hypermixolydian), which Ptolemy ignored because for him it was the same as the first. The shift to the upper part is clearly seen by the darker box indicating the position of the mesē.

To note that with respect to Ptolemy's, as in Cleonides, the Dorian tonos starts with D_3 instead of A_2, one fourth higher.

1.5 Harmony of the Spheres

From the very beginning of their theoretical elaboration, the Greek scholars firmly believed that the study of harmonics was not primarily concerned with producing pleasant melodies. Rather, they believed, as do most modern Western scientists, that there are general ordering principles, perhaps even powers, that act on all beings. The purpose of harmonics was to discover such principles.

The Pythagoreans claimed that the harmonies they discovered in sounds were merely the tip of the iceberg of a larger harmony that affected all of creation: the harmony of the celestial spheres. They believed, for example, that the planets followed orbits governed by ratios of consonances, and that their motion produced a celestial music in which the fastest (to them, the most remote) planets emitted the most acute pitches. Humans could not hear this music because they were exposed to it from birth and were therefore addicted to it. But in principle it was possible to hear it, and the divine Pythagoras had been able to do. This idea was opposed by Aristotle in the *De caelo*. Here, more interesting than the refutation, is an extended and interesting description of the Pythagorean thesis [2].[105]

Little, however, is known with certainty about the original form of the doctrine of planetary music, since neither Pythagoras nor his immediate successors, including Philolaus, left any written record. The first allusion to the doctrine appears in Plato's *Republic*, where Socrates refers to the Pythagorean view that "astronomy and harmonics are sisters" [62].[106] But it is in the myth of Er that Plato gives the first detailed account of the harmony of the spheres:

> Now the whole spindle has the same motion; but, as the whole revolves in one direction, the seven inner circles move slowly in the other, and of these the swiftest is the eighth [the sphere of the fixed stars]; next in swiftness are the seventh, sixth, and fifth, which move together; third in swiftness appeared to move according to the law of this reversed motion the fourth; the third appeared fourth and the second fifth. The spindle turns on the knees of Necessity; and on the upper surface of each circle is a siren, who goes round with them, hymning a single tone or note. The eight together form one harmony; and round about, at equal intervals, there is another band, three in number, each sitting upon her throne: these are the Fates, daughters of Necessity, who are clothed in white robes and have chaplets upon their heads [62].[107]

The myth of Er escapes a strictly musical reading. The harmony of the heavens derives from the music of the sirens, each for each planet, and the singing of the fates, but the association is not easy; for some comments see [61].

There are eight concentric whorls that fit into each other and represent, from the outer to the inner orbit, the seven planets and the spheres of the fixed stars, in the order Saturn, Jupiter, Mars, Venus, Mercury, Sun and Moon. According to the model of the Pythagoreans and Eudoxus, they move faster the closer they are to the sphere of the fixed stars The spindle as a whole moves to the right, while the whorls of the planets move in the opposite direction. Plato also referred to the harmony of the soul as derived from music. In the *Timaeus* he applied the division of the diatonic tetrachord to the human soul, see Sect. 1.3.3.

In the earliest surviving texts on harmonics, those of Aristoxenus and Euclid, there is no reference to the harmony of the spheres. This circumstance does not mean that

---

[105] 2.9.290b-291a.

[106] 530d.

[107] 617a-c.

harmonic scholars with a solid mathematical training did not take the harmony of the spheres seriously. Although the harmony of the spheres was largely commented on by philosophers, in the following we will present the positions of some mathematicians, from Nicomachus to Boethius, who considered the harmony of the spheres as a very important argument to be dealt with.

1.5.1 Nicomachus of Gerasa

One of the most interesting scolar of harmonics is Nicomachus of Gerasa (c. 50–150) CE, three to four centuries after Euclid. A native of ancient Gerasa (corresponding to modern Jerash, now in Jordan), he was an important figure in the popular Neo-Pythagorean movement of the first and second centuries. His *Enchiridion harmonies* (Introduction to harmonics) would later serve as a model for Boethius and a standard textbook for the medieval Christian and Islamic worlds. In his own day and for generations to come, Nicomachus with his *Arithmetic* [56], was to arithmetic what Euclid was to geometry. Indeed, it was even said of Nicomachus: "Arithmeticam Samius Pythagoras invenit, Nicomachus scripsit (Pythagoras of Samos invented arithmetic, Nicomachus composed it)" [57].[108]

He included in his *Enchiridion harmonies* everything in the matrix of the *tetraktys* (1 : 9 : 8 : 6): the diatonic scale; the motion of the planets and the nature of the soul. Below we will comment only on one writing about the harmony of the planets. Nicomachus proposed an interesting (though improbable) etymology of the names of musical notes, according to which they come from the planets (stars) that cross the heavens and move around the earth. Assuming that all bodies necessarily produce sounds when they are vibrated, and that the sounds differ from one another in loudness and pitch either because of the weight of the bodies or their particular speeds, he commented on the situation of the solar system as follows

> The movement of Kronos is the farthest up from us, the lowest pitched note in the diapason was called hypatē, for hypatōn means the highest up. On the other hand, from the fact that the course of the Moon is the lowest of all and is situated nearer to the earth, the name neate [nētē] was derived, for neaton signifies the lowest. The term parhypatē was derived from the position of Zeus, being below Kronos on one side, while at the other side, the position of Aphrodite, being above the Moon, occasioned the name, paranētē. The term mesē was derived from the Sun's most central position, this being the fourth from either end, since the mesē in fact stood at a distance of one fourth in the ancient heptachord. The term hypermesē, also called lichanos, was derived from one of the positions on either side of the Sun, that of Ares, which was assigned the sphere between Zeus and the Sun. On the other side of the Sun between Aphrodite and the Sun, the position occupied by Hermes provided the name paramesē [43, 57].[109]

Figure 1.17a shows the geocentric cosmology of Nicomachus, with thc association of the notes sung by the planets with their distance form the earth.

---

[108] Introduction.

[109] pp. 10–11; 45–46.

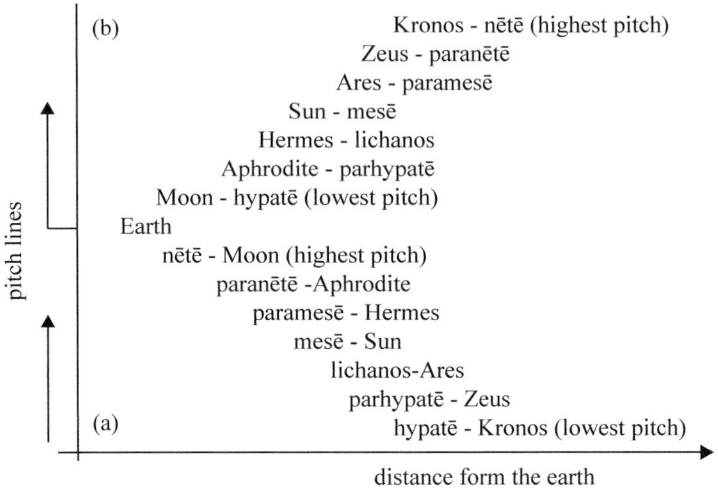

Fig. 1.17 Pitches versus distances of planets. Adapted from [57], p. 51

We are dealing with two conjoint tetrachords (synēmmenōn-diegzeumenōn)[110] after a proslambanomenos (the Earth).

It should be noted that in some fragments Nicomachus told a different and opposite tale of the harmony of the planets, possibly preceding the one reported above, based on the view shared by Pythagoras and his followers, the planets move faster the closer they are to the fixed stars [57],[111] see Fig. 1.17b. Remember that the central position of the Sun in the order of the planets was not asserted before the Alexandrine period. In Plato, for example, the order of the planets was Moon, Sun, Venus, Mercury, Mars, Jupiter, Saturn. A similar treatment of the harmony of the stars is found in Boethius.

> At this point it would seem proper to add concerning the strings of a tetrachord and the celestial order. The hypatē mesōn is assigned to Saturn, whereas the parhypatē is like the orbit of Jupiter. The lichanos mesōn is entrusted to Mars. The Sun governs the mesē. Venus holds the tritē synēmmenōn. Mercury rules the paranētē synēmmenōn. The nētē is analogous to the orbit of the moon [19].[112]

Boethius comparison of strings to the disposition of the heavenly spheres agrees with those assumed by Nicomachus, apart from an inversion between Venus and Mercury.

[110] Indeed paramesē in Nicomachus can be equated to trite [57], p. 48.

[111] p. 54.

[112] p. 219. Translation into English in [20].

1.5.2 The Astrologer Ptolemy

A very detailed description of the functions of the harmony of the spheres is contained in Book III of Ptolemy's *Harmonica*, which is entirely devoted to the subject. Given the strictly technical approach of the first two books, the modern reader is astonished to find so much space devoted to what is now considered at most a fanciful subject. But this was not so for Ptolemy, nor would it be for Kepler, more than a millennium later. In any case, Ptolemy's treatment eliminated many of the Pythagorean notions that tended toward mysticism. For example, Ptolemy did not say that the planets made potentially perceptible sounds.

The list of titles of the chapters of Book III gives a clear idea of Ptolemy's conceptions on the harmonics of the spheres:

1. How through the whole system we might both use and judge the ratios by means of the fifteen-stringed canon.
2. Methods for dividing up to the double diapason with only the eight notes.
3. In which genus must we place harmonic function and our knowledge of it?
4. That the function of the harmonic exists in the natures of all the more perfect things and that it appears particularly in both human souls and heavenly motions.
5. How the consonances are in accord with the first differences of the soul and their species.
6. A comparison between the harmonic genera and those of the first virtues.
7. How harmonic modulations resemble the circumstantial modulations of souls.
8. On the similarity of the perfect system and the circle through the middle of the Zodiac.
9. How the harmonic consonances and dissonances resemble those in the Zodiac.
10. That succession in notes resembles the longitudinal motion of the stars.
11. How the stellar motion in altitude compares with the harmonic genera.
12. That modulations of tonoi are like stellar crossings in latitude.
13. On the similarity of the tetrachords and the aspects of the Sun.
14. By what first numbers might the fixed notes of the perfect system be compared to the first spheres in the cosmos.
15. How the ratios of their motions are calculated by the numbers.
16. How the combinations of the planets should be compared to those of the notes [66].[113]

Here are just a few of the situations commented on.

Ptolemy stated that all living things in nature, in so far as they are healthy and flourishing, are so in virtue of their maintenance of the proper harmonious proportions in their motions and among the elements of which they are composed. The most complete harmony are found in the motions of the heavenly bodies, and those of

[113] pp. 82. Translation into English in [67].

human souls, most since it is "only to each of these that there belong not only the primary and complete sort of movement, that in respect of place, but also the characteristic of being rational [66].[114]

But the astronomer-astrologer, Ptolemy, considered the motion of the cosmos—as studied by astronomy—as that which mostly follows the power of harmonics:

> It is so in the most rational of the relevant sciences, namely Astronomy, which is concerned with the sight and the movement of heavenly bodies which are only visible, and Harmonics, which is concerned with hearing and the movement of things which are only audible, namely sounds. Both employ as uncontradictable helpmeets Arithmetic and Geometry, in order to measure the quantity and quality of the principal movements. They are, so to speak, cousins, for they stem from the sister senses, sight and hearing, and have as their foster mothers Arithmetic and Geometry, to which they are related [66].[115]

A Comparison Between the Harmonic Genera and those of the First Virtues.
The human soul, according to Ptolemy, is made of matter, but unlike ordinary physical objects, it cannot be perceived by the senses. In the *Harmonica* he assigned three faculties to the human soul, in different ways: either according to the Aristotelean-Stoic tradition, for which they are the power of thinking, of feeling and living, or according to the Platonic tradition, for which they are reason, emotion, and desire. The Aristotelian-Stoic division is the more developed.

The connection between the soul and harmonics is easy to find, since the soul has three faculties "there are three first species of homophones and consonances: both the homophone of the diapason and the consonances of diapente and diatessaron" [66][116] The diapason, then, may be compared to the power of thought—for in both there is simplicity, equality, and equivalence—the fifth to the power of feeling, and the fourth to the power of life. The fifth is closer to the diapason than the fourth and sounds better.

On the Similarity of the Perfect System and the Circle Through the Center of the Zodiac.
Ptolemy considered the connection between the 12 divisions of the Zodiac and the divisions of the greater perfect system, the prototype of the Greek harmonics, which had a compass of two diapasons and 15 musical notes. According to Ptolemy, the comparison immediately assumes that both pitches and stars (planets) occur only through intervallic motion. Moreover, the rotational motions of the stars are all circular and regular, and the motions within the note system can be similarly arranged. One can imagine pitches distributed along a closed circle, and thus one grasps the true nature of musical notes, without a starting point.

To place the notes of the greater perfect system on the zodiacal circle, one can start from the constellation of Aries and fold the double diapason into a circle, connecting nētē hyperbolaiōn and proslambanomenos.

---

[114] p. 95.21–24. Translation into English in [67].

[115] p. 94.11-18.Translation into English in [43].

[116] p. 96.1. Translation into English in [67].

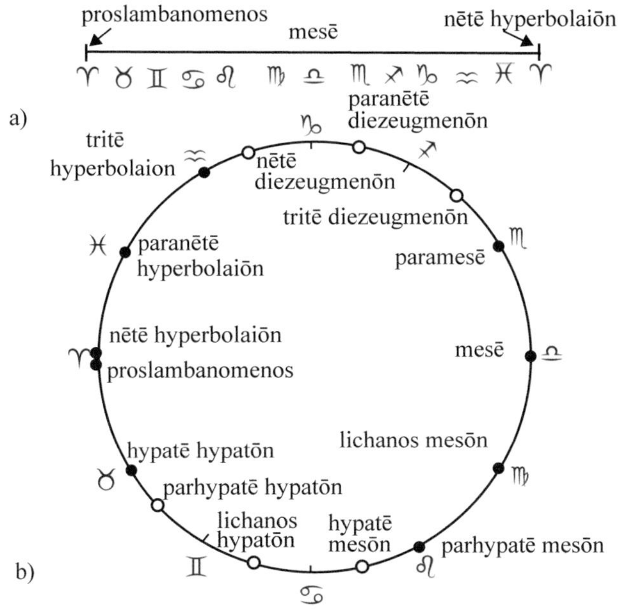

Fig. 1.18 The greater perfect system and the Zodiac. Adapted from [43], p. 30

The relationship between the 12 constellations of the Zodiac and the 15 notes of the greater perfect system is not completely defined, however. The notes must be arranged in a continuum, straight or circular, in which it is assumed that there are 12 positions equally occupied by the constellations. The only criterion given by Ptolemy is that notes that are one diapason apart must be diametrically opposed in the zodiacal circle, since "the effect of the planets in opposition is at its strongest when they occupy diametrically opposed positions in the Zodiac, and a similar relationship obtains between notes that are one diapason apart." [66][117]

A possible arrangement of the notes of the greater perfect system on the zodiacal circle is shown in Fig. 1.18. Here a diatonic genus is assumed; the sequence of tones and semitones, starting from proslambanomenos, is the standard one: T, S, T, T, S, T, T, T, S, T, T, S, T, T. Some notes are in the place occupied by a constellation, they are marked by black circles; others, marked by with empty circles, are between two constellations.

In the present time, when the diapason is naturally divided into 12 equal parts with the equal temperament, the agreement between the signs of the Zodiac is easier, since it is possible to establish a one-to-one correspondence between the constellations and the notes.

How the Harmonic Consonances and Dissonances Resemble those in the Zodiac. In this section, Ptolemy sought further analogies between the greater perfect system

[117] p. 101.18. Translation into English in [67].

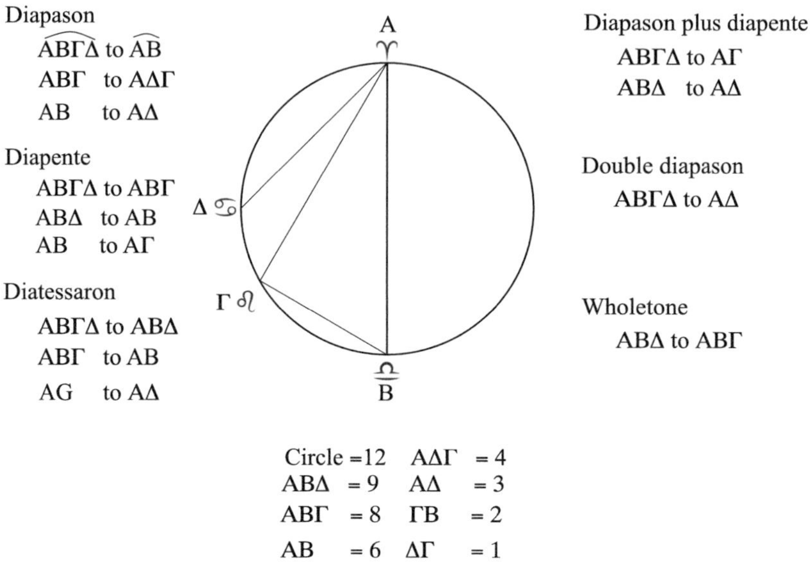

Fig. 1.19 Harmonic consonances and dissonances resembling those in the Zodiac. Adapted from [66], p. 103

and the Zodiac. The consonant intervals can always be divided into four parts; for the largest, the double diapason, is expressed by a fraction whose numerator is four times the denominator (4 : 1), and the smallest, the diatessaron, by the fraction (4 : 3). Similarly, the harmonic aspects observed in the Zodiac are determined by dividing the circle into four parts. Let us draw a circle AB (Fig. 1.19), and from the point A divide it by AB into two equal parts, by AΓ into three, by AΔ into four, and by ΓB into six. Then AB will produce the opposition, AΔ the tetragonal, AΓ the trigonal and ΓB the hexagonal. The proportions of the arcs detached from point A are the same as the consonant intervals, which can be seen by dividing the circle into 12 equal segments (since 12 is the least common multiple of 4 : 1, 2 : 1, 3 : 2, and 4 : 3). The situation is illustrated in Fig. 1.19.

1.5.3 Censorinus

Little is known of Censorinus (fl. 230) common era, except that he was a Roman grammarian and miscellaneous writer. The only writing that has reached us is *De die natali*, possibly written in 238 CE for the birthday of his patron. The contents are of a varied character: the natural history, the influence of the stars, music, religious rites, astronomy, the doctrines of the Greek philosophers, and antiquarian subjects. Below we report Censorinus' ideas on the relationship between harmonics, astronomy and embryology. Censorinus' idea is, of course, laughable today, but the way in which he joined three different disciplines—astronomy or astrology, embryology, and music—is the epitome of his philosophy.

> Hippon, on the other hand, writes that the infant is formed in 60 days; that in the fourth month the flesh solidifies, in the fifth the nails and hair grow, and in the seventh the being is complete. Pythagoras says, more credibly, that there are two kinds of birth: one in the seventh and the other in the tenth month, corresponding to their respective numbers of days. He holds that the periods during which semen is converted into blood, blood into flesh, and flesh into the human form, are proportionate to one another in the same way as those musical tones called symphonoi [26].[118]

Quite interesting is the conception of the harmony of the planets (stellas), which in Table 1.15 gives a view of astronomy in the Roman Empire, after Ptolemy.

> Now as Eratosthenes calculated by geometrical method (ratione) that the greatest circumference of the earth is 252,000 stades[119] [...]. Pythagoras believed the distance from the Earth to the Moon to be about 126,000 stades, and this to be the interval of a tone. Then from the Moon to the planet (stellam) Mercury, called *Stilbon*, is the half of that distance, or a semitone. From Mercury to *Phosphor* or the planet Venus is about the same, i.e., another semitone; thence to the Sun is thrice as far, making a tone and a half. 4. The star (astrum) of the Sun is thus three and a half tones distant from the Earth, making a fifth, and two and a half tones from the Moon, making a fourth. From the Sun to the planet Mars, whose name is *Pyrois*, the interval is the same as from the Earth to the Moon, i.e., a tone; from Mars to the planet Jupiter, called *Phaethon*, it is the half of that, a semitone. From Jupiter to the planet Saturn, called *Phaenon*, the distance is another semitone, and thence to the highest heaven where the signs of the Zodiac are, a semitone again. So from the highest heaven to the Sun the interval is a fourth (two tones and a half), and from the Earth's highest point to the same heaven is six tones, which make the consonance of one octave [26].[120]

Note the reference to Eratosthenes (possibly a contemporary of Censorinus) in the previous quotation, and the relatively small distance estimated for the Earth and Moon, about 3 terrestrial diameters (D), compared to the value proposed by Aristarchus of Samos (c. 310–230 BCE) (20 D) and the essentially correct values of Hipparchus of Nicaea (60 D) (c.190–120 BCE).

Col. 1 shows the order of the planets with respect to the Earth. Note that, unlike Pythagoras (and Plato), the Sun has a central position, and that Mercury and Venus have the same order as in Boethius (480–524 CE) and different from Nicomachus (50–150 CE).[121] Col. 2 shows the relative distance of the planets, measured in tones. Col. 3 the distance of the planets from the Earth, stil in tones. As you can see, the tones and semitones follow an Aristoxenian trend, with a diapason made up of exactly six whole tomes.

The order of the tones and semitones is T-SS1.5T-T-SSS. After the proslambanomenos there is a chromatic tetrachord (SS1.5T), followed by a second disjoint

---

[118] pp. 22–23. Translation into English in [43].

[119] An Italian stadium, which should be the case here, is about 190 m.

[120] pp. 31–32. Translation into English in [43].

[121] Modern measurements have shown that the mean distance of Mercury from the Earth is less than the mean distance of Venus, even though this last planet has the smallest absolute distance from the Earth. Thus, the relative position of Venus and Mercury can be chosen *ad libitum*.

Table 1.15 Position of the planets according to Censorinus

Stars	Relative intervals	Absolute intervals	
Earth			
Moon	1.0 tone	1.0 tone	
Mercury	0.5 tones	1.5 tones	
Venus	0.5 tones	2.0 tones	
Sun	1.5 tones	3.5 tones	Diapente
Mars	1.0 tone		
Jupiter	0.5 tones		
Saturn	0.5 tones		
Fixed stars	0.5 tones	6.0 tones	Diapason

anomalous tetrachord (SSS). The most natural solution would have been to have two disjoint chromatic tetrachords, but this would have resulted in exceeding the diapason. A possible explanation for Censorinus' choice can be found in [42].[122]

References

1. Alypius of Alexandria (1895) Alypi isagoge. In: Jan K (ed) Musici scriptores graeci. Aristoteles, Euclides, Nicomachus, Bacchius, Gaudentius, Alypius et melodiarum veterum quidquid exstat, Teubner, Leipzig, pp 357–406
2. Aristotle (2022) De caelo, http://classics.mit.edu/Aristotle/heavens.html, The Internet Classical Archive. Translated into English by Stocks JL. http://classics.mit.edu/Aristotle/heavens.html
3. Aristoxenus (1902) The harmonics of Aristoxenus. Clarendon Press, Oxford, Edited and translated into English by Macran HS
4. Aristoxenus (1954) Aristoxeni Elementa harmonica. Publicae Officinae Polygraphicae, Rome, Edited and translated into Italian by Da Rios R
5. Aristoxenus (1989) Elementa harmonica. In: Barker A (ed) Greek musical writings (vol 2), Cambridge University Press. Cambridge, Cambridge
6. Artusi GM (1600) L'Artusi ovvero delle imperfezioni della moderna musica. Vincenti, Venice
7. Barbera A (1977) Arithmetic and geometric divisions of the tetrachord. Journal of Music Theory 21(2):294–323
8. Barbera A (1984) Placing Sectio canonis in historical and philosophical contexts. The Journal of Hellenic Studies 104:157–161
9. Barbera A (1984) Octave species. The Journal of Musicology 3(3):229–241
10. Barbera A (1991) The Euclidean division of the canon: Greek and Latin sources. University of Nebraska, Lincoln and London
11. Barker A (1978) Music and perception: a study in Aristoxenus. The Journal of Hellenic Studies 98:9–16
12. Barker A (1989) Greek musical writings (2 vols). Cambridge University Press, Cambridge
13. Barker A (1994) Ptolemy's Phytagoreans, Architas, and Plato's conception of mathematics. Phronesis 39(2):113–135

[122] p. 125–126

14. Barker A (2006) Scientific method in Ptolemy's harmonics. Cambridge University Press, Cambridge
15. Barker A (2007) The science of harmonics in classical Greece. Cambridge University Press, Cambridge
16. Barker A (2014) Ancient Greek writers on their musical past. Serra, Rome
17. Barker A (2020) Chapter nineteen. Harmonics. In: Lynch TA (ed) A companion to ancient Greek and Roman music. Wiley, Hoboken, A comprehensive guide to music in classical antiquity and beyond (Ebook), pp 262–277
18. Boethius (1491-1492) Opera (2 vols). De Gregoris
19. Boethius (1867) Boetii De institutione musica. Libri Quinque. In: Friedlein G (ed) Anicii Manlii Torquati Severini Boetii De institutione arithmetica libri duo. De institutione musica libri quinque, Teubneri, Leipzig, pp 175–371
20. Boethius (1989) Fundamentals of music. Ancius Manlius Severinus Boethius. Translated by Bower CM. Edited by Palisca CV. Yale University Press
21. Boethius (c. 1490) De institutione musica. manuscript, https://colenda.library.upenn.edu/catalog/81431-p3pw86, Kislak Center for Special Collections - Manuscripts Oversize LJS 47. Accessed 12 January 2025
22. Bower CM (1967) Boethius' the Principles of musica. An introduction. PhD thesis, George Peabody College for Teacher
23. Bower CM (1984) The modes of Boethius. The Journal of Musicology 3(3):252–263
24. Busch O (2004) Logos Syntheseos-Die Euklidische Sectio Canonis. Aristoxenos und die Rolle der Mathematik in der antiken Musiktheorie, Olms, Hildesheim
25. Capecchi D (2015) Epistemology of Greek Harmonics. Advances in Historical Studies 4:155–171
26. Censorinus (1845) De die natali liber. Edited by Jahn O. Reimer, Berlin
27. Chrysippus (1964) Chrysippi fragmenta logica et physica. In: Arnim H (ed) Stoicorum veterum fragmenta, vol 2. Teubner, Stuggart
28. Clagett M (1961) The science of mechanics in the Middle Ages. University of Wisconsin Press, Madison
29. Cleonides (1950) Harmonic introduction. In: Strunk O (ed) Source readings in music history, Norton & Company, New York, pp 34–46
30. Cohen H (1984) Quantifying music. Reidel, Dordrecht
31. Comotti G (1991) La musica nella cultura greca e romana. E.D.T, Turin
32. Creese D (2010) The monochord in ancient Greek harmonic science. Cambridge University Press, Cambridge
33. Crocker RL (1963) Pythagorean mathematics and music. J Aesthet Art Critic 22(2):189–198
34. Düring I (1930) Die Harmonielehre des Klaudios Ptolemaios. Cambridge University Press, Cambridge, Cambridge
35. Düring I (1956) Die Harmonielehre des Klaudios Ptolemaios. Porphyrios Kommentar Zur Harmonielehre Des Ptolemaios. PhD thesis, University of Göthenburg
36. Enrico E (2012) Octave-species and the key. A study in the historiography of Greek music theory. PhD thesis, University of Montana
37. Euclid (1916) Sectio canonis. In: Heiberg I, Menge H (eds) Opera omnia (9 vols), vol 8. Teubneri GB, Leipzig, pp 157–183
38. Euclid (1945) The optics of Euclid. Translated into English by Burton HD. Journal of the Optical Society 35(5):357–372
39. Euclid MTJ (1975) An annotated translation of Euclid's "Division of a monochord". Journal of Music Theory 19(2):236–258
40. Fortlage K (1847) Das musikalische System der Griechen in seiner Urgestalt aus den Tonleitern des Alypius zum ersten Male entwickelt. Breitkopf and Härtel, Leipzig
41. Franklin JC (2005) Hearing Greek microtones, https://api.semanticscholar.org/CorpusID:194841090, accessed 12 January 2025
42. Godwin J (1987) Harmonies of heaven and earth. Inner Tradition International, Rochetser

43. Godwin J (1993) The harmony of the spheres. A sourcebook of the Pythagorean tradition in music, Inner Tradition International, Rochetser
44. Gogava A (1562) Aristoxeni musici antiquiss. Harmonicorum elementorum libri III. Cl. Ptolemaei harmonicorum, seu de musica lib. III. Aristotelis de objecto auditus fragmentum ex Porphyrij commentarijs. Valgrisi, Venice
45. Hagel S (2010) Ancient Greek music. Cambridge University Press, Cambridge, A new technical history
46. Huffman CA (2019) Aristoxenus of Tarentum: The Pythagorean precepts (How to live a Pythagorean life): An edition of and commentary on the fragments with an introduction. Cambridge University Press, Cambridge
47. Laloy L (1904) Aristoxène the Tarente et la musique de l'antiquité. Societé Française d'Imprimerie et de Librarie, Paris
48. Landels JG (2000) Music in ancient Greece & Rome. Routledge, New York
49. Litchfield M (1988) Aristoxenus and empiricism: A reevaluation based on his theories. Journal of Music Theory 32(1):51–73
50. Manolova D (2014) Discourse of science and philosophy in the letters of Nikephoros Gregoras. PhD thesis, Central European University, Budapest
51. Marchetti CC (2009) Aristoxenus Elements of rhythm. Text translation and commentary. PhD thesis, State University of New jersey
52. Mathiesen TJ (2008) Apollo's lyre. Greek music and music theory in antiquity and the Middle Ages. University of Nebraska Press, London
53. Monro D (1894) The modes of ancient Greek music. Clarendon Press, Oxford
54. Mueller I (1981) Philosophy of mathematics and deductive structure in Euclid's Elements. Dover, New York
55. Najock D (1972) Drei anonyme griechische Traktate über die Musik. Eine kommentierte Neuausgabe des Bellermannschen Anonymus, Bärenreiter, Kassel
56. Nicomachus (1926) Nicomachus of Gerasa. Introduction to arithmetic. Translated into English by D'Ooge M L, Macmillan, New York
57. Nicomachus (1994) The manual of harmonics of Nicomachus the Pythagorean. Translated into English by Levin FL, Phanes Pres, Grand Rapids, MI
58. Nowacki E (2020) Greek and Latin music theory. University of Rochester Press, Rochester, Principles and challenges
59. Palisca C (1985) Humanism in Italian Renaissance. Yale University Press, New Haven, Musical thought
60. Palisca C (1994) Studies in the Italian music and music theory. Clarendon Press, Oxford
61. Pelosi F (2018) Eight singing sirens. Heavenly harmonies in Plato and the Neoplatonist. In: Jacomien P, Maude V (eds) Sing aloud harmonius spheres. Routledge, London, pp 17–23
62. Plato (2022a) Republic, https://classics.mit.edu/Plato/republic.html, The Internet Classical Archive. Translated into English by Jowett B
63. Plato (2022b) Timaeus, http://classics.mit.edu/Plato/timaeus.html, The Internet Classical Archive. Translated into English by Jowett B
64. Pöhlmann E (1960) Griechische Musikfragmente: Ein Weg zur altgriechischen Musik. Hans Carl Verlag, Nürnberg
65. Pöhlmann E (2020) Chapter eighteen. Acoustics. In: Lynch TA (ed) A companion to ancient Greek and Roman music. Wiley, Hoboken, A comprehensive guide to music in classical antiquity and beyond (Ebook), pp 252–261
66. Ptolemy (1982) Die Harmonielehere des Klaudios Ptolemaios. Hildsheim, Olms
67. Ptolemy (2000) Ptolemy Harmonics. Translated into English and commented by Solomon J, Brill, Leiden
68. Riemann H (1893) Vereinfachte Harmonielehre oder die Lehre von den tonalen Funktionen der Akkorde. Augener, London
69. Simpson PL (2015) Ancient Greek musical modes, https://www.academia.edu/19677694/The_Ancient_Greek_Musical_Modes, accessed May 2024
70. Solomon J (1984) Towards a history of tonoi. J Musicol 3:242–251

71. Steinmann C (1539) Nachklänge: Instrumente der griechischen Klassik und ihre Musik (Materialen und Zeugnisse von Homer bis heute). Schwabe, Basel
72. Various (2025) New Grove Dictionary of Music and Musicians. https://doi.org/10.1093/gmo/9781561592630.article.27643, accessed 1 January 2025
73. Walker D (1978) Studies in musical science in the late Renaissance. Brill, Leiden
74. West ML (1992) Ancient Greek music. Clarendon Press, Oxford
75. West ML (2001) Documents of ancient Greek music: The extant melodies and fragments. Clarendon Press, Oxford
76. Winnington I (1936) The modes in ancient Greek music. Cambridge University Press, Cambridge
77. Wooldridge H (1901-1905) The Oxford history of music (6 vols). Clarendon, Oxford

Chapter 2
Music in the Middle Ages and Renaissance

Abstract Middle Ages music scholars were confronted with two distinct traditions. That of the Church's liturgical chant, the origins of which are not entirely known, and that of Greek harmonics, which was purely theoretical in nature. The two traditions merged in the 9th century during the Carolingian revival. This marked the birth of modern Western music. The centuries that followed saw some fairly significant changes, but the original core remained. The purely diatonic music of the early Middle Ages was enriched with chromaticism, and new consonances such as thirds and sixths, major and minor, were accepted in addition to the fourths and fifths used until then. In this way, Pythagorean intonation was surpassed, leading to the so-called just intonation. The adoption of this richer intonation, however, brought with it a number of disadvantages that were resolved with the introduction of temperament. Monody was flanked by polyphony, which then became counterpoint, in which different melodic lines intersected in a very complex way; finally, there was a reaction to this complexity and a return to Greek music, in which more attention was paid to the text: thus opera was born.

2.1 Ancient Fonts

2.1.1 Middle Ages Heritage

In the Middle Ages, performed music was greatly expanded by its incorporation into Christian rites. It was based on a legacy of songs, namely the sacred chants of the Church, which had their own traditions and were independent of Greek speculative music. In contrast, the development of speculative music was minimal. After the 'publication' of Boethius's *De institutione musica* in the 6th century CE, no relevant treatise on speculative music was published. With the advent of the Carolingian Renaissance in the 9th century, a new wave of speculative treatises, inspired by Greek harmonics, appeared with the aim of placing Church chant on a more rigorous

D. Capecchi and G. Capecchi, *Joseph Sauveur and His Time: Music and Mathematicians*,
https://doi.org/10.1007/978-3-032-04707-6_2

footing. This marked the genesis of what is now recognized as Gregorian chant (for its origin see [21, 51][1]).

The most consulted texts on speculative music were *De nuptiis Philologiae et Mercurii libri 9* by Martianus Capella (fl. 5th century CE) [19, 20]—the ninth book is entirely devoted to harmonics, including elements of rhythmics and the harmony of the spheres [97]—and the *De institutione musica* of Severinus Boethius. New treatises began to be written, but with a pronounced focus on the music actually produced.

Since it was difficult to find musicians who could master the theory of proportions necessary for a full understanding of texts on harmonics, the massive use of the monochord began around the year 1000. Using suggestive terminology, it can be said that the era of the monochord began [48].[2]

The monochord offered a precise aid in learning pitches, intervals, fundamental consonances, knowledge of modal scales, and explanation in a simple way of some aspects of the theory of proportions. In the 12th or 13th century, with the emergence of the universities, when music entered as a discipline of the quadrivium, the use of the monochord was accompanied by the study of musical treatises written to teach the elementary rules of polyphony. The use of the monochord would remain important throughout the Renaissance; Fig. 2.1 gives an idea of how this instrument might be used.

In the universities, the basic text remained *De institutione musica* of Boethius, which was not only a text on music, but also on mathematics. As geometry dealt with magnitudes, arithmetic with integers, harmonics dealt with ratios and their composition [1]. Only from the second half of the 14th century did traditional speculative music experience a substantial change, under the stimulus of a new awareness of Greek music and new developments in mathematics.

2.1.1.1 Notation

Until the beginning of the first millennium of the common era, instrumental and vocal music followed an oral tradition. As musical complexity increased, some form of written representation emerged. Among them the *neumes*—from the Greek neūma, meaning sign—a non-alphabetical form of representation [56].

In the beginning, neumes were very simple signs, accents to be precise, placed above the text of the melody, as shown in Fig. 2.2.

The information on rhythms and intervals contained in them (the earliest surviving documents date from the 9th century) is still not well known. What is certain is that neumes could only help people who already knew the music to sing.

With time, neumes developed into the so-called *diastematic neumes*, which gave relevant information about interval sizes and rhythm. The introduction of the staff at the turn of the 11th century expanded the possibility of writing music, allowing the

[1] vol. 1.

[2] p. 123.

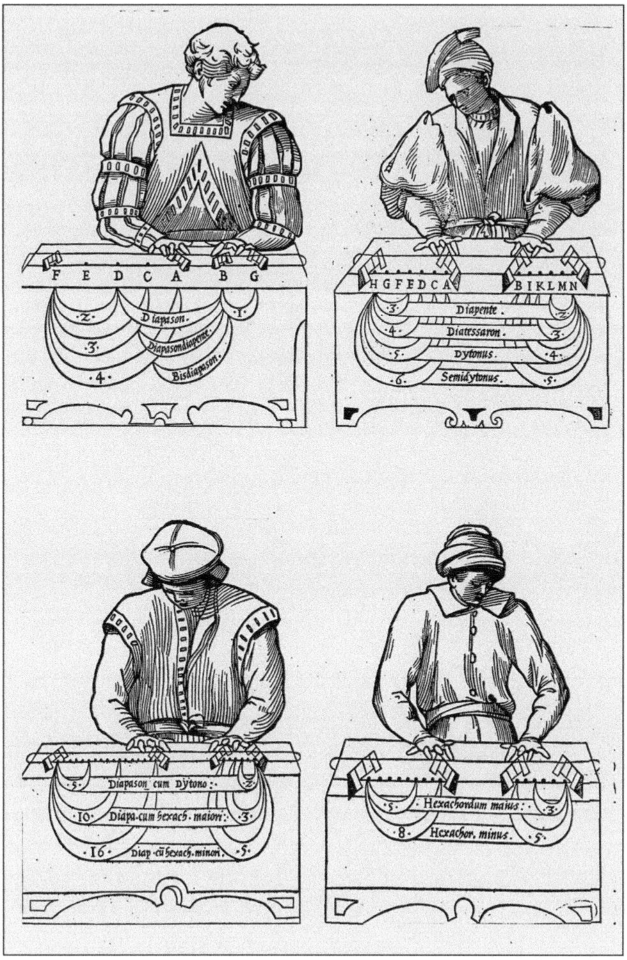

Fig. 2.1 Various uses of the monochord. Adapted from [39], ff. 12v, 14r

performance of music never heard before [4]. In any case, the oral tradition remained fundamental.

In the early centuries, alphabetical notations were also developed which had its roots in ancient Greece and was transmitted to the Latin world by Boethius. This system, which used the Latin alphabet letters (from A to P) to represent pitch, was adopted by Carolingian theorists to record and explain complex musical concepts. It spread primarily in Norman monasteries reformed by Guglielmo da Volpiano (962–1031), demonstrating its importance in the transmission and preservation of medieval music [15]. But it can be concluded that this system of alphabetical notations was used mainly in a theoretical environment. Interesting compromises combined letters to indicate pitches and neumes for rhythm and vocal inflection [66, 94]. In

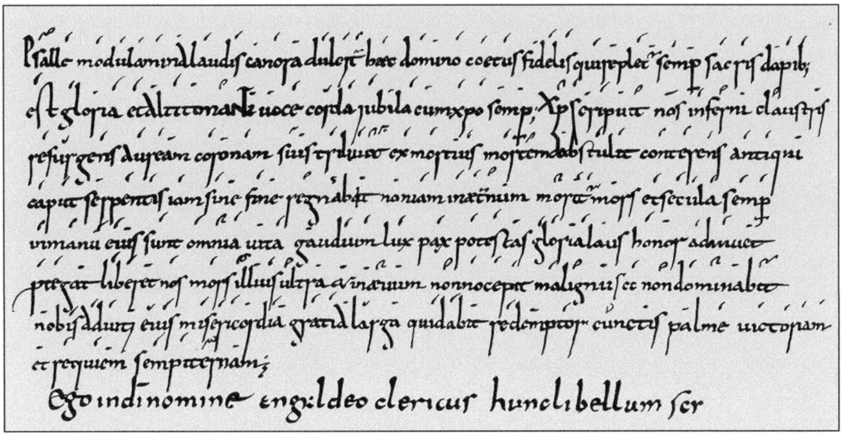

Fig. 2.2 First example of neumes. Redrawn from [2], f. 199v

the following, we will only consider the development of alphabetic notation; for an extensive history of non-alphabetic notation, see [5].

A first appearance of the alphabetical notation in a 'theoretical' treatise is recorded during the recovery of Greek harmonics in the 9th century, with Hucbald of Saint-Amand (c. 850–900) and his *De harmonica institutione* [23, 103]. Here letters were used to indicate the pitch of notes, as shown in Table 2.1.

Hucbald, on his own admission, drawn these notational signs from the list of Greek symbols found in Boethius [12][3]; he urged they should be used in conjunction with neumes. Hucbald's scale started from the bottom with the note proslambanomenos, and the intervals followed the sequence T-S-T as in the modern minor scale. By comparing Figs. 2.1 and 1.16, it is evident that the association of Greek names of notes and Greek signs of pitches comes from the Lydian mode [15].

Another step to alphabetical notation can be found in the *Dialogus de musica*, formerly attributed to Odo of Cluny (878–942), but later found to have been written by an anonymous Lombard author, here cited as the Pseudo-Odo [25].[4] Table 2.2 shows the division of the *monochord*,[5] based on the *Dialogus de musica* [86]. Compared to Hucbald's proposal, a note called Γ is added, which introduces the T-T-S pattern of the modern major scale.

Pseudo-Odo dismissed the tetrachord as the basic system of notes to adopt their arrangement in the octave. There are seven notes in the octave, as in the Greek diapason formed by two disjoint tetrachords, for which Pseudo-Odo used the seven letters A, B, C, D, E, F, G of the Latin alphabet. To emphasize the concept of octave organization—which results both from the acoustic consonance of the octave itself

[3] Book IV, p. 313.

[4] p. 339.

[5] Monochord here and elsewhere can be translated with scale. It can also be understood as the representation of musical intervals on a vertical straight line (an immaterial monochord indeed).

Table 2.1 First example of alphabetical notation; from [58], pp. 921–922. The symbols on the right are Hucbald's original [80], p. 38

V	Nete hyperbolæon itaque habet I iota extensum sic	V	
Π	Paranete hyperbolæon Π græcum extensum	Π	
Y	Trite hyperbolæon Y simplex	Y	
N	Nete diezeugmenon X contractum	N	
O	Paranete diezeugmenon ω quadratum	ʊ	
E	Trite diezeugmenon E simplex	E	
=	Paramese Π græcum jacens	=	
	Nete synemmenon eadem, quæ paranete diezeugmenon	ʊ	
	Paranete synemmenon eadem, quæ trite diezeugmenon	E	
θ	Triste synemmenon θ græcum	θ	
I	Mese I simplex	I	
M	Lichanosmeson M simplex	M	
P	Parypatemeson P græcum (ro) simplex	P	
C	Hypatemeson C sigma græcum simplex	C	
F	Lichanos hypaton F digamma simplex	F	
B	Parypate hypaton B beta simplex	B	
Γ	Hypate hypaton Γ gamma simplex	Γ	
F	Proslambanomenos F dasian rectum	F	

Table 2.2 Pseudo-Odo's monochord; based on [86], p. 253

nētē hyperbolaiōn	aa	
paramētē hyperbolaiōn	g	
tritē hyperbolaiōn	f	
nētē diezeugmenōn	e	
paranētē diezeugmenōn	d	paranētē synēmmenōn
tritē diezeugmenōn	c	
paramesē	♮	
	♭	tritē synēmmenōn
mesē	a	
lichanos mesōn	G	
paryhypatē mesōn	F	
hypatē mesōn	E	
lichanos hypatōn	D	
paryhypatē hypatōn	C	
hypatē hypatōn	B	
proslambanomenos	A	
	Γ	

and from the recurrence of identical intervallic patterns around notes—he used the same letters for the notes of each octave, but with different graphic conventions that remained in use well beyond the end of the Middle Ages (in fact, they are still widely accepted): the uppercase, lowercase, and doubled lowercase letters, as shown in Table 2.2. In the second octave, instead of the letter b, as one might expect, there are two symbols for two different pitches: the b *quadratum*, written ♮, our b-natural (one octave higher than B), and the b *rotundum*, written ♭, our b-flat, corresponding to the tritē synēmmenōn of the unchanging system (the modern symbols for flat and sharp are derived from this).[6]

Actually, Pseudo-Odo's original text did not present the diagram of Table 2.2, but simply the following brief writings:

> Then voices, figures and letters we pose as in the following:
> Γ. I. A. II. B. III. C. IIII. D. V. E. VI . F. VII. G. VIII. a. IX . I.♭. IX. II.♮ X. c. XI. d. XII. e. XIII. f. XIIII. g. XV. aa [86].[7]

Guido of Arezzo (c. 995–1050) retained all this, and in his *Micrologus* of 1026—a treatise that competed for primacy in the Middle Ages with Boethius's *Institutioni musicae*—added four more notes above aa, notated as ♭♭, ♮♮, cc, and dd [27].[8] The note ee was later added to complete a hexachord, as shown in the first column of Fig. 2.3.

Guido did not divide the scale into tetrachords; instead, as Pseudo-Odo, he pointed out its division into octave segments. Because of the dominant role assigned to the octave and the idea of arranging the notes on a staff, the adoption of a new denomination of intervals was natural; diatessaron and diapente became fourths and fifths, respectively (and up to this point we are only talking about a translation from Greek to Latin), but then ditones became thirds and hexachords sixths, tones seconds. The diapason became the octave (the seventh was not mentioned) [27].[9]

The note ♭ has long been considered an extraneous element by most theorists; Guido himself was particularly adamant on this point. Others, however, especially German theorists writing around 1100, were less suspicious and even urged the inclusion of a low ♭, a synēmmenōn grave [25].[10]

Guido is generally credited by historians with having covered all the pitches of the so-called *musica recta* or *musica vera*, that is, the basic system of the diatonic scale, with a series of hexachords, with the structure T-T-S-T-T, whose notes had the names Ut, Re, Mi, Fa, Sol, La.

This account is not entirely true, and Guido was not the only figure in the development of the technique of the hexachord, called *solmisatio*—the modern English solmization—in the 14th century; he had both precursors and heirs [69].[11] What is certain is that around the year 1032, in his *Epistola ad Michaelem* Guido presented a new method of sight-singing based on the six syllables Ut, Re, Mi, Fa, Sol, La, which corresponded to the first syllables of each verse of the hymn of St. John *Ut*

---

[6] Though the *Dialogus de musica* is the earliest extant treatise to report letters as we still use them today, there were other systems of letters that show reduplication at the octave [14, 54], p. 183.

[7] p. 253.

[8] Chap. 2, 10–12.

[9] Chap. 18.

[10] p. 341.

[11] p. 2.

		hard	natural	soft	hard	natural	soft	hard
	ee							La
	dd						La	Sol
superacutae	cc						Sol	Fa
	♮♮							Mi
	♭♭						Fa	
	aa					La	Mi	Re
	g					Sol	Re	Ut
	f					Fa	Ut	
	e				La	Mi		
acutae	d			La	Sol	Re		
	c			Sol	Fa	Ut		
	♮				Mi			
	♭			Fa				
	a		La	Mi	Re			
	G		Sol	Re	Ut			
	F		Fa	Ut				
	E	La	Mi					
graves	D	Sol	Re					
	C	Fa	Ut					
	B	Mi						
	A	Re						
	Γ	Ut						

Fig. 2.3 Systems of hexachords. Light gray boxes, natural hexachord; gray dark boxes, semitones

queant laxis [24, 28]. The six syllables covered the major sixth C-a, an interval with only one semitone E-F in the middle, corresponding to Mi-Fa of the hexachord.

The role of the hexachord in Guido's teaching is emphasized in the following quotation:

> It is a striking paradox of music history that the pedagogical innovation responsible for Guido's quasi-saintly status in the modern era – the introduction of the Ut-La syllables – plays a thoroughly negligible role in the four treatises securely attributable to the Aretino. The syllables are nowhere to be found in three of them (*Micrologus*, the *Prologus in antiphonarium,* and the *Regulae ritmicae*, according to a plausible chronological sequence), and are briefly introduced only in an early section of the *Epistola ad Michaelem* that makes for about twenty lines [...] out of a total of almost 400 lines. Of course, such quantitative analysis is hardly a reliable sign of textual or doctrinal significance, and it utterly fails to explain the enormous interest generated by those syllables in later centuries. It does provide a prima facie indication, however, of the thoroughly marginal role of the syllables in Guido's pedagogy [69].[12]

[12] p. 30.

Table 2.3 Mutation of hexachords

		D	E	F	G	a	b	c	d	e
Natural	D-A	Re	Mi	Fa	Sol	(La)				
Hard	G-e				Ut	Re	Mi	Fa	Sol	La

At the end of the 11th century, Guido's six syllables were transposed to the segment G-e (with Mi-Fa on B-C) and to the segment F-d (with Mi-Fa on a-♭). The hexachord thus covered all the pitches of the musica recta, as shown in Fig. 2.3.

The first hexachord in the figure covers the notes ΓABCDE (a semitone between B and C) and is called a *hard* hexachord. The second covers the notes CDEFGa, with the semitone between E and F, and is called a *natural* hexachord; the third hexachord covers the notes FGabc, semitone between a and ♭, and is called a *soft* hexachord. The hexachords are repeated in the upper octave. The term hard refers to the hexachord that has the semitone between ♮ and c, soft if it has the semitone between a and ♭, the natural hexachord, which is Guido's original, has the semitone between e and f.

Although in Middle Ages music the compass of a melody was not very large, a single hexachord could not be enough to play it. To sing a melody with a compass wider than a single hexachord, one had to switch between two or more hexachords at appropriate points. Usually, this involved choosing a step of the melody common to both the old and the new hexachord as a bridge between them. For example, suppose one has to sing the ascending octave D-d. A possible solution is shown in Table 2.3.

Starting with the natural hexachord, we could sing the first notes of the ascending D-E-F-G as Re-Mi-Fa-Sol, then make a mutation or shift on the fifth note a to the hard hexachord on G, and sing the remaining notes (a-b-c-d-e) as Re-Mi-Fa-Sol-La. The note a is common to both hexachords and serves as a *bridge* between the two. Other example of mutation can be found in [52].[13]

Some treatises suggested that, at least as a training exercise, singers might sing the bridge note of a mutation with the syllables of both the old and the new hexachord, e.g. in the example above one might say that the passage occurs on the note La-Re. There were guidelines as to where and how to make a mutation; however, the choice of the mutation was probably more of a fine art than an exact science [91].[14] In the case of a coir, it was the magister to indicate hexachord mutations with the use of the so called *Guidonian hand* (see [72]).

Figure 2.4 shows two different representations of the Guidonian hand, the phalanges of which are marked with the solmization syllables. Figure 2.4a is taken from an Italian treatise of the early 16th century; Fig. 2.4b from a German treatise of a century later. The assignment of phalanges to syllables is quite different. In the first case the syllables follow a spiral drawn as a dashed line, in the second case they go from bottom to top.

[13] p. 17 unnumbered.

[14] Chap. 1.

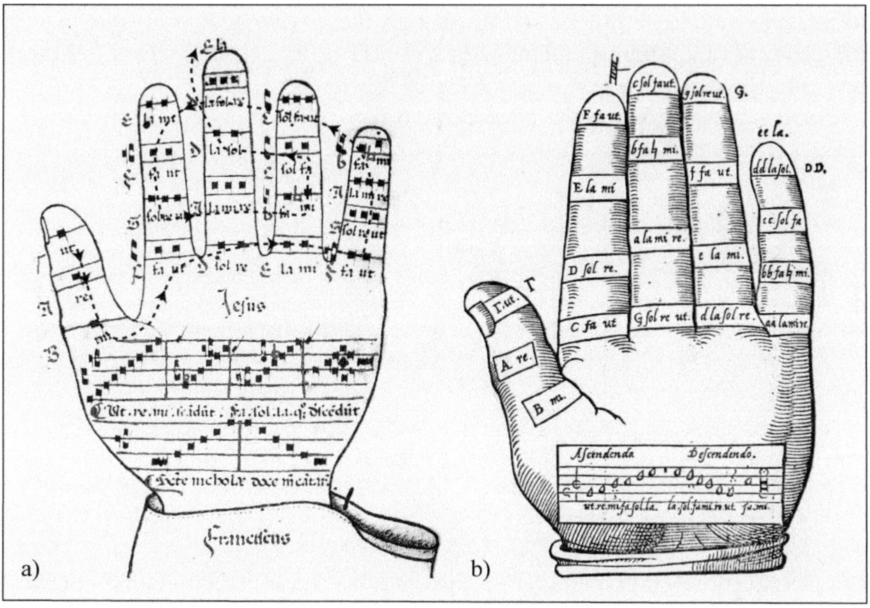

Fig. 2.4 Two different representations of the Guidonian hand. **a** A drawing of the 16th century; redrawn from [3]; **b** a representation of a century later; redrawn from [52], p. 10 unnumbered

Since the syllables had pitches one degree apart, the singer could easily associate intervals with couples of voices by memorizing the hymn. The hexachord had two main functions:

1. Help to locate the semitones of a melody.
2. Help in reproducing rising or falling intervals.

In the beginning, the syllables of the hexachord had no pitch associated with them. Only with time, after the Renaissance, in Latin-speaking countries, musicians began to associate the notes of hexachords with pitches. This association occurred for the normal hexachord with the correspondence:

C	D	E	F	G	A	(B)
Ut	Re	Mi	Fa	Sol	La	(Si)

The note B remained outside this association until the transformation of the hexachord into a heptachord with the introduction of an additional syllable, Si, beginning in the 16th century, to form the modern seven-note scale.

The alphabetic letter name (Γ, A, B, …) of each note [*littera* or clavis] of the musica recta was combined with Guido's syllables [*vox*] (Ut, Re, …) in a composite note name; thus, for example, the lowest note was called Γ-Ut, hence the name *gamut* for a scale. Because the seven hexachords overlap at various points, most pitches in the system have two or three different *voces* in their composite name; for example,

C Fa Ut, D Sol Re, G Sol Re Ut. Among all the voices there is at least one that corresponds to the *littera* in the modern scale UT-ut. For example, in G Sol Re the correspondence is in the first *vox*, in C Fa Ut it is in the second.

Guido is attributed with the innovation of the staff. Until recently, historians of music believed that he was the first to propose:

1. Lines of different positions to indicate the height of the sound.
2. Different colors for some of these lines.
3. Clefs to indicate the pitch space [100].

Indeed we know that he had used the lines, colors, and clefs in his *Antiphonarius*, probably written around 1030, but now definitely lost. What remains is his *Prologus in antiphonarium* [29], though not in the original and thus possibly altered by copyists. Here we find very succinct but meaningful statements, such as:

> The voices are therefore arranged in such a way that each note, no matter how often it is repeated in the melody, is always placed in the same and unique position. To make this position easier to recognize, thick [spissae] (or narrow) lines are drawn. The position of the sounds will sometimes be on the lines themselves, sometimes in the space between them. All sounds placed on the same line or in the same space will be identical [29].[15]

However, according to a more recent history of music staffs, Guido did not invent the staff, which was rather the product of many nameless monks [53], and adapted the music layout from the tradition of the *Musica enchiriadis*—a 9th-century treatise previously attributed to the Pseudo-Odo—in an ingenious way.

2.1.1.2 Church Modes

The Western Church (or Ecclesiastic or Gregorian) modes, whose origins remain somewhat unclear—they date back to at least the 9th century—were employed to organize and classify Gregorian chants with the objective of facilitating their learning and memorization. These modes consist of eight ascending diatonic scales of seven different notes, separated by Pythagorean tones and semitones. Musicians of the Middle Ages assumed that their modes were modeled on the Greek tonoi, but this was a misunderstanding. It was only after the Renaissance that the difference between the Greek and the Church modes was recognized, with Girolamo Mei in 1572 [67].

According to Carolingian theorists, there were four pairs of Church modes, authentic and plagal (i.e., secondary). The four authentic or higher modes begin with the four pitches D, E, F, G in order. They were assumed to have a primary role "because [...], they actually have a single tetrachord of low notes below and two tetrachords of high notes above" [27].[16] The modes cover the interval of an octave, i.e.: D-d, E-e, F-f. G-g, although this is not always the case: "But we often deviate from this rule by going up to the ninth or even the tenth" [27].[17]

---

[15] l. 44–46.

[16] Ch. 11, 23–26.

[17] Ch. 11, 24.

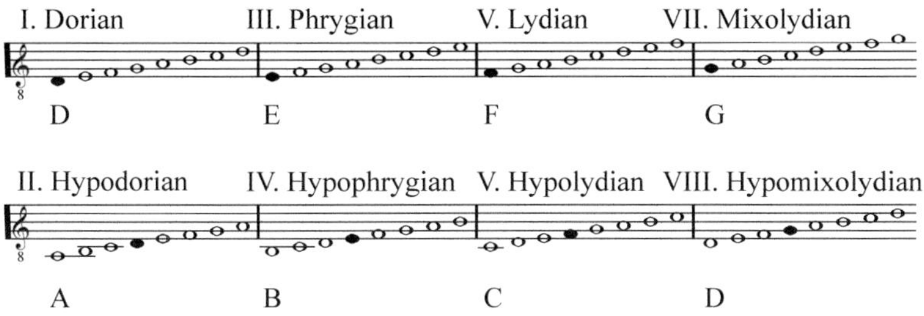

Fig. 2.5 Church modes; black notes are finals

The initial notes (D, E, F, G) were called finals (finales) because they were found at the end of a melody played according to the corresponding mode; this is similar to the tonic in modern tonal music.

The four plagal modes, or lower modes, start from the four notes A, B, C, D. They share with the authentic modes the final and the four notes above it, and complete the octave with a fourth below the final.

Each mode has, in addition to its final, a *repercussio*, the interval around which the melody of the modes revolves. The interval of the repercussio proceeds by ascending from the final to another note of particular importance in the modal scale, called *repercussa*. In France in the 16th century the latter took the name *dominant*, though only for authentic modes [68].[18]

The repercussa of all authentic modes begin one fifth above the final, apart from the case it is B, when it can be changed to C. The repercussa of the plagal modes begin a third above the final. However, during the 10th and 11th centuries the repercussa of some modes was raised; in the second authentic mode the repercussa was on c, in the second plagal mode on a, and in the fourth plagal mode on c.

Since the theory of Church modes is not central to our purpose, and it is very difficult to follow their description before the 11th century with the introduction of new concepts and symbols for notes and scales, we have decided to use the modern notation for notes, which can also be arranged in a staff as in Fig. 2.5. As you can see, there is no alteration; only an accidental is sometimes used—B can be lowered a half-tone to B♭.

In the figure, the compass (ambitus) is limited to one octave above the finals for the authentic modes and to one fifth above and one fourth below for the plagal modes, although in practice, as mentioned above, it may be greater. This limitation is the basis for an alternative classification of modes to that presented above, introduced from the 10th-11th century, called pseudo-classical, in which modes are associated with species of the octave (see Sect. 2.2.1).

To easily understand the difference between Greek tonoi and Church modes, imagine playing them on a piano. Church modes can be obtained by playing only the (diatonic) white keys (for a perfect analogy, of course, the white keys of the piano

[18] Chap. 2.

should be tuned according to the Pythagorean system). For the first mode, the Dorian, start from D and play the next seven white keys; for the second mode, the Phrygian, start from E and play seven more white keys, and so on.

On the other hand, to play Greek tonoi, as well as modern major and minor keys, you have to keep a fixed sequence of tones and semitones, which leads to the introduction of sharps and flats. So, on the piano, you should also play some black keys, except for the C major and A minor scales, whose notes are the same as those of the Hypolydian and Hypodorian church modes.

The names of the modes were not always the same. They had Latin names, such as protus, deuterus, tritus, tetrardus, transliteration from Greek: first, second, third, fourth; they were also simply indicated by Roman numbers, from I to VIII, with odd numbers for the authentic modes and even numbers for the plagal modes. Guido of Arezzo objected to this last nomenclature, because it lost the similarity between authentic and plagal modes: "But it is an abuse of the Latin view to say first, second instead of authentic protus and plagal protus, third and fourth instead of authentic deuterus and plagal deuterus" [27].[19]

Church modes also received Greek names; the following table gives the names in use in the 10th century, as commonly reported in music history handbooks:

Protus authentic	Dorian	Tritus authentic	Lydian
Protus plagal	Hypodorian	Tritus plagal	Hypolydian
Deuterus authentic	Phrygian	Tetrardus authentic	Mixolydian
Deuterus plagal	Hypophrygian	Tetrardus plagal	Hypomixolydian

For a better understanding, Church modes are also reported in Table 2.4. Dark gray boxes are for finals, while light gray boxes are for repercussio. As the table clearly shows, the Church modes have different distributions of tones and semitones.

In the following centuries the number of modes was increased and different theorists introduced different names and interval distributions. In 1547 the Swiss music theorist Heinric Glarean (1488–1563) in his *Dodecachordon* [49, 92], consistent with the new trend in music, added four modes, arriving at 12 modes. The new entries were the authentic Aeolian with a final on a and the authentic Ionian with a final on C, and the corresponding plagal modes. They are shown in Fig. 2.6 [85].

As you can see, apart from octave transpositions, they have the same notes as the Hypodorian and Hypolydian modes, but their finals are different, and modes are characterized more by finals and repercussio than by the scales of notes. Since the finals and repercussio of the Aeolian and Ionian modes are the same as the tonic and dominant of the modern minor A key and major C key, respectively, they are commonly associated with them.

Glarean avoided adding an authentic mode on B because it was irregular. In fact, the octave of all authentic modes is made up of only a fifth and a fourth. This would not be the case for the mode on B defined by the sequence BCDEFGAB, because the fifth BCDEF (S-T-T-S) is a diminished fifth. In any case, in the 18th century another mode was introduced on B, for the sake of completeness even denying its use in

[19] Ch. 12, 13–16.

Table 2.4 Notes and intervals for Church modes

Dorian	Hypodorian	Phrygian	Hypophrygian	Lydian	Hypolydian	Mixolydian	Hypomixolydian
d	a	e	b	f	c	g	d
				e	b		
c	G	d	a			f	c
b				d	a	e	b
	F	c	G				
a	E	b		c	G	d	a
			F	b			
G	D	a	E		F	c	G
				a	E	b	
F	C	G	D				F
E	B			G	D	a	E
		F	C				
D	A	E	B	F	C	G	D

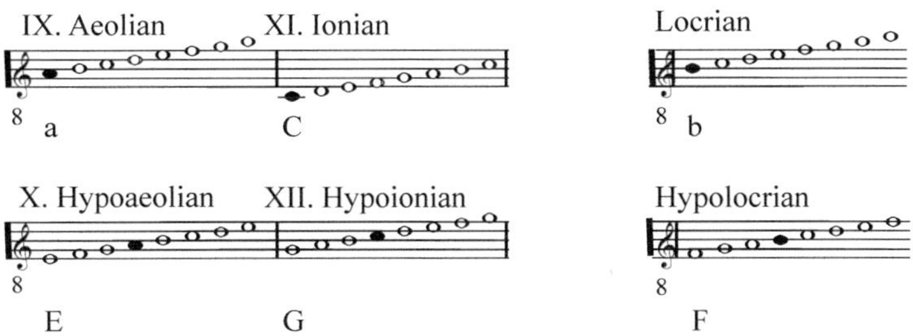

Fig. 2.6 Additional church modes

chant, called Locrian (see Fig. 2.6), with its plagal Hypolocrian, thus arriving at 14 modes. Different from the other modes, the repercussio of the Locrian was a sixth higher than the final [85, 88]. For a comprehensive examination of the evolution of musical styles from the Middle Ages to the Renaissance, see [68].

2.1.1.3 Polyphony and the Art of Counterpoint

Polyphony is a special way of making music in which voices sing different parts together. It was typical of Western musicians who performed it long before there were any written records. A first form of polyphony was the *parallel organum*, in which the original melody was the principal voice and other voices, the organ voices, moved in exact parallel motion (usually a fifth or a fourth above). The parallel organum evolved into the discant organum, in which the organ voices moved more freely [16].[20] The fact that notes of different pitches were played simultaneously changed the perception and even, in some ways, the concept of consonance. The musician of the time was well aware of this. There is a sure reference of the 10th century in Hucbald's treatise *De harmonica*, limited to the organum:

> *Consonance* [consonantia] is the calculated and concordant blending [concordabilis permixtio] of two sounds, which occurs only when two simultaneous sounds from different sources combine into a single musical whole, as happens when a man's and a boy's voices sound at the same time, and indeed in what is usually called "making organum". There are six of these consonances [consonantiae], three simple and three composite; they are designated by specific names: diapason, diapente, diatessaron, diapason-plus diapente, diapason-plus diatessaron, and double diapason [57].[21]

The shift from melodic intervals to simultaneous dyads was not immediately digested by theorists, and many of them continued to conflate the two faces of consonance: the melodic and the dyadic.

The 15th century saw the evolution of polyphony toward contrapuntal techniques that were later adopted by the masters of Renaissance polyphony, including Palestrina, Baroque music, including Bach, and 20th century Schömberg. One of the earliest precise definitions of counterpoint came from the 14th century, when it was already widely used. Pietro dictus Palma ociosa (fl. 1330) wrote in the *Compendium de discantu mensurabili* of 1336:

> So it should be noted that any simple discantus[22] which is nothing but a point against a point [punctus contra punctum] – or a note emitted by natural instruments against another note – can simply be composed and arranged from the unison, the minor third, the major third, the fifth, the major sixth and the octave. For a description of these aforementioned musical intervals of the discantus, in order to understand them more clearly, it is necessary to know what the voice is in music, and what the tone and semitone are [83].[23]

Thus, counterpoint is defined simply as "point against point" or "note against note", composing one note with another note to form the intervals of unison, minor third, major third, and so on. This definition is the one that gave modern polyphony its name, and remained the most widely used, at least until Zarlino.

A fundamental role in the development of counterpoint was played by Johannes Tinctoris (c. 1435–1511). Considered the most important theorist of his time, he wrote countless treatises on speculative and practical music, although only two of them were published during his lifetime. For this reason, Tinctoris' posthumous reputation was not so great in the following generation. The *Liber de arte contrapuncti* (1477) is

[20] pp. 85–99.

[21] p. 19.

[22] The Latin word *discantus* [English discant] came into being in the 12th century from the Greek *diaphonia*, either by direct translation or by analogous word formation. From the beginning, the word carried the implication of singing apart. The term counterpoint was first used for discant in the treatises of John de Muris (c. 1290–1344). This term, developed from the earlier definition of discantus, soon became widespread [38].

[23] p. 6/43.

Fig. 2.7 Example of diminished counterpoint; from [62], unnumbered pages, p. 86

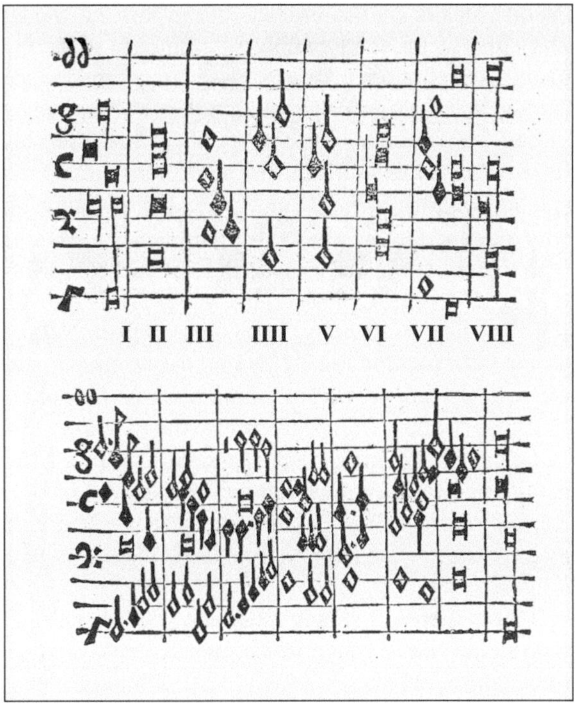

probably his most famous work. Here he contrasted two types of counterpoint: *super librum*, a reference to improvised vocal polyphony, and *res facta*, by which he meant written composition. He saw another important dichotomy in counterpoint: *simplex* and *diminutus*:

> Counterpoint is simple when the note placed against the given note has the same value as it. Counterpoint is diminutus when several notes are placed against one of the same or different duration, which is called floridus [98, 99].[24]

While counterpoint simplex is the traditional counterpoint note against note, counterpoint diminutus concerns the *diminution* of the value of the notes of simple counterpoint into notes of shorter duration, set against the same note. This type of counterpoint was also called floridus, because the diminished notes in the staffs resemble ornamental curlicues and flowers [70].[25] The 'diminution' also concerned the use of dissonances, the treatment of which became important in counterpoint at the end of the 15th century.

In Fig. 2.7 we see a comparison of simple and diminutus counterpoint. At the top we see a ten-line system, which we can call a proto-staff. The Greek-Latin letters on the left correspond to modern clefs. The Roman numerals below the first figure, on

---

[24] Book 2, Chap. XX; p. 474.

[25] p. 177.

the other hand, are the numbers of the bars, i.e. the groups of notes. The notes are in black and white, the two colors being used to distinguish the different voices, four in all. At the bottom is the floridus version; the modern reader, even a non-musician, can see that there is a rich arrangement of notes against notes.

The introduction of counterpoint with the presence of thirds and sixths, which became important as the music of northern Europe, especially England, led to important innovations [70].[26] Singers tended to temper Pythagorean thirds and sixths, which in polyphonic pieces, forming intervals of notes played together, clearly presented their dissonance, which did not bother in medieval monophonies (which followed the teaching of Boethius, ideal for monophonic musical compositions and some musical polyphony). The temperament of the Pythagorean major and minor thirds, (81 : 64) and (32 : 27), lend to the simpler and more pleasing to the ear intervals, (5 : 4) and (6 : 5). The same was true for the Pythagorean sixths (27 : 16) and (128 : 81), which were replaced by the two intervals (5 : 3) and (8 : 5).

In many 15th-century polyphonic situations, the fourth was treated as dissonant, while the new intervals, known as *just intervals*, were classified as pseudo-consonances and considered acceptable: the major third (5 : 4), minor third (6 : 5), major sixth (5 : 3), and minor sixth (8 : 5), though all outside the normally accepted sequence of 1, 2, 3, 4.[27]

Music theorists slowly began to accept the new musical practice, but it met with stubborn resistance from some scholars, including the humanist and musician Franchino Gaffurio (c. 1451–1522), who continued to defend Pythagorean intonation. Among the protagonists of the new trend is Bartolomé Ramos de Pareja (1440–1522), who proposed tuning with major and minor thirds [8].[28] Then, first Giovanni Spataro (1459–1541), a *protégé* of Ramos, and then Lodovico Fogliano (1490–1548)—a humanist able to read Greek musical sources who reacted to Boethius's numerological approach by considering the ear as the final judge of consonances and dissonances of intervals—became the champions of the new diatonic system. Zarlino, an admirer of Fogliano, accepted the new trend and managed to systematize it.

2.1.2 The Role of Humanists

At the end of Middle Ages, music underwent a new development known as *Ars nova*. The term *Ars nova* was first used by medieval music theorists in a specific sense to denote a new technique for notating polyphonic vocal music that was developed in France in the first half of the 14th century. In the 20th century, the term came to be

---

[26] p. 132.

[27] Major and minor thirds were usually called ditones and semiditones, respectively, while sixths were called hexachords.

[28] pp. 89–90.

used as a label for an European musical epoch distinct from that of the preceding *Ars antiqua*.

In the 14th century, new forms of vocal music emerged indeed, particularly in the domain of vernacular polyphony. Additionally, new stylistic developments occurred in forms and genres that had been passed down from the 13th century, including the motet and liturgical polyphony. The new notational techniques, forms, and styles of the 14th century were closely associated with the period of activity of known named composers, particularly in France and Italy [32].

In the Renaissance, the Greek paradigm of speculative music, not only in the form expressed by Boethius, regained strength. The more or less mystical philosophical perspectives on the microcosm and macrocosm, which had begun to disappear in the Middle Ages musicians, also reappeared. However, speculative music was no longer an end in itself, but was somehow connected to practical music, which was now so rich and important that all the major theorists of the time also produced texts on practical music, which, regardless of their titles, were in fact theoretical treatises in the modern sense of the word, dealing in particular with the rules of counterpoint.

The duality of speculative music-practical music can be seen in the titles of some treatises, such as *theorica musica* and *practica musica*; in other cases, the same treatise contains a preliminary section on speculative music, followed by sections on notation, Guidonian hand, and the rules of counterpoint. Renaissance theorists also had to take into account the changing sensibilities of the time.

Of great importance for theoretical music was finding and translating Greek treatises. which occurred several years after the translation of the philosophical texts. The first classical author to be rediscovered was Boethius. The shocking fact that his *De institutione musica* was of Greek origin was discovered by Johannes Gallicus de Namur (c. 1415–1473) [77].[29]

But the fundamental contribution to the rediscovery of Boethius was made by Gaffurio. His first published work, *Theoricum opus musice discipline* of 1480, had the distinction of being the first printed book on music theory in Italy and served as an early version of his later theoretical works. Gaffurio's central achievement as a music theorist, however, was the series of three treatises on the speculative and practical dimensions of music: *Theorica musicae* (1492), *Practica musicae* (1496), and *De harmonia musicorum instrumentorum opus* (1518) [42]. This last treatise did not, as the title might suggest, deal with actual instruments, but rather with details of the author's understanding of Greek harmonics, based on his newly acquired knowledge of Greek sources. Through his treatises, especially the widely read *Theorica musicae*, Gaffurio put his interpretation of Boethius and the ancient authors into the hands of hundreds of musicians [104].[30]

After Boethius, other Greek authors were rediscovered. Among them were Aristoxenus and Ptolemy. Aristoxenus was first known from the *Eisagoge harmonike* of Cleonides, which was translated into Latin in 1497 by Giorgio Valla (1447–1500).

---

[29] p. 5.

[30] p. 74.

Then, in 1562, Antonius Gogava (1529–1569) published his Latin translation of *Elementa harmonica* along with a translation of Ptolemy's *Harmonica* [50].

2.2 Music in the Renaissance

By 1500 a secular musical genre such as the madrigal had become very popular at Italian courts. Texts by famous Italian poets, including Petrarca, were set to music. The origin of the word *madrigal* is uncertain; it may derive from *madre lingua* (mother tongue), meaning compositions based on vernacular texts as they actually were, or from *mandria* (herd), meaning bucolic subject matter.

This genre of music could be sung by groups of friends gathered around a table. It had the characteristic of maintaining a strong adherence to the text; basically, if the text said "I am dying", the music could be sad or dissonant; if it said "I am escaping", there was a canon, i.e., one voice chasing the other, and so on.

In Florence, the Bardi Camerata was founded, following the very goal of the Florentine Renaissance: to restore the Greek theater. According to the members of this Camerata (including Vincenzo Galilei), Greek tragedy should be performed with declamation approaching singing, hence "recitar cantando". The voice was then accompanied by an instrument, such as the lute or archlute, to recall the Greek aedi. From this quest came the—accompanied—monody and the opera [81]. In the opera, the text and its declamation took on a fundamental role, and composers began to take licenses to be more expressive, adding musical rhetorical figures and affects to make the music as expressive as possible.

In 1605, when Claudio Monteverdi (1567–1643) did not respect the rules of counterpoint in his fifth book of madrigals [71], a controversy broke out with Giovanni Maria Artusi (c. 1540–1613), a faithful follower of Zarlino and strict counterpoint. Thus was officially born the *seconda pratica*, which considered the harmony commanded and not commander, and for mistress of harmony placed the oratione [40, 78].

Vocal music was increasingly accompanied on polyphonic instruments, just as instrumental music in general began to emerge. Initially, the ideal was to play the score as faithfully as possible. Over time, the independence of the individual parts was limited to the highest and lowest voices. With the increasing number of polychoral works, the extracted lower part made it easier for the organist to play along: the figured bass was introduced [41].[31]

Figure 2.8 shows an example of a monody in which there are only two voices, the soprano and the figured bass (an organ, harpsichord, or archlute). The numbers above the bass indicate the other voices in intervals that the organist should play above a given note; for example, a fourth (4) and a sixth (6).

This change in practical music also influenced speculative music because of the new role dissonances had gained. In particular, the question of the nature of conso-

---

[31] Vol. 1, p. 13.

Fig. 2.8 Example of monody; from [22], p. XXI

nances became more pressing, and many musicians, especially those with a mathematical background less influenced by the two millennia of Western music, began to think that the perception of consonances was not only a matter of acoustics but also of custom.

In the following, we will consider only the contributions of three of the most important scholars of speculative music: Gioseffo Zarlino (1517–1590), Vincenzo Galilei (1533–1591), and Giovanni Battista Benedetti (1530–1590). The first is generally regarded as the most important music theorist of the1500s representing the apogee of classical speculative music and providing a seal to the contrapuntal practice of what will be known as the *prima pratica*; the second is an exponent of the new way of making music, both in terms of speculative aspects, in which he introduced an experiential approach, and in terms of practical aspects, putting himself forward as one of the representatives of the *seconda pratica*; the third is (one of) the first great mathematician to turn his active interest to speculative music after more than a thousand years.

2.2.1 *Gioseffo Zarlino and the Senario*

Gioseffo Zarlino (1517–1590), a pupil of Adrian Willaert (c. 1490–1562), was one of the leading musicians in Italy and Europe at the close of the Renaissance. From 1565 until his death, he held the position of choirmaster at St. Marco's in Venice, one of the most prestigious positions in Europe. As a member of the *Accademia veneziana* he could also interact with mathematicians and musicians. For his position he had at his disposal a vast corpus of ancient treatises on music; besides Boethius's *De institutione musica* he could see the works of Ptolemy, Aristides Quintilianus, Aristoxenus, Euclid, and many others.

Zarlino's fame as a music theorist rests on his monumental *Le istitutioni harmoniche* (hereinafter *Le istitutioni*) divided into four parts. First edition in 1558 [105], followed by a second edition in 1562 [106], and many more until 1589. It has been translated into English, but in separate parts [26, 109–111]. In 1571 Zarlino wrote another important treatise, the *Dimostrationi harmoniche* [107], a text with a decidedly mathematical slant and a very formal approach, full of definitions, postulates and theorems, but which added essentially nothing new, at least from our point of view. In the following, therefore, we will comment in some detail only on *Le istitutioni*.

Zarlino clearly saw the need to combine the speculative principles of a musical science with practical musical activity: "Even if speculation in itself does not seem to require practical implementation, a theoretician would not be able to apply what he has discovered without the help of a performer. On the other hand, the performer would never achieve a perfect performance without the help of reason. Consequently, these two fields are so closely connected that they cannot be separated" [105].[32]

Part I of *Le istitutioni* serves as an introduction to music, particularly to its foundations as a mathematical discipline of proportions in the tradition of Boethius; Part II reviews the Greek conceptions of music. Beginning with Part III, Zarlino turned his attention to contemporary music, especially counterpoint. Here he began his study with a detailed review of intervals, then moved on to an extensive treatment of practical composition, presenting one of the most comprehensive discussions of how a musician composes in the early history of music. A major significance of this part lies in the lavish attention to practical issues of composition, making the treatise a model for the study and analysis of music of his time.

In particular, Zarlino named the intervals differently from what he had done in the first two parts, adopting what he called the numbering of music practitioners: "The simple ones are all those intervals that are smaller than the diapason, that is, the unison, the second, the third, the fourth, the fifth, the sixth, the seventh, and the octave, that is, the diapason and all those that are larger than it, i.e., the ninth, the tenth, the eleventh, the twelfth, and the others in order" [108].[33]

But while there is only one diapason, there can be several fifths and fourths (etc.). Zarlino in particular commented on the fourth (and on the second):

> The second is of two kinds for practitioners, that is, the tone and the semitone: hence it may be called the second major and the second minor. And the fourth is of three kinds, that is, the diatessaron consonance, the tritone, which is a composition of three tones; and the semi-diatessaron, which is a composition of one tone and two semitones, with the extreme intervals that are dissonant. The same could be said of the fifth, etc. [108].[34]

In Part IV, Zarlino discussed the 12-mode system first articulated by Glarean. He combined the humanist respect for ancient tradition inherited from the Greeks with the practical aspects of music as it was performed in the late Renaissance.

In the following, we will briefly consider mainly the first two parts of *Le istitutioni*. Although not entirely original, they are very useful for our purposes because they contain the mathematical procedures of the 16th century. The first part is divided into 44 short chapters. After a brief discussion of the differences between secular and

---

[32] p. 20.

[33] Part III, p. 149.

[34] Part III, p. 156.

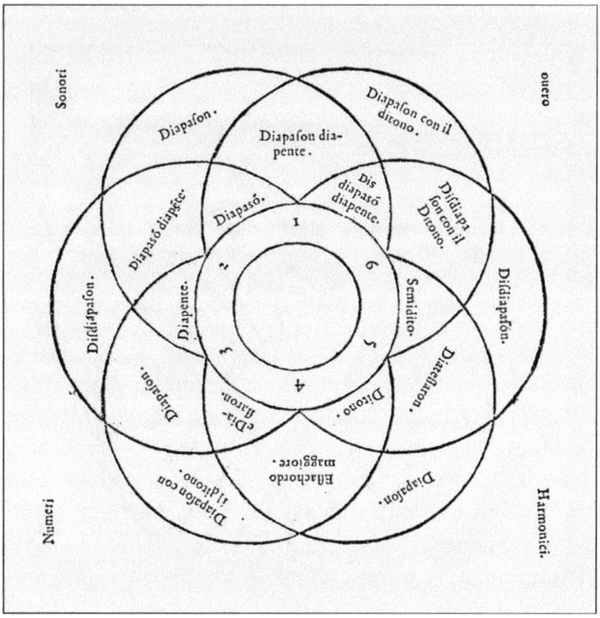

Fig. 2.9 Possible combinations of the senario; from [108], p. 25

human music, and between speculative and practical music, it moves on to a mathematical treatment that is essentially aligned with Boethius. Chapter 13 defines the different kinds of numbers: even, odd, prime, compound, square, cube, and perfect, and dwells on the latter. Perfect numbers are those that are equal to the sum of their submultiples. The first perfect number is 6, $(1 + 2 + 3 = 6)$, the second is 28, and then comes 496. The number 6 is called *senario*, derived from the Latin senarium, meaning composed of six parts. So senario is both a number and a group of numbers.

The properties of the senario, according to Zarlino, are truly extraordinary (Chap. 14); six are the planets, six are the possible types of motion, six are the transcendent entities, six are the modes of the propositions, six are the properties of the elements, and so on. And in music, six are the consonances, of which five are simple: diapason, diapente, diatessaron, ditone (major third), semiditone (minor third); the latter is the unison, the principle of them, although it is only improperly called a consonance. Figure 2.9 clearly shows all the possible combinations of the first six numbers.

Chapter 16 gives the distinction between simple and compound consonances. Simple consonances are those defined by ratios of consecutive numbers that cannot contain an intermediate number, such as (3 : 2). Compound consonances can contain intermediate number, such as (5 : 3), which can contain 4. Compound consonances, on the other hand, can be obtained by adding two simple consonances, e.g. the major hexachord (5 : 3) is composed of the diatessaron (4 : 3) and the ditone (5 : 4).

The minor hexachord (8 : 5), does not fit into the senario, and in fact its consideration would invalidate all of considerations of Zarlino, but he did not give up and used a rhetorical device by inserting the minor hexachord into the senario, as it is composed of the diatessaron (4 : 3) and the semi-ditone (6 : 5).

In Chap. 19, Zarlino defined the sonorous number, showing his Pythagorean attitude:

> The sonorous number [numero sonoro], and not the sonorous body, is the true subject of music, for although all sounding bodies are capable of producing sounds, yet not all of them are capable of producing consonances; unless, that is, they are proportioned to one another and included under a completed form, that is, under the reason of harmonic numbers [105].[35]

Zarlino took the name and the concept of proportion for granted, fitting it into the tradition of the time, and using this name to indicate the ratio between two terms. In the following, we will follow Zarlino's usage, also because there is no ambiguity.

Proportions are divided into rational and irrational, arithmetic and geometric. Arithmetic is between integers and is always rational; geometric is between quantities and can be either rational or irrational, as in the case of the side of a square and its diagonal. The arithmetic proportions, which in the following are represented in the form $(n : m)$, with $n > m$, are classified into several genera:

1. Multiple genus: $n = km$, with $k \geq 2$. It can be dual, triple, etc.
2. Superparticular genus: $n = m + 1$, hence proportions in the form $(m + 1) : m$. Species of this genus are defined by the prefix sequi, in the form sesqui-m-th. For $m = 2, 3, 8$ we have sesquialtera (3 : 2), sesquithird (4 : 3), and sesquioctave (9 : 8), respectively.[36]
3. Superpartient genus: $n = m + k$, with $k \geq 2$, hence proportions in the form $(m + k) : m$. The designation is super-k-th-partient-m-th. For example, $m = 5, k = 2$ super-bipartient-fifth (7 : 5).
4. Multiple superparticular genus: $n = km + 1$,with $k \geq 2$, hence proportions of the form $(km + 1) : m$. The designation is k-sesqui-m-th. For example, $m = 2$, $k = 2$, double-sesquialtera (5 : 2); $m = 4$, $k = 3$, triple-sesquithird (13 : 4).
5. Multiple superpartient genus: $n = km + p$, with $k \geq 2$ and $p \geq 2$, hence proportions in the form $(km + p) : m$. The designation is k-th-super-p-partient-m-th. For example, $m = 4, k = 4, p = 3$, quadruple-super-threepartient-fourth (19 : 4).

Then, starting with Chap. 31, the operations on proportions are introduced: multiplication, addition, subtraction, division. For simplicity we will use a modern approach, identifying the proportions with the corresponding fractions.

1. Multiplication. Multiplication between proportions, as presented by Zarlino, is an indispensable operation for representing multiple proportions on a line, such

[35] p. 30.

[36] In what follows, we suggest a naive translation of the Italian Zarlino terms into English. In [73], more specialized terms are used; for example, *sesquithird* is rendered as *seqsuitertian*.

Table 2.5 Multiplication of proportions. Redrawn from [108], p. 42

3	4		5	6
2	3		4	5
12	8	6		
60	40	30	24	
360	240	180	144	120

as a monochord, using integers. The example he gave is the multiplication of the proportions $3/2$, $4/3$, $5/4$, $6/5$; the operation is commented with reference to Table 2.5.

Let us first consider the multiplication between $3/2$ and $4/3$. We start by multiplying the two major terms together, which gives 12, and the minor term of the first proportion with the major term of the second, which gives 8. Then we multiply the two small terms together, giving 6; these numbers are written below in the order $12 : 8 : 6$. This gives us three integers such that the ratio of the first and second reduced to the radical terms is $3/2$, while that of the second third is $4/3$. Now we multiply by the fraction $5/4$. We multiply the previous three numbers by 5, which does not change their ratio, then we multiply the last number, 6, by the smallest value of the ratio $5/4$, which gives us 24. It is clear that the sequence 60, 40, 30, 24 contains the three considered ratios in sequence. Now we multiply with $6/5$. First we multiply all the numbers in the second row by 6, then we multiply the last number in the second row by 5, and we obtain the numbers 360, 240, 180, 144, 120, which allow us to obtain, in sequence, the five ratios whose product has been made.

2. Sum. The sum of several proportions is the one traditionally used in music for intervals. Zarlino presented it simply by saying that the sum of several proportions is obtained by considering the ratio between the product of the major terms and that of the minor terms of each proportion, as shown in Table 2.6.
3. Subtraction. The subtraction of several proportions is obtained from a sequence of subtractions of two proportions at a time. Figure 2.10, which is classic for musical treatises until the 17th century, shows the algorithm. In modern terms, the difference between $3/2$ and $4/3$ shown in the figure is obtained by dividing the fraction associated with the first ratio, $3/2$, by the fraction associated with the second ratio, $4/3$.
4. Division. To divide a proportion is to divide it into two parts by finding a middle term. The term proportionality is introduced here to denote two successive proportions. There are three types of division: arithmetic, geometric, and harmonic; the value that divides a proportion into the three different types is called arithmetic mean, geometric mean, and harmonic mean, respectively.

    a. Arithmetic mean. The intermediate number in a sequence of three numbers with a constant difference between them. In the proportion $(3 : 2)$ there is no

Table 2.6 Sum of proportions. Redrawn from [108], p. 44

primo	3 Sesquialtera 2	modo	3 Sesquialtera 2	360 Tripla 120
	4 Sesquiterza 3		4 Sesquiterza 3	
	5 Sesquiquarta 4		5 Sesquiquarta 4	
	6 Sesquiquinta 5		6 Sesquiquinta 5	
	360 Tripla 120		Secondo modo	

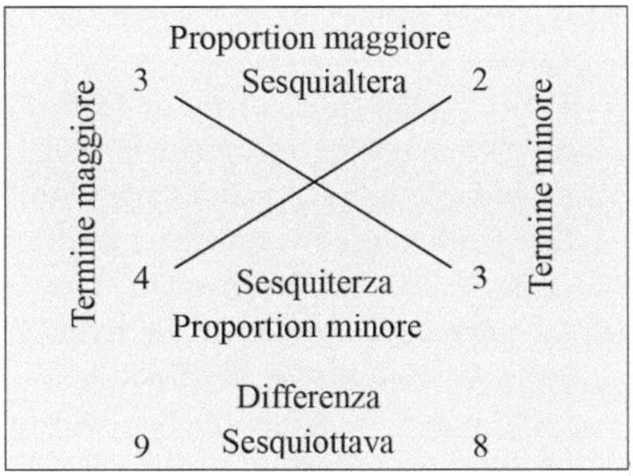

Fig. 2.10 Difference of two proportions. Redrawn from [108], p. 45

intermediate term, so for the arithmetic division we consider the equivalent proportion (6 : 4), which gives the sequence 6, 5, 4, whose terms have unit difference between them. The intermediate term 5 is the arithmetic mean.

b. Geometric mean. The intermediate number in a sequence of three numbers that have the same proportion (ratio) between them. Consider the proportion (4 : 1). The geometric proportionality is expressed by the sequence 4, 2, 1, and the geometric mean is 2.

c. Harmonic mean. The situation in this case is much more complex, and Zarlino only provided an algorithm to find the mean without going into explanations. In modern terms, the harmonic mean b of the proportion $a : c$, contained in the sequence a, b, c, is given by the the inverse of the arithmetic mean of the inverse of a and c, i.e.:

$$b = \left[\frac{1}{2}\left(\frac{1}{a} + \frac{1}{c}\right)\right]^{-1} = \frac{2ac}{a + c}$$

For example, the harmonic mean of the ratio (3 : 2) is $b = 12/5$. Multiplying everything by 5 gives the sequence 15, 12, 10.

According to Zarlino, the harmonic mean is the most natural, even though in the tradition of music theory the most natural was the arithmetic proportion. He probably took this idea from Fogliano, who wrote in his *Musica theorica* of 1529:

> Although in mathematics there are more proportionalities, such as arithmetic proportionality, geometric proportionality, harmonic proportionality, and many others, nevertheless, because I strive to be brief, I shall bypass all those that have no relevance whatsoever to the musician to speak here only about harmonic proportionality, according to which, as will be seen below, all harmony is arranged [39].[37]

Fogliano justified his position because Ptolemy's syntonic diatonic system can be reconstructed using the harmonic mean.

The second part of *Le istitutioni*, divided into 51 chapters, deals mainly with Greek music, but also considers aspects of mondana and humana music. Zarlino concentrated on the tetrachords mentioned in Ptolemy's *Harmonica*, especially on the diatonic genus and its syntonic shade. In Chap. 39 he showed how to find all the consonances of the senario contained in the diapason in the monochord built on the basis of Ptolemy's tetrachords.

From the harmonic division of the diapason $(2 : 1)$ we obtain the diapente and the diatessaron. The sequence of the intervals of the harmonic division of the diapason is 6, 4, 3. If we take these numbers to represent the interval of the diapason, starting with the lowest note, we have first a diapente and then a diatessaron. And this is the most correct choice, according to Zarlino, since the widest interval must be in the lowest part. If one had arithmetically averaged the diapason—$(4 : 2)$ in non-radical form—one would have obtained the sequence 4, 3, 2, in reverse order, that is, with the diatessaron on the lowest part.

The harmonic division of the diapente gives the ditone and the semiditone. Zarlino did not report the sequence of intervals 15, 12, 10 containing the ditone $(15 : 12) = (5 : 4)$ and the semiditone $(12 : 10) = (6 : 5)$, probably because in this case the numbers in the sequence are not contained in the senario. The sequence 15, 12, 10 is known today as *major third chord*. If arithmetic division had been performed, we would have had the sequence 6, 5, 4, which still contains the ditone and the semiditone, but in reverse order; in modern terms, it is a *minor third chord*. Since the major chord has a larger interval in the lower position, it should be preferred according to Zarlino. This preference will be discussed for a long time after Zarlino. Below is a modern evaluation of Zarlino's approach:

> Zarlino was one of several composers and theorists who were increasingly concerned with describing and explaining the vertical structure of music based on triads in the language and concepts of music as a horizontal tradition from the Greeks and chant. Thus in his description of musical structures, we currently refer to as major or minor triads, Zarlino used the vocabulary of ancient Greek musical traditions to describe the harmonic or arithmetic division of the fifth or diapente. In other words, in reading Zarlino, the modem musician can see how the basic chordal structures of music were described in the conceptual language of mathematical proportion, giving the relatively new idea of the triad a rightful place in the most ancient traditions of music theory. In the humanistic conventions of his time, Zarlino

[37] f. 9v. Translation into English in [87].

> anchored his world to ancient learning by citing various classical authors and authorities including Plato, Aristotle, Boethius, and Martianus Capella [104].[38]

The ditone can be divided harmonically to give the major and minor tones. The major semitone is the difference between the semiditone and the major tone. For a discussion of the way Zarlino reasoned about the harmonic division of the diapason, see [87].

Figure 2.11 shows the monochord associated with Ptolemy's diatonic syntonic tetrachords. The numbers between 864–216 (a compass of two diapason) represent the distance from B; the greater the number the lower the pitch. These numbers are obtained by multiplying all the intervals of the diatonic system. The monochord begins with the proslambanomenos, so Zarlino did not use the extension of the note Γ. The figure shows the inclusion of the tetrachord synēmmenōn, which enriches the monochord with two notes, ♭ and *d*.[39] The soft note ♭ is a major semitone (16 : 15) from the mesē, while the distance between d and *d* is a syntonic comma (81 : 80). The introduction of the note *d* is of some importance because it causes the diatonic syntonic tetrachord to consist of 17 strings instead of 16 as in the ditonic diatonic tetrachord.

Zarlino, like most musicians of his time, recognized that a temperament was necessary for modern keyboard instruments because maintaining the consonances of the senario would require a very large number of keys. However, he did not realize, or at least did not fully realize, that even in vocal music, the adoption of the senario presented insurmountable difficulties, which will be well pointed out by Vincenzo Galilei and Giovanni Battista Benedetti.

In fact, Zarlino's motivation for a temperament was quite abstract: to eliminate the gap of the d-*d* comma. In his opinion, this could be done by distributing this comma, after having divided it into small parts, among all the remaining intervals of the octave a-aa. He thought that seven parts would be enough not to change the consonances significantly. This was not a trivial task, however.

The problem of dividing a musical interval, especially of the superparticular type, into equal parts had always interested music theorists. Since ancient Greece, for example, it had become clear that in the field of arithmetic the division could not be made by equal rational proportions.

In geometry, the division into two equal parts, i.e. the search for the geometric mean, was already well known to musicians; it consisted of a very simple construction proposed by Euclid and repeated by Zarlino [96].[40] With reference to Fig. 2.12, consider the diapason consonance (2 : 1) between the strings *ac* and *cb*. To find a geometric mean that lies between these two, proceed as follows: first, describe a

---

[38] p. 201.

[39] The inclusion of a ditonic diatonic tetrachord synēmmenōn in a ditonic diatonic greater perfect system would have given only the note ♭. Introducing a diatonic syntonic tetrachord synēmmenōn into a diatonic syntonic greater perfect system give both ♭ at the distance of a major semitone (16 : 15) from the mesē and *d* at the distance of a fourth from the mesē and a syntonic comma from d.

[40] Book VI, prop. 9, p. 111. This proposition is now commonly known as Euclid's second theorem.

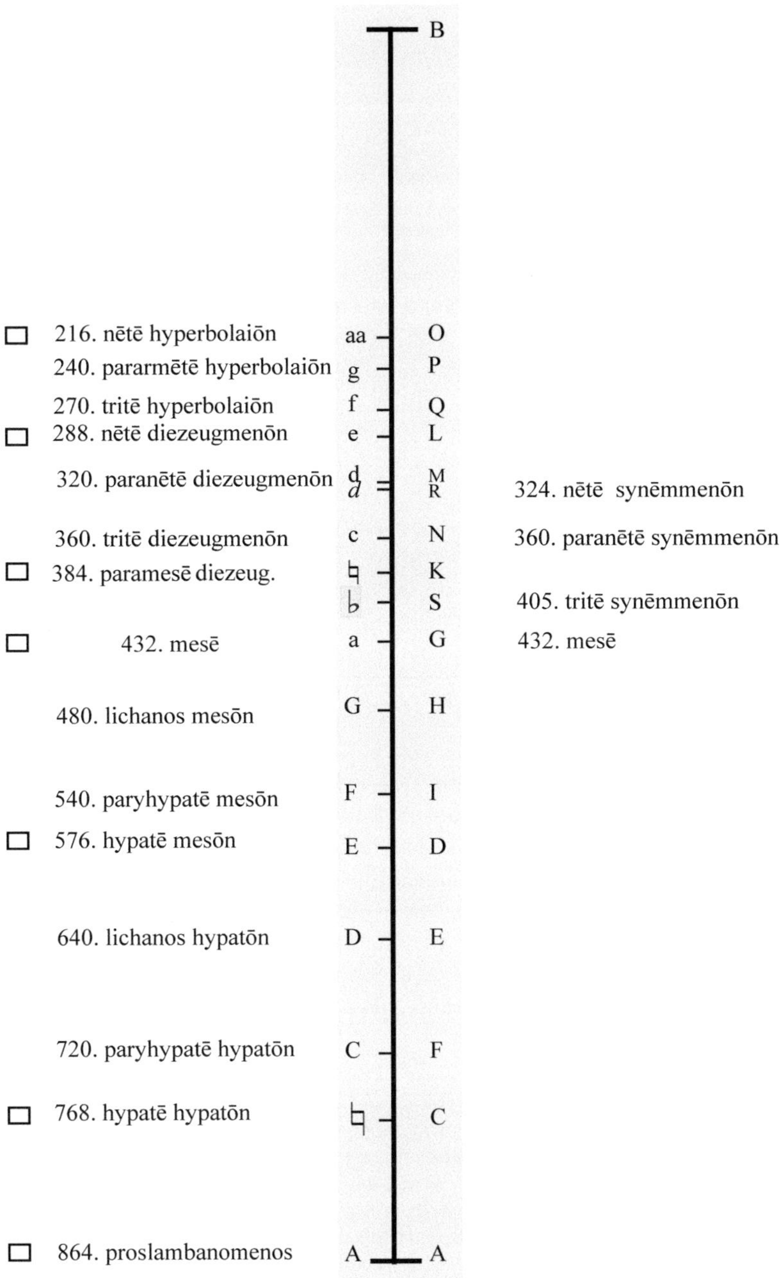

Fig. 2.11 Ptolemy's diatonic syntonic monochord. Redrawn from [108], p. 124

Fig. 2.12 Division of an interval into two equal parts. Redrawn from [108], p. 94

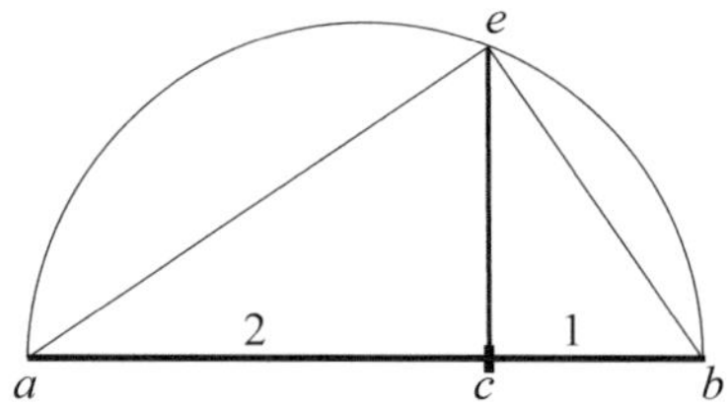

semicircle of diameter ab. Then draw a line from c perpendicular to the diameter, which intersects the circumference of the semicircle at e. The segment ce is what is being searched for; it is indeed $ac : ce = ce : cb$.

When Zarlino wrote his book, algebraic theory had made great progress; a few years later, after the second edition of *Le istitutioni*, Simon Stevin would solve the problem of dividing the diapason into 12 equal parts. Zarlino was far from this, and in any case the division of an interval into seven equal parts was very difficult before the introduction of logarithms and a different conception of proportions. But also Zarlino's contemporary Michael Stifel (1487–1567), among others, was able to provide an algebraic approach to the division of musical intervals. This division is presented in a very clear form in the *Arithmetica integra* of 1544, in chapter IX, entitled *De musicis progressionibus*. For example, the division of the major tone is represented in the sequence $9 : \sqrt{72} : 8$, where the middle term which gives the half-semitone, is an irrational number that can be approximated as you like with a rational number [93].[41] But this move, very natural for a modern reader, was not so for Zarlino, who, like many of his mathematical colleagues, could not make it, because it required a profound change in the logical status of the concept of proportion, which since the Greeks had been associated with integers [84].[42]

To divide into parts greater than two, Zarlino proposed the use of another geometric method, founded on the mesolabion, whose use will be described in Chap. 3. Despite the impossibility of providing a numerical representation and description of the temperament, he proposed an accurate description of it, really very interesting for the time, to the point that we will report it in full:

> But it must be considered that if one wants to make the temperament or *partecipatione* with a good reason and with a solid ground, it is necessary to divide the comma, contained between the strings RB and MB, into seven equal parts, and to distribute them among the seven intervals, contained in the eight strings of the diapason; so that we can reduce the two mentioned strings, containing the comma, to only one. But this should be done in such a way that the intervals remain as much as possible in their just form, so that the ear is not offended, and that every consonance, both in the bass and in the treble, and every other interval, however small, is equally increased or decreased by a given and determined amount in all those intervals which are similar in proportion. This will be very useful, if we act in such a way that every diapente is diminished and imperfect by two sevenths of a comma, and that the diatessaron is increased by the same amount; and it is necessary: since the diapason

[41] p. 79.

[42] Chap. 4.

> must remain unchanged, and in its true and natural proportion, and since it is composed of these two parts, what is subtracted from the one, must necessarily be given to the other; so that by their addition, at the ends, the perfect diapente will be heard. The ditone is made imperfect by a seventh, and the semiditone is diminished by the same amount. Since these two consonances contribute to the formation of the fifth, which is diminished by two sevenths, it is necessary that this amount be divided between these two intervals: as if the third were made imperfect by one seventh, and the minor third by the same – which together make two sevenths – these two consonances, which are part of the fifth, are diminished by the same amount by which their whole is diminished. But the parts of the ditone, that is to say, the major and the minor, will be made imperfect as follows: four sevenths of the comma will be taken away from the former, and three will be added to the latter; and so together they will have the same imperfection that their whole has, that is, they will be imperfect by one seventh. Then the major semitone will be augmented by three sevenths: since it is the smaller part of the semitone, and the major is the larger, these two parts will have the same diminution as in the semitone, that is, they will be diminished. The major hexachord and the minor hexachord will both be augmented by a seventh, since the former is made up of the diatessaron and the ditone, and the latter similarly of the diapente and the semiditone: since the diatessaron is augmented by two parts, and the ditone and the semiditone are both augmented by a seventh, the hexachords will indeed be augmented by this amount. After the instrument has been proportioned in this way, every consonance and every interval, *with the exception of the diapason and the minor semitone* [emphasis added], will be out of its true proportion, but not far from its true form, so that the ear will still be satisfied [105].[43]

The temperament proposed by Zarlino is known as the $2/7$ of a comma temperament (see Sect. 3, below). It should be considered a milestone in the history of temperaments for several reasons; it was the first time that a form of temperament was accurately and correctly described, and it is one of the first times that the idea of dividing an interval into smaller parts was concretized.

In the fourth part of *Le istitutioni* Zarlino discussed ancient and modern modes. This part marks a new stage in Zarlino's thinking. Whereas the third part on counterpoint was largely rooted in the practice taught by his master Willaert, the section on modes is largely the result of Zarlino's own research into sources somewhat alien to the rest of his treatise. Before any considerations, he specified what he meant by mode:

> It should be noted that the word *mode* in addition to all its many other meanings, properly means *reason*, namely, that measure or form which prevents us from going too far in anything we do, making us act in all things with a certain temperateness or moderation [...]. Since each mode had something intrinsic in its tune, and was accompanied by different rhythms, they called some modes grave and severe, some bacchanal and wild, some honest and religious, and others lascivious and bellicose [105].[44]

For ancient modes, Zarlino did not attempt to discover or devise a unified system; his presentation is a compilation of opinions and statements drawn from various ancient authors. Despite the thorough nature of Zarlino's study and his knowledge of the contents of many ancient treatises on music, the modern reader sometimes feels that the author did not fully understand some of the basic concepts of ancient Greek music theory. For Zarlino, a synthesis of the ancient modes is of little relevance

---

[43] p. 126. Translation into English adapted from [111].

[44] pp. 293, 295. Translation into English in [110].

however, since their use had died out in the course of time, making it impossible to find any definite trace of them in contemporary music: "for their use was completely extinguished, that we could not find any trace of them. Nor should we wonder at this, since time consumes all created things [108].[45]

Zarlino could thus present modern modes, especially his own. He referred to the *Dodecachordon* of Glarean, though he did not quote it. His modes, as Church modes, are based on the division of the diapason into diapente and diatessaron into two (apparently) different ways. In the first way he combined a species of diapente (below) with a species of diatessaron (above), and viceversa, for all possible combinations. We have three types of diatessaron, such as (A-D, B-E, C-F) and four types of diapente, such as (D-a, E-b, F-c, G-d), where each note can be raised or lowered by one octave, going from a capital to a small letter and vice versa.[46] By joining the first diapente D-a with the first diatessaron a-d in the acute, we have D-a + a-d = D-d, for the first mode. By joining the same diapente with the same diapason in the grave, we have A-D + D-a = A-a, for the second mode. Proceeding similarly for the other types of diapente and diatessaron, we get 12 modes.

The first mode is called principal or authentic, the second secondary or plagal. For Zarlino, the difference between these two kinds of modes lies in the different way the diapason is divided; the primacy of the authentic mode depends on the more sonorous arrangement of the diapente (D-a) below the diatessaron (a-d).

In the second way to form modes, Zarlino considered six species of diapason (the one beginning with B is avoided because it cannot be divided harmonically [108][47]) and divided them first by a harmonic mean and then by an arithmetic mean, to give still 12 modes.

The twelve modes of Zarlino are shown in Fig. 2.13, with C and F clefs [108][48]; the final is marked with a black note (our addition). Odd numbers are for principal modes, even numbers for plagal modes. In contrast to Glarean, Zarlino did not give any names to his modes; they were simply identified by numbers. And if in *Le istitutioni* the numbering corresponded to that of Glarean, in the *Dimostrationi* they did not [107].[49]

At the turn of the 16th century, the idea spread that a piece of music had to conform to the subject treated in the text, and that the best way to achieve this goal was to choose an appropriate mode. In point of fact, the ethos of modes represented a humanist fashion that endured only until the second half of the sixteenth century because of its inconsistency and the evolution of music. The theorists of the time found a mixed association of ethos with modes coming from ancient Greek and

---

[45] p. 300.

[46] Note that the species of diapason and diapente are defined only by the sequence of the tones and semitones. For example, not only A-D is representative of the first species of the diatessaron, but so is D-G, since both have the sequence T-S-T.

[47] p. 311.

[48] p. 149.

[49] p. 301.

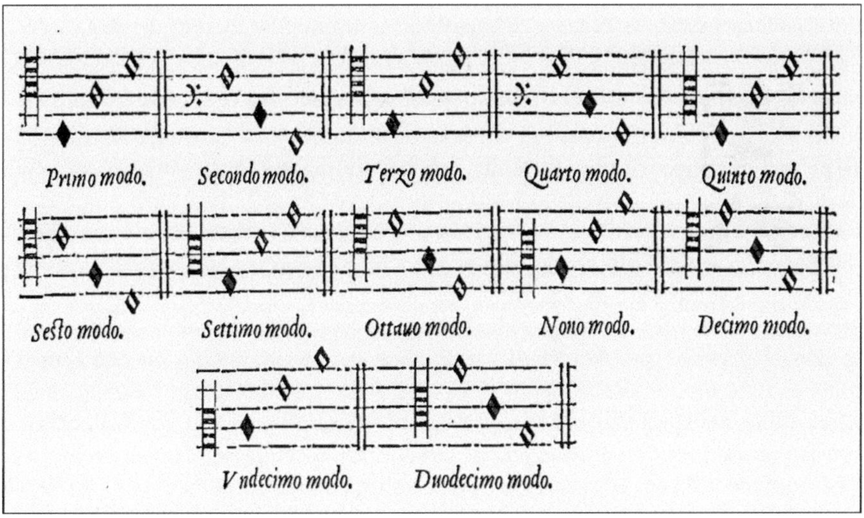

Fig. 2.13 The twelve authentic and plagal modes of Zarlino. Redrawn from [108], p. 311

medieval writers, with the false assumption that they were the same, and associated contrasting ethos with modes.

Zarlino proposed a characterization of his own, arriving at a careful classification, summarized in a long table in [95].[50] He based his classification not only on tradition, but also on musical qualities. For example, this is what he said about the first mode:

> The first mode has a certain effect, between sadness and joy, because of the semiditone that is heard in the counter-punctual relationships above the extreme notes of the diatessaron and diapente, and because of the absence of the ditone in the deep part [of the diapente]. The nature of this mode is religious and devotional, as well as a little sad; therefore, it is best used with words full of gravity that deal with lofty and uplifting subjects [105].[51]

2.2.2 *Vincenzo Galilei's Criticisms Toward the Senario*

> No study of Vincenzo Galilei would be complete without some insight into his rather enigmatic, equivocal personality. It has been said that he stood half in the sixteenth and half in the seventeenth century with regard to his musical perception. As a madrigalist and lutenist, he was a veritable master of sixteenth-century practice. It was, moreover, his desire to benefit the younger generation of composers through his rich experience. On the contrary, he promoted expressive music, and as a pioneer of monody, he envisioned the ideals of the seventeenth century, the era of accompanied solo music. One finds in Galilei's *Dialogo* a manifestation of these monodic tendencies, at least in theory, if not in actual practice. Although this

[50] pp. 113–114.

[51] p. 322. Translation into English adapted from [82].

> noble Florentine did not show how to implement his neo-Hellenic ideas, he did state clearly rudimentary conceptions of the so-called "seconda pratica" [55].[52]

Vincenzo Galilei (c. 1520–1591) probably came to Florence around 1540 to study music. His ability to play the lute attracted the attention of Giovanni Bardi (1534–1612) and members of the Camerata, who encouraged him to study theory. With Bardi's support, Galilei went to Zarlino in Venice. In 1572 he began a correspondence with Girolamo Mei (1519–1594), a philologist interested in Greek music, which would profoundly change his approach to music. At this time he was writing a treatise influenced by Zarlino's teachings, probably for his own students, but discussions with Mei led him to abandon the project because he no longer shared the master's views. In 1581 Galilei published the *Dialogo sulla musica antica et moderna* (hereinafter *Dialogo*) [44, 55], which attacked the ideas of Zarlino and the more traditionalist musicians. His writing provoked a harsh reaction from Zarlino, reported in the *Sopplimenti musicali* of 1588 [108], essentially a collection of quotations from the *Dialogo* with accompanying commentary. Galilei replied in 1589 with the *Discorso intorno all'opere di messer Gioseffo Zarlino da Chioggia* [45], (hereinafter *Discorso*), which bitterly attacked Zarlino and opened a debate that would last until Galilei's death.

In the last year of his life, Galilei completed a treatise on counterpoint, without managing to have it printed, in which he reported his most important findings; in the meantime, he had been working on subjects that we would now call musical acoustics, with a series of experimental observations on the behavior of vibrating strings, organ pipes, and also other objects; these studies are reported in some manuscripts dated around 1589; they will be dealt with later.

2.2.2.1 Theory and History of Music in the *Dialogo*

Galilei is best known for his rejection of Zarlino's senario, showing its arbitrariness, because the intervals Zarlino referred to were those obtainable from a string whose cross-section, material, and tension were considered fixed, with only the length being varied. Thus the octave is characterized by the ratio (2 : 1); but there is no reason to choose the length as the parameter to be varied, Galilei argued. If, for example, the tension were chosen, the octave would be characterized by the ratio (4 : 1).

But Galilei also pursued much more interesting topics from a strictly musical point of view, and if there has not been a satisfactory evaluation of his contribution to the history of music—historians should reflect better on this fact—it is because some of his works have remained in manuscript form and were therefore known only to a very small circle of contemporaries.

In the following, we will consider in short Galilei's contributions to two aspects of practical and speculative music:

---

[52] pp. 17–18.

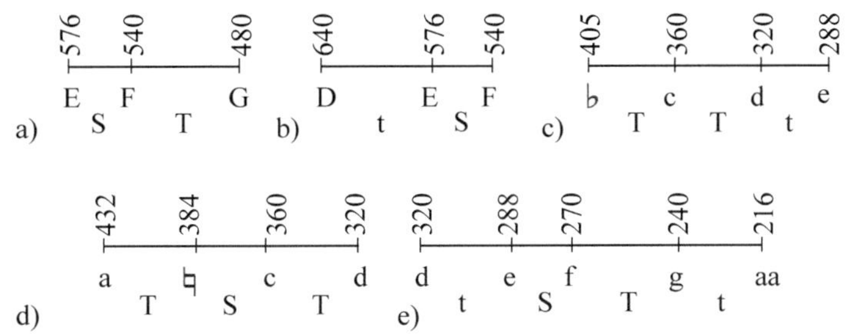

Fig. 2.14 Consonant and dissonant intervals in the diatonic scale

1. Highlighting the problematic nature of the diatonic syntonic system (just intonation).
2. Re-evaluation of Greek music and development of monodic music.

For the sake of space we will avoid comments on his contribution to the *seconda pratica*.

Criticism of the Syntonic Diatonic System.

Galilei referred to Zarlino's senario, as to the syntonic diatonic system, which he identified with that proposed by Ptolemy. It is contrasted with the ditonic diatonic system, characteristic of the Pythagorean division of the tetrachord. Today Ptolemy's syntonic diatonic system is generally referred to as *just intonation system*.

In the first part of the *Dialogo*, Galilei listed some of the inconveniences of the diatonic syntonic system that make it impossible to use in practice, contrary to what most theorists have stated. These difficulties are due to the presence of two tones, major and minor, of different lengths, which cause intervals of the same name, second, third, fourth, etc., to have different ratios. Galilei commented on this fact as follows: "Although the senses, according to what has been said, accept it [the difference] without difficulty, we will nevertheless prove it to the intellect in this way" [44].[53] With reference to Fig. 2.14a, b—showing some windows of Fig. 2.11 of Zarlino's monochord—consider, for example, the minor thirds E-G and D-F. The first minor third is formed by a major semitone and a major tone; it is therefore a just minor third $(16 : 15) + (9 : 8) = (6 : 5)$. The second third is formed by a major semitone and a minor tone, and covers an interval $(16 : 15) + (10 : 9) = (32 : 27)$, which is a semi-ditone of the Pythagorean system $(256 : 243) + (9 : 8) = (32 : 27)$. A similar argument can be made for the major third ♭-d (81 : 64)—Fig. 2.14c—which differs from the just c-e. The fourth a-d—Fig. 2.14d—and the fifth d-aa—Fig. 2.14e—are also not just and therefore dissonant.

These problems make it difficult to use the diatonic syntonic system proposed by theorists, which, according to Galilei, is actually not followed by practitioners.

[53] p. 10.

In the *Dialogo*, an imaginary discussion between two real characters, Pietro Strozzi and Giovanni Bardi,[54] Galilei expressed his conclusions:

> Bardi. I find through long observation that natural voices and instruments made by art do not actually play or sing in this modern music practice any of the nine ancient diatonic species in their simplicity. Our practicing contemporaries inadvertently use only three of them today, mixed together in different ways. These are the intense [diatonic] of Aristoxenus, the ancient ditonic diatonic, and the diatonic syntonic of Ptolemy. Among stringed instruments, I consider that the viola d'arco, the lute, and the stringed fretted lyre play the intense diatonic of Aristoxenus. Hearing and seeing in these instruments the uniformity of tones equally divided into two identical semitones impels me to believe that this is true, for the aforementioned intense diatonic of Aristoxenus was distributed in such a way, as you will understand in due course. However, the organ, the gravicymbalo, and the modern harp – modern only in terms of the newly added strings and not concerning the instrument in its original form, which I believe is very ancient – more or less detach themselves from those other species. [...] The wind instruments like direct flutes, transverse flutes, cornets, and other similar ones have power, due to the distribution of their holes and aided by the fine facility of their wise, expert players, to adjust to one [species) or another according to the need and their [player's] wishes. This is equally true of voices, but only when they are not willing to go against their nature and adapt to those [species]. Moreover, concerning the composing and singing of today, I am persuaded, because of what I told you and am about to tell you, that the ditonic diatonic is mixed with the syntonic of Ptolemy [44].[55]

Music of the Ancient Greeks.

Galilei devoted considerable space to Greek music, although his references are sometimes imprecise, reflecting the lack of knowledge at the time. He identified three schools in the Greek musical tradition associated with the names Pythagoras, Aristoxenus, and Ptolemy, attributing great importance to Aristoxenus, whose writings he knew through an Italian translation of the *Elementa harmonica*. Galilei attributed 13 modes and the equal temperament to Aristoxenus. The latter attribution, quite common in the Renaissance, is derived from the intense diatonic genus associated to Aristoxenus, defined by the sequence S-T-T, in which the semitone S is exactly half the tone T, see Table 1.3.

Quite interesting, and all in all faithful to the original, is the presentation of the Greek modes, those of Cleonides-Aristossenus, Ptolemy and Boethius. In commenting the ethos that can be associated with the various modes, Galilei referred mainly to the different pitches they cover. To Pietro Strozzi, who asked in the *Dialogo*, "Who was the author of this difference of high and low modes? Why were these modes assigned to one nation more than another" [44]?[56] Giovanni Bardi replied:

> Nature was the author of the three original modes which are the Dorian, the Phrygian, and the Lydian, because, when one nation or the other uttered words naturally while singing and speaking, or [doing something in between the two], there existed between their sounds (with regard to highness and lowness) the difference which you have learned [...]. This thing is

[54] The *Dialogo* is in fact a text written in dialogic form with only two characters, the Count Giovanni Bardi (1534–1612) and Piero Strozzi (b. 1550), both members of the Camerata de' Bardi.

[55] p. 30. Translation into English in [55].

[56] p. 71.

> seen and heard by whoever considers well what happens every day in many other provinces, and particularly in those of Italy, because the Lombards generally speak and sing in a deeper tone than the Tuscans do, and the people of Liguria speak in a higher voice than these [...]. If this, then, results from food, water, air, or climate, we will leave such disputes to the inhabitants [of these places] [44].[57]

This difference in pitch is illustrated by the story told by Plutarch that Sappho, the illustrious poetess, invented the Mixolydian mode, a semitone higher than the Lydian, because, as a lady, she could not comfortably sing her own poems in the Lydian mode.

2.2.2.2 Experiment and Experience

There is still no common opinion on Vincenzo Galilei as an experimenter, to the point that some historians doubt that he actually performed any experiment at all. The situation is discussed in [37]; to signal however that Stillman Drake, one of the most important historians of Galileo Galilei, claimed that Galilei father performed true scientific experiments [33].[58] Before analyzing his attitude we present some reflections on the possible ways of acquiring knowledge through practical activity, from the point of view of the 16th and 17th centuries.

It is common to distinguish between experience and experiment. The first would be Aristotelian physics, the second the new physics. In reality, things are more complex. First of all, the two terms experiment and experience, or rather their Latin counterparts—*experientia* and *experimentum*—have long been used without distinction of meaning [90], so it is difficult to clearly distinguish two different approaches to empirical knowledge of the external world in ancient writers. However, the current literature assumes that this is possible and gives the two terms conventional meanings, which somehow justifies their widespread use [31].

With experience it is indicated the way to relate to empirical phenomena typical of the schools and the academic world of 16th century Europe. In this context, experience is defined by general statements about how things are or how they usually are; i.e., to represent the normal course of nature. According to the Aristotelians, experience, derived from sensations by means of memory, is not necessarily individual [31];[59] in fact, it usually is not. It is derived from observations generally accepted by some weighty authority. Even what is written in authoritative texts, Aristotle for philosophy, Galen for medicine, becomes experience. And somehow more or less fantastic accounts of mythological animals can become experience. Regardless of the reliability of the sources, a problem not well appreciated in the Middle Ages, the distinguishing feature of experience is that it comes in static form, an acquired and defined baggage.

Experiment, on the other hand, is the questioning of an event. It can be of two types, contrived or observational, both involving the direct observation of a phe-

---

[57] p. 71. Translation into English in [55].

[58] p. 497.

[59] p. 22.

nomenon and the reporting of its history. There is an observational experiment when one refers to a natural phenomenon with the aim of testing a more or less explicit hypothesis. There is a contrived experiment when one refers to an artificially created phenomenon, with the possibility of replicas. The difference between a contrived experiment and an observational experiment is not so great. Consider, for example, a given body placed in a given fluid; it is an observational experiment if one limits oneself to a single experience; it is a contrived experiment if the observer repeats the experience by, for example, changing the specific weight of the body. But if one makes different observations of different bodies as they occur in nature, is it an observational or a contrived experience? In addition to experimentation by educated people, it is sometimes suggested to consider knowledge acquired out of practical necessity, as in the case of craftsmen. In this case, it is called artisanal epistemology.[60]

In any case, experiment is a dynamic category in which knowledge is constantly expanding. Experiment often has a quantitative approach, so measurements are expected. It is reported in written reports, precisely, so that it is repeatable.

The term experiment is also retained for an activity that has no obvious relation to the empirical world; this is the case of thought experiments. They have a remote origin and can also be found in classical Greece. Interesting are the thought experiments of the Middle Ages, around the 13th century, by philosophers who used the omnipotence of God to imagine falling bodies in a vacuum, usually denied, the earth in motion, usually denied, and so on. From a philosophical point of view, it is natural to ask the question: how can we learn about reality just by thinking? How can thought experiments enable us to acquire empirical knowledge? One possible answer is suggested in [65].[61]

While there is a large literature on experiments in early modern mechanics, less is known about experiments on harmonics. We have seen in Chap. 1 the experimental activity carried on by Ptolemy with the use of the monochord. In more recent times, the experiments of Galileo Galilei (see Chap. 3), Vincenzo's son, are worthy of mention. A first experiment concerned the scraping of a brass plate with an iron chisel. In correspondence of a clear whistle, looking on the plate, Galileo said, it was registered a long series of small stripes, parallel to each other and equidistant. Moreover, the distances of the stripes were the smaller the greater the height of the sound [43].[62] A second experiment concerned the resonance of a glass filled with water. By rubbing the ball of the finger on the rim, a musical note is produced and at the same time the water undulates regularly [43].[63]

Vincenzo Galilei used in his writings the words *sperienza* and *esperienza* with different shades. However, it seems to us that in any case his was not the Aristotelian experience, but he was always concerned with direct observations, although in some cases it can be that he was performing a thought experiment without declaring it. For example, he wrote:

---

[60] There is a large literature on this argument; see, for example, [37], footnote 3.

[61] p. 28.

[62] pp. 144–145.

[63] pp. 142–143.

> I am now coming to say with greater strength, that the fifth in the 3 : 2 proportion is the most perfect, sweeter than it is in any other ratio, as I have judged by ear after many many experiences [*sperienze*] (for I know no better means for attaining certitude) [45].[64]

but also:

> For among coins of the same goodness of material, of the same weight, cavity, thickness, and height, and also among little copper bells poured from the same mold, I have often found a difference of a whole tone. This same variety of the ratios of the diapason may be found also among strings of the same material, equality of length, and goodness, but of uneven thickness, when the same quantity of weight is suspended, and in other ways that I have experimented with many times [46].[65]

In the first case, according to [81], it would not be an experiment but an attentional observation, and the judgment would not be about a fact but about an "aesthetic observation"; in the second case, however, it would be repeated trials with the same material and procedure; it would then be a contrived experiment [79].[66]

Consistent with what we have said about the acquisition of experimental facts, we do not think there is any real difference at the epistemological level between the two cases. Certainly, if one is referring to modern experimentation, there is a difference. In the second case, the experiment could be carried out in the laboratory, using equipment capable of measuring frequencies, without having to rely on the sensory abilities of the observer; it would then be a classical acoustic experiment. In the first case, the observer cannot be avoided. Moreover, the claim that in the first case we are not dealing with a fact but with an observation related to personal taste does not hold up in the light of modern epistemology.

In his *Discorso* of 1589, Galilei had made a somewhat resounding statement: the account attributed to Pythagoras on the origin of consonances contains manifest falsehoods. Pythagoras had declared:

1. Weights suspended from a string produce consonances if they were in linear proportion; analogously, organ pipes produce consonances if their height were in linear proportion.
2. Only very special and multiple ratios could produce consonances.

According to Galilei, both statements are false [45].[67]

In fact, Galilei blamed the error not on Pythagoras, but on his uncritical followers: "I wish to denounce two erroneous opinions which have arisen in men who have been persuaded by the writings of some, as happened to me. But lately I have ascertained the truth by the means of the experience master of all things" [45].[68] Galilei, referring specifically to the octave, showed that it is obtained by quadrupling, not doubling, the weights. Instead, pipes sound at the octave when their volumes are in triple proportion.

---

[64] p. 117.

[65] f. 47r. Translation into English in [81].

[66] p. 114.

[67] p. 104.

[68] p. 104.

Galilei returned to the subject with a work unpublished in his lifetime, probably dated shortly after 1589, *Discorso particolare intorno alla diversità delle forme del diapason* [46].[69] It began by recalling the claims made in the *Discorso*, stating with greater force and against Zarlino that consonances are not determined by the ratio of numbers as numbers (numbered numbers), but by the ratio of the numerical values of certain physical or geometrical quantities (hence numerating numbers) measured on the bodies that produce the sounds, be their lengths, weights, or volumes. The proof of this is that different quantities give different ratios for the same consonance. In particular, Zarlino's senario has no absolute value; it was obtained by chance because he focused only on the length of the strings instead of the other parameters that influence their behavior.

This is how Galilei presented the possibility of obtaining the same consonances for ratios other than direct proportionality:

> The second way of hearing the same diapason, which I said by way of example was through surfaces, is between two strings made of the same material, equal in length, thickness, and goodness[70] from which weights are hung related to each other not as the duple [ratio], as some have said, but in quadruple proportion. From the sound of these strings when they are struck at the same time a diapason will be heard. The same thing will happen if equal weights are suspended from two strings the thicknesses of which are in quadruple proportion, provided the length and goodness are the same. Or they may be in quadruple proportion of length, the thickness and goodness being equal [46].[71]

Note the reference to strings of different sections; while it had always been known that strings with larger sections produced lower pitches, no precise rule had been given. According to Galilei, the octave could be obtained by keeping the length, quality and weight of the strings the same, while varying the area (thickness) in fourfold proportion. A final remark about the possibility of obtaining the octave by keeping the weight, thickness and quality constant and quadrupling the length should be considered provocative. Galilei could not have been unaware that in this way he obtained the double octave, or 15th. He apparently believed that he could also speak of an octave, given the similarity between the octave and the double octave.

Galilei also considered the case of pipes:

> The third and last way to hear the diapason is through the sound that issues from concave bodies when the volumes contained are in octuple proportion and not duple, as some have thought. This will be observed always when two organ pipes that are in any similar ratio to each other [both] the diameter and length of which are in duple proportion, are played together, when an octave will be heard [46].[72]

This statement has been taken by some historians as evidence that Galilei did not carry out any experimental tests on pipes, and perhaps not even on strings, allowing

[69] The manuscript has been transcribed and translated into English in [81].

[70] Palisca translates *bontà* with *goodness*, with the meaning of regularity, but warns that the term could also mean material, specific weight.

[71] ff. 46v, 47r. Translation into English in [81].

[72] ff. 47r, 47v. Translation into English in [81].

himself to be carried away by questions of aesthetics or numerology in order to obtain results that would allow him to polemicize with Zarlino. For example:

> It is evident that Galilei did not do any experiments, since the pitch of a pipe is a function of its length and not of its cubic capacity. In his unpublished *Discorso intorno alla diversità delle forme del diapason*, he asks the question: what interval would be given by two pipes of the same diameter but one of which is double the length of the other? and answers that it could be an equally tempered major third, which, by his own erroneous rule, is correct. We can see that here the mathematical scheme has been elaborated without even the most rudimentary empirical check, and it has gone wrong [101].[73]

The thesis is supported by the fact that Galilei devoted less space to the description of experiments than to (perhaps) numerological considerations.

For example, Galilei argued that the ratio (8 : 1) is the true expression of the octave, because all consonances, including fifths and fourths, are contained in the sequence 1, 2, 3, 4, 5, 6, 7, 8, in the sense that, for example, the fifth, characterized by the ratio (3 : 2), and the fourth by (4 : 3) are obtained by numbers belonging to the sequence, while in the case of the octave measured by the ratio (2 : 1), the fourth and the fifth are expressed by the numbers 3, 4, which are outside the sequence. In addition, the sequence from 1 to 8 also contains the number 7, which for Galilei is very important in music because it allows to obtain the replicas of the different intervals with a simple addition (for example, adding 7 to the fifth gives the twelfth, that is, the replica of the fifth).

There may be some truth in what Galileo's critics say about the pipes, but it should be pointed out that the behavior of pipes is not so easy to grasp, or at least not as easy as that of strings. By the end of the 17th century it was accepted by scholars—and also by Sauveur—the possibility that the frequency (a concept then acquired) of pipes varies inversely to their length; although for a justification on mechanical grounds one has to wait until 1727 with Euler [36]. But this is true under ideal conditions; in practice, the diameter of the pipe also has its importance, and instead of the *geometrical length*, one must consider the *acoustic length*, which is always greater than the geometrical length, and can even be much greater if the ratio of diameter to length is not small. Moreover, in order to obtain pipes of different color, organ builders vary not only the height but also the width, and if the octave is not produced by the value of the ratio 8 : 1, it is not even produced by the ratio 2 : 1, but by an intermediate value.

If Galilei had only considered similar pipes, then the thesis that the octave is produced by a ratio of 8 : 1 would also be empirically justified, at least approximately. In fact, in similar pipes, halving the length to obtain the upper octave, the volume decreases by a factor of eight; unfortunately, he meant the triple ratio in general, as is clear in cases where he compares the sounds of two pipes with the same section but twice the length of each other, where there is a double ratio of volume: "What sort of interval would two pipes of the same diameter but double the length produce? A major third of the intense tuning of Aristoxenus, which is actually the third part of the octave" [46].[74]

---

[73] p. 24.

[74] f. 50v.

Galilei was referring here to Aristoxenus' intense diatonic tetrachord, in which there are two whole tones and one half tone, which he identified with equal temperament. An octave is made up of six tones, and a third is made up of two tones, each of which is $\sqrt[6]{2}$ (modern notation), and two tones are $\sqrt[3]{2}$, which corresponds to the third part of the octave.

However, we believe that instead of focusing on Galilei's errors, one should consider why he came to undermine the established order of numerology by going so far as to challenge even the 'holy' Pythagoras. There is no evidence that other scholars—at least we are not aware of any—suggested the idea to Galilei, nor are there any theoretical considerations that could lead to this result.[75]

Certainly, the reading of Aristoxenus, whose writings had recently been published, played an important role, with its focus on empirical aspects, along with the loss of importance that the principle of authority experienced in the Renaissance, where experience in the Aristotelian sense is questioned in favor of direct experience. But these are the necessary conditions that made Galilei's idea possible, but do not fully explain it. The fact that Galilei was first and foremost a lutenist and not a theorist was certainly the most important stimulus. While tuning the lute, he was touching the change in pitch of a string as the tension varied, and it seems likely that he was driven by curiosity to see what would happen if he increased the tension beyond what was necessary for tuning, first to the fifth and then to the octave. When he saw that the Pythagorean order had been subverted for the strings, he probably doubted the result for the pipes as well.

That Galilei was also guided by an aesthetic principle, which is very likely, does not invalidate his experimentalism. Certainly, his experiments cannot be compared to modern experiments, where one has refined measuring instruments with the possibility of even measuring frequencies precisely, and there is a different sensitivity to experimentation.

Galilei's experimental approach is even more evident in his work *Discorso particolare intorno all'unisono* [47], which does not deal directly with consonances but illustrates Galilei's critical spirit and the fundamental role he entrusted to experience. The subject of this paper was unison, and it was not so much a discussion of whether it was a consonance or not, but rather of the different ways in which unison is realized, with the understanding that it can never be perfect. Galilei considered unison from two different points of view:

1. Direct and empirical, according to which, for example, two strings are in unison when they produce two sounds of the same pitch.
2. Indirect and mechanical, according to which two strings are in unison when they are of the same material, length, diameter, and tension.

[75] In [6], p. 158, the hypothesis is suggested that, at least as far as pipes are concerned, Galilei might have been inspired by the Aristotelian writings, *Problemata* and *De audibilibus*, which he most likely knew, according to which the pitch of the sound increases with the speed of the air expelled, which in turn implies that the pitch increases as the volume of the pipe decreases. But this is only a very general indication.

Galilei believed that in neither case could perfect unison be achieved in practice. However, even if two strings were perfectly in unison, but made of different materials, they would produce sounds of different colors, and thus it can be said that they are either in unison or not in unison:

> It is possible to stretch to a unison two strings of the same material, length, and goodness, but if the thickness is different, their sound will depart from a true unison to the extent that there is an inequality. When the sound of these strings reaches the hearing-at least that of persons versed in such matters-the hearing will recognize an audible difference [46].[76]

Galilei suggested an interesting experiment to show how it is essentially impossible to tune in unison two strings one of steel the other of gut. It is admitted that when the strings are plucked in their whole length the ear perceives them perfectly in unison; when they are divided with frets into short portions, as in the lute, the minute differences that previously were not perceived are now accentuated:

> If I place on a lute one gut string and one steel string and stretch them to be in unison in their way; then if, for example, I position seven frets and I pluck the open strings, or if I position twelve frets (and then pluck the open strings), they will not be in unison [...]. The same happens to the ear as happens to the sight when two lines are so close to being parallel that in the space of a hundred paces they depart only by a fourth of an arm's length. This difference divided by one hundred is so minimal that it would not be detectable by the sense, nor would the sense discern it in the space of a single pace, but in one hundred, yes [46].[77]

2.2.3 *Giovanni Battista Benedetti's Criticism Toward Just Intonation*

Very little is known about the life of the mathematician and natural philosopher Giovanni Battista Benedetti (1530–1590), and only a little more about his scientific works (for a biography see [13]). As a creative mathematician, he belonged to the generation that had rediscovered and appreciated Archimedes. He thought of mathematics as a branch of philosophy and a superior form of knowledge [13].[78]

In 1585, Benedetti made an important contribution to the ongoing discussion of Renaissance music about the structure of the scales with his famous treatise *Diversarum speculationum mathematicarum & physicarum liber*. The last part of the work consists of a large collection of letters (*epistulae*), arranged according to the importance of the addressees, including some who were already deceased. Among the letters are two written to Cipriano de Rore (c. 1515–1565), concerning music.

Benedetti and de Rore were both members of the Farnese court in Parma for a number of years (Benedetti was in Parma for about eight years from at least 1560; de Rore, a generation older, was at the same court from at least 1561 until his death, except for a brief interlude in Venice) [74]. It is therefore reasonable to assume

[76] f. 56v. Translation into English in [81].

[77] ff. 59r-59v. Translation into English in [81].

[78] p. 635.

that the two were acquainted. Beyond any possible personal connection or direct esteem, however, Benedetti may have had an additional motive for his choice of addressee. De Rore was one of the most famous musicians of his time, an innovator. His madrigals had influenced Monteverdi and the Florentine circle around Giovanni Bardi; for many, he was the forerunner of the second practice, the art of letting text and music affect and reflect each other [101].[79]

Beyond the two letters discussed in this section, there is no evidence that Benedetti had any particular interest in music. However, the two letters show that he must have had some training and knowledge of the art, to the point that he may have been a composer himself, which would not be too surprising given Benedetti's status and education, and the value placed on the discipline of music at the time.

The letters were read and commented upon by scholars immediately following Benedetti; it is certain, for example, that they were known to Isaac Beeckman, who quoted Benedetti in his diaries and specifically mentioned his letters [9].[80] They also had (and still have) significant importance for music historians because they were written by a leading figure of the new mathematics [34, 35, 63, 75, 76]. Comments and translations of certain parts of the letters considered relevant have been published (see [75, 76],[81] [34, 35]).

On the other hand, the letters have received less attention from historians of science, probably because they deal with topics of musical theory far removed from the disciplines of mechanics and mathematics in which Benedetti is considered a leading figure. However, they have not been completely ignored, as the last lines of the second letter deal with questions of 'mechanics'.

As for the content of the letters, it can be said that Benedetti's work was original in its time, and many historians consider that it directly led to revolutionize and overthrow of the then dominant system of just intonation, due in no small part to the problems that this system of intonation created in vocal music. In truth, it must be said that no convincing argument has ever been made by historians as to the role of Benedetti's letter in the abandonment of just intonation. Apparently, the letters were largely ignored by contemporary musicians.

The first letter to de Rore deals with just intonation in polyphonic vocal music, and the measure of the size of tones and semitones in this repertoire. The second letter discusses the migration of pitches in a just intonation temperament, the temperament and tuning of keyboard instruments, the law of vibrating strings, and proposes an explanation for the value of the pitches of notes.

2.2.3.1 Practice of Just Intonation and Migration of Pitches

The letters were written at a time when music theory and practice had recently adopted just intonation over Pythagorean intonation. In this transitional period, music theo-

[79] p. 79.

[80] vol. 3, p. 275.

[81] pp. 213–223; pp. 257–264.

rists (such as Ramos, Gaffurio, Spataro, Fogliano, and Zarlino) had not yet reached agreement on how to construct a musical scale and how to position and adjust the various tones and semitones. In this context, the first letter is of great importance, not so much for the results obtained, but rather for the method used to achieve these results. In contrast to music theorists who used a prescriptive approach to define the values of tones and semitones (usually by evaluating them as harmonic or arithmetic means of other intervals), Benedetti used a more observational approach (that of Archimedean mixed mathematics); starting from a limited number of physical principles, he used them to analyze existing musical pieces and thus establish intervallic values. The principles—grounded in acoustic experience and considered certain—are based on the recognition of some musical intervals as consonances: the octave and the fifth (perfect consonances), the fourth, the major and minor thirds, and the major and minor sixths. A similar conclusion was reached by Beeckman, for whom the intervals of any scale should be derived from the difference between consonances, and not from some mathematical principle assumed a priori, as in Descartes, for example [9].[82]

In the first letter, Benedetti's approach was applied to compositions inspired by a passage from Cipriano de Rore's madrigal *Hellas comment* in *Il primo libro de madrigali* [30], a characteristic example of new music, chromatic and with a false harmonic relationship. With his analysis he detected the musical intervals of the major (9 : 8) and minor (10 : 9) tones and the presence of the three semitones (27 : 25), (16 : 15) and (25 : 24).

Benedetti's second letter deals with four main topics: the tendency of pitches to move up or down in vocal performance, the temperament of keyboards and fretted instruments, the law of vibrating strings and a physical explanation of consonances. Only the first topic will be discussed here. It has been treated extensively and in a generally satisfactory manner in [75],[83] [17, 18, 35]; therefore, we will emphasize only the basic aspects and with a different point of view, using the methodology developed in the first letter.

It was well known to Renaissance musicians that when several consonant intervals are played in succession, both in Pythagorean and just intonation, notes are obtained that cannot be contained in any scale, however rich, and therefore cannot be reproduced by fixed-pitch instruments such as keyboards. However, these notes can be reproduced by the human voice, albeit with inconveniences that cause the pitches to wander throughout a piece. The consequences of this (somewhat embarrassing) phenomenon were ignored by the theorists of the time.[84] Benedetti's goal was to fully understand this phenomenon by considering three-voice musical pieces that follow just intonation; however, his considerations and conclusions apply to any polyphony

[82] vol. 3, p. 136.

[83] pp. 263–264.

[84] For example, Zarlino wrote: "But when the voices were proportionate to each other, and well united, without any impediment; and were uttered by the cantors with some discretion, & with good sense, so that one voice did not exceed the other; I take it firmly, that such intervals would be perfectly heard, & that the listeners would take no little pleasure, & satisfaction with the chants, which they heard" [106], p. 136.

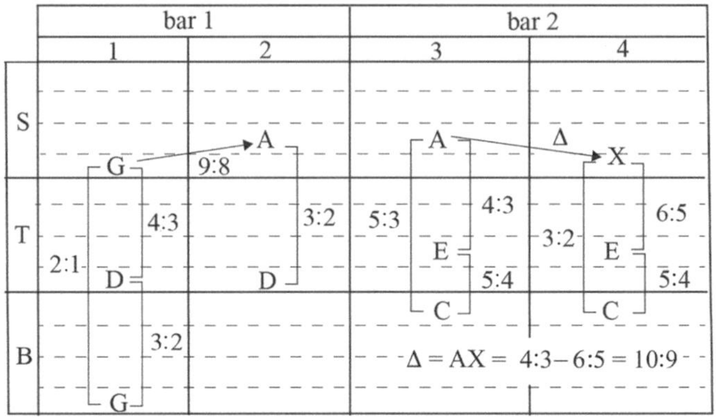

Fig. 2.15 Pitch migration in the first example. A schematization from [10], p. 278

performed in just intonation. To achieve his goal, Benedetti considered two pieces that he had composed specifically to highlight the critical issues of just intonation. In accordance with the customary practice of the time, he assumed the vertical concordance of the three voices, bass (B, lower voice), tenor (T, middle voice), and superius (S, higher voice); and then deduced the consequences of their performance in just intonation. In the first piece, the pitch of the notes is gradually shifted higher, and in the second, they are shifted lower.

To follow Benedetti's reasoning, it is sufficient to look at the four *chords* (here *chord* is use not with the technical modern meaning, but simply to indicate two or more notes played simultaneously) of the first two bars of the first piece, which are shown in Fig. 2.15. If we follow the higher voice, we can see that it goes from a G in the first chord to an A in the second, thus rising by a major tone $(9:8)$ as it goes from a fourth to a fifth in relation to the tenor. The higher voice keeps the A in the third chord, forming a fourth with the tenor, and then drops to the note 'X' for the fourth chord, forming a minor third with the tenor. It is not difficult to see that X is not the same as the G of the first chord, although it is very close. In fact, if we calculate the distance of X from A, starting from the tenor E, we can see that $XA = EA - EX = (4:3) - (5:5) = (10:9)$, which is a minor tone. Since the distance of G from A is that of a major tone $(9:8)$, X is higher than G by the interval $(9:8) - (10:9) = (81:80)$, that is, by a syntonic comma.[85] This upward migration is also reproduced in the following four chords, and could easily be reproduced for the other subsequent chords, resulting in an upward migration of 10 commas or more, so that X eventually differs only slightly from A. This migration also affects all voices, not just the highest. Thus, the pitches of all notes and harmonies in a piece can easily and unintentionally be raised by one comma through their performance in just intonation, even though they remain unchanged in the written music.

[85] Since the difference is small, X can still be called G, as Benedetti did. For a formal definition of a note as an interval of pitches not very different from each other, see [64], p. 20.

A modern reader may be surprised that Renaissance music theorists, advocates of just intonation, did not emphasize this pitch migration before Benedetti, given the evidence for the phenomenon and how well known and documented it is to any modern student of music theory (see [11]). Note that this migration occurs not only in polyphony, but also in monophonic music [60, 89].[86]

There are several plausible explanations for the mystery of how it was possible to use just intonation, despite its instability and the impossibility of practicing it in principle. One, and perhaps the most convincing, is that singers instinctively made the micro-adjustments necessary to avoid migration, and thus the phenomenon was not noticed in practice. This is the explanation Huygens gave in his *Cosmotheoros* [59].[87] Another compelling argument can be made for the possibility that singers simply did not sing with just intonation [7, 61]. In [102] it is shown that essentially just intonation can be maintained in choral music with the precise introduction of syntonic commas in a few places [102]. Two interesting papers [34, 35], suggest some strategies for solving what they call Benedetti's conundrum, modifying only one or two pure intervals rather than the entire system, as temperament does. In the end, it can be added that many composers lacked the theoretical background necessary to recognize the problem in the first place.

References

1. Abdounur OJ (2002) Compounding ratios and intervals: An education/historical approach in mathematics and music. Tech. rep., Max Planck Institut für Wissenschatsgeschichte. Preprint 205
2. Anonymous (820-840) Psalla modulania. München, Bayerische Staatsbibliothek, Clm 9543 199 p. Copied Regensburg, St. Emmeram, abbazia OSB
3. Anonymous (Unknown) Gudonian hand, https://digital.bodleian.ox.ac.uk/objects/4cde97a3-6a4f-4eeb-9232-6e1b0b04fb7e/surfaces/1bdd1a2e-d014-4f60-b168-f6cfd5b2e2cb/, Beginning 16th century. Oxford, Bodleian Library MS. Canon. Liturg. 216. Accessed 12 January 2025
4. Apel W (2010) The notation of polyphonic music 900–1600. Oxford City Press, Great Britain
5. Atkinson CM (2009) The critical nexus. Tone-system, mode and notation in early medieval music. Oxford University Press, Oxford
6. Barbieri P (2007) The Jesuit acousticians and the problem of wind instruments (c.1580-1680). Analectica Musicologica 38:155–204
7. Barbour JM (1938) Just intonation confuted. Music Lett 19(1):48–60
8. Barbour JM (1951) Tuning and temperament. A historical survey. Michigan State College Press, East Lansing
9. Beeckman I (1939-1953) Journal tenu par Isaac Beeckman de 1604 à 1634 (4 vols). Nijhoff, The Hague
10. Benedetti GB (1585) Diversarum speculationum mathematicarum & physicarum liber. Bevilaqua, Turin
11. Blackwood E (1985) The structure of recognizable diatonic tunings. Princeton University Press, Princeton

[86] vol. 20, p. 77; p. 207.

[87] p. 755.

12. Boethius, (1867) Boetii De institutione musica. Libri Quinque. In: Friedlein G (ed) Anicii Manlii Torquati Severini Boetii De institutione arithmetica libri duo. De institutione musica libri quinque, Teubneri, Leipzig, pp 175–371
13. Bordiga G (1985) Giovanni Battista Benedetti. Filosofo e matematico veneto del secolo XVI. Istituto Veneto di Scienze, Lettere e Arti, Venice
14. Browne AC (1979) Medieval letter notations: A survey of the sources. PhD thesis, University of Illinois at Urbana-Champaign
15. Browne AC (1981) The A-P system of letter notation. Music Discip 35:5–54
16. Burkholder JP, Grout DJ, V PC, (eds) (2014) A history of Western music. Norton, New York
17. Capecchi D, Capecchi G (2025) Two letters on music of Giovanni Battista Benedetti to Cipriano de Rore. Part I: Introduction and commentary. Adv Hist Stud 14(1):180–197
18. Capecchi G, Capecchi D (2025) Two letters on music of Giovanni Battista Benedetti to Cipriano de Rore. Part II: Transcription and translation into English. Adv Hist Stud 14(1):198–217
19. Capella M (1977) Martianus Capella and the seven liberal arts. Translated into English by Johnson, R and Stahl, WA (2 vols). Columbia University Press, New York
20. Capella M (1983) De nuptiis Philologiae et Mercurii libri IX. Teubner, Leipzig
21. Carrozzo M, Cimagalli C (2008) Storia della musica occidentale (3 vols). Armando, Rome
22. Cavalieri E (1600) Rappresentazione di anima e corpo. Mutii, Rome
23. Chaier Y (ed) (1995) L'oeuvre musicale d'Hucbald de Saint-Amand. Bellarmin, Ottawa
24. Chailley J (1984) "Ut queant laxis" et les origines de la gamme. Acta Musicologica 56(1):48–69
25. Cohen D (2008) Notes, scales, and modes in the earlier Middle Ages. In: Christensen T (ed) The Cambridge history of western music theory. Cambridge University Press, Cambridge, pp 307–363
26. Corvin L (2000) "Le istitutioni harmoniche" of Gioseffo Zarlino, Part 1: A translation with introduction. PhD thesis, The City University of New York
27. d'Arezzo G (1904) Micrologus. Modern reproduction. Amelli, Rome
28. d'Arezzo G (2005) Epistola ad Michaelem. In: Rusconi A (ed) Guido d'Arezzo. Opere, Edizioni del Galluzzo, Florence, pp 130–154
29. d'Arezzo G (2005) Prologus in antiphonarium. In: Rusconi A (ed) Guido d'Arezzo. Opere, Edizioni del Galluzzo, Florence, pp 116–124
30. De Rore C (1569) Il primo libro dei madrigali. Gardano, Venice
31. Dear P (1995) Discipline & experience. Cornell University Press, Ithaca
32. Desmond K (2001) Ars nova. https://www.oxfordmusiconline.com/grovemusic/display/10.1093/gmo/9781561592630.001.0001/omo-9781561592630-e-0000001360, new Grove Dictionary of Music and Musicians. Accessed 12 July 2024
33. Drake S (1970) Renaissance music and experimental science. J Hist Ideas 31(4):483–500
34. Duffin R (2006) Just intonation in Renaissance theory and practice. J Soc Music Theory 12(3):1–16
35. Duffin R (2017) Cipriano de Rore, Giovanni Battista Benedetti, and the just tuning conundrum. J Alamire Found 9:239–266
36. Euler L (1727) Dissertatio physica de sono. Thurnisiorum, Basel
37. Fix A (2019) Esperienza, teacher of all things Vincenzo Galilei's music as artisanal epistemology. Nuncius 34:535–574
38. Flotzinger R, als (2001) Discant. https://www.oxfordmusiconline.com/grovemusic/display/10.1093/gmo/9781561592630.001.0001/omo-9781561592630-e-0000007839, new Grove Dictionary of Music and Musicians. Accessed 12 July 2024
39. Fogliano L (1529) Musica theorica. De Sabio, Venice
40. Fox T (2004) Explain what Monteverdi meant by seconda pratica and show how this 'second practice' is reflected in three of his madrigals. https://web.archive.org/web/20061211022239/, http://www.musicalresources.co.uk/Resources/Monteverdi%20Essay.pdf, accessed 12 May 2024

41. Freiberg I (2004) Der Frühe italienische Generalbass dargestellt anhand der Quellen von 1595 bis 1655 (2 vols). Olms, Hildesheim
42. Gaffurio F (1518) De harmonia musicorum instrumentorum opus. Gotardum Pontanum, Milan
43. Galilei G (1638) Discorsi e dimostrazioni matematiche sopra due nuove scienze. Elsevier, Leiden
44. Galilei V (1581) Dialogo di Vincentio Galilei nobile fiorentino della musica antica et della moderna. Marescotti, Florence
45. Galilei V (1589) Discorso di Vincentio Galilei nobile fiorentino intorno all'opere di messer Gioseffo Zarlino da Chioggia. Marescotti, Florence
46. Galilei V (1589) Discorso particolare intorno alle forme del diapason, URL Biblioteca Nazionale Centrale di Firenze, Manoscritti Galileiani, Anteriori a Galileo, III, 1589 c., ff. 44r-54v, https://bibdig.museogalileo.it/tecanew/opera?bid=18011&seq=90, accessed 3 July 2023
47. Galilei V (1589) Discorso particolare intorno all'unisono, Biblioteca Nazionale Centrale di Firenze, Manoscritti Galileiani, Anteriori a Galileo, III, 1589 c., ff. 55r-61v, https://bibdig.museogalileo.it/tecanew/opera?bid=18011&seq=90, accessed 3 July 2023
48. Gilles R (2004) La formazione musicale nell'ambito del Quadrivium. In: Nattiez JJ (ed) Enciclopedia della musica (10 vols). Einaudi, Turin, pp 118–129
49. Glarean H (1547) Dodekachordon. Petri, Basel
50. Gogava A (1562) Aristoxeni musici antiquiss. Harmonicorum elementorum libri III. Cl. Ptolemaei harmonicorum, seu de musica lib. III. Aristotelis de objecto auditus fragmentum ex Porphyrij commentarijs. Valgrisi, Venice
51. Grout DJ (ed) (1980) A history of Western music. Third edition with Palisca CV. Norton, New York
52. Gumpelzhaimer A (1591) Compendium musicae (first edition). Schönigk, Augsburg
53. Haines J (2009) The origins of the musical staff. The Musical Quaterly 91(3–4):327–378
54. Herlinger J (2008) Medieval canonics. In: Christensen T (ed) The Cambridge history of Western music theory. Cambridge University Press, Cambridge, pp 168–192
55. Herman RH (1973) Dialogo della musica antica et della moderna of Vincenzo Galilei. PhD thesis, North Texas State University
56. Hodes K (1979) Der gregorianische Choral. Eine Einführung, Wissenschaftliche Buchgesellschaft, Darmstadt
57. Hucbald (1978) De harmonica institutione. In: Palisca CV (ed) Hucbald, Guido, and John on music: three medieval treatises. Yale University Press, New Haven, pp 13–44
58. Hucbald (2006) De harmonica institutione. Documenta Chatolica Omnia, vol MPL132, pp 905-958b
59. Huygens C (1888-1950) Cosmotheoros. In: Societé Hollandaise des Sciences (ed) Oeuvres complètes de Christiaan Huygens (22 vols), vol 21, Nijhoff, The Hague, pp 677–821
60. Huygens C (1888-1950) Oeuvres complètes de Christiaan Huygens (22 vols). Nijhoff, The Hague
61. Keller J, Roten E, Kirschenbaun O (2022) Just intonation in the Renaissance, https://www.earlymusicsources.com/youtube/justintonation, Accessed 2 October 2022
62. Lampadius (1537) Compendium musices. Apiarius, Bern
63. Lindley M (2001) Just intonation, https://www.oxfordmusiconline.com/grovemusic/view/10.1093/gmo/9781561592630.001.0001/omo-9781561592630-e0000014564, Grove music on line. Accessed 25 July 2022
64. Lindley M, Turner-Smith R (1993) Mathematical models of musical scales. A new approach. Verlag für systematische Musikwissenschaft, Bonn
65. Mach E (1919) The science of mechanics: A critical and historical account of its development. Translated into English by McCormack TJ. Open Court, Chicago
66. McGee TJ (1991-1992) Music notation and its influence in the Middle Ages. Scripta Mediterranea 11-12:97–108
67. Mei G (1989) Letter to Vincenzo Galilei, 8 May 1572. In: Palisca CV (ed) The Florentine camerata. Documentary studies and translations. Yale University Press, New Haven, pp 56–77

68. Meier B (1974) Die Tonarten der klassischen Vokalpolyphonie/nach den Quellen dargestellt. Oosthoek, Scheltema & Holkema, Utrecht
69. Mengozzi S (2010) The Renaissance reform of the Medieval music theory. Guido of Arezzo between myth and history. Cambridge University Press, Cambridge
70. Menke J (2015) Kontrapunkt I: Die Musik der Renaissance. Laber, Laber
71. Monteverdi C (1608) Il quinto libro dei madrigali. Amandino, Venice
72. Murray RE, Weiss SF, Cynthia C (eds) (2010) Music education in the Middle Ages and Renaisance. Indiana University Press, Indianapolis
73. Nicomachus (1926) Nicomachus of Gerasa. Introduction to arithmetic. Translated into English by D'Ooge M L. Macmillan, New York
74. Owens J, Schiltz K (eds) (2016) Cipriano de Rore: New perspectives on his life and music. Brepol, Turnhout
75. Palisca C (1985) Humanism in Italian Renaissance. Musical thought. Yale University Press, New Haven
76. Palisca C (1994) Studies in the Italian music and music theory. Clarendon Press, Oxford
77. Palisca C (2006) Music and ideas in the sixteenth and seventeenth centuries. University of Illinois Press, Chicago
78. Palisca CV (1956) Vincenzo Galilei's counterpoint treatise: a code for the "Seconda Pratica". J Am Musicol Soc 9(2):61–96
79. Palisca CV (1961) Scientific empiricism in musical thought. In: Rhys H (ed) Seventeen century science and the arts. Princeton University Press, Princeton, pp 91–137
80. Palisca CV (ed) (1978) Hucbald, Guido, and John on music: three medieval treatises. Translated into English by Babb W. Yale University Press, New Haven
81. Palisca CV (ed) (1989) The Florentine camerata. Documentary studies and translations. Yale University Press, New Haven
82. Palisca CV (2013) The ethos of modes during the Renaissance. In: Cochrane T, Fantini B, Scherer KR (eds) The emotional power of music: multidisciplinary perspectives on musical arousal, expression, and social control. Oxford University Press, Oxford, pp 102–114
83. Palma-ociosa (1336) Compendium de discantu mensurabili, https://shs.hal.science/halshs-04231783v1/document, Raphael Picasos. Édition bilingue de travail Latin-Français. Accessed 15 February 2024
84. Pesic P (2014) Music and the making of modern science. MIT, Cambridge (Massachusetts)
85. Powers HS (2001) Mode. In: Sadie S, Tyrrel J (eds) The new Grove dictionary of music and musicians, 2nd edn. Macmillan, London, pp 378–450
86. Pseudo-Odo (1784) Dialogus de musica. In: Gerbert M (ed) Scriptores ecclesiastici de musica sacra potissimum (3 vols), San-Blasianis, Unknown, pp 247–303
87. Rivera BV (1995) Theory ruled by practice: Zarlino's reversal of the classical system of proportions. Indiana Theory Rev 16:145–170
88. Rockstro WS (1880) Modes, the Ecclesiastical. In: Grove G (ed) A dictionary of music and musicians (AD 1450–1880), by eminent writers, English and foreign, vol 2. Macmillan, London, pp 340–343
89. Sauveur J (1707) Methode generale pour les systêmes tempérés de musique, et du choix de celui qu'on doit suivre. Mémoires de l'Académie Royale des Sciences de Paris pp 203–222
90. Schimitt C (1969) Experience and experiment: a comparison of Zabarella's view with Galileo's in De motu. Stud Renaiss 16:80–138
91. Schulter M (2000) Hexachords, solmization, and musica ficta, http://www.medieval.org/emfaq/harmony/hex.html#intro, Accessed 15 May 2024
92. Senyshyn TL (2000) The theory and practice of *A*-modes in Glarean's *Dodecachordon* (1547). PhD thesis, University of Western Ontario
93. Stifel M (1544) Arithmetica integra. Petreius, Stuggart
94. Strayer HR (2013) From neumes to notes: the evolution of music notation. Music Offer 4(1):1–14
95. Sukljan N (2019) Praetorius versus Zarlino De Musica Disserenda 15(1–2):105–124
96. Tartaglia N (1543) L'Euclide Megarense. Roffinelli, Venice

97. Teeuwen M (2002) Harmony and the music of the spheres. The ars musica in ninth-century commentary on Martianus Capella. Brill, Leiden
98. Tinctoris J (1477) Liber de arte contrapuncti. http://earlymusictheory.org/Tinctoris/texts/dearte23/#pane0=Edited, bologna, Italy - Ms. 2573, formerly RP 194. Accessed 20 August 2024
99. Tinctoris J (1875) Tractatus de musica. Edited by E. Coussemaker. Lefebrve-Ducrocq, Lille
100. Waesberghe SJ (1951) The musical notation of Guido of Arezzo. Music Discip 5:15–53
101. Walker D (1978) Studies in musical science in the late Renaissance. Brill, Leiden
102. Walker J (1996) Intonational injustice: a defense of just intonation in the performance of Renaissance polyphony. Soc Music Theory 2(6):1–12
103. Weakland R (1956) Hucbald as musician and theorist. Music Q 42(1):66–84
104. Williams DR, Balensuela MC (eds) (2007) Music theory from Boethius to Zarlino. A bibliography and guide. Pendragon, Hillsdale (NY)
105. Zarlino G (1558) Istitutioni harmoniche. The author, Venice
106. Zarlino G (1562) Istitutioni harmoniche (second edition). Senese, Venice
107. Zarlino G (1571) Dimostrationi harmoniche. Senese, Venice
108. Zarlino G (1588) Sopplimenti musicali. Senese, Venice
109. Zarlino G (1976) The art of counterpoint. Part three of the *Le istitutioni harmoniche*, 1558. Translated into English by Guy MA and Palisca CV. Norton, New York
110. Zarlino G (1983) On the modes. Translation into English of the fourth part of the *Le istitutioni harmoniche* by Cohen V. Yale University Press, New Haven
111. Zarlino G (2003) *Le istitutioni harmoniche*. 1558, part II. English translation by Perretti L, http://www.tonalsoft.com/monzo/zarlino/1558/zarlino1558-2.aspx, accessed 10 May 2024

Chapter 3
New Mathematics and Physics

Abstract In the late 16th and early 17th centuries, a new generation of physical theories were developed in line with the emerging scientific paradigm. Sound was conceived as a periodic sequence of pulses propagating through the air, and the idea was accepted that the degree of high and low sound was related to the period of these pulses. Thus, the coincidence theory could be stated to explain consonances. Fundamental contributions in this direction were made by Giovanni Battista Benedetti, Galileo Galilei and Isaac Beeckman. The improved understanding of the nature of sound led to study in determining the frequency of vibration of a taut string, which is the prototype of the prince instrument of music, the monochord. Also of great interest were innovations of a mathematical nature, especially the development and use of logarithms in music. This made it possible to put the problem of temperament on a new footing. Already in the 16th century, musicians had dealt with this subject, although not always precisely. The forms of temperament used at that time were closely examined by the new mathematicians, who clarified many of the still unclear aspects.

3.1 Physical Theories

The degree of high and low in a sound, i.e. its pitch, is the most important and elusive physical quantity in the whole science of harmonics. Since antiquity, attempts have been made to assign a measure to it using various approaches; the most fruitful was that associated with the monochord, where the pitch depends on the length of the string; the shorter the string, the higher the pitch. Although the monochord could only provide a measure of the intervals between two pitches given by ratios of lengths, and could not in itself provide a physical explanation of the nature of pitches, the science of harmonics relied heavily on it, from classical Greece at least until the 17th century.

There were various physical explanations for the nature and transmission of sound. It was generally accepted that sound was due to the vibrations of a body, which set the surrounding air in motion, reaching the ear and causing the sensation of sound.

D. Capecchi and G. Capecchi, *Joseph Sauveur and His Time: Music and Mathematicians*,
https://doi.org/10.1007/978-3-032-04707-6_3

However, the nature of air motion had no simple and convincing description; natural philosophers presented two competing explanations.

1. The theory of waves. It is admitted that there is no movement of air toward the ear due to the vibrating body, but that the propagation of sound is a phenomenon similar to the transmission of waves produced by a stone left on the surface of a pond.
2. Theory of missiles. The particles of air move from the vibrating body to the ear.

Originally, wave theory posited that air, the medium through which sound propagates, was a three-dimensional continuum. Conversely, the missile theory postulated that air possessed a corpuscular structure. The first theory, which was closer to empirical evidence, was the most popular, although there was some difficulty in accepting transmission in air without displacement, and some believed that the air somehow moved from the source to the ear (see Sect. 1.4.2.2). The second theory was supported by atomists, who spoke of sound transmission mechanisms similar to those of light; while for light the *idola* (images of external bodies made up of atoms detached from them) moved in a straight line from the object to the eye, for sound it was the words that moved while maintaining their identity to reach the ear and could even follow a curved path if necessary [81].[1] In 17th century atomism, both modes of sound propagation were explained by assuming the corpuscular nature of air. The wave theory was justified by imagining collisions between elastic particles that oscillate about a fixed equilibrium position while transmitting motion to adjacent particles. The missile theory, renouncing the atomists' idea of word transmission, assumed that sound propagates by means of air atoms of different size and speed that start from the source and reached the ear, transmitting the sound sensation by their impact.

In both cases of transmission of sound, motion in air or motion of air, we can speak of pulses; in the wave theory, a pulse corresponds to the maximum amplitude of the wave; in the missile theory, each incoming particle of air can be counted as a pulse.

Pitch was associated with speed. This assumption, attributed to Archytas of Tarentum, is also found in the Aristotelian bodies on music [26].[2] In *De sectio canonis* of Euclid [10][3] and in *De institutione musica* of Boethius [18][4] there are passages which seem to relate pitches to frequency, but these passages are not entirely clear. Gerolamo Fracastoro (c. 1476–1553) spoke of the compression and rarefaction of air, but there was no explicit reference to pitches [92].[5]

The pitch-speed association persisted even when its inconsistency was recognized; for if different sounds of different pitches traveled at different speeds, speech, an association of sounds of different pitches, would be distorted, the more distorted

[1] 4.526–527; 4.595–614.

[2] p. 288.

[3] vol. 2, p. 292.

[4] p. 293.

[5] p. 255.

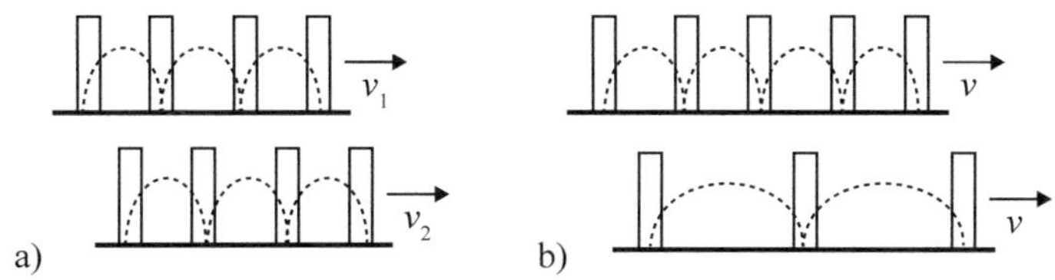

Fig. 3.1 Propagations of pulses of air

the farther one moved from the sound source, and eventually speech would be unrecognizable; which is not the case. Moreover, even consonances could not be explained with the blending of simultaneous sounds, since the more acute sounds reach the ear before the less acute sounds and blending cannot occur.

Considering sound formed by pulses, one can imagine the two different models of propagation shown in Fig. 3.1. In the first model (a), pulses are allowed to be at the same distance; they travel at different speeds, the higher the speed, the higher the pitch. In the second model (b), on the other hand, pulses travel at the same speed and are allowed to be at different distances, the shorter the distance the higher the pitch. In both cases, the ear perceives sounds as a different number of pulses per unit of time, and in both cases a listener can think in terms of frequencies.

In modern times, the difference between these two different modes of transmission has gradually been clarified. For example, Benedetti probably thought of the second model—but he was not quite explicit—while Galileo clearly accepted it and used it to explain consonance with the coincidence theory.

3.1.1 The Theory of Coincidence. Pitches as Frequencies

The theory of coincidence posits that the consonance of two notes is determined by coincident (i.e. simultaneous) pulses of the two notes. Pulse theory leaves the state of the air undefined in the interval separating each pulse; this creates problems in defining simultaneity. One can admit, as Benedetti did, that the pulse is distributed over the entire interval, as shown by the dashed curve in Fig. 3.1. Or one can admit, as Galileo did, that the pulses are separated by an undisturbed stretch. In the first case, the coincidence theory requires that the coincidence be recorded when there is a moment of rest common to the two sounds, the only potentially observable phenomenon; in the second case, when the two pulses strike the ear together.

The theory of coincidence has Greek roots and can be found, for example, in the *Problemata physica* [10].[6] In the literature it is assumed that Nicomachus pointed to the coincidence of pulses to explain consonances [6].[7] But this is far from clear.

---

[6] vol. 2, p. 96.

[7] p. 104.

In the modern era, one of the first proponents of the theory of coincidence was Francesco Maurolico (1494–1575) [6], who devoted a few words to the problem in his *Musicae traditiones carptim collectae* of 1575 [82]. The following quotation is particularly interesting:

> 11. The consonances consist in commensurable proportions. In fact, it is impossible for incommensurable sounds to be consonant, just as it is impossible for vibrations of incommensurable speeds to correspond, since concordance, or rather consonance, is derived from the correspondence of strokes [82].[8]

Here Maurolico assumed that the pitches depend on the speeds of the air pulses, and that consonance can exist only from commensurable values of the speed of two different pulses.

Maurolico's coincidence theory does not require the periodicity of the sound, but simply that the speeds of two pulses that meet in the ear are in the ratio of simple integers. However, it is not difficult to introduce concepts such as period and frequency without evoking the oscillatory phenomenon of the modern theory of sound. Period can be understood as the time interval between one pulse and another; if T_1 and T_2 are the periods associated with two particles of velocities v_1 and v_2, we have $v_1/v_2 = T_2/T_1$. The frequency can be understood as the number of pulses hitting the ear in unit time; if f_1 and f_2 are these frequencies, we have: $v_1/v_2 = f_1/f_2$.

After Maurolico, many other mathematicians and philosophers of nature took up the subject of coincidence, either in manuscript or in printed form. Table 3.1 lists some of them. In the following, we will discuss only the roles of Benedetti and Galileo. The former because he was the first to understand the relevance of the theory of coincidence; the latter because, thanks to his fame and his lucid prose, a full understanding of the theory of coincidence was achieved.

3.1.1.1 The Theory of the Sound of Giovanni Battista Benedetti

Benedetti, in his second letter to Cipriano de Rore (see Sect. 2.2.3), approached the study of the nature of sound with a series of assumptions, partly empirical, partly derived from the natural philosophy of the time; partly mechanical, partly acoustic. They are:

1. The vibration of a string is periodic, let T be that period.
2. The pulses of air produced by the vibration of such a string follow one another at intervals equal to the period T.
3. The shorter the string, the shorter the period T.
4. The shorter the string the higher the pitch.
5. The periods T_1 and T_2 of two consonant notes should be in a simple ratio when there is consonance.

Assumption (1) is a mechanical assumption, and also has an experimental character, but in Benedetti's time its verification was very difficult, especially when observing

---

[8] p. 151. An English translation of part of Maurolico's text can be found in [109], vol. 1, pp. 341–378.

Table 3.1 Mathematicians of the 17th century dealing with the theory of coincidence

1585	Giovanni Battista Benedetti	*Diversarum speculationum mathematicarum & physicarum liber* [15]
1614	Isaac Beeckman	*Journal tenu par Isaac Beeckman* [14], vol. 1
1618	René Descartes	*Musicae compendium* [31]
1621	Niccolò Cabeo	*In quatuor libros meteorologicorum Aristotelis* [22], vol. 2
1623	Marin Mersenne	*Questiones celeberrima in genesim* [89]
1637	Pierre Herigone	*Cursus mathematicus*, vol. 5 [60]
1638	Galileo Galilei	*Discorsi e dimostrazioni matematiche sopra due nuove scienze* [51]

shorter strings. Although it can be said with relative certainty that Benedetti did not personally experiment with vibrating strings (at least there is no evidence of this), he had been involved in some engineering problems at the Savoy court, as well as in the construction of scientific instruments [19],[9] and therefore had a mechanistic view in the interpretation of physical phenomena. It is probably a sum of experiences that led Benedetti to the statement of assumption (1). This is made explicit by him at several points in the text, with locutions of the kind "interval of oscillation" (period) or "number of intervals" (frequency) [15].[10] However, Benedetti was probably unaware of the fact that assumption (1) also implies the isochronism of oscillations.

Assumption (2) can easily be derived from the acoustic theories of the time. It was generally believed that the vibrating body produced a disturbance in the air corresponding to it. Benedetti's novelty was to make precise this assumption accepting explicitly the periodicity of the pulses.

Assumption (3) is easily derived from the apparently trivial (for a modern) statement that a smaller body moves faster than a larger one, as also found in Aristotelian thought (*Problemata physica*, 4.23, see [10][11]; and Aristotle's *De generatione animalium*, 786b, [10]).[12] In the case of periodic motion, it can be postulated that the period is inversely proportional to the speed of the motion. Therefore, it can be inferred that a shorter string will have a lower period.

While this assumption works very well for Benedetti's arguments, it is generally incorrect:

---

[9] p. 599.

[10] p. 283.

[11] vol. 2, p. 94.

[12] vol. 2, pp. 80–84.

a. It is not true that strings move more easily the smaller [shorter] they are; on the contrary, they move more easily the larger [longer] they are, as simple experiments clearly show.
b. It is not generally true that the greater the speed, the greater the frequency, because the speed depends not only on the frequency but also on the amplitude of the vibrations.

Assumption (4) is based on empirical evidence and therefore falls within the domain of musical acoustics. In conjunction with assumption (3), it suggests that the pitch of a sound produced by a vibrating string is dependent upon its period of vibration; more precisely the lower the period the higher the pitch. Assumption (5) is the coincidence theory.

Benedetti embraced the theory of wave, which he called "air waves" [15][13]; it is not clear if he also knew the missile theory. Whatever the theory, the vibrating body produces disturbances in the air in accordance with its vibration. The acceptance of the periodicity of the air pulses makes it possible to use the coincidence theory to explain the consonance: "Nor can I fail to see in the established position, a way of speculating on the generation of simple consonances, that there really is some equality of percussion, or [seu] an equal concurrence of air waves, or [vel] rather of their termination" [15].[14]

Note that Benedetti's period of string vibrations (and air pulses) may not be the physical quantity of the same name used in the modern theory of sound waves, but simply the time interval between two pulses (see Sect. 3.1.1), and possibly Benedetti saw speed as the true cause of pitch and period as a secondary property.

The connection between period and pitch, derived from stringed instruments, can also be considered valid for sounds produced by different objects, such as wind or percussion instruments, thus arriving at a general theory of the difference between high and low pitches and consonances: the value of pitches depends on the period of vibration of the air, consonances occur when these periods are in simple ratios.

This kind of justification, implicit in Benedetti, was made explicit by Galileo Galilei in the *Discorsi e dimostrazioni matematiche sopra due nuove scienze* of 1638, as will be shown below, but it was also taken up by immediate successors, such as Mersenne and Beeckman. Both Mersenne, Beeckman and Galileo reflected more than Benedetti on the mechanism of wave propagation, and by assuming a constant speed for sound (this is certainly true for Mersenne) were able to associate the frequency of the cycle of rarefaction-compression of the air near the eardrum, with pitches.

At the very end of his second letter, it is generally accepted that Benedetti proposed (if somewhat meekly) a criterion for judging the pleasantness of the various coincidences, based on the ratio of the lengths of the strings to the frequencies, for any given interval, "these numbers are not without an admirable regularity" (so the

---

[13] p. 283.

[14] p. 283. Note that the equal concurrence of the air pulses and the equality of their moment of rest are assumed to be equivalent; this last condition leaves less ambiguity in the application to waves of different shapes.

association of admirable regularity and pleasantness is not explicit.) [15].[15] The various consonances, octave, fifth, fourth, major third, minor third, etc., expressed by the ratios in radical form (2 : 1), (3 : 2), (4 : 3), (5 : 4), (6 : 5), suggested to Benedetti that the product between the denominator and the numerator could be used to establish an order of pleasantness; in the cases considered above, the numerical values are 2, 6, 12, 20, 30. Note that using this criterion, the thirds would be 'objectively' less pleasant than the fourths; an idea that has been questioned by most contemporary musicians.

Although Benedetti was not very clear about the "admirable regularity" of ratios, it is worth remembering that it was generally accepted by his contemporaries that the ratio of small numbers produces more pleasing consonances. Benedetti probably was of inspiration to Leonhard Euler who in his *Tentamen novae theoriae musicae* of 1731 assumed as a parameter influencing the suavitas of a ratio $(n : m)$ in its radical form the least common multiplier of m and n, i.e $n \times m$ which he referred to as the *exponent* of the ratio [42].[16]

3.1.1.2 Experiments on Sound by Galileo Galilei

Galileo Galilei had a musical education far beyond that of most mathematicians who studied music. His father, Vincenzo, was a gifted theorist and performer, and many other members of his family practiced music. No wonder, then, that he was a good musician, even as a performer. Below is the witness of his biographer Vincenzo Viviani:

> His most gratifying pleasures were practical music and the touch of the keys and the lute, in which, by the example and teaching of his father, he attained such excellence that he found himself several times in competition with the first professors of those times in Florence and Pisa, being very rich in this instrument of invention, and surpassing his own father in the gentleness and grace of touching it, which gentleness of manner he always preserved until his last days [114].[17]

In addition to his ability as a performer, Galileo studied speculative music extensively. His library included, besides the works of his father and Girolamo Mei, a number of works by contemporary musicians such as Fabio Colonna, Giovanni Battista Doni, and Mersenne's *Harmonicorum libri*; classical texts included Aristoxenus' *Elementa harmonica*, pseudo-Aristotle's *Problemata*, and Ptolemy's *Harmonica* [45].

It may sound strange to a modern reader, but in his time Galileo was considered an important scholar of music; Descartes provocatively declared that it was in music that he made his greatest contribution. Mersenne also held him in high esteem [43].[18] Interest in the scientific aspects of sound, testified in various writings [44], led Galileo

---

[15] p. 283. However it is also suggested that "Benedetti's concern was to demonstrate that the ratios of the consonant intervals are also to be found in the proportion of their products" [27], pp. 76–77

[16] Chapter 4.

[17] p. 602.

[18] p. 167.

to write a small pamphlet, now lost or perhaps incorporated into the *Discorsi e dimostrazioni matematiche sopra due nuove scienze* of 1638 (hereinafter simply *Discorsi*) [43].[19] Here Galileo discussed sound at the end of the first day [52].[20] We are dealing with only a few pages, essentially decontextualized, that Galileo wanted to include in his last masterpiece, probably to leave nothing of his writings unpublished. Despite these circumstances, Galileo's writing is important because it very clearly and explicitly expounds an embryonic form of the wave theory of sound and introduces the idea that the pitches of notes are related to frequencies—the term frequencies will only later become widespread—of air vibrations, thus replacing definitively the numerological paradigm of the monochord with a physical paradigm.

Galileo's contribution has not escaped the attention of music historians, although it has partly escaped that of the historians of science. An extensive commentary on it can be found in [13, 27, 38, 39].

Galileo's (Salviati) argument in his *Discorsi* began with an explanation of the phenomenon of resonance based on an analogy between the oscillations of a pendulum and the vibrations of a string, each characterized by its own mode of vibration/oscillation.

> *Salviati.* First of all, it is necessary to note that each pendulum has its own time of vibration, so limited and fixed in advance that it is impossible to move it in any other period than its own unique and natural one. Take in hand any string you like, to which a weight is attached, and try the best you can to increase or diminish the frequency of its vibrations; this will be a mere waste of effort. On the other hand, we confer motion on any pendulum, though heavy and at rest, by merely blowing on it. This motion may be made quite large if we repeat our puffs; yet it will take place only in accord with the time appropriate to its oscillations. If at the first puff we shall have removed it half an inch from the vertical, by adding the second when, returned toward us, it would commence its second vibration, we confer a new motion on it; and thus successively with more puffs given at the right time (not when the pendulum is going toward us, for thus we should impede the motion and not assist it), and continuing with many pulses, we shall confer on it impetus such that much greater force than a breath would be needed to stop it.
> *Sagredo.* As a boy, I observed that with pulses given at the right time, one man alone could ring a very large bell, and in trying to stop it later, several men would take hold of the rope and all of them would be lifted up into the air; nor could many men together [immediately] arrest the impetus that one man alone had conferred on it by regular pulls [52].[21]

With this premise, Galileo could begin to explain the phenomenon of resonance in music. When a string of a violin or harpsichord is touched, it also begins to vibrate, causing the air around it to vibrate, and waves spread out over a large space and go to strike the string stretched in unison with it, which is ready to make its vibrations at the same time. With the first pulse, the string begins to move a little; with the second, the third, and so on, it finally receives the same vibration as the first string [52].[22]

Having justified resonance, Galileo began to reason about the nature of sound, starting with a simple experiment. If one rubs the ticker string of a viola with a bow

---

[19] p. 164.

[20] pp. 141–150.

[21] p. 141. Translation into English in [53].

[22] p. 142.

so as to produce a loud sound, and places a glass filled with water near the viola, it happens that when the sound of the viola is in resonance with the glass—the resonant frequency (a modern concept for simplification) of the glass can be varied by varying the amount of water with which it is filled—the glass begins to vibrate.

This effect is best seen by placing the glass in a container filled with water. If you rub the edge of the glass with your finger, you will create waves of water around the glass, which will spread out smoothly. It also happens that when the sound goes up one octave during the rubbing, the water waves split into two and become denser, showing that the distance of the waves is related to the pitch of the sound. This experiment by Galileo, which had already been performed independently by Beeckman [14],[23] finds the (2 : 1) ratio for the octave, already associated with halving the length of the strings, in a more general situation (in modern terms, halving the wavelength).

At this point, Galileo turned to the monochord to argue about its pointlessness in explaining pitches. Although Galileo's readings on the phenomenon of the vibrating string are unknown—in particular, it is not known whether he was aware of Mersenne's studies on the subject—he was certainly familiar with his father's experiences. And so he was quite familiar with the behavior of strings, which depends on length, tension, and thickness, here meant as area and not diameter. He later replaced thickness with weight per unit length, based on both experience and theory:

> Here note that the heaviness of the moveable is more resistant to speed than is its thickness, contrary to what one might at first suppose, since it seems reasonable that speed should be more retarded by the resistance of the medium to being separated by a thick but light moveable than by a heavy and thin one; yet in this instance, the contrary happens [52].[24]

Length, tension and weight per unit length affect the nature of the sound emitted by a vibrating string in different ways. If, for simplicity sake, we refer to the octave, we have that we go up one octave if we halve the length of a string, if we increase the tension by four times, and if we decrease the weight per unit of length by four times. Thus as, his father, Galileo can conclude that associating the octave with the ratio (2 : 1), only because this ratio is found in the length of the strings, cannot be taken as an explanation because the ratios (1 : 4) or (4 : 1) due to tension and weight could also provide an explanation.

> These being true experiments, I saw no reasons why wise philosophers should have established the form of the octave as the double [ratio] any more than as the quadruple, or that of the fifth as the sesquialtera rather than the double sesqui-fourth [52].[25]

Galileo therefore suggested the possibility that the pitch of the sound of a vibrating string has to be found in the number of vibrations it makes in a given time.

> But inasmuch as it is quite impossible to count the vibrations of a sounding string, since it makes so many of them, I should have remained always in doubt whether it was true that the higher string of the octave makes double the number of vibrations in the same time as

---

[23] vol. 4, p. 159.

[24] p. 146. Translation into English in [53].

[25] p. 144. Translation into English adapted from [53]. A double sesqui-forth has the ratio (9 : 4), the double of (3 : 2).

> the lower string, had it not been that the waves persist as long as you like in the sounding and vibrating goblet. This showed me sensibly that at the very moment at which the tone is sometimes heard to jump one octave, smaller waves are seen to be generated that precisely bisect those that were there before [52].[26]

At this point, Galileo made an important connection: the pulses produced by the vibrations of a sonorous body bring similar vibrations through the air to the ear, and if one could count the vibrations of the sonorous body, one would also know the vibrations of the air that strain the ear. In this way, although Galileo did not say so explicitly, the wave nature of sound is admitted. What allows this conclusion is precisely the experiment with the glass of water; if going up one octave makes the waves in the water more dense, the waves in the air must also be more dense, leading to an increase in the number of pulses per unit of time.

To support his idea Galileo presented a phenomenon similar in some ways to that which occurs in the glass of water, but with the advantage of leaving a permanent trace. The discovery of this phenomenon, Galileo pointed out, was made by accident.

If you scrape a brass plate with a sharp iron chisel to clean it, after many strokes you will hear a rather loud and clear hissing sound caused by the chisel leaving nicks in the plate. If there is no hissing sound when scraping, then there are no nicks. Repeating the experiment, scraping at a higher or lower speed will produce a higher or lower sound; in the first case the nicks will be more dense, in the second case they will be more sparse. Sometimes, by gradually increasing the speed of the chisel, one hears the sound become more acute and the nicks thicker.

When the strings of a harpsichord are placed next to the brass plate, it happens that some of them resonate with the noise produced by the chisel. Galileo referred to two harpsichord strings a and b that formed a fifth and resonated with hissing. If one compares the distances between the nicks associated with hissing in resonance with the string a and those in resonance with the string b, one sees that they are in the ratio (3 : 2), typical of a fifth. Galileo could thus conclude:

> Going back to our original purpose, I say that the length of strings is not the direct and immediate reason behind the forms of musical intervals, nor is their tension, nor their thickness, but rather, the ratio of the numbers of vibrations and impacts of air waves that go to strike our eardrum, which likewise vibrates according to the same measure of times [52].[27]

After formulating his theory of sound, Galileo tried to explain the phenomenon of consonance with the coincidence theory. In the following we give Galileo's explanation of consonances, restricting for simplicity to the case of the octave, although the approach can easily be extended to all pairs of sounds with rational ratios. Let AB of Fig. 3.2 be a line representing the 'expansion' of the vibration of one string and CD that of another string half the length, i.e. in the upper octave. Suppose the two vibrations, the acute CD and the grave AB, begin together. When the acute vibration arrives at D, it hits the eardrum, Galileo said, but not the grave vibration, which arrives only at E. When the acute vibration returns to C, the grave vibration is at B; at

[26] p. 144. Translation into English in [53].

[27] p. 146. Translation into English in [53].

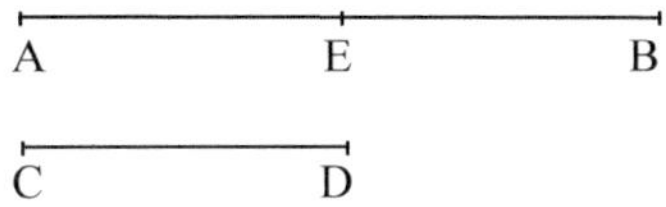

Fig. 3.2 Coincidence for the octave interval. Redrawn from [52], p. 148

that moment, the two vibrations hit the ear simultaneously, producing a consonance and a pleasant sound. Galileo was not bothered by the fact that the simultaneity of the hits occurred on two parts located on different sides of the expansion (he like most of his contemporaries, and like most people of the 18th century, saw the period of an oscillation as the transition from one extreme to the other, and not as an outward and return to the initial position, as is done today). His period is therefore half of ours. Continuing the analysis with another cycle, both vibrations strike on the same side (A and C).

Galileo concluded his analysis of consonance by changing the sense from hearing to sight; for this purpose he imagined the oscillations of pendulums of such lengths that their periods were in the ratio of 2 : 1, 3 : 2, 4 : 3, etc. By swinging the pendulums from the same vertical position, he said he saw that they met periodically, whereas if the lengths were chosen at random, there would be random oscillations.

Two aspects are overlooked in Galileo's analysis: (a) The possibility that the two sounds are out of phase was not considered. This 'amnesia' was common to all his successors and predecessors. Although the presence of the phase shift infects the coincidence theory, it was systematically ignored. (b) No criterion is given to judge consonances, although in a commentary entrusted to Sagredo it is said that unison and octave are the most perfect consonances.

3.1.1.3 Criticisms of the Coincidence Theory

The theory of coincidence, as clearly stated by Galileo, found, besides many appreciations, also immediate criticisms. Very interesting are those of Orazio Ricasoli Rucellai (1604–1674), two generations younger than Galileo; he put his criticism in the manuscript *Dialoghi sopra il Timeo* attributed to 1665–1673 [6]. Rucellai resumed circumstances, well known to musicians of the 18th century, in which the theory of coincidence cannot explain empirical observation.

1. Consonance is also registered for ratios between the frequencies of two sounds that are not expressed by simple integers. This was known, for example, by Vincenzo Galilei, who said in his *Dialogo* that on the lute the tempered fifths were more pleasant than the pure ones [54].[28] Rucellai essentially repeated Galilei's words: "When two harpsichords are tuned together, on one of which a fifth is tuned exactly, with the perfect rationality of the sesquialtera [i.e. = 3:2], and on the other a different, irrational [= tempered] fifth, according to the wishes of the ear, when they are played together, every undisturbed ear will find that the

[28] p. 55.

irrational sounds incomparably better than the rational" [6].[29] The phenomenon is partly explained by continuist theories of consonance [7].

2. Consonance is also registered when two sounds that form a consonance are emitted with different phase. Acoustic experiments show that the ear cannot easily distinguish the phase shift of two sounds. In the limit, even when sounds are completely out of phase, as when a fingernail grazes the center of a vibrating string and each of the two halves into which it is divided emits two octaves out of phase, consonance is observed.

   > The successive pulses of one sound are constantly bisecting the intervals between the pulses of the other And yet the harmony of the unisons is nonetheless *perfectly* agreeable to the ear, as I have often experienced [102].[30]

   Of course, this occurs because man has two ears, and so the phenomenon is more complex than that of the interference of two waves in phase opposition. Wallis tried to save the coincidence theory by arguing that two series of pulses that are not completely in phase at first can become so after a while [115].[31] In this he seems to have anticipated the *Mitnahme-Effekt*, for which there may be a nonlinear coupling effect that tends to synchronize two different oscillators that are not perfectly in phase [103].
3. There are consonances that have fewer coincidences than others, but are more pleasing. This raises the controversial problem of the consonance of the interval of the seventh. The phenomenon can only be explained by saying that the perception of consonances also depends on the way music has developed historically, and what we now call dissonance might have been considered a consonance in another time or place. For a discussion of this point, see [4].[32]

3.1.2 Missile Theories of Sound

The missile theories of sound of the 17th century, when atomism was generally appreciated, are well represented by Isaac Beeckman (1588–1637) [14].[33] His interest in music is documented in letters to Mersenne and in his diaries, which, however, were not published until 1939 [14]; in 1642 a small treatise, *Mathematico physicarum meditationum, quaestionum, solutionum centuria*, reported in Latin and Dutch some extracts from the diaries [58]. In the diaries, in which his interest in music is documented since 1610's, he confessed to have been the worst student his master of music had ever had, and that he was not able to distinguish consonances from dissonances. This explains why his contribution to music was mainly in what we now

[29] p. 215.

[30] pp. 105–106.

[31] p. 132.

[32] pp. 329–355.

[33] vol. 1, p. 92.

call musical acoustics. It must be noted, however, that Beeckman studied speculative music enough and also made an attempt at composition [27].[34] He also introduced Descartes to music, having met him in 1618 when he was in the Netherlands in the army of prince Mauritius.

Beeckman's theory of sound is not easy to understand, partly because it changed over time. In essence, Beeckman conceived of sound as the propagation of atoms, or better, air bubbles. Any vibrating source cuts the surrounding air into small (spherical) bubbles that are sent off in all directions. When the bubbles reach the sense of hearing, they give rise to the sounds heard: "Sound [in the ear] is as air was in the mouth of the speaker" [14].[35]

In a first phase, the different characteristics of notes, pitch and loudness, were individuated in the size of the bubbles and in the amount of air struck respectively, while color remained unexplained. The identification of pitches with size is not strange; it is suggested by the current acoustic theories, for which pitches were associated with speed. Smaller bubbles indicate a more effective action or speed of the vibrating body, such as a string, on the surrounding air from which they originate. The greater the speed of the string vibrations, the greater the speed of the bubbles—as they follow the string—and the smaller their size. Conversely, the smaller their size, the greater the speed and the higher the pitch. The consonance was related to the ratio between the diameters of the moving bubbles of air; for example, in the octave the ratio of diameters was (2 : 1).

Later, Beeckman explained the properties of sound in a different way; pitches were directly related to the speed of the bubbles, while color was related to the diameter. The loudness was still related to the amount of air moved.[36]

In both cases, the motion of the air bubbles is periodic, as is the motion of the string, with a maximum of activity corresponding to the maximum extension of the string and a phase of rest when the string reaches the undeformed horizontal position. The consonance was associated with the contemporary presence of a momentary rest in the vibrating air.

For Beeckman, a pure consonance could be achieved through unison alone; all the others consist of a mixture of concordance, when the speed of the air is zero for both sounds, and discordances, when the speed is zero in one case and maximum in the other: "Harmony and consonance are nothing but a mixture of unison and dissonance" [14].[37]

Beeckman also dealt with the phenomenon of sympathetic resonance, or the excitation of a potential source of sound (a string) by another, distant source of sound (another string), set into vibration by an external cause. This phenomenon, which has been known to musicians since ancient times, is observed with strings of the same properties and length (unison). It can be easily experimented with side by side monochords or with the strings of bowed instruments.

---

[34] p. 191.

[35] vol. 1, p. 92.

[36] The explanatory scheme changed over time, however.[27], pp. 126–127.

[37] vol. 1, p. 160.

Beeckman attributed the phenomenon to the fact that the air moved by the externally excited string reaches and excites the resting string. In the case of two equal strings:

> I say that the touched string partly pushes away from itself the untouched string in the process of dispersing particles of air, partly, through (the vibration of) its own body, moves the air nearest to itself, whose motion moves the following (layer of air) up to the untouched string: and when the touched string returns, it [the untouched string] is moved only by the motion that commonly, but incorrectly, is called 'avoiding the void' [14].[38]

A similar explanation was the standard one among musicians; for example, it can be found in the writings of Girolamo Fracastoro (c. 1476–1553) in his *Liber unus de sympathia et antipathia rerum* of 1546 [47].[39]

Beeckman also sought to explain resonance for strings of different lengths. The air pulses generated by the excited string of length l_0 set in motion any string of length l in its vicinity, but only if l is an integer fraction of l_0 can resonance be more or less efficiently engaged, because only then can the stresses on the string of length l be maintained. For the octave, $l = {}^1\!/_2 l_0$, Beeckman said that the two strings return in concordance every two cycles, and In this case, there is an impulse to move from the excited string to the string that is not excited.; when the strings are in opposition, there is a damping effect, but it does not impede motion. For the higher intervals, the smaller the damping effect, the greater the activation of resonance. Here, for example, is what Beeckman wrote about resonance for the interval (5 : 1), or the resonance of strings whose lengths are in the ratio (1 : 5):

> For the ratio (1 : 5), when the upper string moves five times in the time the lower string moves once, the first, middle, and last strokes concur with the lower string: admittedly the second and fourth [stroke] run counter to it, but the second and fourth do not appear to overcome the first and the fifth as much as the third alone the second and the fourth; thus for the whole stroke the concurrence is stronger [14].[40]

An examination of Fig. 3.3, which refers to a modern interpretation in terms of periodic vibrations, shows the amplitude of the midpoint of strings of different lengths versus time. In Fig. 3.3a the reference is to strings of length l_0 and ${}^1\!/_2 l_0$; in Fig. 3.3b the reference is to strings of length l_0 and ${}^1\!/_5 l_0$. Looking at the figure, it is easy to see that the resonance for the interval (5 : 1) is easier to activate than the resonance for the interval (2 : 1). The activation of the resonance depends on the coincidence of sound with sound and silence with silence. In the case of the resonance (2 : 1), there is opposition of motion in half of the period, and the damping effect is maximum. In the case of the resonance (5 : 1), the damping effect is active for a shorter time.

Beeckman's mechanistic explanation for higher-order resonances is convincing, but only because it maintains a qualitative level. In fact, if the vibrating externally

[38] vol. 4, pp. 206–207. Translation into English in [27], adapted.

[39] Chapter 11, pp. 71–72. Fracastoro's writings, however, should not be read as the work of an originator [110], pp. 22–23.

[40] vol. 4, pp. 206. Translation into English in [27].

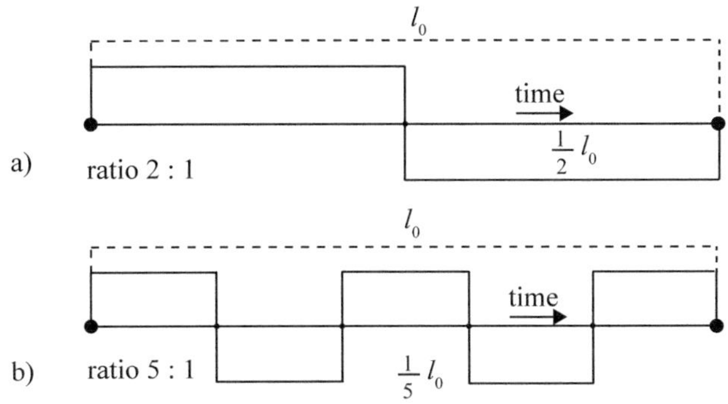

Fig. 3.3 Vibration of of a point of strings with different lengths versus time. Redrawn from [27], p. 135

excited string emitted a pure sound, as Beeckman said, one might have resonance only in unison, but not in the other case.[41]

3.1.3 *Wave Theories of Sound*

Huygens is undoubtedly the great theorist of wave phenomena. A wave is, generally speaking, a motion of matter which has the property of propagating. Matter itself moves only around its point of equilibrium; it is its motion (and not matter) that propagates, continuously deforming the medium. Huygens, tackled the problem of sound in the first chapter of his *Traité de la lumiere* of 1690 [64] because he needed a mechanical model to explain light propagation:

> We know that by means of the air, which is an invisible and impalpable body, sound spreads all around the place where it is produced, by a movement which passes successively from one part of the air to another, and that the extension of this movement, which is equally visible from all sides, must form spherical surfaces which are always expanding and which come to strike our ear [64].[42]

Huygens still had some difficulty in commenting these 'spherical surfaces':

> It will follow that this movement imparted to matter is successive, and that consequently it is extended, as is that of sound, by spherical surfaces and waves; for I call them waves after the resemblance of those seen to form in water when a stone is thrown into it, which represent such a successive extension in a circle [64].[43]

---

[41] This is a result of the modern theory of vibrating strings, according to which the resonance of shorter strings is due to the presence of higher harmonic sounds in the longer string. This point will be discussed in the following Chap. 5.

[42] p. 463.

[43] p. 463.

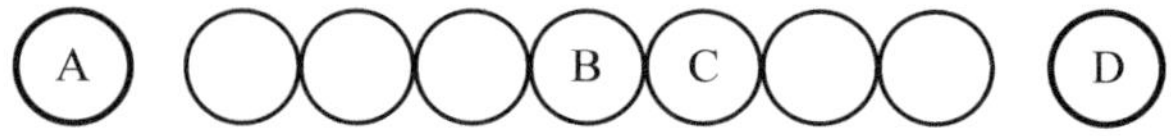

Fig. 3.4 Propagation of sound in a line. Redrawn from [64], p. 474

Continuing his parallel with light, which is the subject of the first chapter, Huygens described the propagation of sound waves in air:

> As for the different ways in which I have said that the movements of sound and light are successively communicated, we can fairly understand how this happens in the case of sound, when we consider that air is of such a nature that it can be compressed, and reduced to a much smaller space than it usually occupies; and that as it is compressed it makes an effort to get back out into the open: for this, together with its penetrability, which it retains notwithstanding its compression, seems to prove that it is made of small bodies that swim and are agitated very quickly in aetheric matter, composed of much smaller parts. So that the cause of the extension of sound waves is the effort made by these small bodies, which collide [64].[44]

Huygens was familiar with Mariotte's and Boyle's work on air pressure and elasticity and had a clear understanding of wave motion in air. Regarding their propagation, he insisted that it is not matter that moves, but its vibrations that propagate:

> It is necessary to observe that though the particles of aether are supposed to be in a continuous motion (as cannot do otherwise) the successive propagation of waves cannot be prevented, because it does not consist in the transport of these parts, but only in a small shaking, which they cannot prevent from communicating to those around them, notwithstanding all the motion that agitates them and causes them to change their place [64].[45]

The theory of sound presented above is essentially qualitative. It assumes oscillatory motion, but does not introduce concepts such as wavelength and frequency at any level. From this point of view, Huygens' theory is not a real theory of waves, but rather a special theory of pulses. This does not prevent him, when speaking of sound, from taking its periodic nature for granted and speaking of frequency ratios, which serves to justify consonances with the theory of coincidence that Huygens attributed to Galileo.

Huygens was aware of the incompleteness of the coincidence theory, since consonances also depend on other factors, such as the harmonic context in which they occur, the taste of peoples and times, and the range of the pitches [63].[46] But nevertheless he accepted the criterion derived from it, in ordering consonances (Fig. 3.4).

One of the defects of the theory of coincidence which was pointed out by musicians of the time was that it indicated that the fourth was more consonant than the third, and this was contrary to the taste of the time. To justify this, Huygens wrote, paraphrasing Mersenne (and Descartes) (see Sect. 4.1.2):

> In general we see the same thing everywhere that in the case of two consonances the one whose first or second replica becomes a multiple ratio appears better than the other. And it

[44] p. 471.

[45] p. 474.

[46] vol. 20, pp. 35–37.

> seems that the reason is that in hearing a given note one supposes and seems to hear somehow its higher octave or even the double octave. And one hears it effectively when sounding some string, or big bell, and even the twelfth and the seventeenth. So that, when the replicas of these consonances are in a multiple ratio for which the coming together of the strokes is more frequent than for the others, one judges the consonance itself by the beauty of its replicas [63].[47]

Thus, while listening to (5 : 4), we simultaneously hear the replicas (5 : 2) and (5 : 1), both of which are more consonant than (4 : 3) because their vibrations coincide more frequently.

3.1.4 *The Frequency Law of Vibrating Strings*

Observing the motion of a vibrating string, one cannot but associate it with some kind of periodicity, at least if the observer has a modicum of practical sense as a craftsman or musician. In the past, however, the phenomenon was not considered interesting because it was not associated with any particular consequence; for example, it was not associated with the pitch of the sound, which was instead attributed to the speed of the string vibrations.

It was not until the 16th century that mathematicians (and to some extent musicians) began to pair pitch of sounds and frequency of the pulsations transmitted to the ear by the air. Thus was born the stimulus to analyze more thoroughly the periodicity of vibrating strings, by either experimental or analytical methods, to obtain a mathematical law that relates period to lengths and other parameters. The knowledge of the mathematical law for the frequency of the string as a function of characteristic parameters, was of a great help for musicians; they could so generate the sound they wanted with no difficulty.

3.1.4.1 A Nonmechanical Derivation

According to modern historians, Benedetti was the first scientist to formulate and demonstrate the law that the frequency of vibration of a string is inversely proportional to its length.

Benedetti commented on the problem of vibrating strings in the last part of the second letter to Cipriano the Rore, see Sect. 2.2.3. A modern reader is at first puzzled by the fact that a law belonging to the dynamics of vibrating strings could appear in a text on speculative music and that the proof of a typical problem of string dynamics is derived outside of mechanics, using assumptions of a musical nature. For Benedetti's contemporaries, the situation was different, because the vibrating string was representative of the monochord, the primary tool of speculative music.

---

[47] p. 36. Translation into English in [27].

Benedetti's approach is very interesting because it combines assumptions of a mechanical and acoustic nature, as resumed at points 1–5 in Sect. 3.1.1.1, repeated below for the sake of clarity:

1. The vibration of a string is periodic, let T be that period.
2. The pulses of air produced by the vibration of such a string follow one another at intervals equal to the period T.
3. The shorter the string, the shorter the period T.
4. The shorter the string the higher the pitch.
5. The periods T_1 and T_2 of two consonant notes should be in a simple ratio when there is consonance.

to which is added an acoustic empirical assumption, for which:

6. The ear says that pairs of vibrating strings with length ratios $(2:1)$, $(5:3)$, $(8:5)$, $(3:2)$, $(4:3)$, $(5:4)$, $(6:5)$ produce consonances.

It is thus possible to derive the law of the vibrating string in a relatively simple and smart way.

Consider two strings of length l_1 and l_2, whose ratio is, for example, $l_1/l_2 = 2$. Under assumption (6), this ratio gives consonant sounds; under the coincidence theory (assumption 5), the periods of the air pulses T_1 and T_2 associated with the vibrations of the two strings must have a ratio expressed by simple numbers. Considering that for assumption (3) $T_1 > T_2$, the most natural choice is $T_1/T_2 = 2$. But for assumption (2) the period of the air pulses is equal to that of the vibrating string, for which the ratio of the period of the vibrating strings is still $T_1/T_2 = 2$, which is equal to the ratio l_1/l_2 between the lengths of the strings. A similar argument can be developed for the ratio $(3:2)$. In Benedetti's words:

> When instead the bridge divides the string so that on one side two fifths remain and on the other three fifths, the consonance of the diapente is generated in these parts, as evident. That is, a vibration interval of the larger portion will be in the same proportion with a vibration interval of the smaller portion [of the longer to the shorter length], that is the time [duration] of the interval [of the longer string] to the time of the shorter will be sesquialtera, for which there will be no coincidence at the same instant except for three perfect intervals [of time, periods] of the minor portion and two of the major [15].[48]

This still particular result is generalized by Benedetti to the case in which the relationship of the strings is any:

> So the ratio between the number of intervals[49] of the minor part to the major part, and the length of the major part to the length of the minor part, will be the same [15].[50]

and for the first time the Pythagorean intervals are given a physical meaning: if there is a consonance for a given ratio m/n for two lengths of string, it is because the ratio of the frequencies of vibration of the two strings, and hence of the air pulses, is n/m.

---

[48] p. 283.

[49] The number of intervals, that is, the number of periods or semi-periods, in a fixed time is proportional to the frequency of vibration of the string.

[50] p. 283.

It is usually assumed that this relation holds for any ratio between the lengths of the strings, rational or irrational. Actually, Benedetti did not seem to be interested in this generalization, which is not proved; he was content to consider it valid only for the rational proportions that give rise to consonant intervals.

3.1.4.2 The First Mechanical Theory

Benedetti's proof is only strict up to a point, since it relies on a number of assumptions that, while convincing, are not necessarily true. For a more interesting proof, based on the laws of mechanics, we have to wait for Isaac Beeckman (1570-1637), a few generations after Benedetti [23].[51] Beeckman read *Diversarum speculationum mathematicarum & physicarum liber* in 1633 and commented it on many sections with appreciation.

The proof of Beeckman dates back to 1614/1615 and was communicated to Mersenne in 1619. For a short commentary see [110]. Below is a quotation explaining Beeckman's approach.

> Let *ab* (Fig. 3.5) be uniformly taut up to *h*. The string *ab* so taut, i.e. *ahb*, is the longest. The string *cb* of half length, which is uniform, is taut in the same way [with the same force)] as *ab* was taut [...]. Since *clb* is half *ahb*—*hb* being equal to *ha* for construction—it follows that *hc* is twice *lm*. Since the string *clb* is affected by [force] neither more nor less than the string *ahb*, both will have the same tendency to move (nixu) toward the reference positions *ab*, *cd* and, having passed them, they will return with the same speed. But since *hc* is twice *lm*, while the point *l* passes twice the position of reference *m*, the point *h* will pass only once the point of reference *c*. And as in *c* and *m* the motion is very fast and very strong – in *h* and *l* the string is at rest [...]. While the string *cb* or *clb* emits the sound twice, the string *ab* or *ahb* emits it once [14].[52]

In some respects, Beeckman's proof is even less convincing than Benedetti's. However, it is more interesting because it contains seeds for the more rigorous proofs that were developed at the turn of the 18th century and later.

Beeckman's proof required two assumptions that were not entirely convincing even to the mathematicians of the first half of the 17th century:

1. Force and speed are proportional.
2. Similarly deformed strings are subject to the same force and *viceversa*.

The first assumption would have seemed unproblematic at first, though not necessarily shared. It expressed a position held by the Aristotelians. A more careful analysis, however, reveals its problematic nature. It was clear to Beeckman that during the motion of the strings there is a point of maximum speed when the string passes through the undeformed rectilinear configuration and a point of rest when the string reaches its maximum deformation. An analysis of the force, at least for a modern one, indicates that maximum speed occurs at zero force and rest at maximum force,

[51] pp. 196–197.

[52] vol. 1, pp. 54-55.

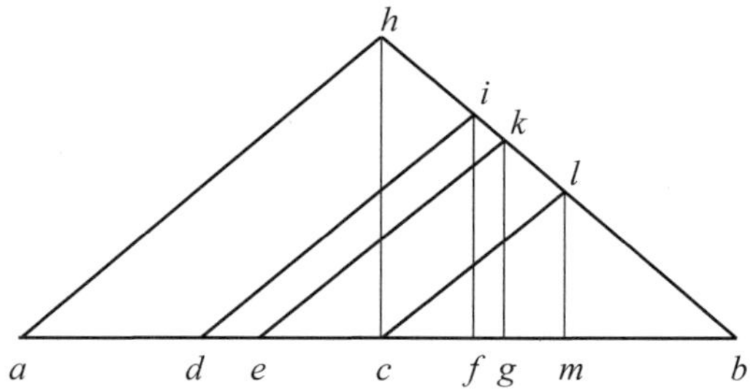

Fig. 3.5 Vibration of strings with different lengths. Redrawn from [14], vol. 1, p. 55

so the Aristotelian axiom is not verified. The axiom can be maintained if the average speed is used instead of the instantaneous speed.

The second assumption raises similar problems; the difficulty in grasping its meaning is due to the ill-defined concept of force it contains. The difficulty is avoided by assuming a static concept of force. When a taut string is subjected to a concentrated transverse force applied at its center (points h, i, k, l in Fig. 3.5), its deformation and tension can be evaluated using the rule of the parallelogram of forces.[53] According to it, for two strings of equal tension but of different lengths, the force required to deflect them is the same if the angle they occupy is the same, that is, if the triangles defined by the baseline and the two parts of the string are similar. Referring to Fig. 3.5, since the triangle ahb and clb are similar, the forces (in h and l) are equal. For the first assumption, force and (mean) speed are proportional, so the velocities are equal, and since lm is half of hc, the oscillation time of the string ahb is twice that of the string clb.

Choosing a particular amplitude for the string vibrations would not limit Beeckman's analysis. In fact, he knew from experience that the vibrations of strings do not depend on amplitude; that is, they are isochronous. This is evidenced by the following quotation, where an acoustic experience is required:

> Since the string comes to rest at last, we must believe that the space through which it moves in the second stroke is shorter than that in the first stroke; and thus the spaces of the strokes diminish. But, since to the ears all sounds seem the same up to the end, it is necessary that all the strokes are always distant from one another by an equal interval of time, and therefore the motions which follow move more slowly [...], since the string crosses a little space in the same time it formerly used to cross a greater one [14].[54]

3.1.4.3 The Law of Mersenne

Marin Mersenne (1588–1648) resumed and perfected, without explicitly mentioning him, Beeckman's reasoning. In particular, he also proved, in some way, that the vibra-

[53] Beeckman should have known the law of the parallelogram as it was proved by Simon Stevin in *Tomus quartum mathematicorum hypomnematum de statica*, published in 1605 [105].

[54] vol. 1, p. 249. Translation into English in [110], adapted.

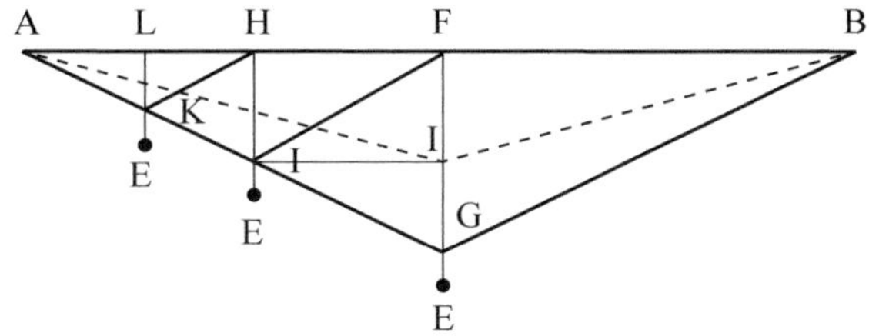

Fig. 3.6 String vibration versus length isochronism. Taken from [91], part 1, *Livre troisieme du mouvement*, p. 157

tions of the string with different amplitudes were isochronous. The proof appeared in the first part of the *Harmonie universelle*, in the *Livre troisieme du mouvement*.

Mersenne proved the isochronism with the assumption that, given a string of a certain length, cross section and axial tension, its 'force' is proportional to the amplitude of the displacement from the position of rest. Again, the word force does not have a precise meaning; one can think of it as the static force applied to the midpoint of the string. The linear dependence of this force, which is correct for small displacements, is taken somewhat for granted by Mersenne. He had the conceptual tools to derive it, the rule of the parallelogram of Stevin, but he did not explicitly call for it.[55]

A given string which has a deflection FI half that of another FG (see Fig. 3.6), has half the force and therefore half the speed (on average), but the space to cover FI is also half that of FG and so the time to reach the horizontal line AB remains the same.

> But it is necessary to notice that the line AB which is displaced to point I will never come back to F as fast as the point I of the line AF will come back to H, because if the path from I to F is equal to the path from I to H, it will take twice as long to come back to F as it will take to come back only to H [91].[56]

Mersenne realized that his arguments had problems when analyzed in depth. The main point, already discussed when dealing with Beeckman, is that if the force pushing the string is proportional to the displacement FG of the point G, then its speed is also proportional to FG, but then the speed of the string should decrease as the string approaches the horizontal position; where it should stop because the force is zero. This goes against the evidence. After some attempts, Mersenne concluded to attribute the cause to air, giving up a clear mechanical explanation, which requires the concept of inertia:

> It seems to me that to explain the phenomenon one must know the different action of the air and the different effort it exerts on the string on which it acts. [The air] can add a new force to push it [the string], which, added to the force it has to reach the center, makes it go on [91].[57]

---

[55] Using the rule of the parallelogram of forces, it is not difficult to show—this is true for a modern reader, but it was also true for Mersenne's contemporaries—that for strings of equal length and tension, the transverse force necessary to represent their midpoints is directly proportional to the displacement.

[56] Part 1, *Livre troisieme du mouvement*, p. 158.

[57] Premiere partie, *Livre troisieme du mouvement*, p. 158.

In the second part of the *Harmonie universelle*, Mersenne explicitly provided the law of the frequency of vibrating strings—taking into account not only the length, but also the cross section, the material, and the tension of the string—commonly, though inaccurately, referred to as the *Mersenne law*. He did this using the language of proportions and rhetorical form, that is, without the use of formulas, by referring to *nine rules* [91].[58] He gave such importance to his rules and claimed that even a deaf man is able to tune a musical instrument by following them: "In any way they may differ, the deaf man could tune them as you like" [91].[59]

Below is the full statement of the nine rules. Note that in them there is no direct reference to the frequency of vibration, although Mersenne possessed this concept (more precisely, he spoke of the period of vibration). What the rules say concerns the condition under which the pitches of two vibrating strings are either in unison, or in the octave, or in the fifth, by varying length, tension, thickness (i.e., area), and specific weight. The rules are largely experimental; except for the inverse dependence on length, which has been 'proved' analytically.

1. First rule. If two strings are equal in length and thickness [and material], and if one makes the sound C-Fa-Ut when stretched with the weight of 1 pound, you must stretch the other with 4 pounds to make it go up one octave, because the weights are in double ratio of the harmonic intervals at which you want the string to go up. The octave interval is 2 : 1, so the double ratio is 4 : 1.
2. Second rule. One must add to the above weight the sixteenth part of the largest weight for the consonance to be just. For example, one must add 4 ounces to the preceding 4 pounds to make the octave just.[60] Consequently, 4¹/₄ pounds against 1, hanging on two equal strings, gave the perfect octave.
3. Third rule. If the strings are of equal thickness [and material] and of different lengths, the tensions stretching them must be twice the ratio of the lengths of the strings in order for them to be in unison. For example, if one string is 2 feet long and the other 1 foot long, and the latter is stretched by a tension of 1 pound, the other must be stretched by a tension of 4 pounds, to which a quarter of a pound must be added according to the above rule.
4. Fourth rule. If the strings are of different thicknesses and of the same length [and material], tensions having the same ratio of thicknesses will bring the strings into unison. For example, if one is 2 in thickness and the other is 3, if the first is

---

[58] Second partie. *Livre troisiesme des instrumens à chordes*, pp. 123–125.

[59] Second partie. *Livre troisiesme des instrumens à chordes*, p. 123. Mersenne's rules translated into modern algebraic language are equivalent to the relations:

$$f \propto \frac{1}{l}\sqrt{\frac{F}{A\rho}} \propto \frac{1}{l}\sqrt{\frac{F}{\gamma}}$$

where f is the frequency of vibration, l is the length of the string, F is the tension (the weight attached to the string according to Mersenne's index), A the area of the cross section of the string, ρ the mass per unit volume, γ the weight per unit length. Note that the frequency is defined less a constant of proportionality.

[60] Remember that the ounce in France in Mersenne's time is $1/16$ of the pound.

stretched by a tension of 2 and the other by a tension of 3, they will be in unison. And if the string were a hundred times thicker, the hundredfold tension would give unison.

5. Fifth rule. If the strings are different in both length and thickness [but of the same material], you must compensate for length and thickness to bring them into unison. Tension follows the single ratio for thickness and the double ratio for length. For example, if one string is 2 in thickness and length and the other is 1 in thickness and length, if you want them to be in unison and the one that is 1 is stretched by 1 pound, the other must be stretched by $6^1/_4$ pounds, because the ratio $6^1/_4$ is made up of the ratio of (1 : 2) to account for the double thickness of the strings and $4^1/_4$ to compensate for the double length.[61]
6. Sixth rule. If the strings are of the same length [and material] but of different thickness, one being 3 and the other 1, and it is desired to make the former rise by a fixed interval in relation to the latter, say by an octave, they must first be tuned in unison according to rule 4, the 3 pound string being tuned to 3 pounds and the 1 pound string to 1 pound. And to make it go up one octave, you have to tension the one of thickness 3 with 12 pounds and 12 ounces, or the sixteenth part of 12 pounds [$^1/_{16}$ of 12 pounds is 12 ounces]. If, on the other hand, you want to make it down one octave, the tension must be a quarter of 3 pounds, or $11^1/_4$ ounces, because you must decrease 12 ounces by a sixteenth, as you had increased it before [$12 - 12/16 = 11^1/_4$] ounces. Finally, if you want to make the string of thickness 1 one octave higher than the string of thickness 3, you must stretch it with a tension of $4^1/_4$ pounds, if the other string is stretched with 3.
7. Seventh rule. When strings are different in thickness and length, you must first align them according to the rule 5. For example, let one string be stretched 1 pound and the other $6^1/_4$ pounds [which are in unison]. If you want to make the second rise one octave above the first, you must stretch it with 26 pounds and 9 ounces, because $26^3/_5$ [26 pounds and 9 ounces] is 4 times $6^1/_4$ (that is 25) pounds plus the sixteenth part of 25.
8. Eighth rule. You must use the same criterion in decreasing the tensions or weights when you want to make the pitch go down as when you want to make it go up by increasing the weights. But you must make the ratios half [square root] to compensate for the different lengths in the former case and double [square] in the latter. That is, you must decrease the tensions in the same ratio by which you increase them; only that to decrease the pitches you must go down, while to increase them you must go up.
9. Ninth rule. If the strings are of different materials, that is, gut, steel, gold, or silver, they must first be brought into unison with the tensions necessary for the purpose; then the previous rules are followed. In order to tune the strings to the same pitch, we can start from experience, which shows that a steel string with a load of 3 pounds will produce a sound as compared to strings of gold, silver or copper [with the same length and thickness], according to the following table:

[61] Here Mersenne is confused, in fact he has to double to compensate for thickness and quadruple to compensate for lengths, which gives 8 instead of 6.

fine gold	fine silver	copper	steel
2886	2160	2025	1920
E	A	B	♮

which shows that gold makes a fourth with silver, silver a tone with steel, and semitone major with copper, which makes a tritone with gold, which makes a fifth with steel.

The letters E, A, B, ♮ indicate that the gold string makes the note E-Mi-La, the silver the note A-Mi-La-Re, the copper the note B-Fa, and the steel the note ♮-Mi.[62] To put them in unison, provided the steel string is loaded with 3 pounds, the gold one is to be loaded with 6 3/4 pounds [they are one fifth a part, so there is a ratio of 1.5 for their interval, which squared gives 2.25, and $3 \times 2.25 = 6.75$] plus one 1/16, that is 7 pounds and 2 3/4 ounces [91].[63]

Notice that in all the first eight rules Mersenne did not make explicit reference to the matter of the string, or better to its specific weight. The 9th rule supplies this lack, but does it in not a general way. There is no explicit statement stating, for instance, that stress should vary in the same ratio of specific weight; even though the four cases reported obey this rule. It is possible that Mersenne was aware of the dependance of frequency with the specific weight only at a later date, possibly by reading Galileo.

3.2 Mathematical Theories

Beginning in the 16th century, the role of mathematics in music took on a new perspective. The fresh mathematics that emerged in the Renaissance differed from the old in its algebraic approach. In this context, the concept of interval as a proportion between homogeneous quantities is replaced by that of ratio between numbers and finally by the concept of fraction. This change of perspective also allowed the use of logarithms in music from the middle of the 17th century. Logarithms made it possible to reduce the sum of intervals from fractional compositions of integers to sums of irrational numbers, and allowed the problem of dividing intervals into equal parts to be solved in an extremely simple way.

3.2.1 The Appearance of Logarithms

As mentioned above, before logarithms could be applied to them, ratios in general, and musical ratios in particular, had to undergo a change in their logical status,

62 Note that Mersenne indicated with B what is now called B♭, and with ♮ what is now called B.

63 Second partie. *Livre troisiesme des instrumens à chordes*, pp. 123–125.

that is, they had to change from the status of proportions between homogeneous magnitudes to that of fractions of the integers measuring those magnitudes. Formally, the transition is simple; given two magnitudes a and b, it is a matter of changing from the notation $(a : b)$ to that of a/b. For example, the interval of fifths $(3 : 2)$ becomes $3/2 = 1.5$. This change, due to the geometric setting of the ratio given by Euclid, came about through a very long process. Still in the 18th century, there was strong resistance to the change of status, even among professional mathematicians who by then had mastered the algebraic language. One example is the great mathematician Isaac Barrow (1630–1677), who asserted in 1664 that ratios were not mathematical entities, especially not numbers [11].[64]

Although the use of logarithms is essentially as old as mathematics, their precise definition did not occur until the 17th century, when they were 'invented' by John Napier (1550–1617) in 1614 (and by Joost Burgi in 1620); the common system of logarithms (base 10) is due to the combined efforts of Napier and Henry Briggs (1561–1630) in 1624. Both Napier and Briggs published tables of logarithms which allowed their practical application [96]. The first application of logarithms to ratios is usually attributed to Johannes Kepler (1571–1630) in his *Chilias logarithmorum* of 1624 [71, 72], but he did not make an explicit connection with music. Later, Pietro Mengoli also (1625–1686) introduced logarithms to measure ratios in his *Geometriae speciosae elementa* [83]. A very weak reference to the application of logarithms to musical intervals, at least according to [116], can be found in Descartes' *Compendium musicae*, published posthumously in 1650 but written in 1618. The treatise contains four circular diagrams representing the division of the octave into a certain number of intervals (see Fig. 4.7). The figures show intervals associated with equal ratios, such as $(360 : 324)$ and $(405 : 360)$, represented by arcs of equal size. If it were not for the logarithmic scale, the former long $360 - 324 = 36$ would be much shorter than the latter long $450 - 360 = 90$.

Disregarding Descartes', other authors used logarithms in music; among them Bonaventura Cavalieri in 1639 for the division of the tone and William Brouncker in 1653, in the English edition of Descartes' *Compendium* [8].[65] But the first extensive and satisfactory representation of musical ratios with logarithms would have to wait for Nicholaus Mercator in 1654-1655 (see below). An example of the application of the logarithms to ratios appears in the following problem:

> To divide one Musicall interval by another. For example, find how often the tonus major 8/9 is found in the diapason 1/2.
> *Rule*. Find the measure of the two reasons given and divide the greater by the lesser.

[64] Lecture 20. "The ratios are not quantities", pp. 366–382.

[65] pp. 282–285.

1	0.000000	8	0.903090
2	0.301030	9	0.954242
	0.301030		0.0051152
The measure of the eight			the measure of tonus major

0.051152) 0.301030 (5.88, so in an eighth there are 5 tonus majores, and 88/100 of a tonus major [117].[66]

The problem to be solved using a modern algebraic language is to find the largest n value such that $(9:8)^n \leq 2$, which by using logarithms reduces to finding the largest n such that[67]:

$$n \log_{10}\left(\frac{9}{8}\right) \leq \log_{10} 2; \qquad n\ 0.05152 \leq 0.30103$$

which gives $n = 5$.

3.2.2 *The Division of the Octave: Tuning and Temperament*

In addition to the evolution in mathematics, one must also consider an evolution in music, analyzed in Chap. 2 and briefly summarized below. Two developments in musical practice dictated that the problem of consonances in a musical scale should be seriously addressed. One was the rise of polyphony; in this situation, the distinction between consonance and dissonance became much more important than in the monophonic practice of music. The other, much more recent development was the emancipation of instrumental music, which was increasingly used not only to accompany vocal music but also independently of it. And while both singers and players of some instruments (the violin family, for example) could adjust the pitch of their notes to some extent, players of instruments with fixed pitches, such as the lute and the organ, and later other keyboard instruments such as the virginal and the harpsichord, lacked such flexibility.

The old problem of finding the musical intervals in the compass of an octave takes on a new guise. First, there is the problem of finding a new division of the octave that includes the new consonances not allowed in Greek music, such as just thirds and sixths (major and minor); then there is the problem of correcting these intervals to eliminate the small and large dissonances that inevitably arise on keyboard and fretted instruments when a melody is developed in one or more voices. These two problems are called tuning and temperament, respectively.

---

[66] pp. 202–203.

[67] In the equation below, the symbol $\log_{10}$ is used to indicate a base 10 logarithm. In the following, this type of logarithm will be referred to simply as log, reserving the symbol ln for the natural logarithm.

The terms tuning and temperament are generally considered to be distinct. Tuning refers to a set of notes and intervals in a scale chosen according to commonly accepted criteria; examples are *Pythagorean tuning* and *just tuning*. Temperament, on the other hand, is an adaptation of a tuning, either to lessen some of its harshness or to make it easier to play. Examples are *meantone* and *equal temperament*.

The difference between tuning and temperament can also be seen by a quick inspection of the ratios that define the musical intervals of each scale. In tuning, these intervals are rational numbers obtained from ratios of relatively simple integers; in temperament, on the other hand, the intervals are irrational numbers obtained from ratios of numbers that are also irrational. Of course, there are borderline cases where it is difficult to make a distinction; for example, in the literature one can find a tuning with a ratio of the kind (164025 : 115921), which is not unlike a ratio between irrational numbers [9].[68]

3.2.2.1 Pythagorean and Just Tunings

In Western music, at least until the 20th century, there are two basic types of tuning, Pythagorean tuning and just tuning. In what follows we will briefly mention them, neglecting for the moment their historical development—which has already been presented in Chaps. 2 and 3 and will be discussed again in Chaps. 5 and 6—in order to reduce the complexity of the subject. We will follow in part the framework suggested in [16].

Diatonic Pythagorean Tuning

We saw in Chap. 1, Sect. 1.3.3, how to divide a monochord according to the diatonic genus, following the Pythagorean approach. This involved starting from a point on the tetrachord and moving forward or backward by fifths, decreasing or increasing by one octave to stay within the octave compass.

The geometric language used there can easily be transformed into an algebraic one, with considerable simplifications. For example, one can simply say that the Pythagorean pitch intervals can be found by the relation:

$$1 \leq \left(\frac{3}{2}\right)^{i} 2^{-n} \leq 2$$

where i and n are either positive or negative integers.

The previous relation can be rewritten more conveniently by introducing logarithms and writing $\overline{v} = \log 3/2$; $a = \log 2$, as:

$$0 \leq i\overline{v} - na \leq 0.30103$$

[68] p. 5.

Table 3.2 Cycles of fifths; from [16], p. 44

$\bar{\imath}$	Modality	Ordinal numbers
6	Augmented fourth	$j = 3$
5	Major seventh	$j = 6$
4	Major third	$j = 2$
3	Major sixth	$j = 5$
2	Major second	$j = 1$
1	Perfect fifth	$j = 4$
0	Octave	$j = 0$
−1	Perfect fourth	$j = 3$
−2	Minor seventh	$j = 6$
−3	Minor third	$j = 2$
−4	Minor sixth	$j = 5$
−5	Minor second	$j = 1$
−6	Diminished fifth	$j = 4$

An even simpler notation can be used by denoting the interval of the i-th iteration of fifths as $\bar{\imath}$, with the understanding that the interval is always within the octave. Since $\bar{\imath}$ varies between 0 and ± 6, we obtain the intervals shown in Table 3.2, where the reference note from which to start the cycle of fifths is currently undefined.

The ordinal numbers j appearing in Table 3.2 represent the order of intervals, starting from the reference note, such that if $j_1 > j_2$, then the numerical value of the interval associated with j_1 understood as a fraction is greater than that associated with j_2. It is easy to verify that the simple relationship exists:

$$2j \equiv \bar{\imath} \pmod 7$$

To get the order number in the staff of the note associated with j, starting from any note, we must add 1. For example, after one cycle, that is $\bar{\imath} = 1$, we have $j = 4$, the number in the staff is 5, which means it is an interval of fifth. For $\bar{\imath} = 2$, we have $j = 1$, which is an interval of second. From this table we can see that the first four notes ($j = 0, 1, 2, 3$), obtained from iterations of fifths, form an augmented fourth or a tritone. So if we want to produce a diatonic scale—i.e. an ordered sequence of unaltered notes with only the two semitones BC and EF and the tritone FGAB—we must start the iteration from the note F. Table 3.3 shows the relationships between iterations of fifths $\bar{\imath}$ and ordinals j with the names of the notes associated with j; Table 3.4, shows notes ordered for increasing j.

The intervals of the diatonic scale starting from C can be evaluated by making the difference of the values of $\bar{\imath}$. They are given in Table 3.5. Thus, the values of Euclid's *Sectio canonis* are found again. The intervals of the Pythagorean scale, derived from Tables 3.2 and 3.5, are summarized in Table 3.6. As it can be seen, only the octave, fifth and fourth intervals are expressed by simple ratios and therefore only they were considered consonants by the Greeks.

Table 3.3 Notes and iterations of fifths

$\bar{\imath}$	0	1	2	3	4	5	6	0
j	0	4	1	5	2	6	3	0
Note	F	C	G	D	A	E	B	F

Table 3.4 Diatonic pythagorean scale

$\bar{\imath}$	0	2	4	6	1	3	5	0	2	4	6	1
j	0	1	2	3	4	5	6	7	8	9	10	11
Note	F	G	A	B	C	D	E	F	G	A	B	C

Table 3.5 Diatonic Pythagorean tones and semitones based on C

Note	C	D	E	F	G	A	B	C
$\Delta\bar{\imath}$	2	2	−5	2	2	2	−5	
Tones	$\frac{9}{8}$	$\frac{9}{8}$	$\frac{256}{243}$	$\frac{9}{8}$	$\frac{9}{8}$	$\frac{9}{8}$	$\frac{256}{243}$	

Table 3.6 Pythagorean intervals

2nd	Third min.	3rd major	4th	5th	6th min.	6th major	7th min.	7th major
$\frac{9}{8}$	$\frac{32}{27}$	$\frac{81}{64}$	$\frac{4}{3}$	$\frac{3}{2}$	$\frac{128}{81}$	$\frac{27}{16}$	$\frac{16}{9}$	$\frac{243}{128}$

Just Tuning

Just tuning is the result of an attempt to remove the harshness of the major and minor thirds and sixths produced by Pythagorean tuning. In just tuning, the consonant intervals are not only the fifth, the fourth and the octave; the major $(5 : 4)$ and the minor $(6 : 5)$ thirds are also introduced; thus, instead of the numbers 1, 2, 3 of the Pythagorean tuning, there is a more complete series 1, 2, 3, 5, which can be extended without loss of generality to 1, 2, 3, 4, 5, 6, a set that Zarlino had called senario.

In the following, instead of giving an independent treatment of just intonation, we start from the Pythagorean one and highlight the differences. Before proceeding, it is worth introducing some well-established definitions and conventions for the pitches of a note. It is now common to measure pitches using the Pythagorean scale, although it is no longer used by musicians.

An example of a Pythagorean scale is represented by the series of notes considered in the previous section, starting from C, ignoring the index of the octave for the sake of notation:

Table 3.7 Tones, semitones and intervals of just intonation

Note	C		D		E		F		G		A		B		C
Tones		$\frac{9}{8}$		$\frac{10}{9}$		$\frac{16}{15}$		$\frac{9}{8}$		$\frac{10}{9}$		$\frac{9}{8}$		$\frac{16}{15}$	
Intervals	1		$\frac{9}{8}$		$\frac{5}{4}$		$\frac{4}{3}$		$\frac{3}{2}$		$\frac{5}{3}$		$\frac{15}{8}$		2

$$C^0\ D^0\ E^0\ F^0\ G^0\ A^0\ B^0\ C^0$$

The exponent ^{0} indicates that the notes are exactly those of the Pythagorean scale, with the intervals between them defined in Table 3.5.

In the just intonation, the Pythagorean scale is modified as follows to make these intervals more pleasing:

$$C^0\ D^0\ E^{-1}\ F^0\ G^0\ A^{-1}\ B^{-1}\ C^0$$

The exponent different from 0 indicates the number of syntonic commas subtracted from the corresponding Pythagorean note; in this case the notes E, A, B are narrowed by one syntonic comma.

It is immediately apparent that the seconds C^0–D^0 and F^0–G^0 are the same as their Pythagorean counterparts and are equal to (9 : 8) This interval also occurs in other positions. For the seconds D^0–E^{-1} and G^0–A^1 the situation is different. The new interval is given by $(9 : 8) - (81 : 80) = (10 : 9)$, known as the minor tone. Table 3.7 gives the tones, semitones and intervals in just intonation for a C major scale.

3.2.2.2 Meantone and Equal Temperaments

Historically, two main approaches have been followed to obtain tempered systems. The first approach was still related to the Pythagorean conception, in which the scales were formed by cycles of fifths; now they were not just, but narrowed by a small amount. The second approach, clearly presented in Chap. 5, consists in dividing an octave into a certain number n of equal parts, requiring that they give rise to five equal average tones (T) and two equal major semitones (S)—also called *diatonic semitone*—so that the tempered tones are intermediate between the major (9 : 8) and minor (10 : 9) tones. By defining the minor semitone (s)—also called *chromatic semitone*—as the difference between the tempered tone and the diatonic semitone, we can see that the octave should be divided into 7 diatonic semitones and 5 chromatic semitones. The number of divisions n of the octave depends on the chosen ratio of sizes of diatonic to chromatic semitones, which is quite arbitrary; for example, for a ratio of diatonic to chromatic semitones of (3 : 2), it is $n = 7 \times 3 + 5 \times 2 = 31$.

Table 3.8 Number of division versus alteration of fifth

n	19	31	43	50	55
p/q	1/3	1/4	1/5	2/7	1/6

In this type of temperament, where the diatonic semitone is larger than the chromatic semitone, the tempered fifth is smaller than the equally tempered value (see below); for this reason they can be called *negative systems* [20].[69] But it is possible to divide the octave in such a way that the fifth is greater than the equally tempered value, by assuming that the chromatic semitone is greater than the diatonic semitone (i.e. the 'major' semitone is less than the 'minor' semitone). For example, for the ratio of diatonic semitone to chromatic semitone of (2 : 3), it is $n = 7 \times 2 + 5 \times 3 = 29$. These systems are called *positive systems*; examples of positive systems, besides the case of $n = 29$, are offered by the equal division of the octave into 17, 41, 53, or 65 parts [20].[70]

It is easy to relate the two different paths to temperament to find a formula that gives the fraction p/q of a comma by which the fifth decreases as a function of the number n of divisions of the octave [9, 99, 102].[71] Table 3.8 gives the values of n corresponding to very high approximations of the most common p/q ratios for negative systems.

One of the earliest representatives of the first approach to temperament taken by mathematicians and music theorists in the 16th century is the so-called meantone temperament, in which the just fifth is narrowed by 1/4 of a syntonic comma (81 : 80). The scale in this temperament is obtained in a manner analogous to that of the Pythagoreans, that is, with a cycle of six forward narrowed fifths, reducing the notes back to the octave when necessary.

If:

$$F^0\ C^0\ G^0\ D^0\ A^0\ E^0\ B^0$$

is a cycle of fifth in the Pythagorean scale, those of the temperate scale is simply:

$$F^{+1/4}\ C^0\ G^{-1/4}\ D^{-1/2}\ A^{-3/4}\ E^{-1}\ B^{-5/4}$$

as for each forward cycle the fifth is narrowed by 1/4 of a comma. In the above sequence it is assumed that C^0 remains unchanged, so in order for FC to be a tempered fifth, F should be increased by 1/4 of a comma.

Note that the two tones C^0–$D^{-1/2}$ and $D^{-1/2}$–E^{-1}, which in the just intonation were a major tone and a minor tone, are now equal, since in the Pythagorean tuning the intervals C^0–D^0 and D^0–E^0 are equal, and those tempered by a quarter comma are still equal because of the equal difference of the exponent. The major third, such as

[69] p. 8.

[70] p. 11.

[71] p. 126, footnote 50.

Table 3.9 Mean tone 1/4 of a comma scale

0	C^0 1.118	100,000
2	$D^{-1/2}$ 1.118	89,443
4	E^{-1} 1.070	80,000
−1	$F^{+1/4}$ 1.118	74,767
1	$G^{-1/4}$ 1.118	66,874
3	$A^{-3/4}$ 1.070	59,814
5	$B^{-5/4}$ 1.118	53,499
0	c^0	50,000

C^0–E^{-1}, is a syntonic comma lower than the corresponding Pythagorean third, and so its interval is $(81 : 64) - (81 : 80) = (80 : 84) \equiv (5 : 4)$, as for the just major third. Since the major third in this temperament is made of two equal tempered tones, the size of each of them is defined by $\sqrt{5/4} = \sqrt{5}/2$. A different view can be obtained by noting that the tempered tone is the geometric mean of the major tone (9 : 8) and the minor tone (10 : 9), given by $\sqrt{\frac{9}{8} \times \frac{10}{9}} = \sqrt{5}/2$.

For this reason, the 1/4 of a comma temperament is called the meantone temperament. It should be noted, however, that the term meantone is often used even when the temperament is different from 1/4; for example, 2/7 or 1/3. This is because in each temperament there is an average tone, a kind of mean value between the maximum and minimum tone, although only in the 1/4 of a comma temperament is the average tone the geometric mean of the two tones.

Table 3.9 gives the scale of the meantone temperament, the first column giving the number of tempered fifths necessary to obtain the note of the scale. The second column gives the note and the alteration (as a fraction of a syntonic comma) with respect to the Pythagorean scale; between the notes are the intervals that separate them. The third column shows the division of a monochord of length 100, 000, so that the value of the table can be compared with that of Table 3.20, obtained by Huygens (see below). Note that there is only one type of tone, with a ratio of 1.118, and one diatonic semitone, with a ratio of 1.070.

While the theory of the meantone temperament, expressed in a modern algebraic language, is rather straightforward, the historical process for its establishment was a long one; it is well described in [9, 79]. This is partly due because Pythagorean tuning was difficult to counteract because of its strong ‘ideological’ motivation, partly because there was no great need for temperament until the 1500s given the type of music in vogue, and partly because the mathematical tools available to musicians until the advent of logarithms were not adequate.

At first, temperament was mainly used for instruments with fixed pitches, such as the keyboard and the lute. Franchino Gaffurio (1451-1522), for example, claimed at the end of the 15th century, that the fifths of organs underwent a small diminution [48].[72]

An accurate understanding of the role temperament played in the formation of musical scales in the past requires an analysis of the interaction between performing and theoretical musicians. An amazing bibliography of tuning, temperament and their history and performance can be found in web page of the Huygens-Fokker Foundation [65].

One of the earliest documented descriptions of temperament, detailed enough to be easily reproduced by any musician, is found in Pietro Aron's *Toscanello in musica* of 1539: "The major third CE must be sound and fair; the fifth CG a little sparse; [...] the fourth must be the opposite of the others named, that it be tempered and raised" [1].[73] Note that Aaron did not give any numerical values for the various intervals; some years later de Salinas provided them in his *De musica libri septem* of 1577 [101].[74]

Zarlino, in his *Le istitutioni* of 1558, presented for the first time in a precise mathematical form, his own form of temperament, known as the $^2/_7$ of a comma temperament: "Each diapente remains diminished by $^2/_7$ of a comma, and a diatessaron is increased of the same amount [...], the ditone [major third] will be imperfect [less] by $^1/_7$ of a comma and of the same amount will be diminished the semi-ditone [minor third] and the hexachord, minor and major will both be increased by $^1/_7$ of a comma" [121].[75]

As already commented in Sect. 2.2.1, Zarlino was not able to find the ratio of the division of a comma into seven equal parts with a numerical procedure. He could, however, use the mesolabium, a simple device for finding a series of geometric means, or equivalently, the terms of a geometric progression, whose invention is usually attributed to Eratosthenes of Cyrene (3rd century BCE) [5].

Etymologically, the word mesolabium is derived from the Greek mesólabos to take in the middle. Such an instrument consisted of a number of equal rectangles of cardboard or wood. A diagonal was drawn on each of them. In Fig. 3.7 three rectangles are shown to find two proportional means between AD and QM, i.e. to find the cube root of the AD/QM ratio. The operation with the mesolabium is simple. Referring to the figure, place the three rectangles so as to obtain the alignment of the points A, O, P, Q. In this way, the three similar trapezoids AOCD, OPGC, PQMG are obtained, from which it is easy to show that PG and OC are the two geometric means of AD and QM.

---

[72] Book 2, Chap. 3.

[73] A not numbered page; second book, chapter 41.

[74] p. 163.

[75] p. 126.

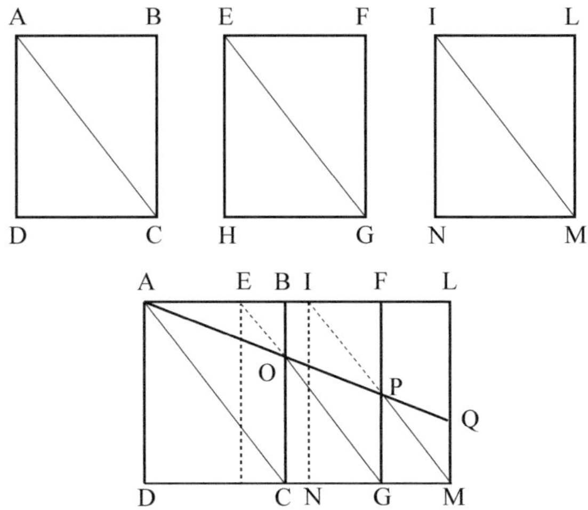

Fig. 3.7 Use of the mesolabium. Adapted from [5], p. 204

In the *Dimostrationi harmoniche* of 1571, Zarlino, named two more forms of temperament corresponding to the $1/4$ and the $1/3$ of a comma. He did not represent them mathematically, as he did for the $2/7$ of a comma temperament. For the $1/4$ of a comma, Zarlino said that it resembles Ptolemy's ditonic diatonic genus, in which all the tones are equal;[76] the hexachord minor and the ditone remain in their pure form: "And this temperament is very pleasing to the ear". While the temperament known as the $1/3$ comma has the major hexachord and the semitone in their natural forms: "It is well true that the temperament made in this way is not as sound as the first two" [122].[77]

The $1/3$ of a comma temperament—which corresponds very closely to a division of the octave into 19 equal parts—may have fifths too much tempered, but it is easy to implement and, most notably, it gives rise to a simple cyclic system. In fact, a cycle of 19 fifths comes back to the fundamental minus a very small interval, about $1/20$ of a syntonic comma (thus in such a keyboard there is no fifth wolf (see Sect. 3.2.2.3). A 19-note keyboard based on the $1/3$ of a comma temperament is such that each tone is divided into three equal parts (e.g. AB is formed by the two equal intervals AA♯ B♭B, with A♯ and B♭, separated by a minor diesis (128 : 125) in just intonation) and each semitone (corresponding to E-F and B-C) is divided into two parts. In 1548 Zarlino ordered from Domenico of Pesaro a two-octave harpsichord with 19 keys per octave, see Fig. 3.8 with the purpose to reproduce both chromatic and enharmonic genus and to reduce the disadvantage of the diatonic system, although he did not refer to the $1/3$ of comma temperament [107, 121][78] (for further reading on the 19-part division see [4][79]).

---

[76] A very strange comment, since Ptolemy's ditonic diatonic system is the Pythagorean system.

[77] p. 221.

[78] p. 141.

[79] pp .298-299

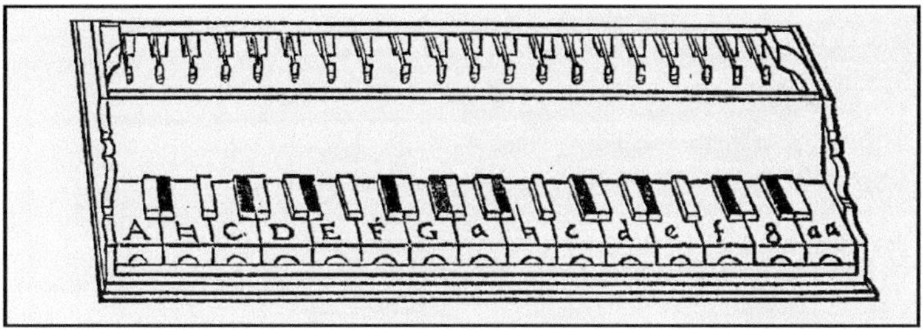

Fig. 3.8 Harpsichord with 19 keys. Redrawn from [121], p. 141

Table 3.10 Regular temperaments of the 17th and 18th centuries. Adapted from [32], p. 74

fraction of comma	$\frac{1}{3}$	$\frac{3}{10}$	$\frac{2}{7}$	$\frac{5}{18}$	$\frac{1}{4}$	$\frac{2}{9}$	$\frac{1}{5}$	$\frac{5}{23}$	$\frac{3}{14}$	$\frac{3}{17}$	$\frac{1}{6}$	$\frac{1}{7}$	$\frac{1}{8}$	$\frac{1}{9}$	$\frac{1}{10}$
	□	○	□	○	■	○	■	○	○	○	■				
year	1577	1749	1558	1749	1539	1666	1697	1749	1751	1751	1650	1758	1758	1758	1758

Zarlino did not claim priority for the $1/3$ comma temperament as he did for the $2/7$ comma, probably because he knew it had already been worked out by Francisco de Salinas in *De musica libri septem*, published in 1577 [80],[80] where he also considered himself the inventor of the $1/4$ of a comma meantone temperament, also qualifying it as "both easier to the intellect and simpler to instruments" [101].[81]

Of the temperaments just considered, $1/4$, $2/7$, $1/3$, the $1/4$ of a comma was the most esteemed and used by theorists and practitioners until the 18th century. For example, Huygens, although critical of any form of temperament, considered it the only one to be admissible (see next sections).

Mathematicians of the 17th and 18th centuries found the problem of temperament as an interesting subject of research, and many, many fractions of the comma with which to narrow fifths were proposed. Table 3.10 shows some of these, with the year of their proposal. Temperaments marked with a black square are the only ones used as a matter of fact. The temperaments marked with a white square or circle, are considered simple proposals, while those without markings are considered academic eculubrations. The temperaments of $1/5$ and $1/6$ of a comma were characterized by less narrowed fifths than for $1/4$ of a comma, at the expense of wider major thirds. Specifically, the $1/5$ of a comma temperament was valued for its proportionality, with fifths narrowed by $1/5$ of a comma and major thirds widened by the same amount, while the $1/6$ of a comma temperament had fifths narrowed by $1/6$ of a comma and major thirds widened by $1/3$ of a comma. These temperaments were used to achieve a compromise between the purity of fifths and thirds while maintaining a certain circularity in the division of the octave.

[80] p. 17 (not numbered).

[81] pp. 140, 154.

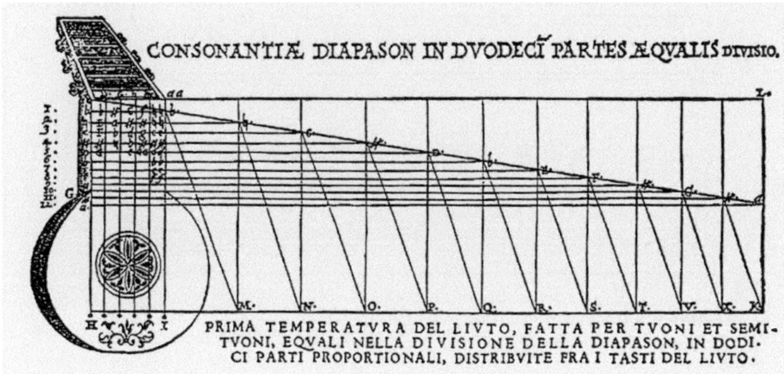

Fig. 3.9 Positioning of equal spaced frets of a lute. Redrawn from [123], p. 209

A special and very important form of temperament is the equal temperament, in which the octave is divided into 12 equal semitones. For a modern reader, the division of the octave into 12 equal parts is a very simple task, since each interval is defined by the ratio $\sqrt[12]{2}$. The actual calculation, however, without the aid of a computer or logarithms, is very tedious, and such was the case for the mathematicians of the 16th (and 17th) century. The interval of the fifth, given by seven semitones, is $\sqrt[12]{2^7} = 1.4983$, just a little below the perfect fifth by the interval $1.5 : 1.498 = 1.00112$, a nearly 1/11 of a syntonic comma, and exactly 1/12 of the Pythagorean comma. It is therefore a much lighter temperament for fifths than the meantone.

Equal temperament appeared very early, at least in approximate form, for fretted instruments such as the lute; according to Vincenzo Galilei, fretted instruments were somehow always tuned in equal temperament. The spread of this type of tuning is documented from at least the middle of the 16th century. Zarlino, for example, reported the case of a lute with 12 equally spaced frets, as documented by Fig. 3.9, where the division into equal intervals is made by means of the mesolabium. However, it is clear that the temperament could not be very accurate with this method, and Fig. 3.9 has more rhetorical than practical value.

From a practical point of view, the rule proposed by Vincenzo Galilei in his *Dialogo* of 1581 is much more convenient and precise. Here he wrote:

> I therefore divide the whole line A B into eighteen parts, and toward the high part (moving away from the low part) where that first part ends, I place the first fret. I divide once more the whole remainder of the same number of parts [i.e. eighteen and from the very same band, I place the first fret under the second. I now proceed to distribute the space which remains beneath the frets, always in that very same order, up to [the number of] twelve, which brings me exactly to [the point] where half of the entire string terminates. The first and lowest octave of this, I find, has been divided into twelve equal semitones and six tones, as Aristoxenus instructed [54].[82]

[82] p. 49. Translation into English in [61].

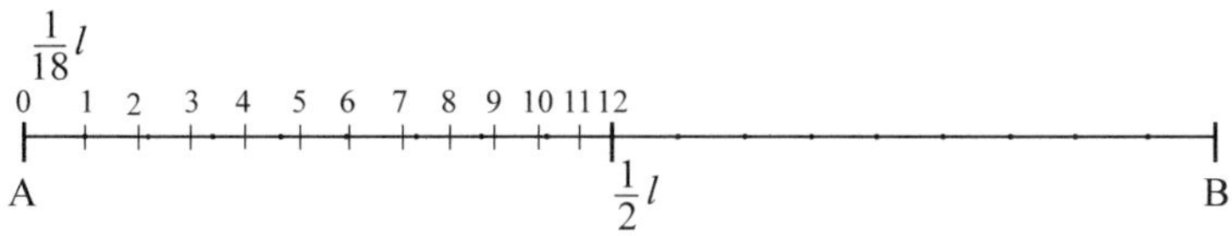

Fig. 3.10 Division of a string into 12 equal intervals for the octave

Galilei's procedure is very clear by looking at Fig. 3.10; the first fret, located at 1, has a straight line of length 17/18 of AB, or 1-B, to its right. The fret 2, has to its right 17/18 of 1-B, i.e. $(17/18)^2$ of AB, and is located at 2. After 12 iterations, fret 12 has at its right side, 12-B, a length of $(17/18)^{12} \sim 0.503$, a little more than half. If we think of fret distances as musical intervals, this means that Galileo divided a little less than an octave into 12 equal semitones of size $(18 : 17)$.

Galilei interrupted his procedure by noting that if he had divided the string AB into a different number of parts, he would not have obtained a correct result. In fact, dividing the string into 19 parts would come close to the octave with more than 12 intervals; conversely, dividing it into 17 parts would come close to the octave with less than 12 intervals. Galilei's procedure is quite clever, but it seems to be essentially empirical in nature; its implementation requires no theoretical knowledge at the mathematical level, apart from the idea of summing intervals.

Galilei's proposal to take the interval $(18 : 17)$ as the semitone measure that roughly divides the major tone into two equal parts was not new. It can be found, for example, in the musical work, *De musica*, by Girolamo Cardano (1501–1576), first draft 1547, second 1574 [25]. Here the semitone $(18 : 17)$ is presented as a possible half major tone $(9 : 8)$, following the division approach that Boethius took in his *De institutione musica*. The interval $(9 : 8)$ is rewritten as $(18 : 16)$, which suggests the sequence $18 - 17 - 16$, from which two semitones can be extracted, the smaller one $(18 : 17)$ and the larger one $(17 : 16)$, not very different from each other; however, Cardano seems to have had no preference between the two.

An early fairly accurate numerical calculation of equal temperament intervals is due to Simon Stevin with his *Spiegeling der singkonst* of the first decade of the 17th century which will be discussed later; but it remained buried until the 19th century.

A similar fate befell the values obtained in China by prince Chu Tsai-yü (Zhu Zaiyu), Ming dynasty (1536–1611) [78]; a brilliant solution, which is surprising since the music of China, as far as we know, had no need of any kind of temperament, although it has a precedent in about 400 CE [9].[83] Table 3.11 reports Tsai-yü values that were not known until their publication in 1779 by the Jesuit Joseph Marie Amiot (1718–1793) [2, 77, 109, 119].[84]

The first printed table of equal temperament in the West is usually attributed to the German mathematician Johann Faulhaber (1580–1635) in 1630 [9].[85] But the

[83] p. 55.

[84] vol. 1, pp. 97–168; vol. 2, pp. 12–15.

[85] p. 78.

Table 3.11 Equal temperament in China. Adapted from [2], p. 143

Name of Chinese notes		Western	Monochord
Hoang-tchoung	Koung	Fa	10,000.000000
Ta-lu			9,438.704312
Tay-tsou	Chang	Sol	8,908.908718
Kia-tchoung			8,408.906415
Kou-si	Kio	La	7,937.000525
Tchoung-lu			7,491.503538
Joui-pin	Pien-tché	Si	7,071.006781
Lin-tchoung	Tché	Ut	6,674.109927
Y-tsé			6,299.600524
Nan-lu	Yu	Re	5,946.003557
Ou-y			5,612.301024
Yng-tchoung	Plen-koung	Mi	5,297.301547

author who contributed most to the knowledge of the numerical values of equal temperament was Mersenne, commented on in the next chapter.

3.2.2.3 Twelve Tone Chromatic Scale and Wolf Fifth

A (twelve tone) chromatic scale is obtained from the diatonic seven-note scale by inserting intermediate notes, called altered notes. They are identified by the symbols ♭ and ♯; for example C♯ is for a note between C and D, while B♭ is for a note between A and B. In the case of C major, the chromatic scale looks like this:

C C♯ D E♭ E F F♯ G G♯ A B♭ B C

Intervals such as FF♯ and E♭E have the same amplitude and are called *chromatic semitone*; the difference between the tone and the chromatic semitone is the *diatonic semitone*. The octave is thus composed of 7 diatonic and 5 chromatic semitones.

In Pythagorean intonation, the altered notes are obtained by continuing the cycle of fifths, adding 5 iterations to the 6 required for the diatonic scale. The sequence of fifths to be considered, starting with F, consists of 9 ascending and 2 descending fifths:

E♭ B♭ F C G D A E B F♯ C♯ G♯

in which the altered notes are separated from the natural notes of the same name by an apotome, which plays the role of a chromatic semitone [16].[86] The diatonic semitone, the difference between the tone (9 : 8) and the apotome (2187 : 2048), is the limma (256 : 243).

By introducing a temperament, the tempered chromatic scale is obtained by cycles of tempered fifths. It is interesting to see what happens to the chromatic and diatonic semitones. A distinction is made between positive and negative temperaments. In

[86] p. 49.

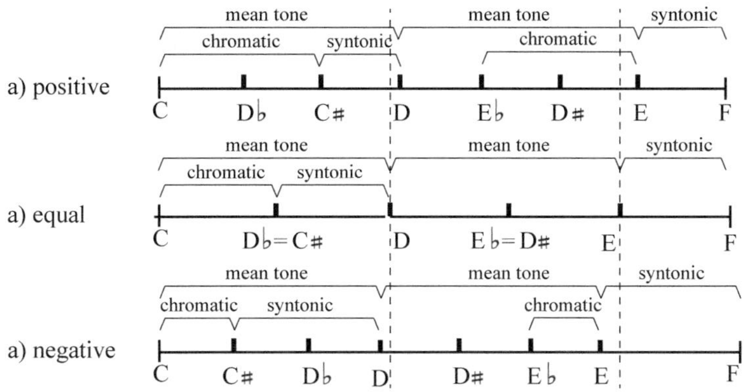

Fig. 3.11 Chromatic and diatonic semitone for different kinds of temperament

positive temperaments, the chromatic semitone is larger than the diatonic semitone; in negative temperaments, the opposite is true. In equal temperament, the two semitones are equal. Figure 3.11 illustrates the situation

Assigning the notes of the chromatic scale to two octaves of a keyboard results in this sequence:

C C♯ D E♭ E F F♯ G G♯ A B♭ B C C♯ d e♭ e f♯ g g♯ a b♭ b cc

from which we see that, although we have proceeded on the basis of tempered fifths, there is one fifth, that defined by G♯ e♭, which is not narrowed as the other. The correct narrowed fifth would actually be G♯ d♯, but d♯ has no place on the keyboard and therefore cannot be played. The G♯ e♭ fifth is generally dissonant and is therefore called the *wolf fifth*. It differs from the correct tempered fifth of the interval d♯e♭, whose amplitude depends on the temperament. In a Pythagorean tuning, this interval is the difference between the apotome and the limma and coincides with the Pythagorean comma; the wolf fifth is narrower than the tempered fifth. In equal temperament there is no wolf's fifth; in positive temperaments the wolf's fifth is narrower than the tempered fifth, in negative temperaments it is wider.

3.2.2.4 Good/Circulating Tempered Systems

The temperaments proposed in the previous section are called *regular temperaments* according to a Bosanquet's definition, for which a regular temperament system is one in which every fifth, or every fifth except one, has the same value [20].[87] This includes equal temperament and the different variations of meantone temperament, as well as temperaments with equal divisions of parts in the octave.

[87] p. 7.

However, there are numerous systems of temperament that do not fall into any of the above classes; they are known as *irregular temperament* [9],[88] and present a variety that seems astonishing. But some of them were proposed by their inventors as modifications of existing systems, and others, though not so designed, are also closely related to regular systems. By making the definitions sufficiently elastic, it is possible to classify each of these irregular systems into one or another of the existing classes.

The irregular temperaments arose when some performers and composers, among them Johann Sebastian Bach, accepted that not all the fifths of a system should be tempered equally. The new irregular systems mixed different amounts of temperament and thus contained different sizes of fifths. By adjusting the amount of temperament and the mixture of tempered and pure fifths, the wolf could be eliminated and more usable chords could be created. This was especially useful when the 18th-century composers began to write more chromatic and complex harmonies.

While the regular temperaments favored one key or another (placing the wolf in a little-used interval of the key), some of the new irregular temperaments could be used in a wide range of keys without adjustment. For these reasons the irregular temperaments were known almost interchangeably as *good (or well) temperaments*, or *circulating temperaments*, meaning that they could be used in any key around the circle of fifths [9].[89] This did not mean that all keys sounded the same, but at least they were all usable, some sounding better than others. The irregular systems became extraordinarily popular as keyboard temperament systems also because they were easier for keyboardists to tune (at a time when harpsichords had to be tuned daily) [40].[90]

An interesting example of a well-tempered system is that of Andreas Werckmeister (1645–1706), the German organist well known to Huygens, music theorist, and composer of the Baroque era. It is usually referred to as *Werckmeister III* to distinguish it from other proposals by Werckmeister himself.

The system is obtained by dividing the Pythagorean comma[91] into four equal parts, and applying a reduction of this amount to only four fifths. Since the Pythagorean comma has an interval defined by $1.5^{12} : 128 \approx 1.013643$, a fifth reduced by a quarter of this amount is given by $1.5 : 1.013643^{1/4} \approx 1.4949$.

Table 3.12 shows the intervals of tones and semitones obtained with the following cycle of 11 fifths:

$$\underline{\mathrm{CG}}\ \underline{\mathrm{GD}}\ \underline{\mathrm{DA}}\ \mathrm{AE}\ \mathrm{EB}\ \underline{\mathrm{BF}\sharp}\ \mathrm{F}\sharp\mathrm{C}\sharp\ \mathrm{C}\sharp\mathrm{G}\sharp\ \mathrm{G}\sharp\mathrm{D}\sharp\ \mathrm{D}\sharp\mathrm{A}\sharp\ \mathrm{A}\sharp\mathrm{F}(\mathrm{B}\flat\mathrm{F})\ \mathrm{FC}$$

where the underlined fifths are narrowed.

The first row of the table gives the names of the notes; here the sharp has no precise meaning; it simply indicates that the note is a variable semitone higher than

---

[88] p. 133.

[89] p. IX.

[90] pp. 50–53.

[91] In the meantone temperaments it was the syntonic comma to be used.

Table 3.12 Pitches and semitones for Werckmeister's temperament; from [59], p. 32

C	C♯	D	D♯	E	F	F♯	G	G♯	A	A♯	B	C
0	7	2	9	4	11	6	1	8	3	10	5	12
1.000	1.053	1.117	1.185	1.253	1.333	1.405	1.495	1.580	1.670	1.778	1.879	2.000
1.053	1.061	1.061	1.057	1.064	1.053	1.064	1.057	1.057	1.064	1.057	1.064	

the previous one. The second row gives the number of fifths, starting with C, needed to obtain the corresponding note. The third row gives the intervals measured from C, and the fourth row the intervals between two successive notes, that is, the value of the semitones.

The advantage of the Werckmeister III temperament is that there are no wolves. The disadvantage is the inhomogeneity of the semitones. The consonance of commonly used intervals is compromised for some keys more than others.

Another well known form of good temperament was due to Francesco Antonio Vallotti (1697–1780), theorist and composer from Padua. He wrote the *Trattato della scienza teorica e pratica della moderna musica*, in three books, which he claimed to have finished by 1728, although only the first book was published in 1779 and the other two (with the second containing the instructions for temperament) remained in manuscript form and were published only in 1950 [112]. The following quotation summarizes the essence of Vallotti's temperament.

> Since, as we have already said, each of the six 5ths is diminished [by 1/6 of a comma], it follows that the major thirds increase by 1/3 and the major sixths increase by 1/2 of a comma: for then the opposite of their respective complements occurs: the fourths increase by 1/6 of a comma, and the minor sixths decrease by 1/3 and the minor thirds decrease by 1/2 of a comma [112].[92]

Vallotti's temperament is a variant of the Werckmeister's temperament.[93] Differently from Werckmeister, Vallotti narrowed the first six consecutive diatonic fifths by 1/6 of a syntonic comma, while the other sixth fifths of the cycle of twelve remained just. Below the same cycle of fifths as that of Werckmeister, with the caution that now the cycle of fifths starts with FC:

CG GD DA AE EB BF♯ F♯C♯ C♯G♯ G♯D♯ D♯A♯ A♯F FC

with the narrow fifths underlined.

Pitches and semitones are given in Table 3.13, starting from C. As above, the first row of the table gives the names of the notes, the second the number of fifths, needed to obtain the corresponding note, the third the intervals measured from C, and the fourth the semitones, whose values are rounded to reduce them to only three classes.

Vallotti's temperament shows a wolf narrowed for a schisma, a rather small amount that was considered negligible by the proponent. This 'inconvenience' was avoided

---

[92] pp. 195–196.

[93] It is possible that a form of temperament close to that of Vallotti was also known by Werckmeister [32], p. 189.

Table 3.13 Notes and semitones for Vallotti's temperament

C	C♯	D	D♯	E	F	F♯	G	G♯	A	A♯	B	C
0	7	2	9	4	11	6	1	8	3	10	5	12
1.000	1.057	1.120	1.189	1.255	1.334	1.409	1.497	1.585	1.677	1.780	1.872	2.000
1.055	1.060	1.060	1.060	1.065	1.055	1.065	1.060	1.060	1.065	1.055	1.065	

by Alessandro Barca, who proposed to narrow the six non-diatomic fifths by 1/6 of a schisma [4].[94] Only a pedantry probably.

A modern form of Vallotti's temperament avoids the wolf most simply by narrowing the diatonic fifths by 1/6 of a Pythagorean comma instead of the syntonic one. Another variation is attributed to Thomas Young (1773-1829) [120]. His cycle of fifths narrowed by 1/6 of a Pythagorean comma differ from Valloti's for a shift of a fifth; it has the form:

CG GD DA AE EB BF♯ F♯C♯ C♯G♯ G♯D♯ D♯A♯ A♯F FC

For this reason, one often speaks of the temperament of Vallotti/Young.

Vallotti/Young's has been until recently one of the most widely known and used temperament to tune harpsichords. Its musical performance and historical relevance are discussed in the literature, showing its advantages [34] and disadvantages [41]. It played a crucial role in the evolution of Italian tuning practices, bridging the gap between meantone temperaments and the more modern equal temperament, and influencing the way musicians approached tuning for a variety of musical styles.

3.2.2.5 The Temperaments of Practitioners

An uncritical reading of treatises from the 16th to 18th centuries can give a misleading impression of contemporary practice. Theorists and mathematicians often presented monochords defined by numbers up to 10 digits, but most did not consider how their results could be useful to practitioners. For this reason, although the problem of temperament was considered important by all musicians, theoretical developments could not be applied by most of them for several reasons: many did not have sufficient mathematical training. Also, without a means of accurately measuring pitches, it was impossible to obtain a temperament that corresponded exactly to the reductions in fifths predicted by mathematics. Finally, the human ear was incapable of distinguishing between temperaments that were not very different. So the results of the theorists, when not ignored, were adapted to practice.

A theorist had the advantage of easily recognizing the main characteristics of musical performance of the temperament he intended to propose; for example, he could specify which intervals remained just and which were more altered. The performing

[94] p. 381.

musician, on the other hand guided only by his ear, when tuning an instrument, could decide to try out a temperament proposed by a theorist or to use his own practical rules.

In the second case, it is the theoretician who has to describe the temperament proposed by the practitioner in mathematical terms, for example by stating the amount of the narrowing of the fifth; he should also be a good musician.

In the first case there are a few possibilities. For example, the practitioner could transfer the mathematical octave division obtained by the theorists to a physical monochord (i.e., a taut string), whose bridge could be moved to the points obtained mathematically, and the sound produced in these positions could be used to tune an organ, for example. This approach was probably used on some occasions, but the movable bridge did not allow for very precise pitches, and in any case the procedure is too cumbersome for a daily tuning.

Performers preferred to tune by ear rather than by mathematics. There were many contemporary descriptions of the tuning procedure and some of them are described in detail in Owen Jorgensen's *Tuning the historical temperaments by ear* [67] of 1977 and in Dominique Devie's *Le Tempérament musical: philosophie, histoire, théorie et pratique* [32], of 1990.

The performer had essentially two ways of helping the ear to tune. The first was based on knowledge of the theoretical expectation of the proposed temperament; e.g. which intervals remained just. After a cycle of a certain number of fifths (or thirds) altered according to personal experience, he could check if there are just intervals. For example, in the $1/4$ of a comma temperament, after a cycle of four narrowed fifths, for instance C-G-D-A-E, he could check if the major third C-E is just or not. If not, he must change the narrowing of the fifth and repeat the process. A detailed description of this approach can be found in Jean Denis' *Trait? de l'accord de l'espinette* of 1643 [93] and Michael Praethorius' *Syntagma musicum* of 1619–1620 [97]. The second approach is to count the beats while tuning in fifths or thirds, considering that each type of tuning is characterized by a certain number of beats per second. This approach, however, could not be followed satisfactorily before the 18th century because of the scarce understanding of the phenomenon of beats; an example of this approach can be found in [67].

3.2.3 *Historical Study Cases*

Below are some examples of octave divisions made by skilled mathematicians who were not always aware of the difference between tuning and temperament.

3.2.3.1 Simon Stevin's *Vande Spiegeling der Singkonst*

The approach of Stevin is intermediate between the Pythagoreans and the Aristoxenians; it is arithmetic in that it defines intervals by ratios like the Pythagoreans, but these ratios are not made up of integers; they are generally irrational, as is natural for numbers expressing the measure of geometric quantities. This position is justified by Stevin's concept of number expressed in his *Arithmetique*:

> Def. 11. Number is what explains the quantity of everything [106].[95]

Underlying Stevin's arguments is the fact that the human ear's perception of consonance has margins of uncertainty. That is, not all people agree that a given interval is consonant or dissonant; and judgment varies not only from person to person, but also from era to era and from country to country. Stevin reported two interesting arguments about the difficulty of perceiving consonance.

The first concerns the interval of the fourth, which in ancient Greece was one of the consonances par excellence, while in Stevin's time it was considered by many to be a dissonance. Stevin had no difficulty in demonstrating the contradictory nature of this position, both theoretically and empirically; he reported that many people, even good musicians, often confused fourths with fifths.

> If anyone considers the fourth a discord, saying that it is unpleasant to his hearing, the contrary and his error are proved to him as follows. Take any two different sounds, such as those of a string and a human voice, or a flute and a string, or a voice and a flute, producing therewith now a fourth, now a fifth, and this several times and also several fourths and fifths higher and lower, each asking his partner what it is that he is singing or playing, then we shall find in practice that, judging thereof without certainty, he will frequently contradict himself, often regarding as a fifth what he previously stated to be a fourth, and again conversely judging that to be a fourth which he had previously stared to be a fifth [106].[96]

Since this is the case in practice, what other words do we need? Who is so unreasonable as to contradict himself by saying, "I don't like the fourth, and I like the fifth a lot"? Stevin asked rhetorically.

To make the example even clearer, Stevin suggested the case of a song for two voices, one playing the flute and the other singing their own part of a given song. If this song is repeated, and the singer goes up one octave, everyone will think it is all right. But where the singer used to be one fifth below the flute player, he will now inevitably be one fourth above him, so if we had a fifth before, we will now hear the fourth as a fifth. As for those who still say that the fourth sounds unpleasant to their ears, but the fifth is very pleasant, Stevin could only conclude that the habit developed over the past century has created a certain bias in them.

Stevin's second argument concerns the difficulty of tuning harpsichords and organs. In Stevin's words: to explain the inexplicable reason why a fifth on the harpsichord cannot be played as well as the natural singing of human voices, it must be understood that not every practical operation can be performed as perfectly as mathematical operations. For example, if a very long piece of cloth is measured by several people, one will find that it is a straw or an inch wider than another, and it rarely happens, and by chance, that all the people arrive at the same result. The same is true of sound, which is the point in question. No sounds can be joined to form

[95] vol. 2, p. 495.

[96] pp. 451, 453.

a fourth, a fifth, a sixth, and so on, so that these intervals are quite perfect, except by chance. But to show this imperfection clearly, place the fret of a lute at the point where you think its string will form a perfect fifth against another string. Then move that fret just the thickness of a hair, up or down, and you will find no appreciable change, although, to be fair, some change has taken place. But then have someone else move the fret so that you don't know if they moved it a hair up or down; or leave it in the same place. If you are asked repeatedly about the goodness of the fifth, you will find that your judgment is uncertain, and you will often say that the good fifth is bad and the bad fifth is good. It is obvious, then, that no human hearing, however acute, is capable of placing two tones with absolute certainty in their perfection. It follows that many of these errors, each of which is imperceptible in itself, in combination produce an appreciable error. For just as in the case of the piece of linen measured by different people, so many small differences in each measuring rod, when added together, make a noticeable difference, so here in the case of tones, if these small imperceptible errors in a fifth are allowed, by repeating the fifth as in the tuning procedure, we eventually arrive at errors that can be considerable. No wonder, then, that masters who carefully tune musical instruments find bad notes that should be good; this is natural, and those who do not understand this do not distinguish between mathematical and practical operations [106].[97]

Stevin's Spiegeling the Singkonst.[98]
Stevin's contribution to music is mainly contained in a short treatise, *Vande spiegeling der singkonst*, written to teach music to prince Mauritius. It was intended as part of the *Hypomenta mathematica*, Stevin's masterpiece. However, it was never published and remained in manuscript, which disappeared soon after his death in 1620, until it was rediscovered and published in 1884 [30].

The manuscript, written around 1605–1608, exists in two different versions. The earlier (version 1) being translated into English and published in Stevin's *Mathematical works* [106],[99] the other version (version 2) is left in Dutch in [30].[100] In the following we will consider version 1, as version 2 has been largely commented on in [27, 28][101] [29]. The manuscript was sent to the organist Abraham Verheyen, who wrote a long letter of strong criticism [30].[102] This, according to [30],[103] would be the reason why Stevin never published his work.

Though little is known of Stevin's musical training, from his quotations it appears that his main reference was Zarlino, whose treatise contains, among other things, relevant hints to ancient Greek music. Another reference was Andreas Papius (1552–1581), for his discussion of consonances, especially the fourth [94].

---

[97] p. 458.

[98] This section is a substantial reworking and expansion of the article [24].

[99] vol. 5, pp. 415–466.

[100] pp. 1–47.

[101] pp. 45–74.

[102] pp. 87–91.

[103] Note 46, p. 265.

The *Vande spiegeling* is organized *more geometrico* with definitions (6), postulates (2) and theorems (1). The first definition introduces the concept of *step*, the name Stevin used for both tone and semitone, as the interval between two successive notes of a musical scale:

> 1st Definition. Step is the next subsequent ascent which one rises in natural singing, of which the smaller variety is called minor step, the larger, major step [106].[104]

The second definition introduces the type of scale that Stevin considered natural in singing; it is defined by the following sequence of major (T) and minor (S) steps T-T-S-T-T-T, which corresponds to the diatonic genus, or in modern terms, a C major scale.

Two definitions (5 and 6) follow that give a name to the intervals in an octave. First, the major and minor steps are redefined as a whole tone and a semitone respectively (no distinction is made between major and minor whole tones), then the intervals between a given note and a reference note are defined in the usual way. For example, a fifth is the interval given by three whole tones and one semitone.

The definitions are followed by two postulates. The first postulate has a physical connotation, and its discussion is the only part of Stevin's treatise in which considerations of the nature of sound are made.

> Postulate. We postulate that as one part of a string is to another, so is the coarseness of the sound of one to that of the other [106].[105]

Note that the postulate does not refer to singing, as has been done before, but to the sound emitted by a vibrating string. Essentially, using modern terms and concepts, the postulate says that the frequency of a sound depends on the length of the string that produces it. In fact, in Stevin's time there was no clear understanding of the concept of frequency of vibration. Stevin spoke of *coarseness*, which is greater for longer strings.

The second postulate states that all tones and semitones in a natural scale are equal.

> 2nd Postulate. All whole tones are equal, and likewise all half tones are equal [106].[106]

The postulate is followed by a very brief explanation and by a table (Table 3.14) with the value and name of the intervals contained in an octave.[107]

---

[104] vol. 5, p. 423.

[105] vol. 5, p. 425.

[106] vol. 5, p. 427.

[107] The way Stevin indicates intervals, $(1 : \sqrt[12]{1/2^n})$, may seem strange to a modern reader who would have written, $(\sqrt[12]{2^n} : 1)$, representing the same ratio. But in Stevin's time, when intervals were measured by string lengths rather than frequencies, the octave was also denoted by the ratio (1 : 2), a ratio less than 1, rather than (2 : 1), a ratio greater than 1. Stevin made a compromise by denoting the octave by $(1 : xx)$, but defining the interval by a ratio greater than 1.

Table 3.14 Division of the octave in equal semitones; from [106], p. 428

As 1 is to — *Theorem.* [2]	So one tone to the other	*Proposition*
1	Selftone	First (prime)
$\sqrt[12]{1/2}$	Semitone	Minor second
$\sqrt[6]{1/2}$	Whole tone	Major second
$\sqrt[4]{1/2}$	One-tone-and-half	Minor third
$\sqrt[3]{1/2}$	Ditone	Major third
$\sqrt[12]{1/32}$	Two-tone-and-half	Fourth
$\sqrt{1/2}$	Tritone	Bad major fourth or bad minor fifth
$\sqrt[12]{1/128}$	Three-tone-and-half	Fifth
$\sqrt[3]{1/4}$	Four-tone	Minor sixth
$\sqrt[4]{1/8}$	Four-tone-and-half	Major sixth
$\sqrt[6]{1/32}$	Five-tone	Minor seventh
$\sqrt[12]{1/2048}$	Five-tone-and-half	Major seventh
1/2	Six-tone	Double-first (octave)

Table 3.14 simply restates the postulate; at least that is the impression of a modern reader. Indeed, the postulate implies that the intervals within a musical scale form a geometric progression, from which the table immediately follows. Assuming that an octave interval is divided into 12 equal semitones, the common ratio of the geometric progression is given by $\sqrt[12]{1/2}$, which is also the ratio of a semitone. A whole tone is given by the ratio $(1 : \sqrt[6]{1/2}) \approx (1 : 0.891)$, slightly smaller than the Pythagorean scale tone: $(9 : 8) \approx (1 : 0.889)$. The interval of fifths, defined by seven semitones, is $(1 : \sqrt[12]{1/128}) \approx (1 : 0.6674)$, a little less than $(3 : 2) \approx (1 : 0.6667)$.

Stevin knew for sure what a geometric progression was, but he probably could not think very easily to find its common ratio. The table of intervals (Table 3.14) is called *vertooch*, which the editor of *Vande spiegeling* translates as theorem. This means that its deduction from the second postulate was probably not as trivial to Stevin as a modern reader might assume. Stevin could have found the tone whose ratio is $(1 : \sqrt[6]{1/2})$ by simply taking one square root followed by one cubic root, this was in his power. But he did not proceed in this way. His starting point seems to be the assumption that a fifth has a ratio of $(1 : \sqrt[12]{1/128})$. Subtracting it from an octave $(2 : 1)$ gives an interval of a fourth, two and a half tones, $(1 : \sqrt[12]{1/32})$, a familiar result for a musician, which subtracted from the fifth gives the tone $(1 : \sqrt[6]{1/2})$, from which the half tone $(1 : \sqrt[12]{1/2})$ is obtained.

The text that follows the postulate is qualified by Stevin as a proof. This is surprising to a modern reader, because to him a postulate cannot be proved, otherwise it would be a theorem. In the 17th and 18th centuries, especially in mixed mathematics, a postulate or principle was simply a starting point for a theory, whether it could be proved or not.

The proof is divided into four sections with the following titles:

1. On ratio in general.
2. Comparison of the geometric ratio with the musical ratio.
3. On the proportions of singable sounds according to the Greeks.
4. On the true proportions of natural tones.

In the first section, Stevin criticized the mathematics of the Greeks, which accepted three different kinds of proportions: arithmetic, geometric, and harmonic. He claimed that this multiplicity was due to a defect in the Greek language (as well as in most modern languages), because it could not allow a clear idea of the concept of proportion. The Dutch language is more precise because its words are composed in such a way that it is easy to understand their meaning [36]. For example, proportion is rendered as equal ratio (everedenheyt), and this makes it possible to exclude arithmetic and harmonic from the list of proportions, leaving only geometric proportions.

In the third section, Stevin claimed that the choice of the value of the intervals of consonances considered by the Greeks is arbitrary and approximate. For example, the fifth is only approximated by the ratio (3 : 2); its correct value is instead $(1 : \sqrt[12]{1/128})$. The error is due to the difficulty of the human ear in detecting the exact ratio: "And though the ancients perceived this fact, they nevertheless took this division to be correct and perfect, preferring to think that the defect was in our singing (as if one were to say: the Sun may lie, but the clock cannot)" [106].[108]

The last section attempted to prove postulate 2 by a simple experiment. It consists of tuning a two-octave keyboard, such as a harpsichord or an organ. In the following, we will simplify Stevin's cumbersome procedure for the sake of space. Refer to the keyboard of Fig. 3.12 with 15 white keys A, B, C, …and 10 black keys for the altered notes. Below the keyboard, Stevin wrote a solmization with a heptachord derived from Guido of Arezzo's hard hexachord, which gives Mi-Fa at the semitone B-C and La-Sa at the semitone EF, with Sa a note a semitone higher than La.[109]

In the tuning procedure, assume key F as the starting note. Then tune a key T_i (A, B, C, …) so that it forms with F an interval of i fifths, with $i = 1, 2, \ldots, 11$, which is maintained within the two-octave compass by lowering this interval by one octave when necessary. The procedure begins by tuning the key eight positions to the right of F to the note f, one octave higher, and one key five positions away, still from F, to the note c, so that Fc is a fifth. The cycle of fifths that follows is given in Table 3.15, where O means an octave and F a fifth.

After tuning all 12 keys with 11 fifths, Stevin performed his experiment by playing the two keys F and T_{11} = H. According to him, he obtained an interval of a fifth:

> This being so, experience shows that H and F make a perfect fifth, and although this is considered a common and certain rule by all those who are skilled in this matter, yet to convince those who should doubt it I thought fit to use the authority of …[106].[110]

---

[108] vol. 5, pp. 431–433.

[109] See Sect. 4.1.2 for a comment on the use of the symbol Sa.

[110] vol. 5, p. 437.

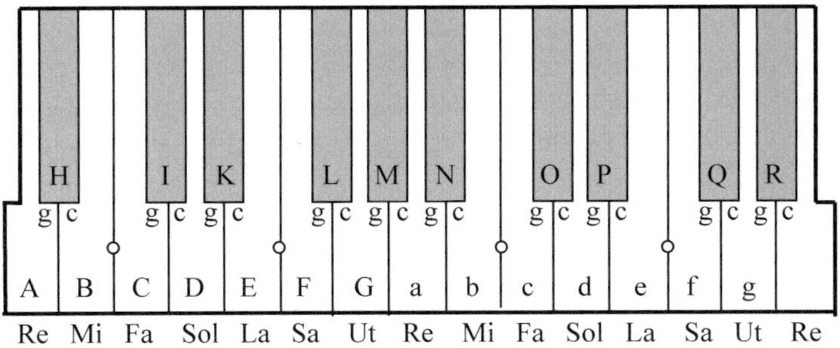

Fig. 3.12 Tuning Stevin's keyboard. Redrawn from [106], vol. 5, p. 436

Table 3.15 Tuning of a harpsichord

1	f	=	F	+	*O*	c	=	F	+	*F*
2	C	=	c	–	*O*	G	=	C	+	*F*
3	g	=	G	+	*O*	d	=	G	+	*F*
4	D	=	d	–	*O*	a	=	D	+	*F*
5	A	=	a	–	*O*	E	=	A	+	*F*
6	e	=	E	+	*O*	b	=	E	+	*F*
7	B	=	b	–	*O*	L	=	B	+	*F*
8	Q	=	L	+	*O*	O	=	L	+	*F*
9	I	=	O	–	*O*	M	=	I	+	*F*
10	R	=	M	+	*O*	P	=	M	+	*F*
11	K	=	P	–	*O*	N	=	K	+	*F*
	H	=	N	–	*O*					

With this result, Stevin was able to prove, by reduction to the absurd, that all semitones, and consequently all tones, are equal.

Before commenting on his procedure, some inconsistencies should be noted. First, Stevin stated here and there in his treatise that the true measure of a fifth is not (3 : 2) but $(1 : \sqrt[12]{1/128})$. One might imagine that from this assumption, which seems a postulate, he could prove the equality of semitones. In fact, he never used this assumption, but rather proved it. Second, it must be said that it is not true that musicians would accept that a cycle of 12 fifths is closed. In fact, assuming the generally accepted value of (3 : 2) for a fifth, 12 fifths, starting from F = 1, gives the interval of the Pythagorean comma $3^{12}2^{-17} = 1.0136$, which is greater than 1. This means that playing $T_{11} = H$ and F gives a wolf fifth shorter than a fifth, highly dissonant.

Table 3.16 compares two monochords. The first column gives the values according to Stevin's division, the second the values obtained with Pythagorean cycle limited to the compass of a fifth. Stevin noted the good approximation for fifths (a differ-

Table 3.16 Pythagorean and equally tempered monochords; from [106], vol. 5, p. 447

10,000	10,000	Selftone	First
9438	9492	Semitone	Minor second
8909	8888	Whole tone	Major second
8404[a]	8437	One-tone-and-half	Minor third
7936	7901	Ditone	Major third
7491	7500	Two-tone-and-half	Good fourth
7071	7023	Tritone	Bad fourth
6674	6666	Three-tone-and-half	Fifth
6298		Four-tone	Minor sixth
5944		Four-tone-and-half	Major sixth
5611		Five-tone	Minor seventh
5296		Five-tone-and-half	Major seventh
5000		Six-tone	Double-first

[a] There is a mistake here, the correct value is 8409 and not 8404

ence of only $6674 - 6666 = 8$ parts) and the relatively unsatisfactory difference for semitones ($9492 - 9438 = 54$ parts).

Now back to the experiment. There is a problem here. In fact, Stevin did not specify how he tuned his keyboard in practice, i.e. how he recognized fifths and octaves. There is no evidence in Stevin's musical writings that his ear was good enough to allow him to tune by ear alone; but he could have gotten help from an experienced musician.

Criticisms of Stevin's Approach.

Stevin's position seems reasonable at first glance. Indeed, there is no reason why consonant intervals should be given by simple ratios between integers. Though the thesis that an octave is made up of 12 equal semitones is fascinating—the same as that implicit in Aristoxenus' geometric approach—the objections that a scholar just after Stevin, or even a contemporary, could make to his position were many:

1. Stevin contradictorily argued that, because of the imperfection of the human ear, experience could not distinguish, for example, that a fifth was defined by the ratio $(3:2)$; but then he believed that the experience of tuning the organ or harpsichord could show that the fifth was indeed $(1:\sqrt[12]{1/128})$.
2. Basically, all musicians considered as consonant only interval expressed by the ratio of integers. In this case, the thesis of the subjectivity of sensations fails because there is agreement among several listeners.
3. By Stevin's time, or shortly thereafter, there were objective theoretical criteria for judging a consonance, and they all tended to favor the Pythagorean thesis. Benedetti, whom Stevin probably did not know, had developed a theory of consonance based on coincidence, the same as Galileo and Beeckman, and the

theory of coincidence explains consonance by the ratio of simple integers, thus rejecting Stevin's thesis.

4. There was also an experimental criterion that could be pursued without any musical skill in tuning musical instruments, and therefore independent of the sensitivity of the human ear. The method is based on the phenomenon of beats. When two notes are played whose frequencies are close to a consonance, a phenomenon occurs in which the amplitude of the sound increases and decreases periodically; this phenomenon is known as beats. Experience with the organ, whose pipes emit sounds that last long enough, suggests that the substantial absence of beats, and thus a better consonance, occurs for simple ratios.

One could certainly refute all these objections, but it would be pointless, since there are no equally stringent arguments to justify Stevin's thesis. If he had an oscilloscope fed with two sinusoidal signals, he would have seen that something extraordinary happens only when the frequencies of the two signals are in simple ratios. Thus, the seemingly implausible Pythagorean thesis that physical phenomena can be described by integers works in this case.

The *Vande spiegeling* is certainly an interesting work and could have provided much stimulus to quantitative music theorists at a time when traditional intonation was in crisis. But it did not. In fact, Stevin's work was read by few and received an unflattering review from them; it was perhaps known only to Beeckman and Huygens, partly because it was written in Dutch. Beeckman, in a letter to Mersenne dated October 1, 1629, wrote that he had first accepted Stevin's ideas and then rejected them because, contrary to his theory of coincidence [14].[111] Huygens, who was the last person to read *Vande spiegeling* before it disappeared, was harshly critical of the theory of irrational consonant intervals: "And those who dared [...] that the fifth does not consist of the ratio (3 : 2), either have not an ear capable of judging, or they believe to have a good reason for it; but they conclude wrongly." [63].[112]

In conclusion, Stevin in fact made no substantial contribution to the history of music, at least until 1884 when his work was published by De Haan (except for a possible indirect influence on Mersenne). Since then, some historians have wanted to see in Stevin a forerunner of the uniform temperament and the first to have provided accurate values for the intervals that define it. This judgment, as extensively documented in [28], is fundamentally incorrect, since Stevin's problem was that of the exact value of consonance, not its best approximation, which was the purpose of a temperament. One also wants to see an influence of Stevin on Schoenberg's dodecaphonic music, but this too may be a forced judgment [33].[113]

---

[111] vol. 4, p. 157.

[112] vol. 20, p. 32.

[113] p. 257.

3.2.3.2 The Ratios of Concords of Nicolaus Mercator

Nicolaus Mercator (c. 1620–1687), not to be confused with the cartographer Gerardus Mercator, was a Danish-born mathematician who spent much of his time in France and England, where he became a fellow of the Royal society in 1666. His most known work in mathematics is the treatise *Logarithmotechnia* of 1668 [88].

There are few relevant musical writings by Mercator: a notebook, possibly from 1654 to 1655, a Latin treatise known as *Theoria musicis*, of unknown date, two short writings in English, from the 1650s and 1660s. All of them are edited by Benjamin Wardhaugh [117], who named the two English writings as *Of musik* (1) and *Of music* (2). In the following, we will focus on *Theoria musicis* and *Of musik* (2), as they contain interesting considerations about the division of the octave.

At least two innovative aspects of Mercator should be emphasized. On the one hand, the massive use of logarithms in operations between intervals, which simplifies their treatment, especially in the case of division. On the other hand, there is a new way of referring to intervals in general and consonances in particular. They are represented by the symbol for fractions. That is, a fifth is represented as $2/3$, instead of $(2 : 3)$;[114] even though, the geometric interpretation of intervals as proportions partially persists. Moreover, pitches are clearly associated with frequencies of vibration. Consonances are thus represented by comparing the frequencies of the two component notes on a time axis, Mercator implicitly assuming the theory of coincidence. For example, for the consonance of the octave, he wrote:

> Suppose now two soundes [Fig. 3.13], the lower sound striking the aire three times at the points of time c, f, h and the higher sound in the same space of time striking six times at the points g, h, i, k, l, d. Thus you see these two sounds would be in a double frequency to one another, and therefore make that concord which we call an eight [84].[115]

Consonant and non-Consonant Intervals.

Among the intervals introduced by Mercator, the first and most important are the superparticular consonances: the octave $1/2$, the fifth $2/3$, the fourth $3/4$, the major third $4/5$, and the minor third $5/6$. This leaves out the sixths, which are treated as complementary intervals, implicitly considered consonances as well. The major sixth $3/5$ is complementary to the minor third, and the minor sixth $5/8$ is complementary to the major third.

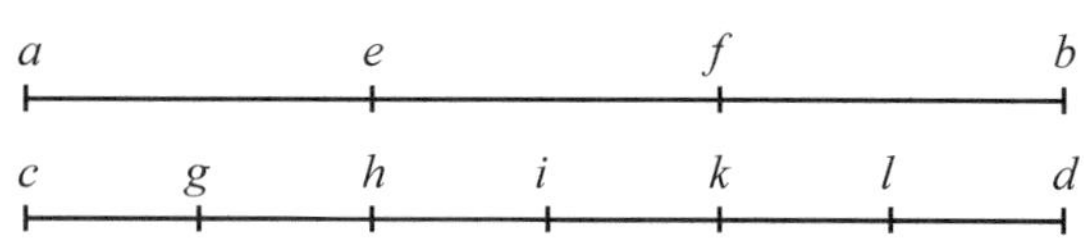

Fig. 3.13 Coincidence of the octave

[114] We use the same convention as Mercator to indicate intervals, with the denominator greater than the numerator.

[115] p. 194.

Smaller intervals, called *concinnous intervals*, are formed by the difference of consonances: The major tone, one fifth less one fourth 8/9; the minor tone, a fourth less a minor third 9/10; the major semitone, a fourth less a major third 15/16; the minor semitone, a fourth less a major third 24/25.

Other intervals are obtained by subtracting the concinnous: the *comma* (80/81), a major tone less a minor tone; the *middle semitone* (128/135), a major tone less a major semitone; the *maximum semitone* (25/27), a major tone less a minor semitone; the *enharmonic diesis* (125/128), a major semitone less a minor semitone.

Discordant intervals are treated separately; they are the concinnous and their complements, including the major seventh, the major semitone's complement, 8/15, the middle seventh, the minor tone's complement, 5/9, and the minor seventh, the major tone's complement, 9/16.

Other dissonant intervals are the *ditonus veterum* (64/81), two major tones, exceeding the modern ditone 4/5; the *pseudo diatessaron* (25/32), a minor third plus a major semitone; the *tritone* (32/45), sum of a minor and two major tones; the *semidiapente* (45/64), complement of the tritone; the *limma* (243/256), a fourth less the ditonus veterum [84].[116]

The Division of the Octave.

Of certain interest is the division of the octave into 53 parts proposed in the *Theoria musicis*. Mercator did not specify whether he thought that this division could find an application for a musical instrument; or whether he saw it as a convenient way in theory to compare intervals with each other. In particular, Mercator did not refer to his division as a form of temperament, as it was considered in [21], and judged a positive temperament in [9].[117]

The starting point of the division of the octave is the observation that the logarithms of the intervals of the just fifth (0.17609) and the octave (0.30103) are very close to the ratio of 31 to 53. This means that by dividing the octave into 53 parts, the fifth is defined by 31 parts, with a really minimal deviation.

Mercator called the *comma* the unit of division of the octave, because it has a value very close to that of both the syntonic and Pythagorean comma (0.00568 for the logarithm of the artificial Mercator comma, versus 0.00539 for the syntonic comma, and 0.005889 for the Pythagorean comma, i.e., a difference of less than 1/50 of a syntonic comma from each): "I consider it not a pure comma, but 1/53 part of an octave" [87].[118]

There is another way to look at the 53 part division. Mercator's goal was to get an almost exact fifth in the division of the octave into m parts; so he could have formulated the problem as: find m, n so that $3/2 \approx 2^{n/m}$, or equivalently solve the equation:

[116] pp. 152–161.

[117] Part I, p. 12.

[118] p. 159. According to Mercator, an octave contains slightly less than 55 syntonic comma [85], p. 226 (he made however a small mistake, as it actually contains almost 56).

Table 3.17 Intervals measured by commas; from [87], p. 133

Comma	1	Sesquitonus	14
Diesis	2	Ditonus	17
Semitonium minus	3	Diatessaron	22
Semitonium medium	4	Tritonus	26
Semitonium majus	5	Semidiapente	27
Semitonium maximum	6	Diapente	31
Tonus minor	8	Hexachordon minus	36
Tonus major	9	Hexachordon majus	39
		Diapason	53

$$\left(\frac{3}{2}\right)^m \approx 2^n, \qquad \text{or} \qquad 3^m 2^{-p} \approx 1, \quad p = m + n$$

This equation has two relevant solutions, $m = 12,\ n = 7,\ p = 19$, which corresponds to the equal temperament, and $m = 53,\ n = 31,\ p = 84$, which is the 53 part Mercator division.

Some historians see a possible clue in the adoption of the 1/53 octave comma in Boethius' writings [9].[119] In fact, in Chap. 8 of Book III of the *De institutione musica*, Boethius spoke of intervals smaller than the tone. Among these is precisely the comma, defined as the difference between the major tone and two 'diesis' [17].[120] It is not difficult to show that Boethius' comma is very close to the 9th part of the tone and the 53rd part of the octave. Of course, Boethius did not actually thought of dividing the octave into 53 comma.

In Table 3.17, the values of the intervals are expressed as multiples of the comma [87].[121] In Table 3.18 are reported the comparison between the logarithms of intervals obtained as multiple of a comma and that of the just intonation; the agreement is good.

In *Of music* (2) Mercator presented a way to find the monochord for 1/4 of a comma temperament, based on iterations of narrowed fifths. Using logarithms, the intervals of the monochord are obtained by a descending cycle of 11 tempered fifths, following an arithmetic progression. Below the way followed by Mercator to solve the problem [87][122]:

[119] pp. 123–124.

[120] Book III, Chap. 8, pp. 277–278. Notice that Boethius called diesis what is commonly called limma (256:243), and his comma is exactly the Pythagorean or ditonic comma, that is the falling to the closure of the circle of 12 fifths.

[121] pp. 133–134.

[122] pp. 228–229.

Table 3.18 Comparison of 53 parts division and just intervals

		Mercator	Just
Comma	1	0.00568	0.00539
Diesis	2	0.01136	0.01030
Minor third	14	0.07952	0.07918
Major third	17	0.09656	0.09691
Fourth	22	0.12496	0.12494
Fifth	31	0.17608	0.17609
Minor sixth	36	0.20338	0.20412
Major sixth	39	0.22152	0.22185

Table 3.19 A descending cycle of logarithms of fifths; from [86], p. 229

0	0.0000000	1000 ×64	64,000	(7)	G♯
1	0.1747425	1495×32	47,840	(2)	C♯
2	0.3494850	2236×32	71,552	(8)	F♯
3	0.5242275	3344×16	53,504	(4)	B
4	0.6989700	5000×16	80,000	(10)	E
5	0.5737125	7477×8	59,816	(6)	A
6	1.0484550	11,180×4	44,720	(1)	D
7	1.2231975	16,720×4	66,880	(7)	G
8	1.3979400	25,000×2	50,000	(3)	C
9	1.5726825	37,390×2	74,780	(9)	F
10	1.7474250	55,900×1	55,900	(5)	B♭
11	1.9221675	83,600×1	83,600	(11)	E♭

The way to divide the chord into tempered fifths.

the terms expressing the fifth	the logarithm	
2	0.3013000	
3	0.4771212	
	0.1760912	the measure of a perfect fifth
the terms expressing a comma		
80	1.9039000	
81	1.9084850	
	0.0053950	the measure of a comma

a fourth of which 0.0013487 is the measure of 1/4 of a comma.

from the perfect fifth	0.1760912
subtract 1/4of a comma	0.0013487
the tempered fifth	0.1747425

Table 3.19 reports the results of the cycle.

The first column indicates the number of fifths that have been stacked; the second the logarithm of such a stack. The third column gives the approximate corresponding string length, together with an indication of the power of two by which it must be multiplied to bring it into the desired octave range (between the two values 40,000 and 80,000); the fourth column shows the result of that multiplication. The fifth column reports the number of orders of the pitches in decreasing order of height, named in the final columns: D, C♯, C, B, B♭, A, G♯, G, F♯, F, E, E♭.

3.2.3.3 Christiaan Huygens's *Cycle Harmonique*

Christiaan Huygens was the most sophisticated in musical matters of the Dutch mathematicians. He learnt the theory and the practice of music from his father, Constantijn. He published only one short treatise on music, a letter to the publisher Henri Basnage de Beauval (1656–1710) in October 1691, intended for publication in the *Histoire des ouvrages des sçavants*, the *Lettre touchant le cycle harmonique*. This work is now known as the *Novus cyclus harmonicus* after its translation from French into Latin in 1724 [63].[123] In spite of his musical ability, only one simple attempt at musical composition by him has come down to us [98].[124]

There are, however, some manuscripts published in the *Oeuvres complètes de Christiaan Huygens* in volume 20, which, together with the editorial notes, cover the space of about 150 pages. They are divided into five sections:

1. Theory of consonance, division of the monochord.
2. Observations on ancient and modern singing.
3. Notes on writings on ancient music.
4. Notes on writings on modern music.

While Mersenne and Stevin had lived in a transitional period from the Renaissance to the Baroque, Huygens was influenced by the latter, in which there was a greater richness of musical themes.

Huygens' approach to music is essentially that of a mathematician; there are very few physical considerations. Even the theory of sound is mentioned only in connection with light transmission.

He fully embraced the Pythagorean approach, for which consonances are expressed by ratios of integers.

> It is a fact of experience, and those who have an ear for music cannot deny it, that consonances that follow the correct proportions are better than those that occur when we depart from these just numerical proportions. And those who dare to argue the contrary, that the fifth is not in the ratio of 3 to 2, either have no ear capable of judging, or think they have a reason. But they conclude wrongly [63].[125]

In what follows we limit to commenting with some extension only on his published work both for reasons of space and because it is the one that surely had the most influence on later developments in speculative music. Of note, however, that unpublished

---

[123] vol. 20, p. 19.

[124] p. 202.

[125] vol. 20, p. 32.

writings are very relevant; for instance Huygens' remarks on the interval of seventh, which he believed should have some space among consonances, even though all past and contemporary musicians were of the opposite opinion [27].[126] To support his thesis, although he was unable to make an experimental verification he argued that in addition to the five higher harmonic sounds identified by Mersenne there were also the sixth and the seventh harmonics. Huygens' considerations on the subject unfortunately remained unknown until the publication of his *Oeuvres complètes* in 1940.

Huygens thoroughly disliked temperament, with a couple of caveats; although he shared some of the motives of his contemporary Andreas Werckmeister (1645–1706) for the need of a temperament, he condemned him as an author "without learning and of little value" [63],[127] and clearly did not share his solution, which was to distribute the Pythagorean comma over a number of selected fifths (a good temperament, see below). For him, equal and good temperaments, while apparently mathematically attractive, must be rejected on the basis of sense experience. For Huygens, a temperament was nothing but an empirical achievement for which no reason or mathematical proof could be found. For a scientist like him, no solution was acceptable that did not in some way combine experience and reason [27].[128]

In the *Novus cyclus harmonicus* Huygens regarded the 1/4 of a comma mean-tone temperament as the best compromise that could be made with respect to just intonation—it is the "veritable temperament" whose priority was disputed by Zarlino and de Salinas—and sought a rational derivation of it. This was achieved by dividing the octave into 31 equal parts. Huygens opposed the opinion of Mersenne and de Salinas, who spoke ill of their division. He contended that the criticism directed towards their division stemmed from the fact that it lacked strictness [63].[129] He denied any derivation from Zarlino or de Salinas, and did not mention that a hint had already been found in the writings of Nicola Vicentino (1511–1572) [3, 113, 118] and Marchetto da Padova (fl. 1300) [63].[130] But, above all, he did not mention Lemme Rossi (1601–1673), who in his important *Sistema numerico* of 1666 explicitly noted the almost perfect correspondence between the monochord based on the division of the octave into 31 equal parts and that based on cycles of fifths narrowed by 1/4 of a comma [100].[131] However, it is worth noting that although Huygens did not mention Rossi in the *Novus cyclus harmonicus* of 1691, he had already achieved his result without publishing it in 1661 [63].[132]

The calculation of the intervals of the octave is made quite simple by using the recently available tables of base 10 logarithms. Huygens called the 31st part of the

---

[126] pp. 212–214.

[127] vol. 20, p. 133.

[128] p. 224.

[129] vol. 20, p. 142.

[130] vol. 20, p. 144.

[131] pp. 86–88. For further reading on the spread of the division of the octave into 31 parts see [4], pp. 299–303.

[132] vol. 20, p. 141.

octave a diesis and calculates its value d by dividing the logarithm of 2, the octave interval, by 31 [63][133]:

$$d = \frac{\log 2}{31} = 0.0097106450$$

The diesis considered in just intonation is usually defined by the proportion $^{128}/_{125}$, which corresponds to a logarithm of 0.0102999566, a difference of 5% with respect to Huygens' diesis. To construct a table of the interval resulting from the division of the octave into 31 parts, he assumed a monochord of length $l = 100,000$. The octave is at $^1/_2 l = 50,000$; the 31 parts p_i are obtained with the relation:

$$p_i = \log 50,000 + id$$

with $i = 1, 2, \ldots, 31$.

The results are reported in Col. 1 of Table 3.20.[134] Col. 2 reports the length of the monochord, Col. 5 the length of the monochord corresponding to the meantone $^1/_4$ of a comma temperament, and Col. 6 the corresponding logarithms. Cols. 3 and 4 list the names of the solmization notes and Greek names, respectively. Notes countersigned by the symbol for sharp, ♯, are one diesis higher than the corresponding note not countersigned; notes countersigned by the symbol for flat, ♭, are similarly one diesis lower than the corresponding note not countersigned. Huygens provided the following 12 notes, commonly referred to as the twelve-tone chromatic scale:

C C♯ D E♭ E F F♯ G G♯ A B♭ B C
Ut Ut♯ Re Ma Mi Fa Fa♯ Sol Sol♯ La Sa Si Ut

The solmization adopted in the table is based on the normal Guido of Arezzo's hexachord, to which the syllable Si is added, indicating its early adoption. In addition, the syllables Ut, Re, Mi, etc. are used not only to locate semitones, but also to mean pitches. Note also the use of the syllable Sa to denote Si♭, as Stevin did, albeit with a different solmization (see Fig. 3.12), and an analogous syllable Ma for Mi♭.

Huygens discussed the difference between the ratios of the consonances of his division and those of the meantone temperament. For example, take the fifth C-G of the meantone temperament. From Col. 6 of the table he obtained a distance in the logarithmic scale equal to $5.0000000000 - 4.8252574989 = 0.1747425011$, which is the logarithm of his tempered fifth. The logarithm corresponding to the just fifth is 0.1760912590, with the difference of 0.0013487583, which is exactly $^1/_4$ of the syntonic comma ($\log_{10} 81/80 = 0.005395031$). The fifth measured from the numbers in Col. 1, corresponding to the Huygens division, is given by $5 - 4.8252083893 = 0.1747916100$. It differs from the meantone fifth by

[133] p. 147.

[134] Our table is slightly different from the one published by Huygens. Some minor errors have been corrected, the symbol ♯ is used instead of *, B♯ has become simply B, and B has become B♭, following modern usage.

Table 3.20 Comparison of the division with 31 diesis versus 1/4 of a comma temperament; from [63], vol. 10, p. 173

31 diesis				1/4 of a comma	
1	2	3	4	5	6
N 97106450					
4.6989700043	50,000	Ut	C	50,000	4.6989700043
4.7086806493	51,131				
4.7183912943	52,278				
4.7281019393	53,469	Si	B	53,499	4.7283474859
4.7378125843	54,678				
4.7475232293	55,914	Sa	B♭	55,902	4.7474250108
4.7571338743	57,179			57,243	4.7577249574
4.7669445193	58,471				
4.7766551643	59,794	La	A	59,814	4.7768024824
4.7863658093	61,146				
4.7960764543	62,528			62,500	4.7958800173
4.8057870993	63,942	Sol♯	G♯	64,000	4.8061799740
4.8154977443	65,388				
4.8252083893	66,866	Sol	G	66,874	4.8252574989
4.8349190343	68,378				
4.8446296793	69,924				
4.8543403243	71,506	Fa♯	F♯	71,554	4.8546349804
4.8640509693	73,122				
4.8737616143	74,776	Fa	F	74,767	4.8737125054
4.8834722593	76,467				
4.8931829043	78,196				
4.9028935493	79,964	Mi	E	80,000	4.9030899870
4.9126041943	81,772				
4.9223148393	83,621	Ma	E♭	83,592	4.9221675119
4.9320254843	85,512			85,599	4.9324674685
4.9417361293	87,445				
4.9514467743	89,422	Re	D	89,443	4.9515449935
4.9611574193	91,444				
4.9708680643	93,512			93,459	4.9706225184
4.9805787093	95,627	Ut♯	C♯	95,702	4.9809224750
4.9902893543	97,789				
4.9999999993	100,000	Ut	C	100,000	5.0000000000

$0.1747425011 - 0.1747916100 \approx 0.00005$, which is less than the 1/100 of a comma [63].[135]

In the letter of 1691 Huygens considered only the difference on the fifth; the motivation of his choice is based on the need to show that his fifth is 1/4 of a comma less than the just fifth, to validate the division of the octave into 31 diesis.

Huygens did not say how he came to choose 31 as the number to divide the octave. He could have set as an objective to get a just major third when dividing the octave into m parts (because in the 1/4 of a comma temperament the major third is not tempered); so he could have formulated the problem as: find m, n so that $5/4 \approx 2^{n/m}$, or equivalently, solve the equation:

$$\left(\frac{5}{4}\right)^m \approx 2^n, \qquad \text{or} \qquad 5^m 2^{-p} \approx 1, \quad p = 2m + n$$

This equation has two relevant solutions, $m = 28,\ n = 9,\ p = 65$ and $m = 31,\ n = 10,\ p = 72$ [37].[136]

Migration of Pitches.

Huygens proposed a temperament because he was well aware of the disadvantages of just intonation. He reported in detail the phenomenon of pitch migration, already pointed out by Benedetti in Sect. 2.2.3. Below are some single voice excerpts showing how, starting from a note and playing a melody by going up and down by just intervals, when one would expect to return to the starting note, it arrives either higher or lower, by a not negligible interval.

> Moreover, even if the voice sings the intervals of fifths and fourths very precisely, it will necessarily deviate from the true tone. For example, when we sing V S R L M V, this M is a whole comma higher than it should be to make the third MV just, assuming that we sing only the fifth VS, the fourth SR, the 5th RL and the 4th LM, so it will happen that to make the third in tune we will sing the last V a comma higher than the first, or else the voice must not sing the just intervals of the consonances. And if we continue to sing by the same intervals, we will raise another comma, and so on and so forth, so that the voices that are not regulated by some instrument will necessarily go astray and will be some false intervals and chords, and this will be seen by experience [63].[137]

To understand the passage, one must remember Huygens' symbology of notes,[138] summarized in the table below:

[135] vol. 10. p. 174.

[136] p. 54.

[137] vol. 20, p. 77.

[138] Huygens' letters stand for the initials of the solmization syllables of the natural Guido's hexachord, except for C, which stands for B-Si. Possibly C is for Ci, one of the names that circulated for the syllable that later became Si, used later by Huygens himself. The use of the initial letters of the normal hexachord to indicate pitches can be found since the beginning of the 17th century.

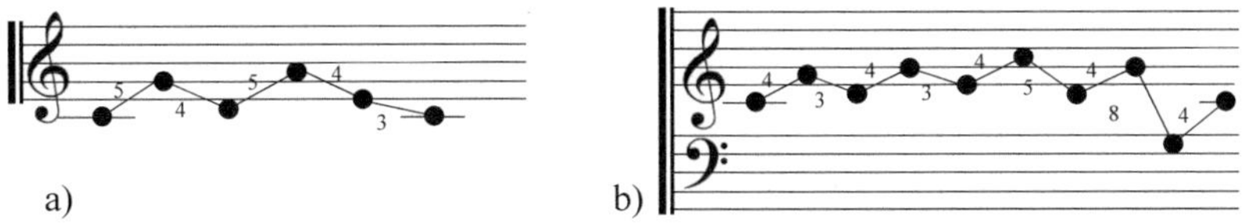

Fig. 3.14 Migration of pitches. **a** Up 1 comma, **b** Down 2 commas

Standard names	A	B	C	D	E	F	G
	La		Vt	Re	Mi	Fa	Sol
Huygens' names	L	C	V	R	M	F	S

So the note sequences V S R L M V can be rewritten as C G D A E C, that is, an ascending fifth, a descending fourth, another ascending fifth, a descending fourth, and a descending major third (Fig. 3.14a). By identifying ratios with fractions, the ascending intervals are $3/2 \times 3/2 = 9/4$, the descending intervals are $4/3 \times 4/3 \times 5/4 = 80/36 = 20/9$. The difference between the ascending and descending intervals is $9/4 : 20/9 = 81/80$, i.e. the last C is one comma higher than the first.

Pitch migration can also occur in the opposite direction, i.e. after a cycle of some just intervals, one or more commas can be found below the starting point. This happens, for example, when the notes rise for fourths and fall for thirds. For example, consider the sequence proposed by Huygens, using standard symbols: C F D G E A D G G C (see Fig. 3.14b). The ascending fourths are CF, DG, EA, DG, GC. The descending (minor) thirds are FD, GE; there is also a descending fifth AD and a descending octave GG. The ascending intervals are given by $(4/3)^5$, for a total of $1024/243$, the descending intervals are $(6/5)^2 \times 3/2 \times 2$, for a total of $108/25$. The difference between the descending and ascending intervals is $1024/243 : 108/25 = (81/80)^2$, or two commas, the value by which the note C descends with respect to the original.

In the following quotation, Huygens again considered a simple example of pitch migration. And gave his idea because this apparently critical drawback of the just system of intonation is not so critical and can be easily and instinctively eliminated by singers who actually perform micro-temperament. Of course, for fretted instruments, the temperament should be transferred to the hardware.

> If someone were to sing the notes that musicians call C, F, D, G, C, raising and lowering the pitch of the voice and keeping the intervals absolutely just, the last C would be lower than the first C from which one began to sing; this difference is called a comma. From the just ratios of these intervals, which are (4 : 3), (5 : 6), (4 : 3), and (2 : 3), we get a ratio of (160 : 162), or (80 : 81), which is exactly the comma. Therefore, if the same chant is repeated nine times, the voice will end up lowering closer to a major tone, which is a ratio of (8 : 9). This is not at all pleasing to the ear, which remembers the initial tone and returns to it. We are thus forced to use a hidden temperament and to sing the intervals imperfectly,

which is actually much less offensive. Singing is almost always in need of such corrections, as can be clearly seen by analyzing the proportions in the way we have explained here. We have therefore endeavored to explain these things for the benefit of those who are interested in these arts and who are not without some foundation in geometry [62].[139]

3.3 The Harmony of the Spheres According to the Mathematicians

In the Renaissance, with the rediscovery of Neoplatonism, the harmony of the spheres again became the subject of treatment also by philosophers, musicians and mathematicians. This phenomenon, though in an increasingly less obvious form, will continue at least throughout the 17th century [55, 57].

The approach is multifaceted, ranging from that of Christian philosophers, in which Neoplatonism is combined with religious elements and the theory of the spheres is imbued with eschatological and hermetic functions in which God's direct intervention is envisaged, to that of mathematicians, who, while retaining the numerology of Pythagoras and thus accepting a basis that is not purely rational, seek to interpret the empirical elements of the world with the harmony found in vocal and instrumental music.

Among the best known and most studied representatives of the first category are Marsilio Ficino (1433–1499), Francesco Giorgi (1466–1540) and Robert Fludd (1574–1637); among those of the second, Gioseffo Zarlino (1517–1590, Johannes Kepler (1571–1630), Marin Mersenne (1588–1648) and Athanasius Kircher (1602–1680), ending with Jean Philippe Rameau (1683–1764) and Giuseppe Tartini (1692–1770).

A postulate of ancient speculative music is that harmonics is ontologically prior to material existence. An interesting and extreme example of this ontology is the idea at the basis of some religions, for which the priority of harmonics is a priority of sound in time [56].[140] Instead of saying "let there be light",[141] they say "let there be sound".

> In the beginning there was nothing but absolute silence in which the first thoughts, inaudible rhythms, arose. Such rhythms were the model for the coming concrete creation [56].[142]

If the harmony of the spheres was seen by Christian philosophers as a manifestation of God, many mathematicians thought that, God or not, the universe must have regularities, and they believed that they could grasp these regularities in ordinary music, where they were easily perceived, in the form of consonances.

We reproduce below a quotation from Giuseppe Tartini, which, although it comes from a musician of a time when the harmony of the spheres was no longer in vogue,

---

139 p. 755.

140 p. 58.

141 Genesis 1:3.

142 p. 58.

expresses very well the idea of the role of vocal or instrumental music in universal harmony.

> This means that the harmony of the universe is the whole tree: music is only a branch, but necessarily of the same nature and root. Proof of this is the music inherent in the human species, which alone is capable of the science of number. In this sense there is in numbers science and nature as divinely understood by Plato; and in this sense we see the possibility of discovering the tree from the branch, the whole from the part [108].[143]

To a modern reader, the idea of universal harmony may seem a bit bizarre, but if he thinks about it, the approach of 21st century scientists is not much different. They too assume that there are harmonies in the universe, but they look for them in more abstract rules, in the form of invariants, symmetries of the mathematical laws that describe the universe. This approach, in which the ontology behind the regularities is barely visible, will probably seem no less curious to scientists of the future than harmony based on musical consonances seems to us today.

In the following, we will only mention the concepts of the harmony of the spheres of Johannes Kepler and Marin Mersenne, two mathematicians who played an important role in the speculative music of the 17th century.

3.3.1 Johannes Kepler's Rational View

Johannes Kepler (1571–1630) authored important works on speculative music and on the harmony of the spheres. His results were summarized in the *Harmonices mundi libri V* (herein after *Harmonices*) of 1619 [69], although he had begun his studies many years earlier, as evidenced by his first printed book, the *Mysterium cosmographycum* (herein after *Mysterium*) of 1596 [70], and by numerous letters.

From an early age, Kepler was immersed in the musical traditions of Württemberg Protestantism, both in an educational and a religious context. From his fifth year onwards, he engaged in the practice of German psalmody, in addition to the Latin sequences and hymns that he later referenced in the *Harmonices mundi*. This daily singing was complemented by weekly lessons in theory. The standard of musical cultivation and theoretical instruction in Württemberg was notably high, encompassing contrapuntal music. Kepler's solid theoretical knowledge and practical skill derived from his primary schooling continued and deepened during his theological studies in Tübingen (1589–1594). Kepler also participated in performances of church music and private festivities [35]; thus he was not only a scholar of speculative music, but also an amateur musician, and had been surrounded by musical performances all his life, participating personally on many occasions [95].

Before Hans Georg Herwart von Hohenburg (1553–1622) gave him a copy of Ptolemy's *Harmonica* in 1607, the only source on ancient music that Kepler had studied was probably Boethius's *De institutione musica*. From the *Harmonica* Kepler was able to delve into other Greek musical sources, such as the texts of Aristoxenus

[143] p. 4.

and Euclid, the musical section of the Aristotelian *Problemata*. In addition to these texts, he was familiar with others, one of the most interesting being Porphyry's commentary on Ptolemy's *Harmonica*, writings attributed to Cleonides, the work of Jacques Le Febvre d'Étaples (1455–1536). As for contemporary authors, in addition to Vincenzo Galilei, Kepler was certainly familiar with *L'arte del contraponto* by Giovanni Maria Artusi (1540–1613), the works of Sethus Calvisius (1556–1615), with whom he had a long correspondence, and the *Monochordum* (1604) by Andreas Reinhard.

If we then consider what Kepler wrote in the pages of Book III of the *Harmonices*, we notice that the references to authors of music, are very few: the names of Vincenzo Galilei and Ptolemy stand out above all others (the former mentioned 13 times and the latter 9 times), followed by Aristotle (6) and Plato (3); all the others—Aristoxenus, Calvisius, Euclid, Porphyry, Proclus—are mentioned only once or twice. There is no mention of important authors such as Boethius, Glarean, Gaffurio, or de Salinas; Zarlino is mentioned only once in the entire work [111].[144]

As far as we know, Kepler was not aware of Benedetti's ideas; however some passages in the *Harmonices* suggest that he was aware of the relationship between the frequency of vibrations and the pitch of sound. Indeed, in Chap. I of Book III, Kepler discussed the "marvelous experience" of sympathetic resonance, already illustrated decades earlier by Gerolamo Fracastoro (1476/1478–1553) [92].[145]

3.3.1.1 The *Mysterium Cosmographycum*

The *Mysterium*, written by a young Kepler, before he had conjectured the elliptical orbits of the planets, has an astronomical imprint decisively framed in the Copernican vision of the universe, at a time when this was considered a dangerous belief by both the Catholic and Reformed churches. The goal was to justify the values of the distances of the planets from the Sun and their ratios.

Kepler speculated that the orbits of the six planets known at the time—Mercury, Venus, Earth, Mars, Jupiter, and Saturn—could be arranged in spheres nested around the five Platonic solids: octahedron, icosahedron, dodecahedron, tetrahedron, and cube. He guessed that these polyhedra arranged in circumscribed spheres for a given polyhedron and inscribed spheres for the next polyhedron gave a fair approximation for the relative sizes of planetary orbits around the Sun. There has long been speculation about the role of this work in the history of science, moving from positions that see it as lacking any scientific foundation, an exercise in mysticism, to assessments, prevalent today, of its complexity.

Leaving aside the theological and metaphysical implications, Kepler's approach can be interpreted as a method of regularizing experimental data (the distances of the planets from the Sun). This is the geometric counterpart of the modern best-fit criteria for lines, parabolas, and circles. And the values of planetary distances obtained

---

[144] pp. 82–83.
[145] p. 255.

with the Platonic polyhedra could be considered more reliable than the experimental results, which are inevitably affected by measurement errors. The agreement between the Platonic solids and the orbits of the planets could also be used to support the Copernican thesis. Today, however, with the discovery of additional planets, Kepler's approach would no longer be possible.

Certainly, Kepler regarded his approach as not merely a regularization procedure but also as an explanation that posits God as the efficient cause, orchestrating the arrangement of the planets in accordance with a (potentially unique) harmonious order, which represents the formal cause.

To give an idea of the union between reasoning based on indubitable statements, which also take into account empirical reality, and those that require the acceptance of metaphysical and religious dogmas, we quote below two short excerpts, the first at the beginning of the treatise, the second toward the end.

> Bear with me now, patient reader, if I trifle for a moment with a serious subject, and indulge in allegories a little. For I think that from the love of God for man a great many of the causes of the features in the universe can be deduced. Certainly at least nobody will deny that in fitting out the dwelling place of the universe God considered its future inhabitant again and again. For the end both of the universe and of the whole creation is man [70].[146]

> Now, friendly reader, do not forget the end of all this, which is the conception, admiration and veneration of the most wise Maker. For it is nothing to have progressed from the eyes to the mind, from sight to contemplation, from the visible motion to the Creator's most profound plan, if you are willing to rest there, and do not soar in a single bound and with complete dedication of spirit to knowledge, love, and worship of the Creator [70].[147]

Kepler did not introduce concepts of musical harmony until Chap. 12, where he discussed the division of the Zodiac into 12 parts. Here he found an initial difficulty, since according to the speculative music of his time there were 7 consonances, while the polyhedra defining the positions of the planets were only 5. He got away with it by arguing that the two sixths, as well as the two thirds, major and minor, were not very different from each other [68].[148]

3.3.1.2 *Harmonices Mundi*

Harmonices mundi was a product of the late Renaissance, an age in which Neoplatonism, hermetics, cabalism, alchemy and magic were appreciated pursuits. German court was a centre where practitioners of the occult sciences met, and they no doubt influenced Kepler's investigations. The treatise was widely read; the section on music must have been studied by the young Isaac Newton, who showed signs of familiarity with it [66].

The Latin text of the *Harmonices mundi* is available, apart from its original print [69], as volume 6 of *Johannes Kepler's Gesammelte Werke* [73]. There is a complete

[146] p. 29. Translation into English in [75].

[147] p. 88. Translation into English in [75].

[148] p. 42.

English translation [76] and a partial translation (Book V) [74]. A basic commentary is given in [104].

The treatise is organized into five books as follows:

1. Introductory, on the origin and constructions of the regular polygons which establish the harmonic proportions.
2. Architectonic, on the congruence of the regular polygons in the plane or solid.
3. Harmonic, on the origin of the harmonic proportions in the polygons, and on the nature and distinguishing features of matters relating to music.
4. Metaphysical, psychological, and astrological, on the mental essence of the harmonies and on the types of them in the world.
5. Astronomical and metaphysical, on the most perfect harmonies of the celestial motions, and the origin of the eccentricities in the harmonic proportions [69].[149]

The five books are followed by an appendix containing a comparison of Kepler's Book III with Book III of the *Harmonica* of Claudius Ptolemy and with the harmonic speculations of "Robert of the floods, called Fludd". The latter severely criticized the *Harmonices mundi* soon after its publication in his *Veritatis proscenium* (1621), claiming that the hermetic or chemical approach is deeper than the mathematical.

Kepler knew about Ptolemy just before writing his treatise, and remained very surprised to find large similarity between Ptolemy's and his work.

> In addition during this intermediate period, in which the extremely laborious restoration of the motions held me in suspense, my appetite was particularly intensified and my purpose stimulated by the reading of the Harmony of Ptolemy, which was sent over to me in manuscript by that great man, born for the advancement of philosophy and indeed of every kind of scholarship, Johannes Georg Herwart, Chancellor of Bavaria. There I found unexpectedly, and to my great wonder, that almost the whole of his third book was given up to the same study of the celestial harmony, one thousand five hundred years before [69].[150]

The first two books of the *Harmonices* are geometrical in nature. Book I deals with regular polygons and their derivations obtained by prolonging their sides, which are called stars.

Book II discusses the congruence of regular polygons, that is, their ability to tessellate the plane, covering it through repetitions without overlapping. Book III is in fact a treatise on speculative music. Books IV and V deal with the harmony of the spheres; specifically, using Boethius' categories, Book IV deals with human music and astrology, Book V with worldly music. In the following, we will only give some details of Book III and Book V.

Book III. A Treatise on Speculative Music

Book III begins with a critique of Pythagoras' numerological approach, which is considered inadequate on both the empirical and metaphysical levels. On the empirical level, because thirds and sixths are not considered consonances, which they are;

---

[149] Frontspice. Translation into English in [76].

[150] p. 178.

on the metaphysical level, for several reasons. Meanwhile, it is not clear why consonances should be produced only by the numbers 1, 2, 3, 4 and not also by 5, 6, 7, etc. Then number cannot be a cause, but only an ordering principle on which natural things are based. Kepler took up Aristotle's criticism of Pythagoreanism and argued that number is neither an efficient, nor a material cause, nor a formal cause, nor a final cause [69].[151]

For these reasons, Kepler declared that he had set himself the task of illuminating this part of mathematics and physics by discovering causes which, on the one hand, would satisfy the judgment of the ears in determining the number of consonances and the other melodic intervals, without going beyond what the ears can bear, but which, on the other hand, would establish a clear and open criterion between the numbers which form musical intervals and those which have nothing to do with the matter [69].[152] Kepler looked for these causes not in integers, but in magnitudes, that is, in geometry.

By ingenious reasoning, which we cannot honestly consider more convincing than the purely arithmetical reasoning of the Pythagoreans, Kepler came to show that in an octave there are seven consonances, the octave, the major and minor sixths, the fifth, the fourth, and the major and minor thirds. This overcomes Zarlino's senario approach, which did not really justify the minor sixth (8 : 5).

Kepler's theory starts from the classification of regular polygons described in Book I. Regular polygons include those that are *provable*, that is, constructible with the use of a ruler and compass alone,[153] and those that are non-provable, such as polygons with 7 sides (heptagon), with 9 sides (ennagon), with 11 sides (hendecagon), etc.

The first three provable polygons, triangle, square, pentagon, to which the diameter is added, are called *radical polygons*. From them all other provable polygons are obtained by doubling the sides. Of the polygon Kepler distinguished between *parts*, a portion of the polygon, provided that it is not larger than a semicircle; the *remainder*, i.e. what remains, which is larger than a semicircle; and the *whole*.

Consonances associated with provable polygons are obtained by examining the parts, the remainders and the whole. For example, in the pentagon we have:

Parts	Remainders	Whole
1	4	5
2	3	5

The consonances to consider are given by the ratios: (4 : 1), (3 : 2), (5 : 4), (5 : 3) (5 : 2) (5 : 1).

By increasing the number of sides of the regular polygons, you also have dissonant parts, which correspond to a number of parts equal to the side of a non-provable polygon; for example, for the octagon, in addition to the consonant parts and remainders

[151] pp. 7–8.

[152] p. 8.

[153] Note that Kepler did not explicitly identify *provable* with *constructible*, but this is a common interpretation.

Table 3.21 Construction of consonances; from [69], Book III, p. 21

Sunt ergo

Consonæ Partes.	Consona Residua.	Dissonæ Partes.	Dissona Residua.	Respectu Totius
1.	1.			2.
1.	2.			3.
1.	3.			4.
1. 2.	3. 4.			5.
1.	5.			6.
1. 3.	5.		7.	8.
1. 3.			7. 9.	10.
1. 5.			7. 11.	12.
1. 3. 5.		7.	9. 11. 13. 15.	16.
1. 3.		7. 9.	11. 13. 17. 19.	20.
1. 5		7. 11.	13. 17. 19. 23.	24.

1, 3, 5, we also have the dissonant remainder 7. Table 3.21 summarizes the conclusions of Kepler's axiomatic theory of musical consonances. The consonances are represented by the ratio $(m : n)$, where $m > n$ can be either the number of remainders or the whole, n can be either the number of parts or remainders; m and n should not be the number of sides of a non-provable polygon, and should be co-prime.[154] For example, in the case of the regular polygon of 10 sides, in addition to the consonant parts 1 and 3 reported in Table 3.21, there are also the consonant parts 4, 5, and 6, which, however, have ratios in non-radical form, that, when reduced, report consonances already defined in the polygons of sides 5 and 2.

Kepler's procedure naturally leads to an infinite number of consonances, since the regular polygons are infinite. However, he succeeded in reducing the consonances to the seven accepted in his time by introducing the concept of harmonic division of a chord.

> Definition. If the whole string is divided into parts such that they are individually in consonance both with each other and with the whole, we shall call the division harmonic [69].[155]

Only seven consonances can divide the string harmonically.

The expansion of the numbers characteristic of the division of the octave occurs in the manner shown in Fig. 3.15, where ratios $(m : n)$ are such that $n > m$. First, the whole is expressed in the form of a fraction, that is, with unity at the top as the numerator and unity at the bottom as the denominator. Then each number is taken

[154] Kepler only reports consonants expressed in radical form.

[155] Book III, p. 22. Translation into English in [76].

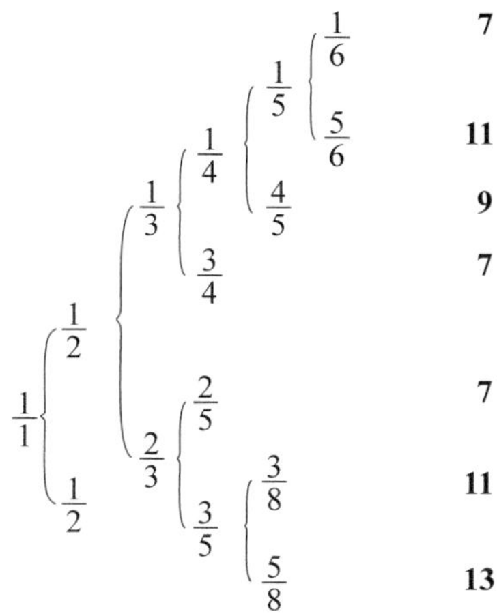

Fig. 3.15 Chain of consonances; from [69], Book III, p. 27

separately as the numerator, and the sum of the two is taken as the denominator. Thus, from any given fraction, two branches are formed, until the sum yields the number which indicates a non-provable polygon [69].[156]

Book V. The Harmony of the Spheres

Book V applies the results of Book III to astronomy. As we have said, Kepler was astonished at the correspondence between Ptolemy's views of celestial harmony and his own; but he recognized that none of the ancients could possibly have known the full truth in these matters. He wrote this at the beginning of Book V:

> To be sure, much was still lacking in the astronomy of that age; and Ptolemy, having begun badly, could plead desperation. Like the Scipio of Cicero, he seems to have recited a kind of Pythagorean dream rather than advancing philosophy. Both the primitive state of ancient astronomy and the wonderfully exact agreement between the two meditations, at an interval of fifteen centuries, confirmed my intention of pressing onward. What more is there to say? The very nature of things was revealed to men through the different interpreters of distant ages; it was the finger of God, as the Hebrews say, which conceived the same thing here in the minds of two men who gave themselves totally to the contemplation of Nature, concerning the structure of the world; since neither of the two led the other into this path [69].[157]

Kepler had much empirical data on the motion of the planets, either obtained by him or by Tycho Brahe. They can be classified as follows:

1. Their distances from the Sun.
2. Their periodic times.
3. Their daily eccentricities.

---

[156] Book III, p. 27.

[157] Book V, p. 178. Translation into English in [104].

Table 3.22 Apparent daily pass at aphelion and perihelion; from [69], p. 198

Planet	Position	Angle	Harmonies	Angle	Ratio
Saturn	At aphelion	1–46	Between	1–48	
	At perihelion	2–15	And	2–15	(5 : 4)
Jupiter	At aphelion	4–30	Between	4–35	
	At perihelion	5–30	And	5–30	(6 : 5)
Mars	At aphelion	26–14	Between	25–21	
	At perihelion	38–10	And	38–10	(3 : 2)
Earth	At aphelion	57–30	Between	57–28	
	At perihelion	61–18	And	61–18	(16 : 15)
Venus	At aphelion	94–50	Between	94–50	
	At perihelion	97–37	And	98–47	(25 : 24)
Mercury	At aphelion	147–00	Between	164–00	
	At perihelion	384–00	And	394–00	(12 : 5)

4. Their daily times spent on their arcs.
5. Their angles to the Sun, or apparent daily arcs to observers, as it were, on the Sun.

Kepler attempted to correlate this data set with the intervals of the musical scale, but found significant correlations only with the data in item 5, which represent the angular velocities of the planets. These are listed in Table 3.22.

Column 1 of the table gives the names of the planets, Col. 3 the daily angles observed by the Sun,[158] Column 5, the same angles slightly modified by Kepler to obtain ratios of integers to have an exact value of consonances. Column 6 gives the musical intervals between aphelion and perihelion corresponding to the values in Col. 5.

Table 3.23 reports the data of Table 3.22 normalized so that they are in the same octave. The numbers in parentheses indicate the number of octaves the values of Table 3.22 are narrowed.

Kepler arranged the normalized speeds of the planets in a staff. He conventionally took the notes associated with Saturn as G. In Fig. 3.16a, the fundamental note is that of Saturn's aphelion. The Earth is G at the octave, the note A is missing (*vacat*), and the sequence of tones and semitones includes B and E, thus being classified by Kepler as *cantus durus*. Figure 3.16b takes as its fundamental note that of Saturn's perihelion. This time it is Jupiter at the octave of Saturn, and what is missing is the note F. The sequence of tones and semitones includes B♭ and E♭ and is thus classified by Kepler as *cantus mollis*.

Figure 3.17 looks at the notes associated with each planet. Or better, the notes at aphelion and perihelion integrated by 'intermediate notes'. Actually, Kepler did not

[158] The sign—of the angle separates minutes from seconds; thus 57–30 is really 57.50 minutes. Note that the Earth makes about 1 degree per day, or about 360 degrees per year.

Table 3.23 Speed of planets in the staff; from [69], p. 203

Position	Minutes	Seconds
Perihelion of Mercury (7)	3	0
Aphelion of Mercury (6)	2	34
Perihelion of Venus (5)	3	30
Aphelion of Venus (5)	2	58
Perihelion of Earth (5)	1	55
Aphelion of Earth (5)	1	47
Perihelion of Mars (4)	2	23
Aphelion of Mars (3)	3	17
Perihelion of Mars (1)	2	45
Aphelion of Jupiter (1)	2	15
Perihelion of Saturn	2	15
Aphelion of Saturn	1	46

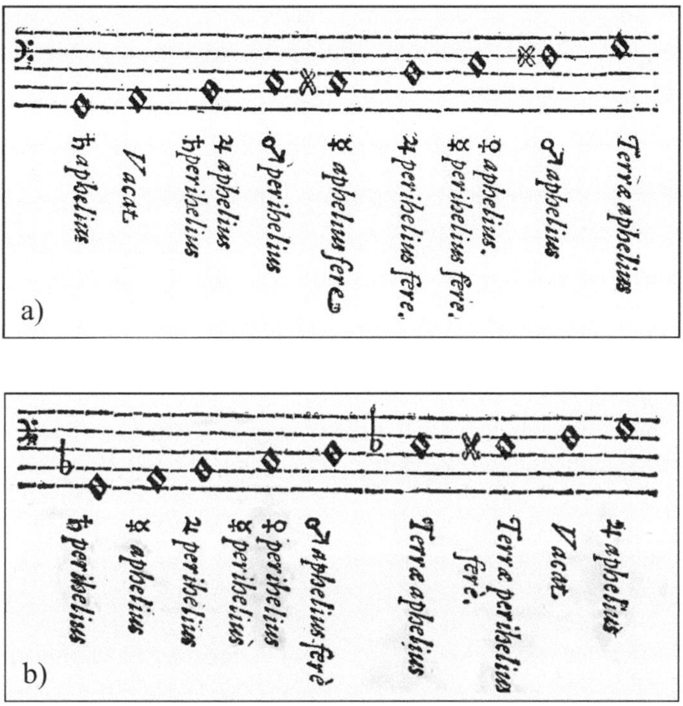

Fig. 3.16 Musical scales associated to planets; from [69], p. 204

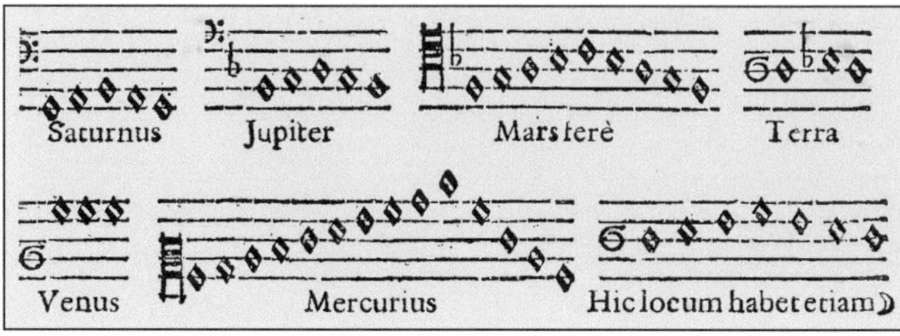

Fig. 3.17 Harmonies of planets; from [69], p. 207

think of notes in the classical sense, i.e. sounds with stable pitches, but of sounds with variable pitches. In his mind, these sounds could be heard if the Heavens were filled with air.

Kepler's *Harmonices* culminated in his so-called third law of planetary motion: "The proportion between the periodic times of any two planets is exactly the sesquialtera proportion of their mean distances" [69],[159] that is the cube of the mean distances are proportional to the square of the periods.

3.3.2 *Marin Mersenne's* Harmonie Universelle

Marin Mersenne, with Kircher can be considered the last great mathematician with a deep knowledge of speculative music to write about the harmony of the spheres. The very titles of some of his treatises testify to his interest: *Traité de l'harmonie universelle* (1627), *Les préludes de l'harmonie universelle* (1636), *Harmonie universelle* (1636), *Harmonicorum libri XII* (1648). The period in which Mersenne lived was marked by an overwhelming development of the new science based on mechanism; the numerological theories of Pythagoras and Zarlino found a formidable competitor in the field of music: the theory of coincidence, made plausible by the conception of sound as a series of pulses periodically following each other through the air. The theory of coincidence, though imperfect and unable to account for many aspects of the harmony of sounds, nevertheless provides a simple physical (mechanistic) explanation of consonance, removing the aura of mystery with which Pythagorean numerology had imbued it. Mechanistic theories also began to be formulated in astronomy, explaining how the movements of the stars could be explained in terms of efficient causes.

The existence of particular causal explanations that differ from each other for the various real or supposed harmonies of the universe undermines the very idea of a single universal harmony. Perhaps the only reason that could support this idea

[159] p. 189.

was associated with the existence of a demiurge who created man and the world by orchestrating many different but coordinate causal mechanisms. This belief certainly united Mersenne with Kepler, although the latter had less reason to doubt a universal harmony.

With the rationalism of the 17th century and the Enlightenment of the 18th, reference to the intervention of a God to explain natural phenomena became increasingly rare, and the ontology that guided traditional speculative music no longer found a following among musicians and scientists, with a few exceptions.

This notwithstanding Mersenne dealt extensively with the harmony of the spheres in the second book of his *Traité de l'harmonie universelle* [90], which begins with an exposition of his idea of harmony.

> The second book of the *Traité* shows that harmony can be found everywhere in the world and teaches how sounds can be used to understand, or at least to remember, the most beautiful things explained in the other sciences. And although it is not necessary for composition, the perfect musician must not ignore what it teaches if he is not to deprive himself of what all scholars have compassionately sought in the harmony of the universe [90].[160]

After a long preface explaining how the inherent harmony of music had been conceived, he listed a series of theorems (15) which are later proved

I The various reasons for consonances are found in rhythms, metrical feet, and verses; consequently, the perfect musician keeps the reasons or proportions of words as those of consonances.

II Sounds and consonances are similar to flavors, smells, and other objects of the senses, such as colors, sizes, and figures.

III Sounds may be compared to weights, to the machines of war and peace, to the beating of the louse, and to the health of body and soul.

VI The consonances may be compared to the four elements, their qualities, and many other similar things.

V The consonances may be compared to the heavens and planets, by reason of their dispositions, aspects, magnitudes, distances, and movements.

VI It is necessary to know the distances between the planets, and their magnitudes, in order to know whether they have the same cause as consonances or dissonances.

VII To determine which are the consonances or dissonances that correspond to the magnitudes, distances, and movements of the planets in the preceding.

VIII We can find several consonances and dissonances in the magnitudes and distances of the stars, if we follow the system of Copernicus.

IX Consonances are similar to the lines, figures, and solids of geometry.

X There is the same proportion of the difference of the sounds to the difference of the strings, considered in their length, as there is of the difference of the weights to the difference of their distance from the center of the balance, or from the support of the lever, which the Greeks call hypomoclion; consequently sounds may be compared to weights, and to moving forces.

---

[160] Book II, Preface, first unnumbered page.

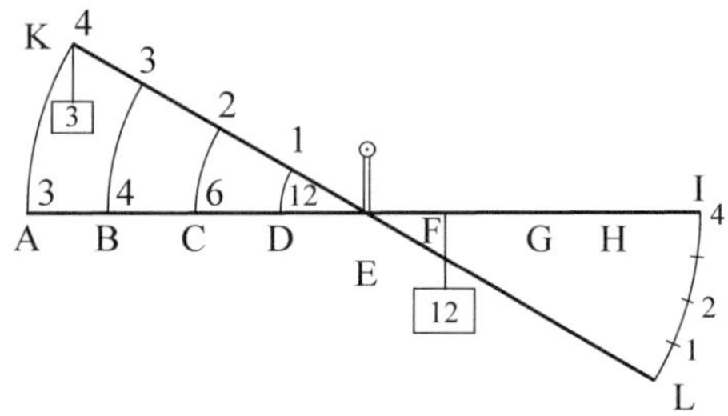

Fig. 3.18 Harmony of a lever; from [90], Book II, p. 398

XI To state and explain what are the principles and principal maxims of the science of mechanics.

XII To determine whether the harmony and consonances which Robert Fludd puts into the intellectual, celestial, and elemental world, are well established.

XIII To explain the harmony that Plato put into his numbers and that Offusius[161] treats in his book *De divina facultate astrorum in larvatam astrologiam* (1750); and show in what Fludd or Robert des Flots is lacking in its harmony.

XIV The consonances of music are similar to the proportions of bodies and to the degrees of active and passive qualities or resistances of the elements.

XV The architecture and its proportions are similar to the consonances and concerts of music: this is particularly noticeable in the structure of the Temple of Solomon, whose; rich and measures are explained [90].[162]

Theorem X concerns the new science, specifically Galileo's statics, as stated in *Le mecaniche* [49], a text well known to Mersenne who published it in French, in 1634, under the title *Les mechanique* [50]. The Galilean statics is based on what is now called the principle of virtual work. In its basic version, it admits that the effectiveness of a weight on a lever, on an inclined plane, and on any of the simple machines, depends on its greater or lesser tendency to move. For example, in a lever, the points further away from the fulcrum have a greater tendency to move, and the weights placed there are more effective, i.e., as Mersenne said, they are heavier.

The proof of the theorem is largely mechanical; only here and there is a reference to the laws of harmony. For example consider the balance AI of Fig. 3.18, divided into eight equal parts. The 12-pound weight suspended in F is in equilibrium with a 12-pound weight in D, with a 6-pound weight in C, a 4-pound weight in B, and a 3-pound weight in A, since, according to Mersenne, the weights are heavier the farther they are from the fulcrum E of the balance. And since the weight in C is twice as far from E as the weight in F, 6 pounds in C is worth as much as 12 in F and D, and 4 in B and 3 in A is worth as much as 12 in D. At this point, Mersenne introduced the idea that the weights on the balance can be associated with consonances: "We will say that the four numbers 12.6.4.3 give us the most important consonances of

[161] Jofrancus Offusius (c. 1505–c. 1570), German astronomer-astrologer.

[162] Book II. Tables des theoremes.

music, because 12 and 6 make the octave, 12 and 4 the twelfth, 6 and 4 the fifth, 6 and 3 [again] the octave, and 4 and three the fourth" [90].[163]

Mersenne saw the forced nature of the analogy (to a modern reader, no more forced than the association of the four elements with the tetrachord, for example), and shortly thereafter wrote that if there is an analogy between the lengths of the arms of a balance and the lengths of a string, it is in reverse order.

> We can add that the string is balanced when the bridge is in the middle, because it sounds the same on either side, and the bridge is like the center of gravity of the two parts of the string, which makes the sound all the sharper and lighter when its end is closer to the bridge. Consequently, these two parts of the string are the opposite of the lever arms, where they are heavier when they move faster. But the parts of the string are the more free and the softer the further they are from the bridge, because the speed of the movement decreases as it moves away [90].[164]

Theorems XII and XIII are a critique of Robert Fludd's conceptions of the harmony of spheres (see for example [46]), very much in vogue when Mersenne wrote the *Traité*. In Fludd's narrative, the earth is a musical instrument whose strings and the harmonies they produce are due to the hand of God: the world has a mathematical and musical order.

Mersenne was less concerned with Fludd's geocentrism, toward which he took a critical position, than with the presence of a hermetic influence of his harmony. Fludd considered truth to be occult in the eyes of mathematicians, but that it could be revealed in the register of symbol, analogy, and image. Fludd's monochord remains unacceptable not only because of its false mathematical, acoustic and astronomical data, but also because of the temptation of its alchemical interpretation, which would contribute not only to an erroneous conception and representation of the world, but also to a distancing from Christian truth [12].

Among Mersenne's technical criticisms of Fludd is that he uncritically accepted the legend of Pythagoras and his hammers.

> This is completely contrary to truth and experience, as I have shown in the fourth book,[165] because all hammers make more or less the same sound when they strike the anvil; and there is no one who can distinguish whether the sound made by the 20-pound hammer is louder than the sound made by the one-pound hammer [90].[166]

Theorems V to VIII discuss the harmony of the stars; of particular interest is the proof of Theorem VIII, which gives in full Kepler's results contained in Book V of his *Harmonices mundi*.

---

[163] Book II, p. 400.

[164] Book II, p. 405.

[165] The reference is to a planned fourth book of the *Traité*. [43], pp. 15–16.

[166] Book II, p. 447.

References

1. Aaron P (1539) Il toscanello in musica. Seffa, Venice
2. Amiot JM (1779) Mémoire sur la musique des Chinois tant anciens que modernes. Imprimerie de Parlment, Paris
3. Barbieri P (1983) I temperamenti ciclici da Vicentino (1555) a Buliowski (1699): Teoria e pratica ‘archicembalistica’. L’Organo 21:129–208
4. Barbieri P (1987) Acustica accordatura e temperamento nell’illuminismo veneto. Torre D’Orfeo, Rome
5. Barbieri P (1996) Il mesolabio e il compasso di proporzione: le applicazioni musicali di due strumenti matematici. 1558–1675. In: Passadore F, Rossi F (eds) Musica, scienza e idee nella Serenissima durante il Seicento. Venice, Edizioni Fondazione Levi, pp 201–220
6. Barbieri P (2001) Galileo’s coincidence theory of consonances, from Nicomachus to Sauveur. Recercare 23:201–232
7. Barbieri P (2002) La nascita delle teorie ‘continue’ della consonanza. La ignorata curva di Draghetti e Foderà, poi di Helmholtz (1771–1837). Acta Musicologica 74(1):55–75
8. Barbieri P (2008) Enharmonic instruments and music 1470–1900. Il Levante, Latina
9. Barbour JM (1951) Tuning and temperament: a historical survey. Michigan State College Press, East Lansing
10. Barker A (1989) Greek musical writings, vols 2. Cambridge University Press, Cambridge
11. Barrow I (1734) The usefulness of mathematical learning explained and demonstrated. Translated into English by Kirkby J. Austen, London
12. Basilico B (2017) L’harmonia mundana au XVIIe siècle. Les critiques de Marin Mersenne adressées au monocorde du monde de Robert Fludd. L’Atelier du Centre de Recherches Historiques 17
13. Baskevitch F (2005) Notion de vibration sonore au XVIIème siècle. https://www.academia.edu/7180028/. Accessed 25 July 2023
14. Beeckman I (1939–1953) Journal tenu par Isaac Beeckman de 1604 à 1634 (4 vols). Nijhoff, The Hague
15. Benedetti GB (1585) Diversarum speculationum mathematicarum & physicarum liber. Bevilaqua, Turin
16. Blackwood E (1985) The structure of recognizable diatonic tunings. Princeton University Press, Princeton
17. Boethius, (1867) Boetii De institutione musica. Libri Quinque. In: Friedlein G (ed) Anicii Manlii Torquati Severini Boetii De institutione arithmetica libri duo. De institutione musica libri quinque, Teubneri, Leipzig, pp 175–371
18. Boethius (1948) De institutione musica. In: Choen M, Drabkin I (eds) A source book in Greek science, McGraw-Hill, New York, pp 291–294; 298–299
19. Bordiga G (1985) Giovanni Battista Benedetti. Filosofo e matematico veneto del secolo XVI, Istituto Veneto di Scienze, Lettere e Arti, Venice
20. Bosanquet RH (1875a) Temperament: Or, the division of the octave (part I). In: Proceedings of the musical association. 1st Sess. (1874–1875), pp 4–111
21. Bosanquet RH (1875b) Temperament: Or, the division of the octave (part II). In: Proceedings of the musical association. 1st Sess. (1874–1875), pp 112–158
22. Cabeo N (1646) In quatuor libros meteorologicorum Aristotelis commentaria (2 vols). Corbelletti, Rome
23. Capecchi D (2018) The path to post-Galilean epistemology. Springer, Cham-Dordrecht
24. Capecchi D, Capecchi G (2023) Music in Holland: consonances according to Simon Stevin. In: SISFA 43th annual conference. Federico II University Press, Naples, pp 91–98
25. Cardano G (1663) Della natura de principii et regole musicali. In: Hieronymi Cardani mediolanensis opera omnia, vols 4 and 10. Huguetan & Rovaud, Leiden, pp 621–630
26. Choen M, Drabkin I (eds) (1948) A source book in Greek science. McGraw-Hill, New York
27. Cohen H (1984) Quantifying music. Reidel, Dordrecht

28. Cohen H (1987) Simon Stevin's equal division of the octave. Ann Sci 44(5):671–488
29. Cohen H, Kursel J (2020) Simon Stevin's music theory revisited: a dialogue. In: Davids C et al (eds) Rethinking Stevin, Stevin rethinking. Constructions of a Dutch polymath, Brill, Leiden, pp 252–272
30. De Haan D (1884) Simon Stevin. Vande spiegeling der singkonst et Vande molens. Deux traité inédits, De Haan, Amsterdam
31. Descartes R (1650) Musicae compendium. Ackersdijck and Zijll, Utrecht
32. Devie D (1990) Le Tempérament musical: philosophie, histoire, théorie et pratique. Société de musicologie de Languedoc, Béziers
33. Devreese J, Berghe G (2008) Magic is no magic. Witpress, Southampton, The wonderful world of Simon Stevin
34. Di Veroli C (2010) Vallotti as the ideal German good temperament. Harpsichord & fortepiano 15(1):9–14
35. Dickreiter M (1973) Der Musiktheoretiker Johannes Kepler. Francke, Oppenheim
36. Dijksterhuis E (1970) Stevin and the Dutch language. In: Dijksterhuis E (ed) Simon Stevin: Science in the Netherlands around 1600, Nijhoff, The Hague, chap 13
37. Dillon G, Musenich R (2009) The Huygens' comma: Some mathematics concerning the 31-cycle. Thirty-One 1:49–56
38. Dostrovsky S (1975) Early vibration theory: physics and music in the seventeenth century. Arch Hist Exact Sci 14(3):169–218
39. Drake S (1970) Renaissance music and experimental science. J Hist Ideas 31(4):483–500
40. Duffin R (2007) How equal temperament ruined harmony (ebook). Norton, New York
41. Duffin RW (Unknown) Why I hate Vallotti (or is it young?). https://casfaculty.case.edu/ross-duffin/why-i-hate-vallotti-or-is-it-young-1-1/. Accessed 25 July 2024
42. Euler L (1739) Explicatio phaenomenorum quae a motu successivo lucis oriuntur. Commentarii Academiae Scientiarum Petropolitanae 11:150–193
43. Fabbri N (2006) De l'utilité de l'harmonie. Filosofia, scienza e musica in Mersenne, Descartes e Galilei. PhD thesis, Scuola Normale di Pisa
44. Fabris D (2011) Galileo and music: A family affair. In: Corsini EM (ed) The inspiration of astronomical phenomena VI, Astronomical Society of the Pacific Conference Series, vol 441, pp 57–71
45. Favaro A (1886) La libreria di Galileo Galilei. Bollettino di Bibliografia e di Storia delle Scienze Matematiche e Fisiche 19:219–293
46. Fludd R (1617) Utriusque cosmi historia. Galler, Oppenheim
47. Fracastoro G (1546) Liber unus de sympathia et antipathia rerum, item, De contagione, et contagiosis morbis, et eorum curatione, lib. III. Giunta, Venice
48. Gaffurio F (1496) Practica musice. Rothomagensem, Milan
49. Galilei G (1634a) Le mecaniche. In: Favaro A (ed) Le opere di Galileo Galilei (National edition) (20 vols), vol 2, Barbera, Florence, pp 147–192
50. Galilei G (1634b) Les mechaniques de Galilée. Traduit de l'Italien par P.M.M. Guenon, Paris
51. Galilei G (1638a) Discorsi e dimostrazioni matematiche sopra due nuove scienze. Elsevier, Leiden
52. Galilei G (1638b) Discorsi e dimostrazioni matematiche sopra due nuove scienze. In: Favaro A (ed) Le opere di Galileo Galilei (National edition) (20 vols), vol 8, Barbera, Florence
53. Galilei G (1974) Two new sciences. The University of Wisconsin Press, Madison, Translated into English by Drake S
54. Galilei V (1581) Dialogo di Vincentio Galilei nobile fiorentino della musica antica et della moderna. Marescotti, Florence
55. Godwin J (1987) Harmonies of heaven and earth. Inner Tradition International, Rochetser
56. Godwin J (ed) (1989) Cosmic music : musical keys to the interpretation of reality. Essay, Inner Tradition International, Rochetser
57. Godwin J (1993) The harmony of the spheres. A sourcebook of the Pythagorean tradition in music, Inner Tradition International, Rochetser

58. Gogava A (1644) Mathematico physicarum meditationum, quaestionum, solutionum centuria. Sloot, Utrecht
59. Grenfell MT (2005) The development of the equal temperament scale. Evolution or radical change? PhD thesis, Danbury, Connecticut
60. Herigone P (1637–1646) Cursus mathematicus, vols 6. Le Gras, Paris
61. Herman RH (1973) Dialogo della musica antica et della moderna of Vincenzo Galilei. PhD thesis, North Texas State University
62. Huygens C (1888–1950a) Cosmotheoros. In: Societé Hollandaise des Sciences (ed) Oeuvres complètes de Christiaan Huygens, vols 22 and 21. Nijhoff, The Hague, pp 677–821
63. Huygens C (1888–1950b) Oeuvres complètes de Christiaan Huygens, vols 22. Nijhoff, The Hague
64. Huygens C (1888–1950c) Traité de la lumiere. In: Societé Hollandaise des Sciences (ed) Oeuvres complètes de Christiaan Huygens (22 vols), vol 19, Nijhoff, The Hague, pp 459–550
65. Huygens-Fokker (Unknown) Tuning and temperament bibliography. https://www.huygens-fokker.org/docs/bibliography.html. Accessed 2 Jan 2025
66. Jeans S (2001) Kepler [Keppler], Johannes, new grove dictionary of music and musicians. https://www.oxfordmusiconline.com/grovemusic/display, https://doi.org/10.1093/gmo/9781561592630.001.0001/omo-9781561592630-e-0000014903. Accessed 20 July 2024
67. Jorgensen O (1977) Tuning the historical temperaments by ear. Northern Michigan University Press, Marquette (Michigan)
68. Kepler J (1596) Mysterium cosmographicum. Gruppenbachius, Tubing
69. Kepler J (1619) Harmonices mundi. Libri V. Ioannes Plancus, Linz
70. Kepler J (1621) Mysterium cosmographicum. Kempfer, Frankfurt
71. Kepler J (1624) Chilias logarithmorum ad totidem numeros rotundos, praemissa demonstratione legitima. Caspar Chemlinus, Marburg
72. Kepler J (1625) Supplementum chiliadis logarithmorum. Caspar Chemlinus, Marburg
73. Kepler J (1937) Johannes Kepler Gesammelte Werke, vols 23. Beck, Munich
74. Kepler J (1939) Harmonies of the world by Johannes Kepler (English by Wallis CG). https://mathship.com/900/Astronomy/kepler_harmonies-of-the-world.pdf, edited by Globalgrey 2109. Accessed Sept 2024
75. Kepler J (1981) Mysterium cosmographicum. The secreto of the universe. Duncan AM (Trans) Abaris Book, New York
76. Kepler J (1997) The harmony of the world. In: Alton EJ, Duncan AM, Field JV (Trans)American Philosophical Society, Philadelphia
77. Kuttner FA (1975) Prince Chu Tsai-Y?'s life and work: A re-evaluation of his contribution to equal temperament theory. Ethnomusicology 19(2):163–206
78. Lam J (2001) Zhu Zaiyu, new grove dictionary of music and musicians. https://doi.org/10.1093/gmo/9781561592630.article.49365. Accessed 12 Jan 2025
79. Lindley M (2001) Temperaments, new grove dictionary of music and musicians. https://doi.org/10.1093/gmo/9781561592630.article.27643. Accessed 12 Jan 2025
80. Lindley M (2022) An historical survey of meantone temperaments to 1620. www.academia.edu/1134182/An_historical_survey_of_meantone_temperaments_to_1620. Accessed 25 July 2022
81. Lucretius Carus T (2009) Titi Lucreti Cari. De rerum natura. In: Munro HAJ (eds) Libri sex. Cambridge University Press, Cambridge
82. Maurolico F (1575) Musicae traditiones carptim collectae. In: Franceschi F (ed) Opuscula mathematica, Franceschi, Venice, pp 145–160
83. Mengoli P (1659) Geometriae speciosae elementa. Ferroni, Bologna
84. Mercator H (2013) Of musik (1). In: Wardhaugh B (ed) The compendium musicae of René Descartes: early English responses. Brepolis, Turnhout, pp 193–219
85. Mercator H (2013) Of musik (2). In: Wardhaugh B (ed) The compendium musicae of René Descartes: early English responses. Brepolis, Turnhout, pp 221–2236

86. Mercator H (2013) Of musik (2). In: Wardhaugh B (ed) The compendium musicae of René Descartes: early English responses. Brepolis, Turnhout, pp 132–140
87. Mercator H (2013) Theoria musicis. In: Wardhaugh B (ed) The Compendium musicae of René Descartes: early English responses. Brepolis, Turnhout, pp 144–187
88. Mercatror N (1668) Logarithmotechnia. Godbid, London
89. Mersenne M (1623) Questione celeberrimae in Genesim. Cramoisy, Paris
90. Mersenne M (1627) Traité de l'harmonie universelle, vols 2. Baudry, Paris
91. Mersenne M (1636–1637) Harmonie universelle, vols 2. Cramoisy, Paris
92. Palisca C (1985) Humanism in Italian Renaissance musical thought. Yale University Press, New Haven
93. Panetta VJ (ed) (1987) Treatise of harpsichord tuning by Jean Denis. Cambridge University Press, Cambridge
94. Papius A (1581) De consonantiis, seu pro diatessaron libri duo. Planini, Antwerp
95. Pesic P (2005) Earthly music and cosmic harmony: Johannes Kepler's interest in practical music, especially Orlando di Lasso. J Seventeenth-Century Music 11(1)
96. Pierce R (1977) A brief history of logarithms. Two-Year Coll Math J 8(1):22–26
97. Praethorius M (ed) (2014) Syntagma musicum. Tomus secundus. Faulkner Q (Trans) Zea, Lincoln (Nebraska)
98. Rasch R (1992) Six seventeenth-century Dutch scientists and their knowledge of music. In: Coelho V (ed) Music and science in the age of Galileo. Springer, Dordrecht, pp 185–210
99. Romieu JB (1758) Mémoire théorique & pratique sur les systèmes tempérés de musique. Mémoires de l'Académie Royal des Sciences de Paris pp 483–519
100. Rossi L (1666) Sistema numerico, overo musica speculativa, dove si spiegano i più celebri sistemi di tutti i tre generi. Laurenzi, Perugia
101. Salinas F (1577) De musica libri septem. Mathias Gastius, Salamanca
102. Smith R (1759) Harmonics or the philosophy of musical sounds. Merril, London
103. Stanzial D, Bonsi D, Gonzalez D (2001) Nonlinear modeling of the "mitnahme-effekt" in coupled organ pipes. In: Proceedings of the international symposium of musical acoustics musical sounds from past millennia. Venice, pp 333–337
104. Stephenson B (1994) The music of the heavens: Kepler's harmonic astronomy. Princeton University Press, Princeton
105. Stevin S (1605) Tomus quartum mathematicorum hypomnematum de statica. Ex officina Patii, Leiden
106. Stevin S (1955) The principal works of Simon Stevin. Crone E et al (eds) (5 vols) Swets & Seitlinge, Amsterdam
107. Sukljan N (2018) Zarlino's harpsichord. A contribution to the (pre)history of equal temperament. Musicol Anual 54(1):5–22
108. Tartini G (1767–1770) Scienza platonica fondata sul cerchio, https://www.discovertartini.eu/Scienza_platonica.pdf, archivio regionale di Capodistria. ms. 233: antica 128, inventario: 4-233; cc. 95. Accessed 18 Sept 2024
109. Tonietti T (2014) And yet it is heard: musical, multilingual and polycultural history of mathematics (2 vols). Birkhäuser, Basel
110. Truesdell CA (1960) The rational mechanics of flexible or elastic bodies. In: Leonhardi Euleri Opera omnia (1911-, in progress), 2, vol 1, part 2, Teubneri GB; [then] Fussli Turici; [then] Birkhäuser, Basel
111. Uggias G (2015) Keplero e la musica. Il Libro III dell'Harmonice mundi (Linz, 1619). Translation and introduction. PhD thesis, Alma Mater Studiorum. University of Bologna
112. Vallotti FA (1950) Trattato della moderna musica Il Messaggero di S, Antonio, Padua
113. Vicentino N (1555) L'antica musica ridotta alla moderna pratica. Barre, Rome
114. Viviani V (1654) Racconto istorico di Vincenzo Viviani. In: Favaro A (ed) Le opere di Galileo Galilei (National edition), vols 19 and 20, Barbera, Florence, pp 597–632
115. Wallis J (2016) John Wallis: writings on music. Cram D, Wardhaugh B (eds) Routledge, New York

116. Wardhaugh B (2008) Musical logarithms in the seventeenth century: Descartes, Mercator, Newton. Hist Math 35:19–36
117. Wardhaugh B (2013) The Compendium musicae of René Descartes: Early English responses. Brepolis, Turnhout
118. Wild J (2014) Genus, species and mode in Vicentino's 31-tone compositional theory. Music Theory Line 20(2):1–19
119. Yoko N (2013) The Jesuit Jean Joseph Marie Amiot and Chinese music in the eighteenth century. In: Luis MRS (ed) History of mathematical sciences. Europe and China: science and the arts in the 17th and 18th centuries, World Scientific, pp 81–92
120. Young T (1800) Outlines of experiments and inquiries respecting sound and light. Philos Trans R Soc Lond 90:104–150
121. Zarlino G (1558) Istitutioni harmoniche. The author, Venice
122. Zarlino G (1571) Dimostrationi harmoniche. Senese, Venice
123. Zarlino G (1588) Sopplimenti musicali. Senese, Venice

Chapter 4
The Mathematicians at the Close of the 17th Century

Abstract Since the second half of the 16th century, the role of speculative music has undergone a major change due to developments in mathematics and natural philosophy that allowed for a more effective treatment of its classical problems: the formation of scales, the understanding of the nature of consonances, the functioning of musical instruments, and the presence of overtones. This last aspect was studied empirically and satisfactorily by various English scientists, including John Wallis. The theoretical analysis was conducted by Daniel Bernoulli, who unfortunately did not yet have the necessary mathematical tools. Music theory was studied in the newly emerging scientific academies, notably the Royal society of London and the Académie des sciences of Paris. Toward the end of the century, Sauveur began his work as a mathematician, pursuing the systematic development of musical acoustics.

4.1 The Legacy of the 17th Century

In the second half of the 17th century, the role of the speculative musician underwent a profound change. Developments in mathematics and natural philosophy, which had by then become physics in the modern sense, made it possible to deal more effectively with the classical problems of music: the formation of scales, the understanding of the nature of consonances, and the functioning of musical instruments, of which there was an increasing variety.

The new knowledge required important background in mathematics and physics; this inevitably led to a gradual process of specialization. Speculative music became the preserve of mathematicians, who often had important musical skills; practical music and the theory of composition, was carried on by people who, with few exceptions, had little scientific knowledge, whom we now clearly call musicians, along with practitioners.

Mathematicians of the late 17th century could count on a small but valuable legacy, easily usable, derived from the musicians and mathematicians of the previous century and the first half of the 17th century. Only the following publications were readily available to the successors of the men whose works we have discussed so far:

D. Capecchi and G. Capecchi, *Joseph Sauveur and His Time: Music and Mathematicians*,
https://doi.org/10.1007/978-3-032-04707-6_4

1. Zarlino's *Le istitutioni harmoniche* (1558), the first part of which is entirely framed in speculative music. Probably one of the last treatises on music written by a musician who also had a scientific attitude.
2. Benedetti's two letters in *Diversarum speculationum mathematicarum et physicarum liber* (1585). This writing was hardly read.
3. Beeckman's *Mathematico physicarum meditationum, quaestionum, solutionum centuria*. This posthumous publication (1644), of one entry from the Beeckman's diary, contained only a handful of varied notes on music.
4. Galileo's ten-page passage on music in the *Discorsi* (1638).
5. Descartes *Musicae compendium*. Published posthumously in 1650.
6. Mersenne's voluminous works, especially the *Harmonie universelle* (1636/1637).

The three latter works were the sources from which the next generation of mathematicians drew most of their information about the science of music [25].[1]

4.1.1 *The* **Compendium** *of René Descartes*

In 1618, at the age of only 22, after meeting Isaac Beeckman, who was then becoming interested in problems of musical acoustics, Descartes wrote a short treatise on speculative music and dedicated it to his Dutch friend, the *Musicae compendium* (hereinafter *Compendium*).The first printing in Latin was in Utrecht in 1650 [29], the year of Descartes' death. It was soon republished in 1656 still in Latin [29]. A French translation was published in 1668 [30]. Some copies of the *Compendium* circulated before its publication, however. Beeckman had his own copy; but so had Huygens, and possibly some other scholars [92].[2] The booklet was well received, mainly because of the author's fame and its conciseness.

In 1653 an English translation appeared as the *Excellent compendium of musik with necessary and judicious animadversions* [28, 92]. As the title suggests, the book was divided into two parts; the first a fairly faithful translation, the second a commentary and criticism, longer than Descartes' treatise, with a timid appearance of the logarithm for measuring musical intervals. Both the translation and the commentary were anonymous. Today there is quite unanimous agreement that the translation of the *Compendium* is due to Walter Charlton (1620–1707), the translator of Gassendi who played a crucial role in transmitting Mersenne's ideas to England, while the animadversions are due to William Brouncker (1620–1684) [93].[3]

The *Compendium* does not present great innovations, in content it is rather in continuity with late 1500s theorists such as Zarlino, who is explicitly mentioned, and de Salinas, essentially limiting to presenting the diatonic systems. This is for example the opinion of HF Cohen:

---

[1] pp. 205–206.

[2] p. XII.

[3] p. 30.

> No music-scientific work of a major participant in the scientific revolution adheres so closely to the Renaissance style of musical theorizing as Descartes' little treatise. Whereas all scientists discussed so far aimed at replacing the old numerological concepts by new, scientifically more acceptable modes of analysis, Descartes' *Compendium musicae* can best be defined as 'Zarlino,more geometrico' [25].[4]

The *Compendium*, however, is very innovative in its presentation. It was written by a mathematician who had become sophisticated by then, so it treats musical themes by giving little space to Renaissance numerology, and explains all the basic concepts in a simple and compact way.

In addition, there are some interesting innovations that influenced music theorists and mathematicians, especially French ones, including Sauveur.

1. An aesthetic theory that replaces Pythagorean numerology in judging the pleasantness of consonances.
2. The introduction of consonances by monochord bisections.
3. The introduction of circular diagrams for octave division.
4. A limited use of the concept of overtone, already known to the Greeks, i.e. that every musical note contains its higher octave(s).

Descartes did not publish anything specific on music beyond the *Compendium*; his later developments of musical ideas are reported in his correspondence with Mersenne, first published nine years after his death, and in some of his philosophical treatises.

In the following, we will deal mainly with the *Compendium*, not so much because it is more interesting, but rather because it is basically Descartes' only musical writing that has been read with much attention. Descartes took a purely mathematical approach in his treatise, and although sometimes referred to the monochord as the *string*, he treated it as a geometric segment, and when he said that the height and depth of sound are related to the length of the segments: "One sound bears the same relation to another sound as one string [nervus] bears to another string" [29],[5] he was following a tradition going back to the Greeks. Only in one place did he make a passing reference to the nature of sound, but without explicitly mentioning its periodicity: "a sound strikes the ears with many percussions or reverberations, and the more rapidly the more acute the sound is" [29].[6]

Chapter 1 of the *Compendium* begins without an abstract definition of sound, as was often the case in earlier treatises on speculative music, simply stating that it is the object of the study of music, following in the tradition of mixed mathematics where useless definitions were avoided. Sound is primarily characterized by rhythm and pitch; note that the idea of introducing rhythm into a treatise on speculative music was not common.

Chapter 2 is entitled *Praenotanda*, an archaic Latin word that has the meaning of general premise. It presents a series of points concerning the aesthetics of sound,

---

[4] p. 163.

[5] p. 5.

[6] p. 13.

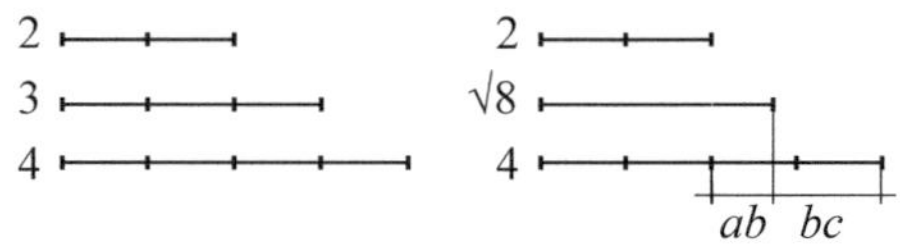

Fig. 4.1 Arithmetic versus geometric proportions

that is, what are the conditions under which sounds can be considered more or less pleasant. Given the brevity and interest of the topic, these general premises are given in full below.

1. Notice, first, that all the senses are capable of some pleasure.
2. Second, that this pleasure of the senses consists in a certain proportion and correspondence of the object to the sense; whence, for instance, the discharge of a musket, or the sound of thunder, would be unfit for music, because it would injure the ear, as the bright glare of the Sun's rays directly injures the eyes of the beholder.
3. Thirdly, this object must be such that it does not appear confusing to the sense, which does not have to work to know and distinguish it. Therefore a figure, however regular, is not pleasing to the eye if it is overloaded with many features, as is the part of the astrolabe called the mother; whereas a figure such as the spider of the same astrolabe, whose parts are more equal and observe more symmetry, disturbs less the eye that looks at it. The reason for this is that the sense is much more satisfied in the latter object than in the other, where there are many things that it cannot receive clearly enough.
4. An object is more easily perceived by the senses whose parts are less different from one another.
5. These parts have less difference between them, of which there is more proportion.
6. This proportion must be arithmetic and not geometric, inasmuch as in this case there are fewer things to consider, the differences are everywhere the same, and therefore the sense does not work so much to know clearly and in detail all that is encountered. For example, with reference to Fig. 4.1, the proportion of the lines 2, 3, 4 is more easily known than that of the lines 2, $\sqrt{8}$, 4.
   Indeed in the first figure the sense is only required to notice the unit of which one line exceeds the other, while in the second figure one must know the parts *ab* and *bc*, which, being incommensurable cannot be perfectly known at the same time by the sense [...].
7. Among the objects of each sense, that which is either very easy or very difficult to perceive by the senses is not pleasant to the soul. But [it is pleasant] what is not so easy to know that it does not leave something to be desired by the passion with which the senses are accustomed to their objects, nor so difficult that it causes the senses to suffer in the work of knowing it.
8. Finally, it should be noted that variety is very pleasant in all things [29].[7]

[7] pp. 1–2.

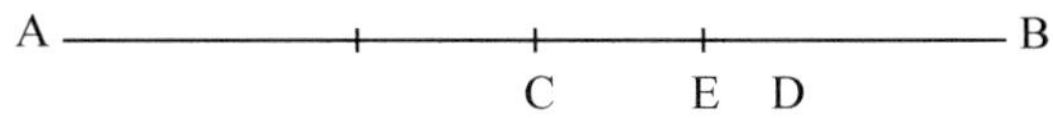

Fig. 4.2 Division of the monochord. Redrawn from [29], p. 5

Historians are particularly interested in the sixth premise, where the superiority of arithmetic division over geometric division is affirmed by reporting a judgment that belongs to the sense of sight over that of hearing.

Chapter 3 discusses rhythm. According to Descartes, rhythm in sounds must be composed of equal parts.

> Because these, as we noted in the 4-th premise, are the easiest to know; or parts that are in double or triple proportion, without going beyond them, because they are the most suitable to be heard distinctly, as we said in the 5-th and 6-th premise. Now if the measures were more unequal, the ear could only with great difficulty and effort apprehend their difference, as experience teaches us; for if I were to set five notes against one, it would be difficult to sing it [29].[8]

Rhythmic scansion must be proportionate in order to arouse the pleasure of the senses. The rhythm must proceed, Descartes prescribed, according to the binary measure, in two equal parts, or ternary, in three.[9] However, when the rhythmic succession is not accompanied by a melody, the scansion can benefit from more articulate patterns. To express the latter concept, Descartes used the example of the drum, "instrument of war [instrumentum bellicum]".

> I will not omit to say that the power of *tempo* in music is so great that it can, even by itself, produce a certain pleasure: as is evident in the drum, instrument of war [instrumento bellico], in which nothing is taken into account but the measure of tempo, which therefore—I think—may be found to consist not only of two or three parts, but perhaps even of five or seven and more. And since, indeed, in such an instrument the sense has nothing else to turn its attention to but tempo, there can then be a greater variety, so that it occupies more of the sense [29].[10]

This chapter deals with the characterization of sound in terms of pitch, introducing the concepts of consonance, degree, dissonance. Chapter 5 is devoted to consonances. Contrary to tradition, Descartes did not consider unison to be a consonance. The term consonance is not defined, as usual, by reference to acoustic sensation, but in a very neutral way, by reference to some ratios of small integers. Only later is the purely arithmetical definition flanked by the acoustic meaning, based on aesthetic criteria rather than numerology: an interval is the more pleasant, the more simple it is.

Consonant intervals are introduced by dividing a geometric segment, the line AB, into equal parts, as shown in Fig. 4.2. Dividing it into two equal parts, AC and CD, gives the octave or diapason; dividing it into three equal parts, AD, DE, EB, gives a twelfth and a fifth. Descartes suggested that AB might be divided into 4, 5, or

---

[8] pp. 2–3.

[9] In the 17th century, rhythm was conceived as today [20]. The binary measure is when there are two accents in a bar, the ternary measure when there are three accents in a bar.

[10] p. 4.

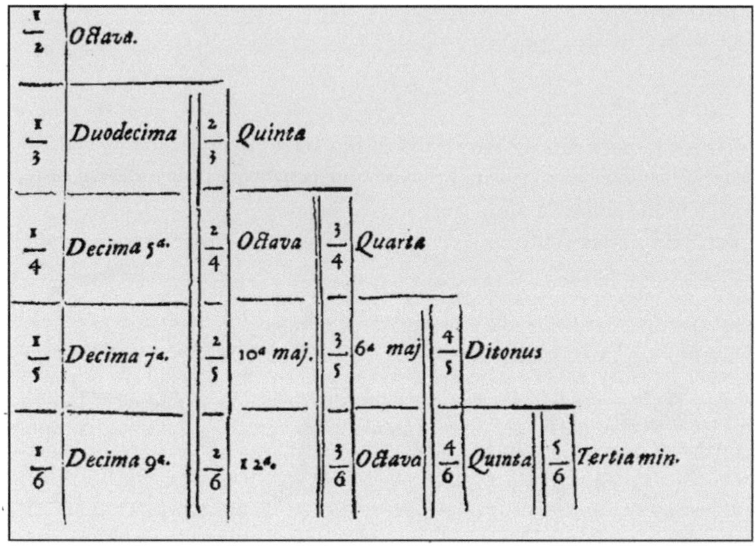

Fig. 4.3 Consonances in the division of a string into equal parts. Redrawn from [29], p. 6

6 parts, but no more, "because of the imbecillitas of the ear" [29],[11] which cannot easily distinguish differences of sound other than those of the above division.

Consonances are shown in Fig. 4.3; the division into two parts gives only one type of consonance, the division into three gives two types defined by AD, a twelfth and a fifth AE, with respect to AB. Descartes distinguished between proper and accidental consonances, obtained by dividing the string into 2, 3, 5 parts (numbers 4 and 6 are ignored because they are composed of the first three numbers). With reference to Fig. 4.2, in the case of the fifth, the proper consonance is given by AD, the accidental consonance by DB, where DB is only a remainder.

Having introduced the consonances, Descartes proceeded to comment on them, also on the basis of his aesthetic criteria. Much space is devoted to the octave, which is considered to be the first and major of the consonances, the one that differs least from unison. It is the only consonance that can be geminated or doubled. One octave added to a consonance gives a consonance: "of all consonances of any kind, there are but three species: one is simple, another is a compound of a simple and one octave, a third is a compound of a simple and two octaves" [29].[12] Other types are not allowed because the proportions would be too large. Figure 4.4 clearly illustrates the types of consonance, listed in orderer of preference.

Here the minor sixth, obtained by subtracting the ditone from the octave, is added and the ditone, i.e. the major third, is assumed to be more consonant than the fourth.

[11] p. 5.
[12] p. 5.

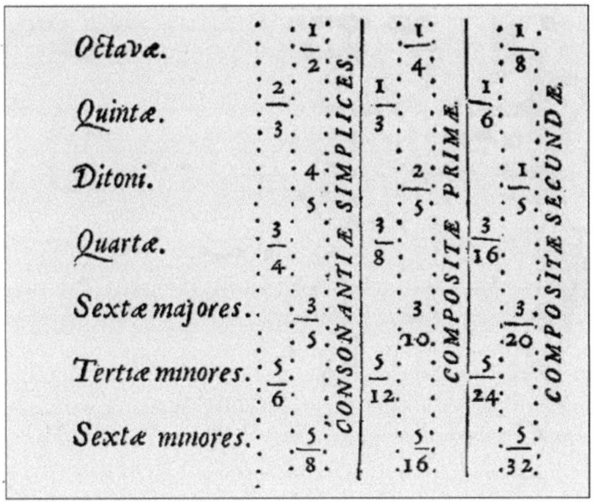

Fig. 4.4 Types of simple and composite consonances. Redrawn from [29], p. 8

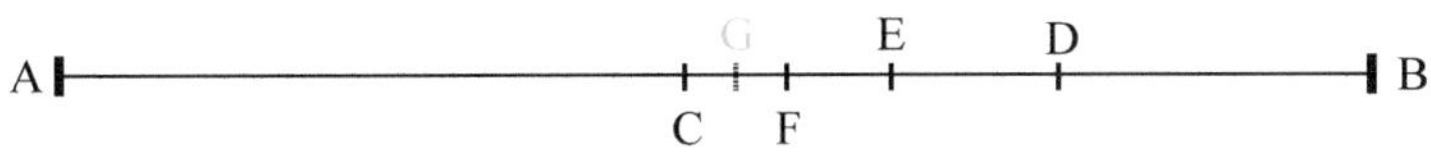

Fig. 4.5 Octave division. Redrawn from [29], p. 8

The primacy of the octave is reiterated by showing the derivation of the various consonances through its dichotomous divisions. Here Descartes departed from Zarlino's senario and adopted a criterion of economy that would become common in the theoretical treatment of just intonation. But there is a flaw in this criterion, which was discovered by Beeckman. He initially appreciated Descartes' approach to obtaining diatonic musical intervals, but later criticized it for providing only major consonances, major and minor tones, but failing to determine semitones [11].[13]

This can be seen in Fig. 4.5. Starting with a sequence of bisections, the points encountered are C, D, E, F, G. AB/AC = $^2/_1$ defines an octave, AD/AC = $^3/_2$ a fifth, AE/AC = $^5/_4$ a major third, AF/AC = $^9/_8$ a major tone. Descartes ended his argument here; but if we continue the bisection, we obtain point G, not indicated in Descartes' original figure, with AG/AC = $^{17}/_{16}$, which defines a semitone, but it is not the major semitone of the diatonic system (16 : 15).

The presentation of consonances continues with the fifth, which Descartes considered the most perfect, specifically the second kind of fifth, namely the 12th, "because it is not so sharp [acriter] as a ditone, not so languid as an octave" [11]. And it is only because music also seeks variety that other consonances should also be used.

The fourth, according to an approach that will become traditional, is seen as the shadow of the fifth, because it is somehow always present with it. In Fig. 4.6, DB is one fifth above AC, EF is the octave above AC, that always accompanies it—Descartes knew it was a pervasive interval—so EF over DB is a fourth.

[13] vol. 3, p. 136.

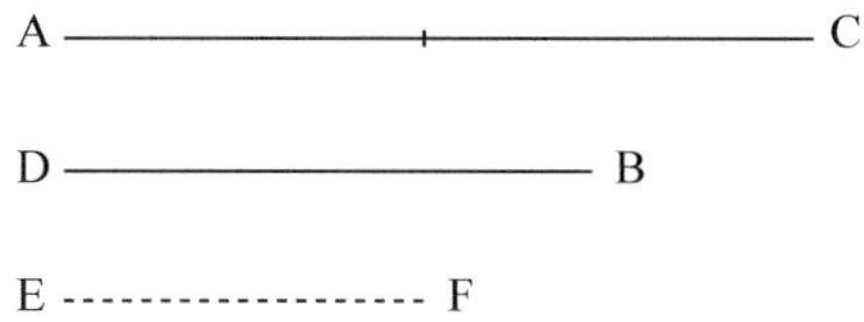

Fig. 4.6 The fourth as a shadow of the fifth. Redrawn from [29], p. 12

The last consonances that Descartes commented on are the thirds and sixths. Here are some interesting considerations about the presence of compounds:

> The perfection of a given consonance depends not only on its qualities as a simple unit, considered in itself, but at the same time on all its compounds. The reason for this is that one never hears a pure consonance without also hearing the resonance of its compounds, since it has been said that a unison also contains the resonance of its higher octaves. Now the ditone, considered in this way, is composed of much smaller numbers than the fourth, and therefore it is also more perfect [29].[14]

A superficial reading of the above passage would lead one to believe that Descartes was thinking of overtones as we understand them. In fact, he was thinking only of the octaves, especially the second and third, which form the compound consonances as described in Fig. 4.4. In the last part of the passage, Descartes gave his explanation for the fact that the major third is more consonant than the fourth—although it presents a smaller number of coincidences—with a reasoning that will be taken up by Mersenne and Huygens (see Chap. 3). In fact, the major third (5 : 4), when composed with two octaves, yields the (multiple) compound consonance (5 : 1), which has a greater number of coincidences than the fourth. Not only, but according to Descartes aesthetics the ratio (5 : 1) is easier to understand than the ratio (4 : 3). Since the consonance (5 : 1) is always present in the major third, it softens it and makes it preferable to the fourth. On the role of octaves in the perception of consonances, see [63].[15]

After consonances, Descartes moved on to degrees, or tones. He considered major (9 : 8) and minor (10 : 9) tones (T and t respectively), and major (16 : 15) and minor (25 : 24) semitones (S and s respectively), as well as the syntonic comma (81 : 80) (c), which he called a schisma.[16]

Figure 4.7a,b, taken from the Latin edition of the *Compendium*, show two different arrangements of tones and semitones in the octave represented by a circle; they differ mainly in the different position of the major tone relative to the minor tone. In Fig. 4.7a, starting with the number 540 and rotating clockwise, there is the sequence of tones and semitones t-T-S-T-t-T-S, in Fig. 4.7b, on the other hand, the sequence is T-t-S-T-t-T-S (note that dark circular sectors represent major tones, composite by schisma and/or semitones). That is, the first sequence begins with the minor tone, the second with the major tone (this point will be commented on in Chap. 5). In Fig. 4.7b,

[14] pp. 12–13.

[15] pp. 35–38.

[16] Usually, this term denotes the difference between a Pythagorean comma and a syntonic comma; what Descartes called a schisma is commonly known as a syntonic comma.

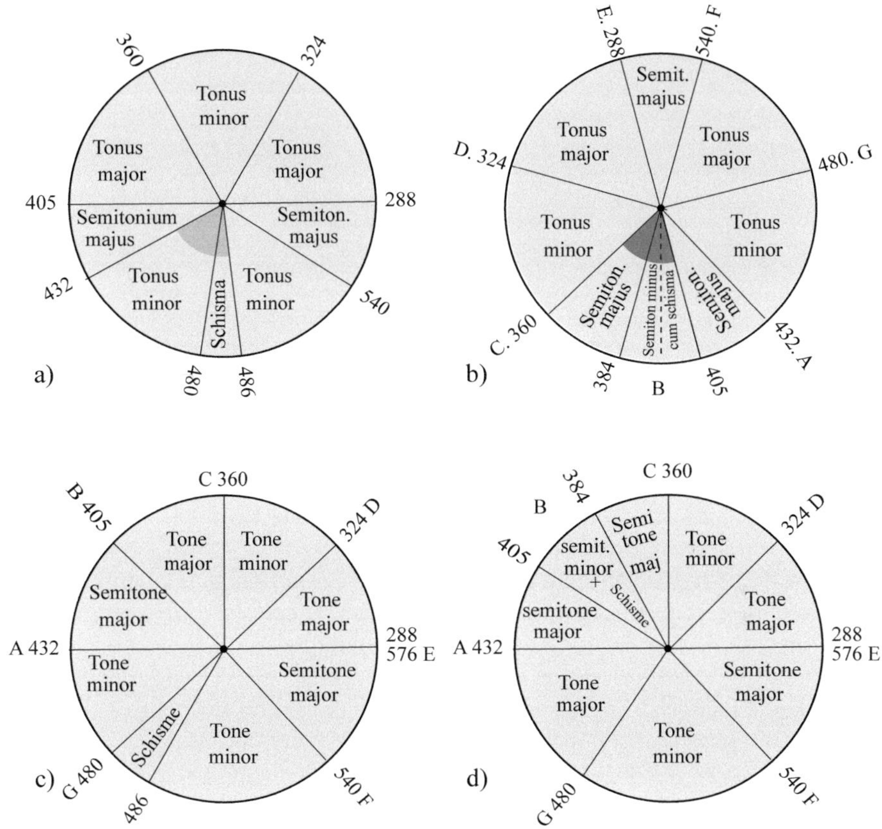

Fig. 4.7 Circular representation of the octave in the *Compendium*: **a–b** Latin edition [29], p. 18; **c–d** English edition [28], p. 32

in addition to ♭, identified by the number 405, the note ♮, identified by the number 384, is also included. The notes ♭ and ♮ are separated by a minor semitone plus a syntonic comma. Figure 4.7c, d come from the English edition of the *Compendium*. They are very similar to the previous figures, though the distributions of spaces in the different circular sectors is quite different. Figure 4.8 shows Guido d'Arezzo's hexachords in three different circles.

Circular representations of pitches were not new with Descartes [56, 91].[17] One of the first occurrences is due to Zarlino, who represented the senario around a circle as shown in Fig. 2.9, drawn form the *Le istitutioni* and especially represented some musical intervals in the *Sopplimenti musicali* [94].[18]

---

[17] p. 34.

[18] p. 296.

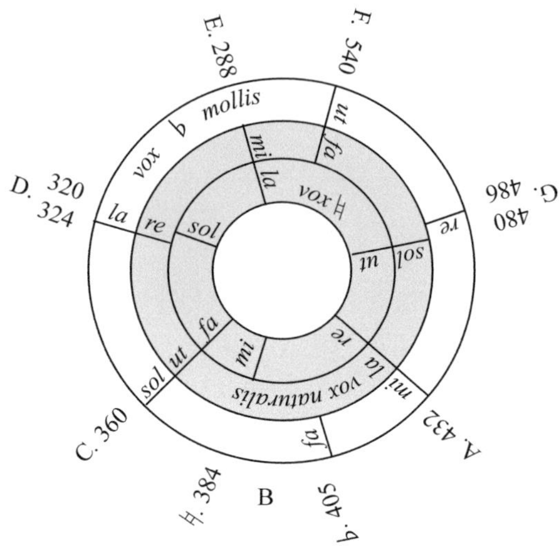

Fig. 4.8 Circular representation of solmization. Redrawn from [29], p. 20

In [56, 93][19] it is maintained that Descartes' diagrams imply a logarithmic representation of the intervals because equal intervals are represented by equal circular sectors instead of sectors proportional to the differences in pitches. The assertion, based in part on visual inspection of the figures, is supported by a comparison of the figures in the Latin and English editions, in which the sectors are not equal.

This is a suggestive but not entirely convincing hypothesis. It is also possible that Fig. 4.7a,b have only a qualitative purpose, since the equality of the intervals and the equality of the sectors can be accepted quite naturally. The difference between the figures in the English and Latin editions, apart from the fact that they have texts in different languages, is not so striking, and one can think that the English figures are simply improved copies of the Latin ones.

The diatonic scale is described in more detail in Fig. 4.9, where three octaves are shown, which Descartes considered sufficient. The letters on the left of the illustration represent the names of the notes. Notice that in the figure, ♭ and ♮, allow for different relative positions of the major and minor tones. For example, in the major third C-E 360-288, both the sequences 360-320-288 (Tt) and 360-324-288 (tT) are allowed.

In the last part of the *Compendium*, a few pages are devoted to dissonances and some rules of composition to end with a very brief discussion on modes.

Descartes returned more than once to questions of musical theory, especially in his correspondence with Mersenne. In the following, we will only report his considerations on nature of sound and consonances. He first clearly recognized the periodic nature of sound:

> Sound is nothing but a certain trembling of the air which tickles our ears, and the turns and returns of this trembling are all the more sudden the sharper the sound is; so that when two sounds are separated by one octave, the lowest will tremble the air only once, while the

[19] p. 30.

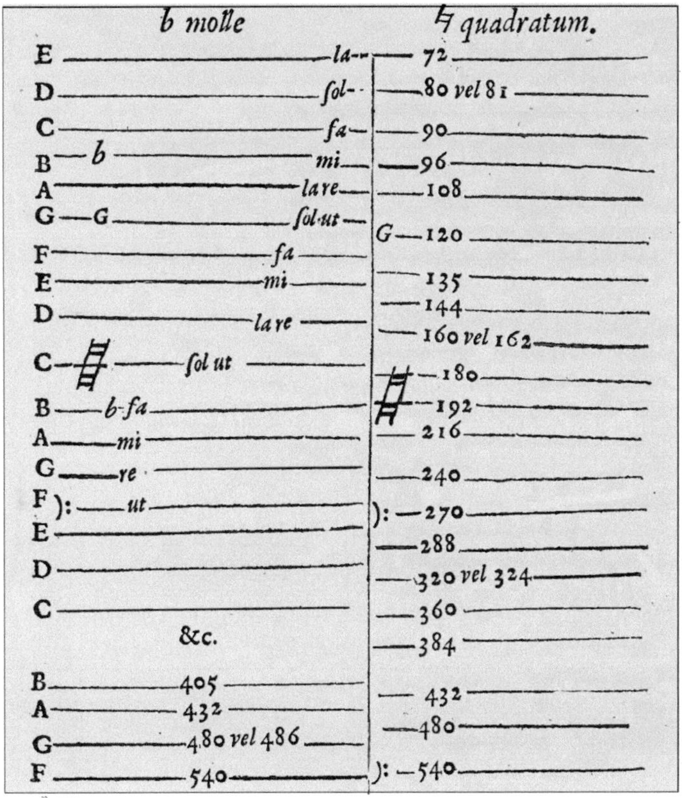

Fig. 4.9 Division of the monochord. Redrawn from [29], p. 18

> highest will make it tremble twice, and so for the other consonances. Finally, it must be supposed that when two tones strike the air at the same time, they will be the more in tune if their vibrations meet each other more often, and if they cause less inequality throughout the body of air [31].[20]

Musical sound is distinguished from noise by its periodicity. In the following quotation, Descartes rather naively introduced periodicity by stating that if a sound does not repeat itself, it cannot be distinguished as a low or high sound. He spoke of a repetition of two or three times. In fact, considering that the frequencies of music are of the order of 100 Hz, two or three repetitions would last only a few hundredths of a second and therefore would not even be perceived:

> For it must be noted that if the sound hits the ear only once, it will certainly be heard as noise and will not be distinguished as a grave or acute sound; for this it is necessary for it to hit the ear at least two or three times, so that from the intervals between the two hits we can estimate how deep or high it is. When you put your finger on a string, immediately after you

[20] vol. 1, Letter to Mersenne, October 1631, pp. 223–224.

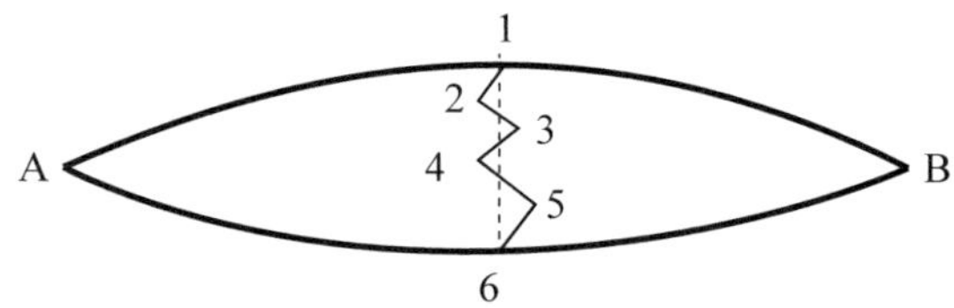

Fig. 4.10 Irregular vibrations of a string due to imperfection. Redrawn from [31], vol 1, p. 267

> have touched it, before it has had time to make several returns, you will hear some noise, but you will not be able to judge whether it is grave [31].[21]

Descartes rejected Beckmann's theory of sound: "To say that it is the same part of the air that goes from the mouth of the person speaking to the ear of the receiver is ridiculous." [31].[22] He rather thought that the cause of sound was in wave due to the agitation of the medium, as in the case of a stone falling into a pool of still water.

The propagation of wave sound, according to Descartes, occurs through phases of contraction and expansion, clearly described in a letter to Mersenne of 1640. When a body moves through the air, that part of the air which was in place cannot leave without condensing a little, and especially it cannot expand to resume its natural place without pressing on all the surrounding air which condenses into the form of a sphere, then it expands compressing more air a little farther away, and so on. A body does not have to move a great distance, it only has to move quickly for a portion of the air to form spheres, and it is these spheres derived by condensation or rarefaction that cause sound. Several sounds or spheres can coexist because the same body is capable of different movements at the same time [31].[23]

In another letter to Mersenne of 1633, Descartes proposed his own theory to explain overtones. He justified his theory with reference to Fig. 4.10. The string AB in its oscillations goes from position 1 to position 6, producing its natural sound, which depends on the length, tension and specific weight of the string. In moving from 1 to 6, the string passes through intermediate positions 2, 3, 4, 5. In the figure they are not shown on the straight line connecting the points 1–6, but rather 'mysteriously' arranged in a zigzag path. Descartes said that because of the irregularity of its parts "after being passed from 1 to 2, it returns to 3, to 4, to 5, and finally to 6, which produces a sound higher than a twelfth of the fundamental" [31].[24] This statement is difficult to interpret; if the numbers were drawn on the line 1–6, one might think of faster oscillations for intermediate positions, but this hypothesis is to be disregarded as Descartes could not have been unaware of the isochronism of the rope oscillations; perhaps the indication of numbering is only meant to indicate some irregularity of motion, but in this way no explanation is given.

While in the *Compendium* the agreeableness of consonances was judged primarily by aesthetic criteria, especially based on the soul's mode of learning, in the corre-

---

[21] vol. 1, Letter to Mersenne, January 1630, p. 107.

[22] vol. 1. Letter to Mersenne, January 1630, p. 109.

[23] vol. 3. Letter to Mersenne, July 1640, pp. 125–126.

[24] vol. 1. Letter to Mersenne, July 1633, pp. 267–268.

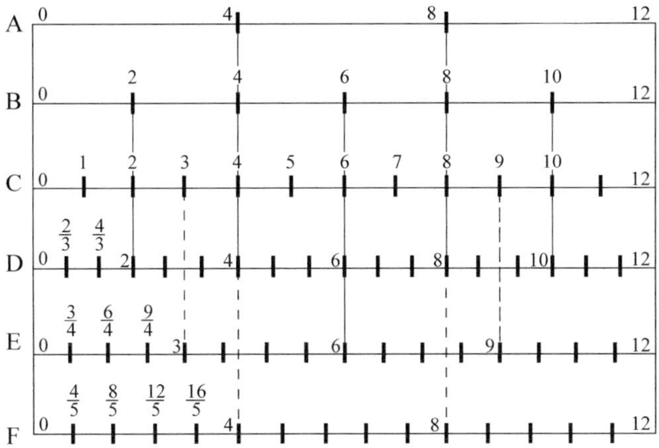

Fig. 4.11 Coincidence for different intervals. Redrawn from [31], vol. 1, p. 224

spondence Descartes got down to the philosophy of nature, adopting the theory of coincidence illustrated in Fig. 4.11. The bold dashes represent the pulses of a sound, the numbers above the dashes are for the time, the first number is for the period of the sound. So the sound in line A has a period of 4, the sound in line B has a period of 2, the sound in line C has a period of 1, the sound in line D has a period of 2/3.

Line A represents a sound one octave lower than that of line B, and consequently each vibration lasts twice as long. Line C, on the other hand, represents the duration of a sound that is one octave higher, so the intervals between pulses are half as long. Line D represents the sound that makes the fifth with C, and the 12th and 19th with B and A. Line E makes the fourths with C, the 11th and 18th with B and A. Line F, which makes the major third with C, makes the 10th and 17th with B and A. It is obvious from this figure that the sounds that make the octaves are those that agree best with each other; those that make the fifths follow them, the fourths after them, and the thirds are the least agreed of all.

Descartes gave value to the theory of coincidence, but he believed that it is one thing to say that a consonance should be grateful, another that it actually is grateful:

> Concerning the sweetness of consonances, two things must be distinguished: that which makes them simpler and more harmonious, and that which makes them more pleasant to the ear. Now, what makes them more pleasant depends on the places where they are used; there are places where even false fifths and other dissonances are more pleasant than consonances, so we cannot say absolutely that one consonance is more pleasant than another. But we can say that, as a rule, thirds and sixths are more pleasing than fourths; that, in happy songs, thirds and major sixths are more pleasing than minor ones; the reverse in sad sounds, etc., so that there are more occasions when they can be used pleasantly. But we can absolutely say which consonances are the simplest and most consonant; for this depends only on the fact that their sounds unite more closely with each other, and that they approach more closely the nature of unison; so we can absolutely say that the fourth is more consonant than the major

> third, though ordinarily it is not so pleasant, as coffee is much sweeter than olives, but not so pleasant to our taste [31].[25]

The pleasantness of a tone, however, depends not only on personal taste, but also on custom.

> For the music of the ancients, I believe, had something more powerful than ours, not because it was more advanced, but because it was less. In fact, those who had a great naturalness for music, not obeying the rules of our diatonic, did more by the sole power of imagination than those who have corrupted this power by the knowledge of theory. Moreover, the ears of listeners not accustomed to such regulated music as ours were much more easily surprised [31].[26]

4.1.2 The **Harmonie Universelle** *of Mersenne*

Marin Mersenne (1588–1648) devoted most of his production to research on sound (mostly music) and for this reason is often considered as a forerunner of acoustics. His results were reported in various treatises, including *Questiones in genesim* of 1623, the *Traité de l'harmonie universelle* of 1627, *Questiones harmoniques* and *Les preludes de l'harmonie universelle* of 1634, the *Harmonicorum libri* of 1648; of some interest is also the manuscript known as the *Livre del la nature des sons*, possibly written after 1626 [36]. But his masterpieces on music are the two impressive volumes of the *Harmonie universelle* of 1636–1637 with a total of about 1500 dense pages [52].

Mersenne studied sound more in depth than he did of optics, and considered it very reliable: "The ear, moreover, never errs as to the acuity or gravity of a sound" [52].[27] The large number of pages devoted to analogies between sound and light, and the prolonged interest in music for over 20 years, in an attempt to establish laws corresponding to those of reflection and refraction of light, testify to Mersenne's efforts not to succumb to his skepticism, which he also expressed to many of his interlocutors.

As a premise to his *Livre premiere des consonances*, Mersenne declared his skeptical epistemology of physics. I don't want us, said Mersenne, to take the meaning of the word *demonstrate*, which is often used at the beginning of propositions, in the same sense and with the same meaning as in geometry, but only with the meaning of knowing, or examining, etc., [...] because I know that it is very difficult to prove anything in physics, if one makes the demonstration with rigor. That's why everyone is free to follow the opinion he wants, according to the most plausible reasons, for example, those who believe that all tones and semitones are identical (see the eleventh proposition of the book of dissonances), as Stevin did at the beginning of

---

[25] vol. 1. Letter to Mersenne, October 1631, p. 223.

[26] vol. 1. Letter to Mersenne, December 1629, pp. 101–102.

[27] Part I, Livre premier de la voix, p. 85.

the first book of his *Géographie*,[28] and the Aristoxenians of Italy with several others, will not lack reason, and it will be difficult to prove to them that the fifth is exactly sesquialtera and the tone sesqui-octave, or if necessary a thousandth part, etc. [52].[29]

The *Harmonie universelle* is considered by historians to be essentially a compilation of the achievements of the science of the time, with an emphasis on the physical and mathematical aspects of music. If this is so, it must be said that it is a very well crafted compilation, which, thanks to personal contributions, is truly original in its completeness. The work is divided into two parts: the first, published in 1636, contains a substantial section on general mechanics as well as strictly musical topics of a speculative and practical nature. The second part, published in 1637, deals mainly with musical instruments.

The *Harmonie universelle* is prefaced by a short text, only two pages, added at the time of printing, the *Abregé de la musique speculative*, in which, according to Mersenne, the main points of the theoretical parts of music are presented. They are all concerned with the physical aspects of music, using a term spread some years later, they are concerned mainly with musical acoustics. It is worth listing them in full:

1. Sound is nothing but a beat of air which the ear perceives when it is touched. Now the two principal properties of sound consist in its strength and in the qualities we have called *grave* and *acute*. Its strength is all the greater when it is produced by a more violent blow of the air: and this blow is all the greater the greater the quantity of air that is struck at the same time.
   As for its gravity, it is all the greater when it is made by slow beats; and conversely, it is all the more acute when it is made by faster beats; for example, if there is a sound in a given time of 50 beats, and another sound in the same time of 100 beats, this last sound will be twice as acute as the first.
2. When two or more sounds are produced together and at the same time, we call them consonances if they agree well and please the ear and the mind. Now the reason for this agreement is taken from the union of the sounds, so that their agreement is sweeter, because they have their union closer and wider, as we find in unison, in octave, in diapente, etc.
3. The unison is the union or blending of two sounds made by an equal number of beats of the air; the octave is the blending of two sounds, the lowest of which is made by one beat and the highest by two; and the diapente is the blending of two sounds, the lowest of which is made by two beats and the highest by three.
   All simple consonances are understood and explained by the first 6 numbers: 1, 2, 3, 4, 5 and 6, because the octave goes from 1 to 2, the fifth from 2 to 3, the fourth from 3 to 4, the ditone or major third from 4 to 5 and the minor from 5 to 6. They represent the number and comparison of their beats.
4. The octave is the sweetest of all, after the unison; because its beats unite more often: for the first beat of the acute motion unites with the first part of the beat of

[28] A rather curious reference.

[29] Part I, Advertisement au lecturer. Before the Livre premier des consonances.

the grave sound, and the second beat with the last part: that is, its beats of 2 beats unite in 2 beats, those of the fifth of 3 beats in 3 beats, etc.
And when the union is equal on the acute side, and unequal on the grave side, the consonance that equally unites its sounds on the part of the one and the other is sweeter: for example, the beats of the fifth unite 3 strokes in 3 strokes of the higher sound, and 2 in 2 of the lower. The twelfth, on the other hand, unites its notes in every stroke of the lower one, which is why it is softer.
5. Since weight cannot raise a string to the octave unless it is quadruple, we can say that the acute sound of the octave is 4 times heavier than the grave sound. But if the strings are of different lengths and of the same thickness and matter, the weight that must raise the string 2 times longer to the octave must be 8 times [actually 16 times], because the quadruple only brings the double string into unison with the subdouble and then the [further] quadruple to the octave.
6. It may be said that 230 toises is the proper measure of the direct propagation of sounds, as they make their way in the time of a second, either the wind favors or it is contrary, and either the sounds are strong or weak, and that 162 toises is the measure of reflected sounds, as a syllable pronounced as quickly as possible will strike the wall 81 toises away, and then return to the ear within a second.[30] But if the sound is made by circles similar to those made by water, it is certain that the motion of the air carrying the sound is 1870 times faster than the motion of water; from this it can be concluded that the air is 1870 times easier to move, more fluid, less resistant, and lighter than water [52].[31]

One of the innovations of the *Harmonie universelle* over the theoretical music treatises of the 16th century is the introduction of the concept, though not the term, of frequency for measuring the pitches of notes. In this way, musical intervals are identified by ratios of frequencies rather than by ratios of the lengths of the strings that produce them. This implies, among other things, a reversal of the roles of arithmetic and harmonic means. For example, in Zarlino's *Institutioni*, the arithmetic mean of the octave, producing the sequence (4 : 3 : 2), places the fourth below the fifth, while the harmonic mean of the octave produces the sequence (6 : 4 : 3), with the fifth below the fourth. Zarlino appreciated the latter circumstance because for him the largest interval had to be in the lower position of the monochord; the harmonic mean therefore has a privileged position. When frequencies are used instead of lengths, the two progressions (4 : 3 : 2) and (6 : 4 : 3) become (3 : 4 : 6) and (2 : 3 : 4) respectively. In this situation, the arithmetic mean enjoys a privileged role instead.

Consonances
The subject of consonances has been of great importance in the treatises on speculative music of all periods. Mersenne followed Descartes on this point, and while he accepted the theory of coincidence, he did not assume that it could definitively

---

[30] There are two conflicting values for the speed of sound, 230 and 162 toises per second.

[31] Shortly after the table of contents.

Table 4.1 Consonances more agreeable according to the ear and to the theory of coincidence; from [52], Part I, *Livre premiere des consonances*, p. 82

I	II
Octave	Octave
Fifth	Fifth
Third major	Fourth
Third minor	Sixth major
Sixth major	Third major
Sixth minor	Third minor
Fourth	Sixth minor

decide on agreeability. Table 4.1 compares the agreeability indicated by the ear (Col. I) with that indicated by the theory (Col. II).

Mersenne tried to explain the discrepancy between theoretical and perceived consonances. The explanation for the greater pleasantness of the major third compared to the fourth is is similar to that of Descartes (see Fig. 4.6); it will be taken up in a more elaborate form by Huygens some years later (see Sect. 3.1.3):

> I say that the fourth must seem less good than the third major, because if you take this third, which is (5 : 4), and the octave of 5, which is 10, you get the tenth major with 4 [(10 : 4) ≡ (5 : 2)]; but if you take the fourth (4 : 3), you get the eleventh with (8 : 3), which is much rougher and less pleasant than the tenth major (5 : 2). For we must judge the soundness of the consonances not only by considering the singles, but also the replicas [higher tones]: the replicas of the major thirds are more pleasant than those of the fourths, because they unite their tones more often, as their lower terms show [52].[32]

Equal Temperament

Mersenne also addressed the problem of equal temperament. Although the idea was not his own, he provided fairly accurate tables and made known to the scientific and musical communities the numerical values of equal temperament, which had been relegated to inaccessible texts.

The *Harmonie universelle* contains several tables with numerical values for equal temperament monochords, given to him by mathematicians of his milieu. Table 4.2, for example, is due to Jean Beaugrand (c. 1595–1640). The first and third columns of the table give the names of the notes, the first starting from A, the second from C. Mersenne made it clear that it does not matter where one starts, since the semitones of equal temperament are equal to each other. The second and fourth columns give the approximate values above and below the correct values, which are given in the fifth column, not present in the original table.

A division with much more digits, due to Jean Gallé (b. 1580), is also present in the *Harmonie universelle*, as reported in Table 4.3, which shows the division for two

[32] Part I, Livre premiere des consonances, p. 77.

Table 4.2 Equal temperament; from [52], Part II, Livre premiere des instruments, p. 38

I	II	III	IV	V
a	100,000	c	100,000	100,000.000
G♯	105,946	♮	105,945	105,946.309
G	112,246	♭	112,245	112,246.205
F♯	118,921	A	118,910	118,920.712
F	125,993	G♯	125,992	125,992.105
E	133,481	G	133,480	133,483.985
D♯	141,422	F♯	141,421	141,421.356
D	149,830	F	149,829	149,830.708
C♯	158,741	E	158,740	158,740.105
C	168,179	D♯	168,178	168,179.283
♮	178,172	D	178,171	178,179.744
♭	188,771	C♯	188,770	188,774.863
A	200,000	C	200,000	200,000.000

Table 4.3 Gallé's table; from [52], Part II, Nouvelles observations physiques et mathematiques, p. 21

NOMBRES DE L'ACCORD EGAL.

	I	II	
I	100000000000	144000	000
II	94387431198	135919	009
III	89090418365	128290	202
IV	84089641454	121089	089
V	79370052622	114292	876
VI	74915353818	107878	109
VII	70710678109	101823	376
VIII	66741992715	96108	470
IX	62996052457	90714	317
X	59460355690	85622	912
XI	56123102370	80817	267
XII	52973154575	76281	243
XIII	50000000000	72000	000

different monochords. Col. 3 serves only to complete the values of Col. 2; all except the value in the third row are found to be correct. The actual value here should be 89 089 871... instead of 89 090 418.... [68].[33]

[33] p. 208.

Overtones

An interesting phenomenon studied by Mersenne was the presence, verified empirically, of more than one simple tone in the sounds emitted by various instruments, including stringed instruments (presence of overtones):

> When an open string is struck and sounded, it produces at least five different sounds at the same time, the first of which is the natural sound of the string, which serves as the basis for the others, and which is only taken into account in singing and in other parts of music, since the other [sounds] are so faint that only the best ears can hear them easily.
>
> [...]
>
> Now these sounds follow the proportions of these numbers (1), 2, 3, 4, 5, because we hear four sounds different from the natural one, the first of which is in the octave above, the second in the twelfth, the third in the fifteenth, and the fourth in the seventeenth, as we can see from the said numbers, which contain the ratios for these consonances in their smallest terms. It is necessary to notice two things: no sound is ever heard lower than the natural sound of the chord, because they are all higher; these sounds follow the same ratios as those of the trumpet [52].[34]

In the last part of the above quotation, Mersenne clearly argued that sounds other than the fundamental are more acute than it. Continuing his analysis, Mersenne wrote:

> In addition to these four extraordinary sounds, I hear a fifth higher sound, which I hear especially at the end of the natural sounds and sometimes a little after the beginning. It forms a major 20th with the natural sound, with which it is like (20 : 3). But I almost always notice that the 12th and the 17th are heard more clearly than the others; from this it follows that one often seems to hear only one of these, or that it is easily mistaken for a fifth or a tenth if one is not careful. When one hears the octave and the fifteenth, it is the latter that is heard more distinctly than the former [52].[35]

It is also likely that instead of the 20th, Mersenne heard the perfect 19th (6 : 1), or perhaps the compound major sixth (10 : 3) between the third and tenth partials [41].[36]

The problem of overtones seemed difficult to explain because, once pitches were associated with frequencies, the existence of several simple tones at the same time seemed to require that a body (e.g. a string) could vibrate at several frequencies at the same time, which was hard to accept. Here is how Mersenne expressed his perplexity:

> [Since the string] produces five or six notes [...], it seems absolutely necessary that it should strike the air five, four, three, and twice at the same time, which is impossible to imagine, unless one says that half the string strikes it twice, while the whole string strikes it once, and that, at the same time, the third, fourth, and fifth parts strike it three times, This situation is contrary to experience, which clearly shows that all parts of the string make the same number of returns in the same time, because the continuous string has a single motion, even though the parts near the bridge move more slowly [52].[37]

---

[34] Part II, Livre quatriesme des instruments, p. 208.

[35] Part II, Livre quatriesme des instruments, p. 209. Translation into English in [41].

[36] p. 41.

[37] Part II, Livre quatriesme des instruments, p. 210.

Some have resorted to the different surfaces of the chord to explain all these sounds, said Mersenne, supposing that the core of the string makes a different sound from that made by the outer surfaces, but there is no reason to believe that the string is divided into several concave cylinders covering the convex cylinder in the middle, like onion skins covering each other. And although the string is made of these different cylinders, it could not make different sounds if it did not strike the air 1, 2, 3, 4, 5 times at the same time: which is no easier to explain by the multitude of these cylinders, than by the other point of view [52].[38]

According to Mersenne, the different tones should be produced by the surrounding air:

> It is more probable that these different sounds are produced by the different motions of the external air rather than by those of the internal air, and that these, when struck by the string, make a quantity of small movements similar to those of the water in a glass, which one makes sound by running one's fingers along the rim, or to those of the water in which one has dipped one end of a monochord [52].[39]

Higher pitches were also identified by Mersenne in the wind instrument. In particular, he noticed that open pipes jump to the octave, while stopped pipes rise to the twelfth (three times the natural frequencies) [52].[40] A phenomenon for which he "cannot find the reason" [52],[41] but which is explained by the modern dynamics of pipes, according to which stopped pipes emit only odd higher harmonics, and the twelfth is the first odd harmonic.

Mersenne also studied the phenomenon of the beats, but was unable to identify the cause.

> But it is difficult to explain this difference which is the cause that pipes distant only a semitone or a quarter of a tone make a movement of air like to a beat of a drum when they play together, because one hears ten beats quite clearly in one second [in both cases] [52].[42]

Solmization

In Mersenne's century, the system of hexachords was gradually abandoned. While it may have been ideal for the solmization of plain chant, it became increasingly unsuitable for part music. Here it was difficult and confusing to deal with a system that used different names for the same pitch. Moreover, the increasing use of chromaticism in the 16th century made the system even more obsolete.

Bartolomeo Ramos de Pareja (c. 1440–1522) proposed the use of eight syllables in his *Musica practica* of 1482 [65, 66].[43] Meanwhile, other more complex systems appeared in Germany and the Netherlands. The syllable Si as the seventh in solmization is signaled at the end of the 16th century in the Flandres, with uncertain

---

[38] Part II, Livre quatriesme des instruments, p. 211.

[39] Part II, Livre quatriesme des instrumens, p. 211.

[40] Part II, Livre sixiesme des orgues, pp. 346–347.

[41] Part II, Livre sixiesme des orgues, p. 236.

[42] Part II, Livre sexieme des orgues, p. 362.

[43] p. 65.

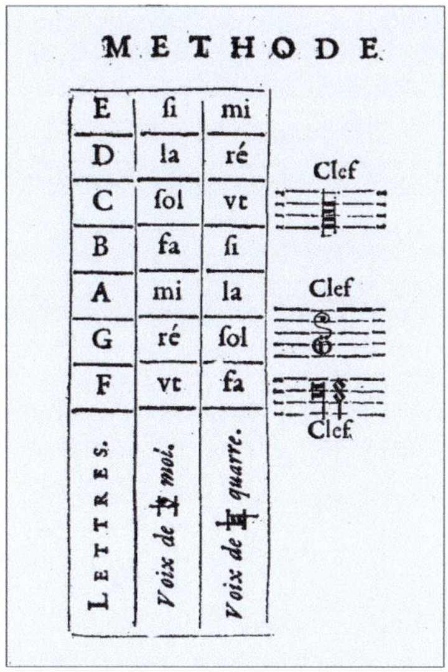

Fig. 4.12 Two examples of solmization with the new syllable Si; from [3], p. 18

attribution [2]. The first printed treatise advocating the use of Si appeared in Germany with Joachim Burmeister (1564–1629) and Sethus Calvisius (1556–1615). But it was in France that Si was first established, to the point that Jean Jacques Rousseau came in 1768 to claim that Si was only known in France [74].[44]

This is evidenced by the musical treatise *Méthode facile pour apprendre à chanter la musique*, whose author is unknown, although two candidates are suggested: Jean Le Maire (c. 1581–1650) and Guillame Gabriel Nivers (1632–1714) [3]. Figure 4.12 shows two different types of suggested solmization, one for ♭, the other for ♮.

This is how the syllable Si is presented in the introduction to the above treatise:

> Of all the methods of learning to sing that have ever appeared, none is easier, shorter, more learned, or more certain than this method, which is called the *Méthode du Si*. It is necessary to renounce reason in order not to yield to the demonstrative proofs made for it, especially in comparison with the common and ancient method called *Méthode des muances*. Here is the cognitive proof: in order to express the different sounds of music, it is proper to give them different names. Now there are seven different sounds in music, so there must be seven different names. Consequently, there is an abuse and a defaut in the ancient method, which admits seven sounds and only six names, and which gives several names to the same sound, and different sounds to the same name [3].[45]

In his *Méthode claire, certain et facile pour apprendre à chanter la music*, first edition 1683, Jean Rousseau (1644–1699) proposed a similar approach. He pointed out that with the introduction of the note Si, solmization is simplified; for example, the old note A-La-Mi-Re becomes more simply A-La-Mi [73].[46]

[44] Article Si, p. 440.

[45] Avant-propos, first page. English translation adapted from [2].

[46] p. 9.

Table 4.4 Different kinds of solmization for the species of octaves; from [52], Part I, Livre troisiesme des genres de la musique, p. 192

I			II			III		
c	FA	DVT	d	SOL	RE	e	LA	MI
♮	MI	CI	C	FA	VT	D	SOL	RE
A	RE	LA	♮	MI	CI	C	FA	VT
G	SOL	SOL	A	RE	LA	♮	MI	CI
F	FA	FA	G	SOL	SOL	A	RE	LA
E	MI	MI	F	FA	FA	G	SOL	SOL
D	RE	RE	E	MI	MI	F	FA	FA
C	VT	VT	D	RE	RE	E	MI	MI

IV			V			VI			VII			
F	FA	FA	G	SOL	VT	a	LA	RE	♮	MI	MI	CI
E	MI	MI	F	FA	BI	G	SOL	VT	A	RE	RE	LA
D	RE	RE	E	MI	LA	F	FA	BI	G	SOL	VT	SOL
C	FA	VT	D	RE	SOL	E	MI	LA	F	FA	BI	FA
♮	MI	CI	C	FA	FA	D	SOL	SOL	E	MI	LA ou	MI
A	RE	LA	♮	MI	MI	C	FA	FA	D	SOL	SOL	RE
G	SOL	SOL	A	RE	RE	♮	MI	MI	C	FA	FA	VT
F	FA	FA	G	VT	VT	A	RE	RE	♮	MI	MI	CI

Mersenne treated the subject of solmization in at least two parts of the *Harmonie universelle*. In the *Livre troisiesme des genres de la musique*, to overcome the difficulty of mutating hexachords, he proposed the introduction of two more syllables; one to denote the note following La by a tone, and the other for the note following La by a semitone: "The syllable Ni has already been invented for the note following La, but since its pronunciation is too similar to that of Mi, I prefer the use of Ci, and Bi for the semitone" [52].[47] Mersenne also mentioned the syllable Dut to indicate the upper octave of ut to avoid duplication of syllables.

Table 4.4 shows the use that can be made of the notes Ci and Bi in the representation of the seven octave species. In the table there are two columns for each type of octave, in the first column is the normal solmization, in the second column is the one with Ci and Bi syllables: Ci when there is a tone after the note La, Bi when there is a semitone.

A little later, Mersenne also reported another type of syllabizing capable of representing all the semitones of the octave. It is clearly illustrated in Fig. 4.13, where the syllables of the tones of the diatonic octave are Ut, Re, Mi, Fa, Sol, Ib; those of the semitones are obtained by reversing the letters: Tu, Er, Im, Af, Los, Bi. Mersenne is not explicit about his source, probably one of his correspondents, the physician

[47] Part I, Livre troisiesme des genres de la musique, p. 192. Note that two syllables, Ba for ♭ and Bi for ♮, were proposed in 1614 by Adriano Banchieri [9], p. 411.

Fig. 4.13 Syllables for the 12-note chromatic scale; from [52], Part I, Livre troisiesme des genres de la musique, p. 193

Table 4.5 Solmization in four octaves; from [52], Part II, Livre cinquieme de la composition de musique, p. 343

1	2	3	4	5	6	7	8	9	10	11	12	13	14	15	16	17	18	19	20	21	22	23	24	25	26	27	28
Γ	A	B	C	D	E	F	G	A	B	C	D	E	F	g	a	b	c	d	e	f	*g*	*a*	*b*	*c*	*d*	*e*	*f*
V	R	M	F	S	L	Z	V	R	M	F	S	L	Z	v	r	m	f	s	l	z	*v*	*r*	*m*	*f*	*s*	*l*	*z*

Christophe de Villiers (c. 1585–c. 1650), "a very excellent philosopher and doctor of medicine" [17].

In the *Livre cinquieme de la composition de musique*, Mersenne explicitly referred to Jean Le Maire, to whom he attributed the introduction of the syllable Za to be placed after the note La: "If we can put Za, invented by M. Le Maire, after the 6th syllable La, all the difficulties of B mol and B square will be reduced to the syllable Za, which will be intoned a tone or only a semitone higher than La, and for this purpose we will always put a diesis when that will be higher by a tone; and if the diesis is not there, it will be only a semitone higher than La" [52].[48] The octave can thus be written with this syllable:

C	D	E	F	G	A	B	C
Vt	Re	Mi	Fa	Sol	La	Za ♯	Vt
V	R	M	F	S	L	Z	V

where the last line contains the initials of the second.

Table 4.5 shows four octaves in which Mersenne used Le Maire's symbol; in the first row are the position numbers of the notes, in the second row are the littera, in the third row the vox, defined only by the initials. The different octaves are distinguished by the different formatting of the characters: uppercase, small caps, normal, italic.

4.2 Mathematical Physics and Frequencies

By the late 1600s, the idea of the periodic nature of sound was well established. The terminology, however, was still uncertain; for example, there was no common use of the term frequency, but only locutions such as speed of vibration, oscillations in

[48] Part II, Livre cinquieme de la composition de musique, p. 343.

a fixed time, etc. The name frequency appeared but was not emphasized in some of Gassendi's writings [10].[49] Aside from the nomenclature, what was missing was the acknowledgement of the true nature of sound. Sauveur, like Galileo and Mersenne, but somehow also Huygens, regarded sound as a periodic succession of pulses whose precise shape was of little importance to the ear. A simple sound (pure tone) was a succession of similar pulses, while a compound sound contained pulses of different kinds. In particular, there was no mathematically grounded wave theory—which would come many years later with Euler—in which, for example, sounds are represented by superpositions of sinusoidal waves [35].

Below we report on analytical and experimental attempts to measure absolute frequency as cycles per second, or Hz, until the early years of the 18th century. For measurement methods of the late 18th and 19th centuries, see [34].

4.2.1 Absolute Frequency

Galileo in the *Discorsi* had highlighted the parameters that influence the frequency of vibration of strings: the length, the weight per unit length, and the tension to which it is subjected. Mersenne presented them again, but in a less precise form than Galileo, because he did not fully understand the dependence on weight. Nevertheless, he discussed the law of the vibrating string at length in his *Harmonie universelle*. This law is essentially empirical, except for the inverse proportionality between frequency and length, which was somehow proved on a mechanical basis. Expressed in a modern formulation, Mersenne's law has the form:

$$f = K\frac{1}{l}\sqrt{\frac{F}{\gamma}}$$

where K is a constant of proportionality, F is the tension (force), l the length and γ the weight for unit length of the string.

However, Mersenne's law does not give the absolute values of the frequencies, but only the relative ones: for example, one can say that the frequency is halved by doubling the length, but one cannot say how much the frequency of a string of a given length is worth.

Mersenne was well aware of the limitations of his law, both because it was incomplete and because it lacked a mechanical justification, and he appealed to the mathematicians of the day to try to solve the problem. In particular, in 1746 he wrote to Huygens to take up the enterprise, but Huygens replied that he had often thought about the matter, but the solution would be very difficult [85].[50]

Although Mersenne was unable to provide a rigorous demonstration or a formula, he did have some success in finding the absolute value of frequency. For this purpose,

[49] pp. 25–28.
[50] p. 47.

Fig. 4.14 Stancari's rotating device. Redrawn from [8], p. 17

he claimed that it was sufficient to determine experimentally the value of the absolute frequency in a particular case, which allowed to determine the constant of proportionality K. Mersenne considered the vibrations of a very long string, about 6 m, so that one could easily count its vibrations in a given time. However, he did not insist much on the experiment, both because he was more interested in the method than the result, and because the experiment was not too easy to perform with accuracy, and therefore the result he obtained was not very reliable. The experiment was repeated many years later, in 1860, and with greater care [34].[51]

Another experimental approach to the measurement of absolute frequency was pursued shortly thereafter by Hooke and Huygens with devices broadly classifiable as sirens. Particularly interesting is the attempt of Huygens, probably in 1682, who succeeded in producing a sound of known frequency in unison with the D of a spinet by spinning a series of wheels suitably linked together [43].[52]

Somewhat more interesting, though later, is the approach of Vittorio Francesco Stancari (1678–1709). In 1713 a posthumous writing, the memoir *De certa soni mensura costituenda*, reported on an experiment of 1706, not unlike those of Huygens and Hooke, but more refined and, above all, pursued with greater perseverance. Stancari said that he got the urge to measure a standard frequency as a result of reading Sauveur's work.

The device he used is shown in Fig. 4.14; it consists of a wooden wheel about one meter in diameter with 200 iron foils threaded around its outer circumference. When the device was spun, it produced a distinct, though not too loud, sound of known frequency. To measure the pitch of a given sound, it was sufficient to bring the sound emitted by the wheel into unison with the sound to be measured.

[51] p. 298.

[52] vol. 19, p. 375.

Stancari, who was not only a mathematician but also a good musician, had the foresight to keep the device running for a long time to minimize the error in counting the revolutions and thus the frequency; moreover, each measurement was repeated several times. With such an instrument he measured the frequency of the tuner of the organ of the Basilica of San Petronio in Bologna, finding a frequency of $A_4 = 386$ Hz [8].[53]

In 1701, Joseph Sauveur had made an attempt to generate sound with known frequency using a completely different approach that cleverly combined experience and theory (see Chap. 5 below). It consisted in counting the beats produced by two organ pipes tuned to the minimum semitone (25/24). Sauveur claimed to be able to produce a standard sound with a frequency of 100 Hz.

In addition to these experimental attempts, theoretical ones were made during the same period. Huygens had published in 1673 his seminal text *Horologium oscillatorum* in which he demonstrated that, in the absence of friction, the oscillation of a heavy body along a cycloid occurred with a period independent of the amplitude of the oscillation, unlike what happened for a simple pendulum moving on an arc of a circle [42].[54]

In a fragment published in the *Oeuvres complètes*, Huygens proved that the gravity of a body—in modern terms, the force acting on the body moving on the cycloid—varies in proportion to the distance from the lowest point of the cycloid [43].[55] This fact, even with the laws of mechanics of the day, enabled him to state that any motion subject to a force proportional to the displacement—as is the case with the vibrating string—is isochronous and harmonic.

Huygens' approach to the law of vibration of strings was a step-by-step process. First, he considered vibrations of strings such as those of Fig. 4.15, in which a lumped mass at the midpoint is the object of a restoring force, due to the tension of the string that varies linearly with the displacement of the mass. For the string of Fig. 4.15b, Huygens arrived at the formula:

$$f = \frac{1}{\pi}\sqrt{\frac{g}{2a}}\sqrt{\frac{K}{G}}$$

where a is half the length l of the string, K is the value of the weight tending the string, G is the weight of the body G.

To compare the previous relation with the modern formula:

$$f = \frac{1}{2l}\sqrt{\frac{gF}{\gamma}}$$

just take $K = F$, $G/l = \gamma$ to get:

[53] p. 18.

[54] pp. 184–186.

[55] vol. 18, p. 489.

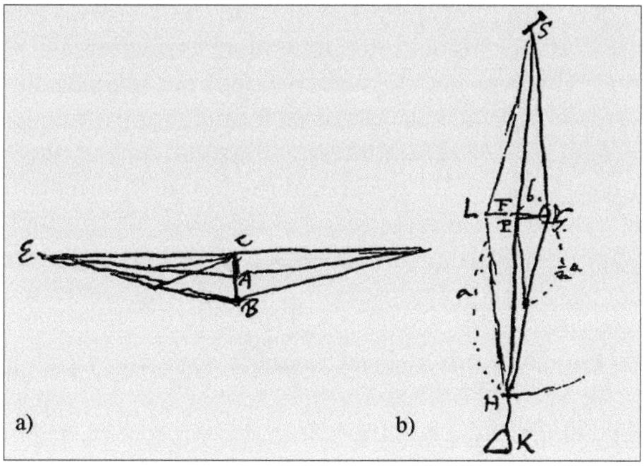

Fig. 4.15 Vibration of a string with a lumped mass. Redrawn from [43], pp. 489–490

$$f = \frac{1}{\pi l}\sqrt{\frac{gF}{\gamma}}$$

The only difference is that the numerical coefficient is π instead of 2; therefore, by concentrating the mass in the middle, we get an error of about 50%. Huygens also showed that the transverse oscillations of the string are isochronous with the circular oscillations of the 'compound' pendulum SLH around the axis SK.

Huygens then considered the case of a mass distributed along the string, or a weightless string loaded by many small weights. But he did not mention the solution. The editor of the *Oeuvres complètes* developed Huygens' suggestion in his own way, following an energy approach, and arrived at the relation [43]:[56]

$$f = \frac{\sqrt{10}}{\pi}\frac{1}{2l}\sqrt{\frac{gF}{\gamma}}$$

which differs by about 1% (excess) from the solution believed to be correct.[57]

Truesdell more or less explicitly suggests the possibility that even in the case of a string with distributed weight, the period of the circular oscillations can be compared with the period of the transverse oscillations. In the former case, each elementary part of the string is subjected to a centrifugal force proportional to its distance from the axis, and in the latter to a restoring force due to the tension of the string, also proportional to its distance from the axis [85].[58]

---

[56] vol. 18, p. 495.

[57] Note that all formulas for frequency consider this quantity as it is considered today, that is, the frequency of a complete round trip of oscillations. Throughout the 18th century, at least in France, the frequency was considered for half a complete cycle, and was therefore double the modern value.

[58] p. 49.

Sauveur, in his memoirs of 1713, took up Huygens' suggestion of assimilating the circular motion to the transverse oscillation and obtained the same result. His, since Huygens had neither developed nor published his results, should be regarded as the first sufficiently accurate formula for the frequency of vibrating strings.

Still in 1713, about the same time as Sauveur, the English mathematician Brook Taylor (1685–1731) published the 'correct' formula for the frequency of vibration of a string and also the shape of the curve—a sinusoid—representing the transverse displacements, though limited only to the fundamental mode (i.e., higher modes were not considered). The memoir with which Taylor presented his results is the *De motu nervi tensi* [83].

Taylor's result is correct, but the rigor with which it was obtained left something to be desired even for his time, so much so that a few years later Johann Bernoulli felt compelled to confirm it into two papers [15, 16].

Shortly after publishing his theoretical work, Taylor performed some experiments to obtain empirical verification, following Huygens' 1682 approach [43].[59] He compared the sound produced by a quill attached to the escape wheel of a pendulum clock for which he knew the frequency of rotation. He found that when the quill rotated 1532 times per second, the sound it emitted was tuned to the A La Mi Re of a harpsichord string, or rather two octaves higher. The string was 12.3 inches long, with a tension of 10 ounces and a weight of 1 grain per foot. The Taylor equation gave a frequency of 383 Hz, exactly the same value obtained by the experiment, provided that 1532 is divided by 4, for lowering it by two octaves.[60]

However, Taylor found that the agreement of his equation with experiments was not always so good. Below is a quotation from an experiment which testifies to such a circumstance.

> March 10, 1712/13 I stretcht a string of the best white packthread 126 feet long [38.4 m], weighing 350 grains, with a weight of 52 ounces or 24,960 grains. This according to my second [theorem in] *De motu nervi tensi* ought to make 4.27 vibrations per second. But by observing the vibrations of it for 5 seconds at a time-thrice repeated, I found it to make but 4.07 vibrations in a second. Which is about 1/20 less than it should be by the calculation. This difference is much less than what it was in the experiments I tried with strings of 24 1/2 feet. Which agreed very well with each other, making all of them between 72 and 73 vibrations in the time they ought to make 100; which is more than 1/4 part less than the calculation. Whence it appears that the more the string is stretched the nearer it agrees with calculation.[61]

Subsequent studies of the vibrating string by mathematicians such as Euler and d'Alembert, and then Lagrange, represent an important milestone for mathematical analysis and continuum mechanics with the introduction and use of partial derivative equations, but they are not central to our purpose, for which we refer to the literature [85].

We would like to emphasize that Alexander Ellis (1814–1890), in his exhaustive collection of information on attempts to produce sounds of known frequency, sub-

---

59 vol. 19, p. 375.

60 All frequencies are doubled, with respect to modern use.

61 Quoted from [21], pp. 19–20.

stantially minimized all the attempts made in the early 1700s. In particular, he did not even mention the proposals of Huygens and Hooke, perhaps considering them only simple suggestions, and did not mention Stancari's interesting experience. Under the category of sirens, his first reference is to the siren of Cagnard de la Tour, which dates from 1819 [34].

4.2.2 Overtones

Although it was known all along that the sound emitted by a string also contained sounds at frequencies higher than the fundamental, it was still not clear where they came from. Only a few thought that they were directly derived from the vibrations of the string, because it seemed impossible that the string could vibrate at more than one frequency. A simple matter of common sense, and also true if instead of saying "one frequency" we say "one way".

Only one of the great mathematicians of the 18th century went against the grain; he was Daniel Bernoulli, perhaps not of the same rank as Euler and d'Alembert, but gifted with a greater sensitivity to more properly physical aspects. In the early 1730s he found that the vibrations of a heavy chain can occur in more than one way, that is, in modern terms, that there is more than one proper mode. In the early 1740s, Bernoulli looked at the vibrations of a bending beam and found that they too had multiple proper modes and frequencies, but that these did not follow the rule of natural numbers, also suggesting the possibility of a coexistence of multiple modes of vibration, providing, if not a proof, at least a plausible explanation [27, 85].[62]

But it was in two memoirs of 1753 [13, 14], that Bernoulli gave his thesis on the simultaneous existence of multiple modes of vibration, limited to the cases of vibrating strings and hanging chains. He began the first of his memoirs by referring to Taylor's theory and stating that it allowed—although Taylor had not found higher modes—that a stretched string "can form its vibrations in an infinity of modes [...]. These modes consist of the number of antinodes that the string can form during vibrations" [13].[63]

Figure 4.16 shows the cases with 1, 2, 3, 4, 5 antinodes (bellies). For each type of oscillation, the transverse deformations can be larger or smaller, but even the largest ones can be considered infinitely small according to Bernoulli. This fact allows him to use what is now known as the principle of superposition of effects, one of the first applications in vibration dynamics. Bernoulli asserted that the types of vibration that can be derived from Taylor's approach are not only abstract in character, but can be verified empirically with, for example, the trumpet marine [13].[64] The absence of any reference to Sauveur's work is remarkable.

[62] pp. 154–199; p. 353.

[63] p. 148.

[64] p. 150.

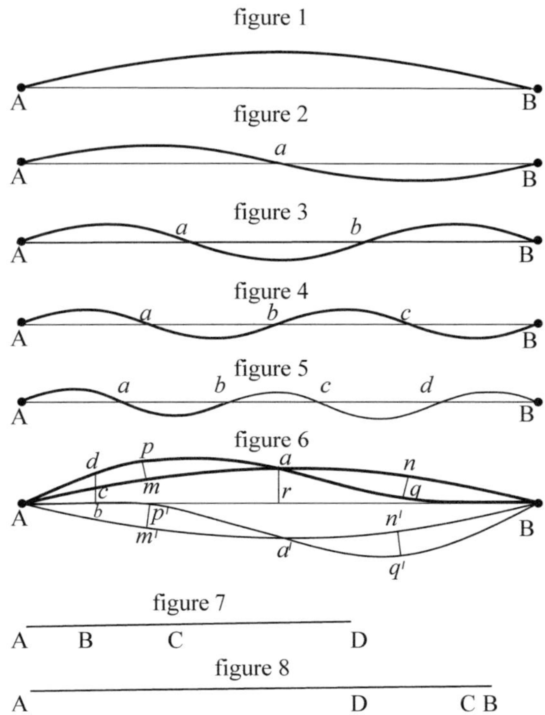

Fig. 4.16 Superposition of modes. Redrawn from [13], p. 552

For Bernoulli, the multiplicity of vibrations is manifest in all sonorous bodies; for example, with a transverse flute one can, by a simple variation of the mouthpiece, produce the fundamental, the octave, the twelfth, the double octave, etc., which stand in the progression of 1, 2, 3, 4, 5, etc. But it should not be believed, Bernoulli argued, that this is a general law, because, for example, in closed pipes we get sounds that follow the progression of the odd numbers 1, 3, 5, 7, etc., while a bar can vibrate with incommensurable frequencies [13].[65]

> My conclusion is that every sonorous body potentially contains an infinity of sounds and an infinity of corresponding ways of making its regular vibrations; finally, that in every different kind of vibration, the bending of the parts of the sonorous body is different [13].[66]

But these different kinds of sounds, Bernoulli said, can not only occur in isolation, but can also occur together: "In fact, all musicians agree that a plucked string simultaneously produces a fundamental and other much higher-pitched sounds" [13].[67] Something similar happens with bells and steel bars. But even air is not exempt from this multiplicity of coexisting sounds, "which proves conclusively how the different waves of air mingle, is when one hears distinctly all the parts of a concert, and that all

[65] pp. 150–151.

[66] p. 151. Translation into English in [85].

[67] p. 152.

the waves caused by the different parts mingle in the same mass of air without interfering with each other, as is the case with the rays of light entering a dark chamber through an opening, which do not interfere with each other" [13].[68]

"What I have just said about the nature of the oscillations of bodies attached to a stretched string", Bernoulli continued, "I do not hesitate to extend to all small reciprocal motions that can occur in nature, provided these [...] are set up by a permanent cause. For every body which is somewhat displaced from its point of rest will tend toward that point with a force proportional to the small distance from the point of rest" [13].[69] In figure 6 of Fig. 4.16, Bernoulli illustrated the superposition between two different modes of vibrations, the first and second. Although the drawing contradicts this, he imagined that the amplitudes of the vibrations were infinitesimal, so that, for example, the curve describing the first mode could be compared with the axis AB of the string.

These are Bernoulli's considerations of the superposition of modes:

1. The ideal curve AmanB makes its oscillations with respect to the axis AB entirely following the simple oscillations of figure 1.
2. Each point on the curve ApaqB has its relative motion with respect to the points on the curve AmanB, as do the absolute motions of the curve shown in figure 2.
3. Each point in figure 2 makes exactly two oscillations, while the same point in figure 1 makes one. It is therefore necessary that all points on the curve ApaqB end their oscillations together with those on the curve AmanB.
4. If a point on the curve ApaqB is outside the curve AmanB, the same point, at the end of its two oscillations, will be inside the latter curve, which will have made only one oscillation; thus, if the curve AmanB assumes the configuration A$m'a'n'$B, the curve ApaqB assumes the configuration A$p'a'q'$B, and this A$p'a'q'$B remains what it was by changing the order of the sides, that is, the same as BqapA.

Bernoulli argued that all solutions could be obtained by superimposing what he called Taylor curves. For this purpose he introduced the functions corresponding to the modal forms in Fig. 4.16.

figure 1	$y = \alpha \sin \frac{\pi}{a} x$
figure 2	$y = \beta \sin 2 \frac{\pi}{a} x$
figure 3	$y = \gamma \sin 3 \frac{\pi}{a} x$
figure 4	$y = \delta \sin 4 \frac{\pi}{a} x$

where a is the length of the string, x the abscissa, and y is the ordinate. Combining all these curves, with coefficients α, β, etc., gives:

[68] p. 153.

[69] p. 152.

$$y = \alpha \sin \frac{\pi}{a} x + \beta \sin 2\frac{\pi}{a} x + \gamma \sin 3\frac{\pi}{a} x + \delta \sin 4\frac{\pi}{a} x + \text{etc.}$$

which contains infinite terms.

> XVIII. I have so far avoided calculations, and have based all my reasoning on principles confirmed by experience, that there can be a mixture of vibrations in a single sound body, which are absolutely independent of each other. To well consider this principle, it is not different from that of the composition of motions; however, to put it in all its light, I thought I must open here to mechanical reflections, and calculations that this matter requires [13].[70]

What is still missing is the study of the evolution in time of the coefficients α, β, γ, δ which govern the superposition of the modes. Bernoulli will complete the study in the last part of his memoirs, which we will not comment on here for reasons of space.

Bernoulli's theses were immediately countered by Euler and d'Alembert with the two arguments:

1. A function expressing the vibrations of a string can have only one frequency.
2. Not all functions representing the shape of a vibrating string can be expressed as a series of sines.

The first argument is mostly physical, the second mostly mathematical, although they are not completely independent.

The answer to the first argument is complex and also depends on one's concept of function; for how can a periodic sound have more than one frequency? This is what Bernoulli thought:

> As to the question whether the new oscillations are really simple oscillations and synchronous for all points, or whether they are rather a mixture of several different oscillations coexisting in the same string, all of different duration, I have spoken of them only to better explain their nature, far from quarreling with such great men about the meaning of certain terms [13].[71]

Bernoulli showed how a string could be deformed by imagining the superposition of different modes, giving a unique form to the function expressing the vibrations of the string and thus to the air pulses reaching the ear.

The second argument was still more complex to discuss and Bernoulli was unable to provide a response; a fully satisfactory answer would come much later with the work of Jean Baptiste Joseph Fourier (1768–1830) with his *Théorie analytique de la chaleur* [40] of 1822. Fourier had developed his work to study thermal processes and had only hinted at its possible application to music; for an explicit application to this field one would have to wait for George Simon Ohm (1789–1854) [45], whose achievements will be commented on in Chap. 6 below.

4.3 Music in the Academies

Since the second half of the 17th century, scientific research was carried out mainly in the various academies that had spread throughout Europe. During this period,

[70] p. 160.

[71] p. 172.

professional mathematicians applied to the study of music with a passion and intensity comparable only to that of the mathematicians of ancient Greece. With different and more powerful mathematical tools, however, and with different goals. There was more of a focus on practical music and learning how to use the many musical instruments that were becoming popular.

4.3.1 Royal Society in Music

From its founding in 1660 to the early 18th century, the minutes of the Royal society of London chronicle the acoustic experiments conducted at its meetings, reporting on studies ranging from string vibration to the medicinal powers of music. At its heart was an attempt to discover the fundamental laws of speculative music. Although most research focused on acoustic studies, on vibrating strings, on the transmission and speed of sound through various media, it often extended into related areas: hearing and the ear, instruments, and even musical composition itself. Indeed, documents from the early years of the Royal society support the view that the organization actively encouraged both scientists and musicians to develop fixed rules for composition and theory, and to reconcile the mathematical principles of acoustics with practical temperament systems. The society was particularly intrigued by the concept of a scientific method of composition proposed by the theorist/composer/teacher John Birchensha (c. 1605–c. 1681), a frequent advisor to the society in the field of music [54].[72]

In the following, we report only the writings of two of the society's greatest exponents: John Wallis and Isaac Newton.

4.3.1.1 John Wallis's Ancient and Modern Music

John Wallis (1616–1703), along with Isaac Barrow, regarded as the greatest English mathematician before Newton, wrote in the spirit of the age on mathematical physics, and his *Mechanica sive de motu* of 1671 is a milestone [86]; he also produced important texts on music.

The first evidence of Wallis' interest in music is the correspondence with Henry Oldenburg (c. 1619–1677) of 1664, related to the musical writing of Birchensha [90].[73] Very interesting for our purposes are his edition of Ptolemy's *Harmonica* of 1682, with a high philological standard for the time, and the appendix to this edition, *The harmonics of the ancients*. But probably his most known contributions to music rests on two letters to Oldenburg of March 14 and 27, 1677, which contain discussion of the presence of higher modes (modern term). To signal also two papers published in the *Philosophical transaction* of the Royal society [88, 89].

[72] p. 63.

[73] Introduction.

Below are brief reflections on the appendix to Ptolemy's *Harmonica*, concerning the main approaches to speculative music, the Aristoxenians and the Pythagoreans. Some space is devoted to the modes of vibration.

The Harmonics of the Ancient
The harmonics of the ancient, is a rather long and interesting treatise on music. As well as illustrating the relationship between ancient and modern speculative music, it is also intended to provide Wallis with a technical forum in which to publish his own ideas about music. In the last part of the Appendix he did indeed rework the ideas he had reported in some letters to Oldenburg many years earlier [90].[74] The appendix covers four main topics: terminology, pitches and their study, Greek harmonics, and modern speculative theory. Many ancient music theorists are mentioned, Euclid, Eratosthenes, Aristoxenus, Ptolemy, Boethius, Henri Estienne, Guido of Arezzo, Jacques Lefèvre d'Étalpes, Heinric Glarean, Gioseffo Zarlino,Vincenzo Galilei; though it is uncertain how much familiarity Wallis had actually acquired with them; he assumed, for example, that Euclid was an Aristoxenian. Among the modern theorists, reference is made to Kepler, Mersenne, Descartes.

All of *The harmonics of the ancients* move mainly on a mathematical level. In a few places, Wallis raised physical considerations: for example, about the validity of Aristoxenus' approach, and when he presented a possible explanation of the nature of consonance by referring to the theory of coincidence.

Wallis discussed Aristoxenus' ideas at length without condemning them. Quite interesting is his distinction between interval (diastēma) and ratio. The former term (interval) would have been used by Aristoxenians because of its reference to geometry. The latter term (ratio) would have been used by the Pythagoreans, and he noted that Ptolemy never used the term intervals. In fact, this assumption has no historical roots, for as we saw in Sect. 1.3.2, Euclid, a Neopythagorean, used the term interval. Wallis can be excused, however, since he considered Euclid to be an Aristoxenian.

Wallis attributed to the Pythagoreans a physical measure of pitch based on the speed of the air reaching the ear: "thus with double the speed the level of pitch is doubled, and they make the octave consonances. And likewise for other consonances" [90].[75] Without historical evidence, however; if it is indeed true that, as was common in ancient Greece, the Pythagoreans associated pitches with greater or lesser air speed, none of them classified consonances by ratios of speed, as Maurolico suggested (see Sect. 3.1.1).

The Pythagoreans compared pitches of sounds to ratios of whole numbers, using arithmetic, Aristoxenus opposed this view:

> [Aristoxenus] and his followers considered this to be excessively precise and had recourse to the judgement of the ears and wished to derive from that source the first principles from where the rest were to be deduced. Crying that those who pursued the details of ratios acted in vain, they asserted that the judgement of the ears showed they were often contrary to

[74] pp. 45–63.

[75] p. 113.

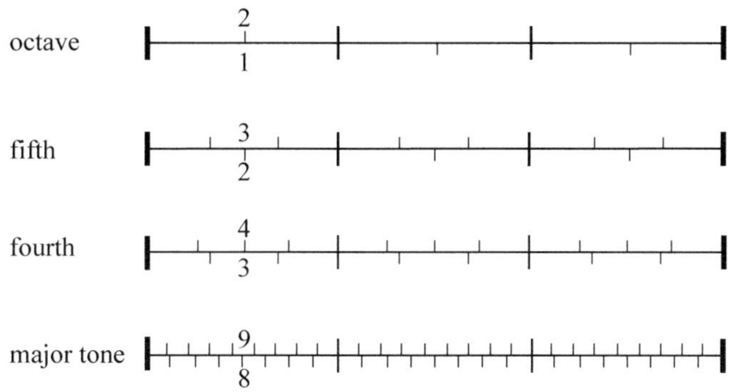

Fig. 4.17 Coincidence theory. Redrawn from [90], pp. 132–133

> the phenomena. And perhaps this was because some of the Pythagoreans had sometimes assigned ratios not as carefully as was right, as Ptolemy sometimes admits [90].[76]

Aristoxenus placed thus sensory judgement in the first place and thence deduced his demonstrations. He put it thus:

> A geometer has no need of the power of sense. But the musician puts the precision of sense in almost the first place. For we judge the sizes of intervals by hearing, but we perceive their force in the end by the intellect [5].[77]

Sometimes, commented Wallis, Aristoxenus seemed to speak uncertainly as to whether a fourth was really exactly two and a half tones, or whether it was close to it, differing only by the smallest amount. Likewise, whether a fourth is to be measured from one of the smaller intervals (the semitone or the diesis), or whether it is in fact incommensurable with all of them: in other words, whether what is called a semitone is exactly half a tone, and thus a fifth of a fourth; whether an enharmonic diesis is exactly a quarter of a tone, and whether a chromatic diesis is a third of it, and so on. And whether a ditone is exactly eight times the smallest diesis or a little less [90].[78]

Wallis, as a mathematician, could not but side with the Pythagoreans, but with some reason. For instance, when intervals differ very little from each other, the precision of the ear can hardly distinguish them, if at all—for the ratio (16 : 15) differs from that of the limma, (256 : 243), and that of the major, (9 : 8), from that of the minor, (10 : 9), by the ratio (81 : 80); which is so small that the ear cannot distinguish it from the unison. In such a case we must judge by reason rather than by sense as to which opinion is to be preferred (since sense will not be unduly troubled by allowing either) [90].[79]

---

[76] p. 113.

[77] p. 189. Free translation into English by Wallis.

[78] p. 115.

[79] pp. 131–132.

Wallis dealt with the cause of consonance in a section entitled *Pythagorean theory placed on a physical basis*, without citing who first proposed a solution to the problem. He credited a Pythagorean theory of sound that actually had no historical basis, which calls for the theory of coincidence to explain consonances [90].[80]

Wallis presented the theory of coincidence assuming that "the more often the coming and going of two strings struck side by side coincide, the more pleasing these strings sound together to the ear; the more rarely (and thus the more often they clash), the less pleasing they sound" [90].[81] This criterion is illustrated in Fig. 4.17, where the coincidence of sounds is shown for the octave, the fifth, and the major tone. Note the inclusion of this last interval, the major tone, which can hardly be considered a consonance, among the examples of coincidence, and the exclusion of the major and minor thirds and sixths, at a time when they were commonly considered consonant.

Interestingly, Wallis assumed either of two cases: the two tones are in phase or they are not in phase. If "they either begin at the same time, or at least coincide at some point, the coincidence of their vibrations will last as long as they continue [to sound]. But if [...] there is no coincidence, a similar alternation will continue: or even, not improbably, each, struck by the strokes of the other (carried by the common air), will be impeded in its own motion by the strokes of the other, until little by little, after a few strokes, they will come together in motion or simultaneous vibration" [90].[82]

However, Wallis was not sure about the coincidence theory: "Whether this is the true physical cause of the pitch of sounds, which the Pythagoreans not without reason, and others after them, suppose; and, further, whether the degree of pitch is proportional to the frequency of vibrations in a given time: I do not know whether this has yet been sufficiently proved by experiment, and therefore I do not wish to affirm it too boldly" [90].[83]

Wallis completed his appendix to Ptolemy's *Harmonica* with some interesting reflections on the aesthetic value of Greek music.

> And so I am completely convinced that neither was the ancients' music more accurate than ours, nor did those prodigious effects which they often recall (by Orpheus, Amphion, Timotheas, etc.) ever touch the minds of men: except, you might say, through the exaggeration of historians, or perhaps they touched an inexperienced people because of the great rarity of music, rather than its excellence [90].[84]

Today, Wallis continued, it is easy to admit this, since musicians play almost exclusively for themselves, to please the soul, rather than, as the ancients seem to have done, to move emotions back and forth. Moreover, their simpler, single-voice music did not obscure the text as our more elaborate music does; so, for example, when

[80] p. 132.

[81] p. 132. The locution "the more often the coming and going" calls for the modern term frequency.

[82] p. 132. The same assertion is found in [60], p. 8; see also Sect. 3.1.1.3.

[83] p. 133. Only here did Wallis use the term *frequency*, but the reference to "a given time" suggests that his frequency is not the same as ours.

[84] p. 130.

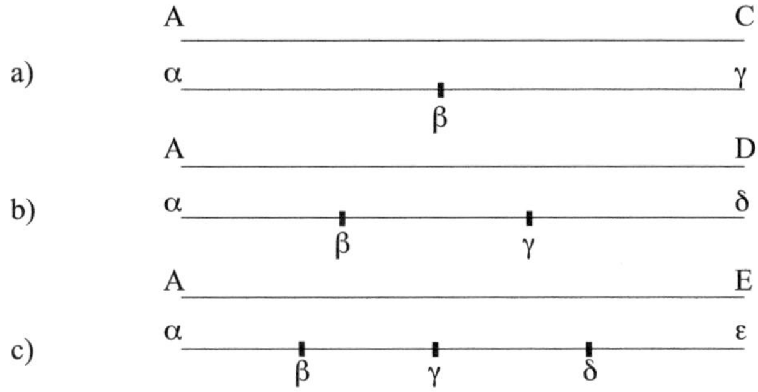

Fig. 4.18 Nodes in a string excited by another string of half length (**a**), one third (**b**), and one fourth (**c**). Redrawn from [87], p. 840

tragic texts are produced with tragic recitation, tragic verse, and tragic sound, which they put together in their music, it is no wonder that they arouse tragic emotions, not unlike us when we recite tragedies: and likewise for other emotions [90].[85]

Nodes in Vibrating Strings. Letters to Oldenburg of 1677

Wallis first interest in the nodes of vibrating strings is attested by a letter to Oldenburg, dated March 14, 1677 [87]. The letter begins with a description of an experiment:

> Whereas it hath been long since observed, that if a viol string, or lute-string, be touched with the bow or hand; another string on the same or another instrument not far from it [...], will at the same time tremble of its own accord [...]. But adde this to the former observation; that, not the whole of that other string doth thus tremble, but the several parts severally, according as they are unisones to the whole or the parts of that string which is so struck. For instance [supposing] AC to be an upper octave to $\alpha\gamma$; & therefore an unisone to each half of it, stopped at β. Now if, while $\alpha\beta$ is open, AC be struck; the two halves of [this] other, that is $\alpha\beta$ & $\beta\gamma$, will both tremble; *but not the middle point at* β [emphasis added, see Fig. 4.18a]. Which will easyly be observed if a little belt of paper be lightly wrapped about the string $\alpha\gamma$, & [re]moved successively from one end of the string to the other [87].[86]

Wallis went on to observe that similar results are found when AD (Fig. 4.18b) has a natural frequency three times that of $\alpha\delta$, that is, when AD is an upper twelfth to $\alpha\delta$. In such a case there are two points that do not vibrate, while if AD is a double octave to $\alpha\epsilon$ there are three fixed points (Fig. 4.18c), and so on. Wallis continued his letter by considering what happens when a string is struck at the potential fixed points. In such a case the string does not sound clear.

Figure 4.18 is misleading to a modern reader. For example, AC in Fig. 4.18a represents a string an upper octave above $\alpha\gamma$; it should be drawn as half of $\alpha\gamma$. Similarly, AD should be a third of $\alpha\delta$ and AE a quarter of $\alpha\epsilon$. This was not a mistake, as it was not uncommon to represent segments that did not match our sense; the same was done by Robert Plot (see Fig. 4.19b below, for example) [64].[87]

Although Wallis is commonly credited as the discoverer of the presence of intermediate nodes in a vibrating string, it must be said that this is an inaccurate attribution.

[85] p. 130.

[86] pp. 839–840.

[87] p. 328.

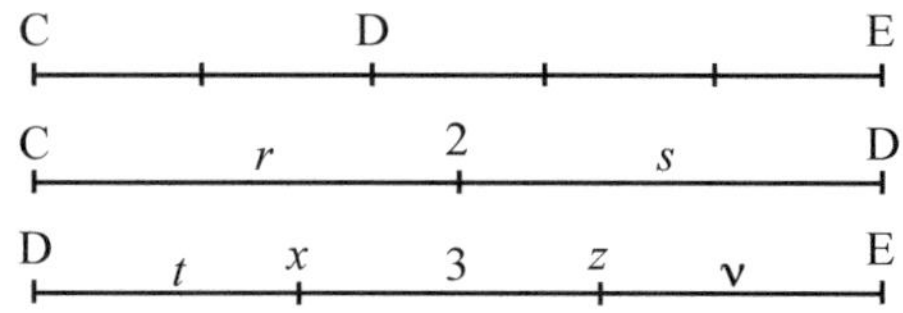

Fig. 4.19 Higher harmonic sounds for a division of a string into five parts. Redrawn from [64], after p. 328

His was not an original work, at least not an original experimental work, since the results had already been found some years before. This is testified, for example, in the *Natural history of Oxford-shire* by Robert Plot (1640–1696) of 1677, the same year as the letter from Wallis to Oldenburg: "Which phaenomena I shall always gratefully acknowledge were first discovered to me by the ingenious Thomas Pigot [...], which have also been observed for about these years by the no less ingenious William Noble. The solution whereof in all their cases, as received from the learned and accurate hand of the Reverend Narcissus Marsh" [64].[88] The discovery is dated 1673, with Noble preceding Pigot. The results of the experiment were communicated to Wallis by Narcissus March (1638–1713) in 1676, as Wallis testified a few years after the publication of the letter to Oldenburg [90].[89]

What Plot wrote in his *Natural history of Oxford-shire* is not very different from what Wallis wrote; in some respects more complete and interesting. This is the incipit of the section on music:

> It hath been lately observed here at Oxford that though viol or lute strings rightly tuned do affect one another, yet most of them do it not in all places alike, as has till now been supposed: for if the lesser of two octaves be touched with the hand or bow, each half of the greater will answer it, but will stand still in the middle, and if the greater of the two octaves be touched on either of its halves, all the lesser will answer it, but if touched on the middle, the lesser will not stir any where at all. So if the lesser string of two fifths be touched on either of its halves, each third part of the greater will answer it, but if on the middle they will not stir; and if the greater of two fifths be touched on either of its thirds, each half of the lesser will answer it, but if in the divisions they will not stir: and so of twelfths, fifteenth, etc. [64].[90]

As you can see, the first part of this quotation is very similar to Wallis' writings; but the second part refers to a more complex experimental situation. Plot considered a string CE divided into five parts and referred to CD as the two parts of fifth. He said that if we strike a point of one of the two parts s, r of CD, DE begins to vibrate in each of its three parts. If CD is struck in its center, DE does not vibrate at all. This phenomenon is much more complex than that shown by Wallis in his Fig. 4.18b and will be explained by Sauveur some years later.

Considering that Plot wrote his considerations independently of Wallis and in the same year, we can say that Wallis' main contribution was probably to publish the experience of Pigot and Noble in the *Philosophical transactions*, a widely circulated journal.

---

[88] p. 288.

[89] p. 10.

[90] p. 288.

Fig. 4.20 The trumpet marine. Adapted from [18], p. 103

The second letter to Oldenburg, that of March 27, deals with the resonance of strings caused by wind instruments: "A [lute-string or] viol-string will thus answere, not onely to a consonant string on the same or a neighbouring lute or violl; but to a consonant note in wind-instruments. Which [was] particularly tryed, on a violl, answering to consonant notes on a chamber-organ, very remarkably" [90].[91]

Wallis (and Plot) considered the presence of higher modes in a vibrating string in an abstract way, as a special case of sympathetic resonance. But the phenomenon has also emerged from a practical situation, and the points of rest were first noticed in the study of the behavior of the trumpet marine. This consists of a bowed monochord, up to two meters long, equipped with a vibrating bridge, which was in common use from the 15th century until the middle of the 18th century. Figure 4.20 shows an engraving that clearly illustrates the way the trumpet marine is played.

In its fully developed form the instrument is capable of sounding all the pitches of the harmonic series up to the 16th partial. A comparison of its tone with that of a modern wind trumpet shows that the latter's partials drop off sharply after the 10th or 11th. The use of a straight mute, however, raises the upper partials of the trumpet so that it begins to exhibit the same configurations as the trumpet marine.[92]

---

[91] p. 73.

[92] The history of the trumpet marine can be divided into two overlapping periods: the first from 1450 to 1650, the second from 1550 to the late 19th century. From the beginning a vibrating bridge was

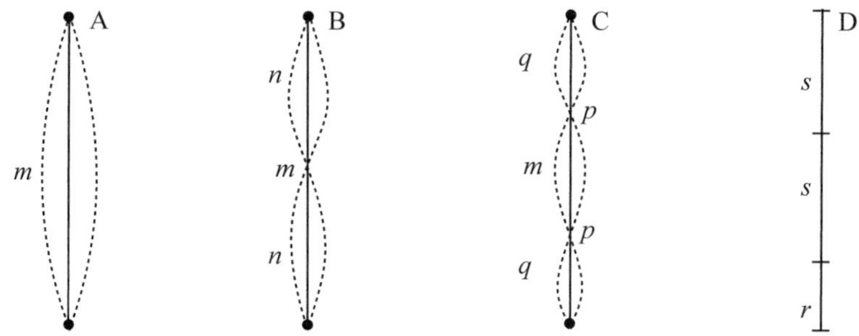

Fig. 4.21 Nodes and antinodes in the trumpet marine. Redrawn from [70], after p. 563

Francis Robartes (1650–1718), in the 1692 *Philosophical transaction* of the Royal society of London, took up the analysis of the presence of intermediate nodes in the string of the trumpet marine. Note that he did not cite Wallis, probably with some justification, since the behavior of the trumpet marine was already known and Wallis, for that matter, had no claim to priority. It is possible that the phenomenon studied by Wallis was not seen as strictly related to that studied by Robartes. The former concerned sympathetic resonance between two different and separate strings, the vibration being induced by pulses of air; the latter concerned vibrations induced by contact between two continuous strings.

Robartes' analysis is well summarized in the following quotation and illustrated in Fig. 4.21:

> Now in the trumpet marine you do not stop close as, in other instruments, but touch the string gently with your thumb, whereby there is a mutual concurrence of the upper and lower part of the string to produce the sound. This is sufficiently evident from that, if any thing touches the string below the stop, the sound will be as effectually spoyled as if it were laid upon that part which is immediately struck with the bow. From hence therefore we may collect, that the trumpet-marine will yield no musical sound but when the flop makes the upper part of the string an *aliquot* of the remainder, and consequently of the whole: otherwise as we just now remark'd of Fig. D the vibrations of the parts will cross one another, and make a sound suitable to their motion, altogether confuse [70].[93]

This quotation emphasizes the role that the position of the finger plays in the functioning of the trumpet. The finger should be placed in such a position that the small part of the string is a submultiple of the whole string, otherwise a confused sound is obtained, as is clear from Fig. 4.21d. Actually, as Sauveur will show, harmonic

---

added as a means of improving the tone or increasing the loudness. The instrument with the vibrating bridge eventually became known as the *Trumscheit* and coexisted with the tromba without it until the mid-16th century, when the latter disappeared [1]. Antonio Vivaldi composed two concertos featuring trumpet marine violins, namely RV 221 and RV 588. Additionally, several operas emploied the trumpet marine as a timbral coloring element, as exemplified by J.B. Lulli's *Serse*. The *Concert for trumpet marine, strings, and timpani* by Christian Gottfried Telonius (1742–1802) is a relatively well-known piece of music [84].

[93] p. 561.

motion can occur even when this requirement is not met, but shorter and longer portions of the string form a ratio of small integers.

4.3.1.2 Isaac Newton's Division of the Octave

Virtually all of Newton's writings on music are contained in a notebook he filled out during his undergraduate years, 1664–1666, which is now catalogued as Ms. Add. 4000 in the Cambridge university library. Only a small portion of the manuscript is devoted to music and is divided into two parts: Part I from f. 104v to f. 113r and Part II from 138v to 143r; the rest deals with mathematical matters. Part I contains disordered minutes, mainly calculations. Part II is the occasion for a small treatise entitled by Newton *Of musik*. It has recently been published, with meager comments, in an article focusing on the famous analogy between the spectrum of light and the musical scale, of no interest here [62]. A fairly accurate edition of both parts can be found in [59] and and interesting comment in [19].[94]

Newton's approach to music was essentially mathematical; the subject was the formation of musical scales and, in particular, the division of the octave into equal parts. There are no references to the theory of sound, which was an important subject of his later period. In general, it can be said that Newton's writings were virtually unknown to his contemporaries and thus did not influence the development of speculative music in the 18th century (for example, the division of the octave into 53 equal parts is usually attributed to Mercator; but Newton himself, more or less at the same time, considered this division in his writings). Nevertheless, the reading of the Add. 4000 is very interesting, as it helps to understand how the highly gifted mathematician Isaac Newton approached the subject of music of his time.

Part I of Newton's manuscript is not easy to decipher because of the lack of comments. Of particular interest is the use of the logarithms in the division of the octave. Table 4.6 shows data related to the division of the octave into 12 semitones that form a syntonic chromatic scale. Col. 1 gives the intervals of the system for which the sequence *a-b-a-c-a-b-a-a-c-a-b-a* is followed from bottom to top, using three different semitones; the major diatonic semitone a ($^{16}/_{15}$), the major chromatic semitone b ($^{135}/_{128}$), and the minor chromatic semitone c ($^{25}/_{24}$). Col. 2 returns the monochord in the range 360–720, corresponding to the intervals of Col. 1, Col. 3 returns the logarithms of Col. 1, Col. 4 the logarithms of a tempered monochord, which is given in Col. 5. The last column gives the base 2 logarithms of the intervals in Col. 1. More precisely, if i is the generic interval of Col. 1, it reports the value $\bar{c} = -12 \log_2 i$. The sign $-$ reflects the fact that the intervals of Col. 1 are defined by the ratios m/n, where $n > m$. With this choice, a seventh column based on tempered intervals would report a sequence of integers of the kind $1, 2, 3, \ldots 11, 12$, noting that $\bar{c}$ is nothing more than the modern cent used to measure intervals (except for a factor 100).

---

[94] pp. 97–129.

Table 4.6 Dividing the octave into 12 equal semitones. Redrawn from [58], f. 105v

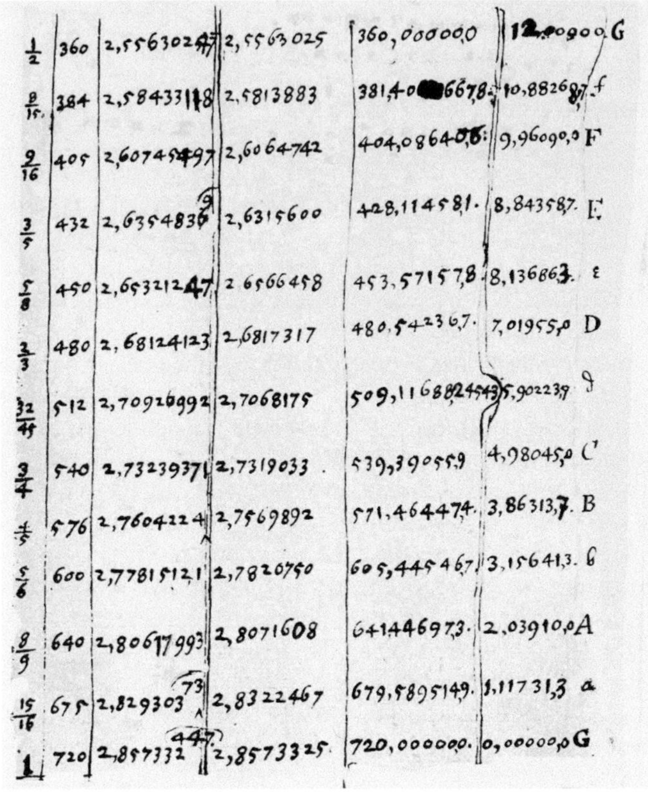

1/2	360	2,55630247	2,5563025	360,0000000	12,000000 G
8/15	384	2,58433118	2,5813883	381,40[illegible]6678	10,882687 f
9/16	405	2,60745497	2,6064742	404,0864006	9,960900 F
3/5	432	2,63548389	2,6315600	428,1145181	8,843587 E
5/8	450	2,65321247	2,6566458	453,5715,78	8,136863 ε
2/3	480	2,68124123	2,6817317	480,5423667	7,019550 D
32/45	512	2,70926992	2,7068175	509,1168824543	5,902239 g
3/4	540	2,73239371	2,7319033	539,3905559	4,980450 C
4/5	576	2,76042224	2,7569892	571,4644747	3,863137 B
5/6	600	2,77815121	2,7820750	605,4454467	3,156413 b
8/9	640	2,80617993	2,8071608	641,4469723	2,039100 A
15/16	675	2,82930373	2,8322467	679,5895149	1,117313 a
1	720	2,857332447	2,8573325	720,0000000	0,000000 G

Table 4.7 Transcription of the note names and numbers

G	*	A	*	B	C	*	D	*	E	F	*	G
G	*	F	E	*	D	*	C	*	B	A	*	G
0	1 + 1/17	2 + 1/51	3 + 1/13	4 − 1/15	5 − 1/102	6 − 1/41	7 + 1/102	8 + 1/15	9 − 1/13	10 − 1/51	11 − 1/17	12

Newton considered the division of the octave into parts even different from 12; in particular, he considered the division into 15, 19, 20, 24, 25, 29, 36, 41, 51, 53, 59, 100, 120, and 612 parts, as shown in ff. 108r-180v of the manuscript.

The division into such a large number as 612 is particularly puzzling. A motivation for this choice is suggested in [57],[95] based on Table 4.7. The first line of the table shows a syntonic chromatic scale starting from G; the second line shows the same scale in reverse order. The asterisk indicates a (major) semitone of just intonation

[95] Part I, f. 208r.

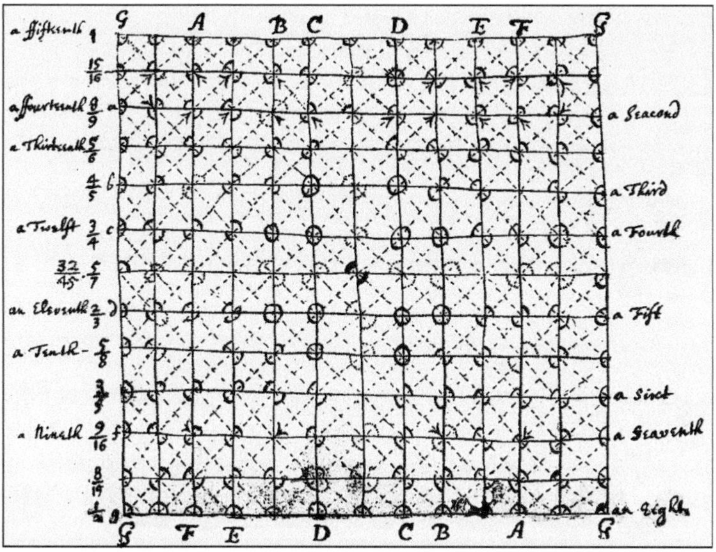

Fig. 4.22 Intervals between two notes of the diatonic scale. Redrawn from [58], f. 107r

(16 : 15). The integers in the third row refer to the number of semitones of the equally tempered scale, while the fractions are the part of a whole tempered tone to be added to the integer to obtain the exact number of semitones for the intervals in just intonation. The more common values, 102, 51, and 17, are also the most meaningful, because while they give good approximations to the minor and major triads FA♭C and DF♯A, they are submultiples of 612. So a division of the octave into 612 equal parts can yield exactly these values.

Still in Part I of the manuscript, to signal Fig. 4.22, which reveals the ingenuity of the young Newton. The figure can be used to find the interval between two notes of the diatonic scale: suppose you want the interval from A to D. From the top A of Fig. 4.22, move to the right along the diagonal to meet the vertical from D. From this point, draw the horizontal to the vertical of the last G; it will read 'a fourth'. If, instead, one moves to the left along the diagonal from the lower A to the lower D and draws the horizontal, one meets on the vertical from G the complementary interval, that is, a fifth. The circles or parts of circles indicate whether or not the intervals exactly match the ratio indicated on the left [59].[96]

To signal the presence in Part I of circular diagrams to represent various kinds of solmization similar to those reported by Descartes, to indicate a most likely reading by Newton of Descartes' *Compendium*, at least in its English edition.

Part II of the manuscript is perhaps less interesting, although it does contain some valuable information. For example, the fact that Newton left room for empirical observation to judge the pleasantness of combinations of sounds. And Newton, as

[96] p. 96.

Table 4.8 Modes and their pleasantness. Adapted from [58], f. 140

	key		*2d* minor	*3d* major	*3d* minor	*4th*	Tritonus	*5th*	*6th* minor	*6th* major	*7th*		*8th*	
4	*o*	·	*p*	·	*q*	*r*	·	*s*	·	*t*	*u*	·	*o*	1
3	*s*	·	*t*	*u*	·	*o*	·	*p*	·	*q*	*r*	·	*s*	2
5	*r*	·	*s*	·	*t*	*u*	·	*o*	·	*p*	·	*q*	*r*	3
2	*p*	·	*q*	*r*	·	*s*	·	*t*	*u*	·	*o*	·	*p*	4
1	*t*	*u*	·	*o*	·	*p*	·	*q*	*r*	·	*s*	·	*t*	5
6	*u*	·	*o*	·	*p*	·	*q*	*r*	·	*s*	·	*p*	*u*	6

he will do in the *Opticks*, resorted to an analogy of light and sound to justify the presence of non-consonant intervals in music:

> But as too suddaine a change from lesse to greater light offfends ye eye by reason yt, ye [spirits rarified by the] augmented motion of ye light too violently stretch ye optick nerve: soe ye suddaine passing from grave to acute sounds is not so pleasant as if it were done by degrees, for there because of two greate a change of motion made thereby in the auditory spirits. And as a man suddainely coming from greater to lesse light, cannot discerne objects thereby so well, as when if he came to it by degrees or as when hee hath staid some while in ye lesser light (by reason yt ye motion of ye spirits in ye optick nerve caused by ye greater light, doth, untill it bee allayed; disturbe & as it were drowne ye motion motion of ye weaker light) soe if the slower motion of ye lower sound immediately succede ye much more smart motion of ye higher its impression on ye auditory spirits—being then less perceptible, ye lower sound must bee more less pleasant yt had if moved ye step had beene graduated [58].[97]

One of the most interesting points in Part II concerns the formation of modes and the gradation of their pleasantness. Table 4.8 lists the first six modes, which, according to Newton, are twelve in number. He designated the notes with the letters o, p, q, r, s, t, u, which are related to the traditional symbology according to the table below:

o	p	q	r	s	t	u	o
G	A	B	C	D	E	F	G

The numbers in the first column of Table 4.8 indicate the order in which the modes are generated, starting from E—with tones and semitones alternating according to the sequence STTTSTT, where there is no distinction between major and minor tones—and continuing in fourths. Thus we have the modes beginning with G, A, D, G, C, F. Note that Newton did not give his modes Greek names. In the last column, the numbers indicate the order of pleasantness, the judgment of which is left to the senses, but the justification of which is attempted on theoretical grounds [59].[98]

Note, moreover, that the mode considered to be the most pleasant, the first one, is in fact a major mode; this would be exactly the case if, instead of the sequence

[97] f. 138r. Transcripted in [59].

[98] f. 140.

GABCDEFG, one used GABCDEF♯G, that is, if one raised F so as to obtain the sequence of tones and semitones TTSTTTS characteristic of major modes. This choice corresponds to the late 17th century tendency in music to prefer the major mode.

4.3.2 *Académie des Sciences of Paris*

Music has been an object of attention in the Académie des sciences of Paris since its foundation in 1666. The earliest references date from 1676, although we must wait until the reform of 1699, when the writing of the *procès-verbaux* became regular, for precise accounts.

The history of music as a discipline of the Académie before the French Revolution is reported by A. Cohen in his *Music in the French Royal academy of science*, of 1981, which is very rich in bibliographical references [24]. The history can be divided into two phases: a first phase with Jean Baptiste Du Hamel (1624–1706) as permanent secretary, which lasted until 1683. There was then a considerable lull in activity due to the death of Colbert, the Academie's protector and organizer. The second phase began in 1691 with a new political climate and a new perpetual secretary, Bernard Le Bovier de Fontenelle (1657–1757), who replaced Du Hamel in 1697. This second phase ended shortly after Sauveur's contributions around the mid-1700s.

In both periods, speculative music received a great deal of attention from mathematicians and natural philosophers, who considered it on a par with other scientific disciplines. Although this was a general process in Europe, a direct consequence of the development of the new science, it found more fertile ground in the French academy than elsewhere.

In the early period, among the academics who dealt with music, we should mention Gilles Personne de Roberval (1602–1675), Claude Perrault (1613–1688), Huygens, Mignot La Voye (c. 1619–1684) [24],[99] to whom should be added Edme Mariotte (c. 1620–1684) and François Blondel (c. 1618–1686). Here is an interesting excerpt concerning Blondel, reported in the procès-verbaux discussing an experiment similar to that of Galileo Galilei in the *Discorsi*: "Mr. Blondel has demonstrated that, after stirring water in a glass by rubbing its edges with his finger, when the sound [produced] reaches one octave, the small circles in the water are doubled" [24].[100]

The second period was more interesting for our purposes because it already saw Sauveur as the main actor. The protagonists were Philippe de La Hire, Denis Dodart (1634–1707), Nicolas Malebranche (1638–1715), Louis Carré, and finally, of course, Joseph Sauveur [24].[101] In what follows, we will discuss in some detail only the contributions of de La Hire and Carré; the former because Sauveur's activity in music began with him, the latter because he was one of his contemporaries.

---

[99] p. 6.

[100] p. 12.

[101] p. 18.

4.3.2.1 Philippe de La Hire and the Trumpet Marine

Philippe de La Hire (1640–1718) was a scholar with an extraordinarily wide range of interests, from physics to natural sciences, meteorology, art; particularly important is his contribution to the statics of arches [46]. He discussed the nature of sound on several occasions at the Académie; his findings are collected in two works *Explication des differences des sons de la corde tendue sur la trompette marine* of 1692 [49] and *Experiences sur le son* [47, 48] published in the memoirs of the Académie.

The Trumpet Marine
The memoir of 1692 on the trumpet marine is rather interesting, not so much for its contributions to acoustics or speculative music, which are modest, but rather for the aid it provides to the understanding of speculative music among Parisian academics at a time when Sauveur was beginning to take an interest in the subject. A modern reader is somewhat confused when reading this memoir, for he cannot tell whether the results reported refer to real or thought experiments, and it is hard not to agree with Truesdell, who considers La Hire's experiments: "haphazardly conceived and vaguely described" [85].[102]

La Hire did not master the theory of sound, for which he referred to Claude Perrault's 'bizarre' (at least to a modern reader) remarks reported in the second volume of his *Essais de physique* of 1680 [61]. According to Perrault, the vibrations of bodies are merely the occasions of sound sensations which actually have different origins. La Hire also showed ignorance of the theory of overtones presented by Wallis in 1677 and its application to the trumpet marine by Robartes in 1692; with the mitigating circumstance that not even Sauveur, at the time he developed his theory of harmonic sounds, knew of these works. His technical language is also very poor and imprecise.

After a brief introduction, La Hire began to study the theory of the trumpet marine, wondering why, if one touches the string a little above or a little below the just division, the tone does not change, and then why the sound makes jumps, passing abruptly from one note to another, more precisely, he said, from one consonance to another.

La Hire began his analysis by considering a string AB stretched tightly and divided into two equal parts at C by placing a finger lightly on one of the two parts, AC, and causing it to vibrate. The vibrations of this part are transmitted to the other part CB because of the interaction between the two parts of the string; they tend to reach unison, although the sound of CB may be very weak at first.

So far, so good; however, the following considerations concerning the case where the finger is placed at the point F, which divides the string of Fig. 4.23 in the ratio (2 : 1), are contradictory. La Hire said:

> If, on the other hand, the string is divided at F by the finger so that the part BF is twice AF, when AF is touched, the two parts will tend to vibrate within one octave of each other, although BF will emit a much weaker sound than AF, its motion being induced by the

[102] p. 126.

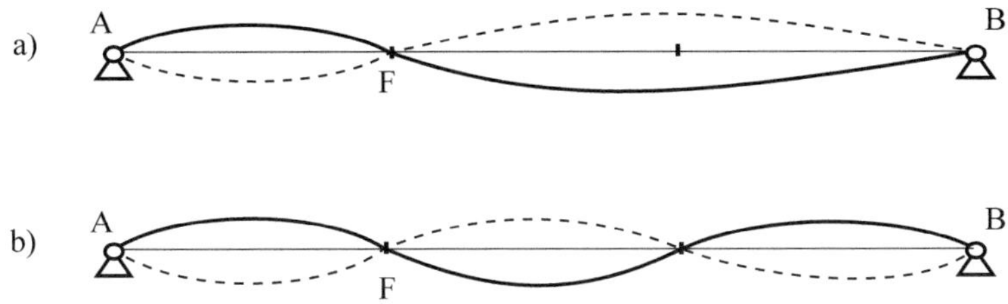

Fig. 4.23 Vibration of the string of a trumpet marine for a ratio of 2 : 1 in the division of the string

> shaking of AF. Since the vibrations of the two strings do not match, they will destroy each other unless they are supported by the bow. If the strings are divided into different numbers, still in concordant proportions, the phenomenon will be the same, but if the string divisions have proportions expressed by large numbers, such as 23 to 37, then the vibrations will die out very easily [49].[103]

So, according to La Hire, if you rub AF vigorously with the bow, the two parts of the string will play one an octave above the other. But this is not the case in practice, because the two parts of the string interact. What actually happens is that FB, after a transient phase, begins to vibrate at the same frequency as AF; this results in the presence of an intermediate node, as shown by Sauveur in his 1701 memoir and illustrated in Fig. 4.23, where (a) indicates the transient phase of vibration and (b) the steady phase.

At this point, La Hire opened a parenthesis to explain that the two parts of the string have their own natural way of vibrating (their proper frequency, in modern parlance), which is not too much affected by the motion of the bow, which only has the task of keeping the vibration alive. The effect of the bow is similar to that of a saw with evenly spaced teeth.

Although the memoir is theoretical and applies to any type of trumpet marine, La Hire described the type used in his day (see Sect. 4.3.1.1). The string of the trumpet is stretched over a board fastened at both ends and supported by a bridge with two feet, only one of which rests on the board while the other is free to strike the board at the same frequency as the vibrating string. When the bow touches the string, in addition to the sound of the string, the sound of the foot of the bridge against the board can be heard.

Somewhat accidentally, he returned to the theoretical aspects and tried to explain why the trumpet marine always produces the same sound whether it is touched a little above or below the right split. If a finger is placed exactly in the middle of the string and its upper part is stressed, La Hire continued, the trembling of this string is transmitted to the lower one, and the free foot of the bridge strikes on the board one octave higher than in the case of the open string. If, on the other hand, one places one's finger a little higher or lower than the center of the string, the part stressed by the bow, although a little longer than the other, will still produce the same sound

[103] pp. 503–504.

as before. In fact, although in isolation the two strings would tend to vibrate at a frequency of one less and one more, the interaction causes them to vibrate in the same way, especially if the bow lends a lot of tension to the string.

La Hire 's explanation is quite satisfactory according to modern theories, and the result is also true. In fact, the two strings sound at the same frequency. If they had actually two different lengths, this would not be possible because each string would vibrate at its own frequency. But the constraint imposed by the lightly resting finger means that the effective length of the strings is not precisely defined, and it actually adjusts so that the two parts of the string vibrate in unison. On the other hand, what La Hire said is inconsistent in the case where the length ratio is approximately (2 : 1), or (3 : 2), as he added: "The same thing happens when the finger divides the string in the ratio of (2 : 1), then the two strings will sound in the octave. If the division is approximately (3 : 2), then the shorter string forms the major third and the longer string forms the minor third. However, it is difficult to hear these two tones" [49].[104] Indeed, in such cases it is not a matter of small adjustments; for example as discussed just above, in the ratio (2 : 1) what happens is not that the two parts begin to vibrate exactly one octave of the other, but rather that they begin to vibrate at the same frequency, as shown in Fig. 4.23.

The synchronization of the oscillations of two elastic bodies with slightly different natural frequencies, brought into contact by a deformable body such as a finger, discussed by La Hire, is an interesting phenomenon, first pointed out in the case of two pendulums by Huygens ("of Zuichen") [49][105] in 1665, who observed that if two pendulum clocks with slightly different periods are hung on the same support of sufficiently elastic material, such as a wooden board, the oscillations of the pendulums influence each other and after a while the two pendulums synchronize. In particular, Huygens noted that whatever the initial conditions, the coupling brought the system into anti-phase synchronization; furthermore, he observed that even by disturbing the oscillations, the pendulums would become synchronized again after a transitional phase [43].[106] The phenomenon has been the subject of much work in an attempt to explain the cause of this synchronization more convincingly than Huygens did. For a comprehensive and interesting work, although not recent, see [12] (Fig. 4.24).

To this experiment of Huygens, La Hire contrasted another. He considered a simple pendulum consisting of a lead weight of five or six ounces with a wire of the length necessary to beat the second. The pendulum is suspended at the end of a thin wooden beam that is sufficiently elastic. Several situations are considered. From the case in which the cantilever is perpendicular to the wall to which it is attached, and the oscillations of the pendulum occur in the vertical plane formed by the cantilever and the pendulum wire, to the cases in which the cantilever is bent in the same vertical plane so as to form an obtuse angle or an acute angle with the pendulum wire. In the first case, the period of the pendulum is the same as that of the original pendulum suspended from a fixed point, in the second case it is longer, and in the third case it

[104] p. 508.

[105] p. 509.

[106] vol. 17, pp. 183–186.

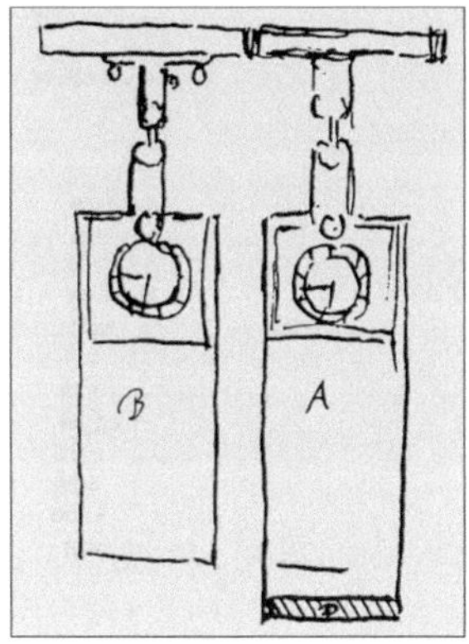

Fig. 4.24 Synchronization of two pendulums. Redrawn from [43], p. 183

is shorter. La Hire attempted an explanation of this behavior which demanded the interaction between the cantilever and the pendulum, and it is as ingenious as it is naive, and for the sake of simplicity will not be repeated here.

On the basis of his (incorrect) analysis, La Hire went then on to consider what sounds could be obtained from the trumpet marine by moving the finger along the string. In this analysis, in comparing the lengths of the two parts of the string separated by the finger, La Hire sometimes considered the mutual ratio of the two parts, sometimes related the two parts to the whole string, probably uncertain how to describe the sounds produced.

Consider, said La Hire, the string AB of Fig. 4.25, if, having set the whole string in motion, you rest one finger on C, the tone goes up one octave. If you imagine that the string AB is divided into 11 parts and you put your finger on D so that AD : DB = (5 : 6) and you rub AD with the bow, the two strings will tune to a minor third. If you then move your finger to C, the pitch jumps of the interval (11 : 10) because the length of the string changes from AD = 5/11 to AC = 5.5/11.

This conclusion is somewhat singular. First, La Hire considered only the sound of the string's AD and its complement DB, and then completely forgot that the two parts of the string form a minor third interval, and compared the two parts AD and AC. Just as the following considerations are singular. If you divide the string into E, so that AE and EB are in the ratio 4 : 5 (imagine AB divided into 9 parts), the two strings are in the ratio of a major third. Putting the finger in C and reasoning as above, we get a jump given the ratio of AE to AC equal to 9 : 8, or one tone.

At this point La Hire rather puzzlingly stated that the considerations made so far were valid only in theory, but that in "practice various phenomena occur which cannot

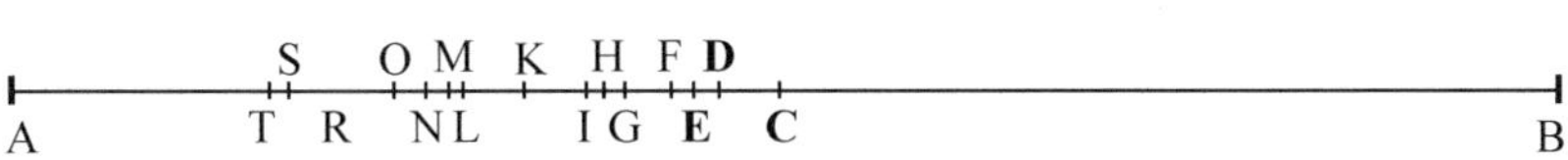

Fig. 4.25 Division of the string of the trumpet marine. Redrawn from [49], p. 512

be predicted a priori" [49].[107] Thus he did not admit that his theory was wrong, but simply believed that in natural philosophy theory and practice could diverge without embarrassment.

When he dealt with what happens in practice, his description of the behavior of the trumpet marine is somewhat closer to reality. However, there are passages that describe behaviors that do not correspond to practice and thus make one doubt whether La Hire had an accurate knowledge of the real behavior of the instrument he was describing.

He recalled that if the string is divided into two parts in a ratio close to a just ratio, the two parts of the string interacting with each other will come to vibrate at the same frequency, provided that the bow acts vigorously on one of the strings. For example, La Hire said, place your finger so that the string is divided into two parts that are in a ratio of (5 : 4). In this case, the two parts of the string tend to vibrate with two different frequencies, but because of the mutual interaction, this is not possible, and the strings come to vibrate with one sound, with an intermediate frequency. And the resulting sound is more or less the same as when you put your finger on C, the middle of the string. More or less, because the intermediate vibration also depends on the way the bow is held.

If La Hire's reasoning were correct, the string would always sound as if the finger were in C, regardless of the ratio of the lengths of the two parts of the string; but we know that this is not true. La Hire also somehow admitted this by saying that what you get in practice is an unpleasant sound.

He went on to consider the case in which the finger pressing down on the string passes out, leaving the string empty; in this case the resulting sound is practically destroyed, because the parts are not in tune, unless the sound before the finger is released is particularly loud. When the whole string and the two parts are in tune, they somehow synchronize to produce a pleasant sound [49].[108] Again, it is not clear whether La Hire is referring to a real experience or just imagining it.

Ondulations

Having finished his theory of the trumpet marine, La Hire moved on to discuss an experience communicated to him by the abbot Jean Gallois (1632-1707), one of the founders of the *Journal de Sçavans* (for a biography, see [37]). As was not uncommon in the late 17th century, the experience involves everyday objects rather than specially constructed artifacts. In this case, it involves studying the sound emitted by two elastic metal arms connected by a ring, such as the tongs used for fireplaces. Depending on how the ends of the arms are handled, the sound changes. It goes from a situation where the ring is suspended from a wire and the two arms at the other end are enclosed by a piece of iron or plastic material, to a situation where the ring is attached to a piece of solid wood and the ends are free.

Experiments show that the sound can be very different in the different cases until there is a substantial silence, but, at least according to La Hire, in all cases

[107] p. 516.
[108] p. 517.

there is evidence of similar vibratory motions. This fact would confirm the theory of Perrault that vibrations are not responsible for the sound, which is instead caused by *ondulations*, while vibrations produce only very faint noises and only by chance. Of course, La Hire never explained how he could say that the vibratory motions remain essentially the same.

To explain what he meant by ondulations, La Hire referred to an elastic solid body that is struck, for example, with a hammer. It first falls back a little where it was struck, but then, because of the elasticity, the affected surface reaches the undeformed state and then passes over it with many forward and backward movements which may be regarded as very rapid vibrations produced around the affected part and very different from the large vibrations affecting the whole body. Then any raising or lowering of the affected part is transmitted to the whole body as in the case of water, "and these deformations are what I call ondulations" [49].[109]

This explanation of the nature of sound, though very curious and very Cartesian, leads to an interesting result, perhaps the only one in the whole work, according to which (elastic) bodies are characterized by natural modes of vibration with their own frequency:

> The ondulations of resonant bodies do not end at the boundary of the body, because if they did, the sound would immediately cease throughout the body. Having arrived at one of the ends of the body, they return as if reflected and return several times along the body without being destroyed, as happens in water, and for this reason the resonant bodies always produce the same sound wherever they are struck. Thus the ondulations will always be the same in the same body if its parts are homogeneous, and you will not have that the intensity of the sound will be increased or decreased when you strike it either with more or less force. [49].[110]

In the case of a string, La Hire said, it seems at first glance that the theory of ondulations cannot be applied because it is not well understood how and whether ondulations can form. But if you think about the wave motion that occurs in a string when you pluck it with a finger or a feather, you can imagine small ripples in the string that can be interpreted as ondulations.

La Hire concluded his memoirs of 1692 by stating that the theory of sound he had expounded did not invalidate his reasoning about the trumpet marine based on vibrations, because the inverse proportionality between the frequencies of the sounds and the length of the strings must be considered an empirical law that holds whatever the actual mechanism of sound production is.

In his memoirs of 1716, *Experiences sur le son*, La Hire presented some interesting experiments and explained, with some modifications, his theory of ondulations, which he here called *fremissements*. Strengthened also by the fact that this theory had been adopted by Louis Carré in his writings on music, in particular *De la proportion que doivent avoir les cilindres pour former par leurs sons les accords de la musique* of 1709 [23] (see below).

The experiment is based on the sound emitted by several wooden cylinders made from different trees and of different sizes. The cylinders are either supported in

---

[109] p. 522.

[110] p. 522.

the middle by fingers or suspended from a rope. When they are struck with a hard body at different points along their length, the sound emitted varies according to the points where they are struck. La Hire rejected the hypothesis that this was due to the heterogeneity of the material, because when the cylinders were struck near an eighth of their length, the sound was always much lower than when the cylinders were struck at other points, including the ends, and this was true for all the cylinders. La Hire repeated the experiment with metal rods and the result was the same.

For him, these results were a confirmation of Perrault's theory, which he had adopted in his memoirs of 1692, that sound does not depend on macroscopic vibrations, but only on microscopic ones, i.e. fremissement. A modern reader's explanation of this experiment is that the vibrations of the cylinders are periodic and characterized by the presence of nodes and anti-nodes. When a body is hit at a potential node, the resulting vibrations are strongly damped.

After describing his experiments, La Hire presented a variant of the ondulation/fremissement theory. Consider an elastic body struck by a hard object; you will hear immediately a sound. Since there are no vibrations yet, this can only be caused by the "fremissements" of the affected body. The movement of the material particles of the body displaced by the shock at the point of impact transmit their displacement to all the successive parts of the body; the vibrations force the air contained in the pores of the body to come out, but immediately these pores, because of their elasticity, expand even more than before and receive new air, so that repeated rapid shocks are determined in the air that reaches the ear and produces the sound. Nevertheless, it cannot be denied that the vibrations of the body also produce a movement of the air contained in the pores of the body, but this movement is much slower than that of the sound and the ear does not perceive it.

> This movement, mixed with that of fremissement, produces different sounds, so that it seems to me that one can say that the fremissements influence the sound produced by the trembling, and thus the different color of the various bodies are explained [48].[111]

Thus, according to Hire, even if the vibration-induced air movement alone is not perceived by the ear, it still affects the quality of the perceived sound.

According to La Hire, the interplay between vibrations and fremissements is also responsible for consonances, as can be seen from the following argument, which seems rather confusing to us:

> If the fremissements of a body can be tuned to its vibrations, one will hear a distinct sound and will have a consonance with the sound of another body struck in a different way, if the fremissements are the same. If the vibrations are different and closely related to each other by (1 : 2), or by (2 : 3), etc.., the fremissements will be in tune with those vibrations and will produce a consonance [48].[112]

---

[111] pp. 267–268.

[112] pp. 267–268.

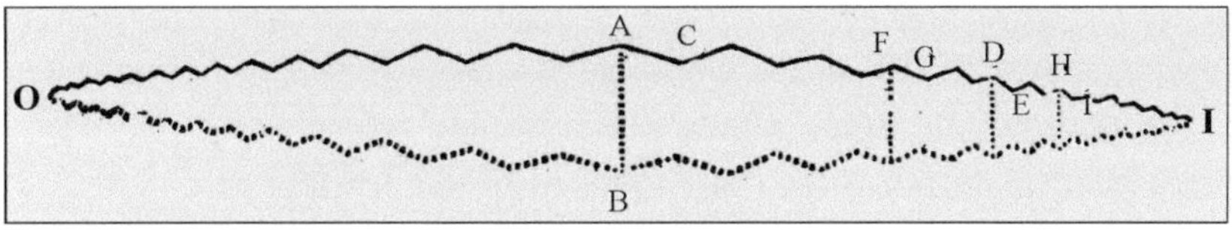

Fig. 4.26 Tremblemens along a string. Redrawn from [61], p. 129

4.3.2.2 Louis Carré and the Fremblemens

Louis Carré (1663–1711), a mathematician of the Académie des sciences of Paris, wrote one of the first treatises on Calculus [38]. Among other subjects he showed a special interest in speculative music and acoustics. The procès-verbaux de l'Académie des sciences contain the titles of some of his writings [24].[113] All that has come down to us of these, however, is the extensive summary *Traité sur le son* of 1703, published in the *Journal des Sçavans* [22], and a memoir *De la proportion que doivent avoir les cilindres pour former par leurs sons les accords de la musique* of 1709, published in the Académie's memoirs [23]. His musical interests included acoustics, musical systems and the mechanics of musical instruments (he was commissioned by the president of the Académie to describe all the musical instruments of France).

Traité sur le Son

The *Traité sur le son*, from what can be gleaned from its summary, would not have been particularly original. Perhaps the most interesting part concerns the exposition of the theory of sound, which is a reworking of Perrault's, adopted by La Hire .

Below is how Perrault explained the way a string produces the vibrations that can give rise to both the overtones and the color of the notes:

> Imagine, then, that the string IAO [of Fig. 4.26], when plucked, is drawn to A, then to B, then back to A, then to B, so that it comes and goes more or less like a pendulum. The relaxation of the string, in addition to this oscillation that gives it a kind of oval shape, causes it to bend again into 100 parts and to vibrate in such a way that each part, such as AC, FG, DE, HI, is also bent into 1000 parts, so that these last are the ones that hit the air particles whose recoil is responsible for the sound, and each of the parts AC, FG, HI [with its tremblemens] produces a particular sound different from the others, so that the parts toward the middle, such as AC, produce a more grave sound than those toward the end, such as DH, because here the string is tighter and stiffer. All the sounds appear to each other to produce the overall sound, and that is what makes up its tone (color) [61].[114]

There is, however, an interesting and important innovation in Carré, his *tremblemens*, corresponding to the *ondulations*, have periods proportional to those of the macroscopic vibrations:

---

[113] p. 23.

[114] pp. 129–131.

> That is, in spite of the fact that these tremblemens are in the same ratio with respect to frequencies as the total vibrations, one can always take these total vibrations for the measurement of all chords [22].[115]

The theory of tremblemens led Carré to a conclusion that is at least curious when applied to sounds produced by vibrating strings of different materials. Comparing the behavior of two strings of equal length, diameter, and tension, one made of gold and the other of iron, Carré wrote: "Gold makes a low sound and iron a high sound; since gold particles, being softer and more flexible than iron particles, have less elasticity, they must have less rapid and less frequent tremblemens" [22].[116] Thus the cause of the lower frequencies of the tremblemens, and thus of the vibrations, is attributed to the lower elasticity of gold, and not to its greater specific weight, as was accepted by all the mathematicians of the time.

Memoir on the Cylinders

The 1709 memoir is strictly experimental in nature and quite interesting for the history of acoustics. It deals with the sounds emitted by cylinders of various materials and sizes when struck by a hard body. It begins with the following sentence:

> The solid cylinders whose sounds form the chords of music are in triple and inverse relation to those of the sounds which express these pitches [23].[117]

This proposition is quite convincing when applied to similar cylinders where the ratio of the volumes is in triple proportion to that of the linear dimensions (for example, if we take two similar cylinders whose diameters and lengths are in the ratio (3 : 2), then with volumes in the ratio (27 : 8), the emitted pitches are in the ratio (3 : 2), that is, they form a fifth, with the larger and longer cylinder at the lower pitch). Carré also wondered whether this would hold even if the cylinders were not similar. The answer, left to experimentation, will be negative

The proposition quoted above must be subjected to systematic verification. Carré called a certain Blanchet, a maker of harpsichords and spinets, as a witness to his experiments.

I In the first set of experiments, four beechwood cylinders A, B, C, D, made from the same piece of wood, are considered, differing only in length, with bases of 15 lines, while A is 4 feet long, B 3 feet, C 2 feet, and D 1 foot. When a harpsichord is tuned to the just diatonic system and cylinder A is played, it is in unison with the G Re Sol of the harpsichord; cylinder B, on the other hand, is in unison with the A Mi La above it. Cylinder C is is in unison with C Sol Ut, so cylinders A and C form the fourth Sol-Do, and cylinders B and C form the minor third La-Do. Cylinder D is in unison with F Ut Fa. Cylinders A and D make a major seventh. A and C and C and D, which are in the same ratio of length, both make a fourth.

[115] p. 116. According to Carré, a chord in music is a mixture of two sounds, one of which is low and the other high in relation to each other.

[116] pp. 116–117.

[117] p. 47.

Table 4.9 Sounds of the different cylinders

A	B	C	D	Cylinders
4	3	2	1	Lengths
G Re Sol	A Mi La	C Sol Ut	F Ut Fa	Pitches

For the sake of clarity, Table 4.9 compares the pitches and lengths of the various cylinders:

From this we can conclude that the chords of these cylinders do not follow any particular rule.

II In a second set of experiments, consider three cylinders and E, F, G from the same piece of wood, each two feet long, whose volumes are in the ratio 4 : 3 : 2. Again, there is no definite rule for the chords.

III A third set of experiences concerns five cylinders H, I, K, L, M made of wild cherry, one of the most sonorous woods. Cylinder H has a base of 16 lines ($1^1/_3$ inches) in diameter and a length of 16 inches. The other cylinders I, K, L are similar. Cylinder I has a base of 12 lines, cylinder K has a base of $10^2/_3$ lines, and cylinder L has a base of eight lines. The last cylinder, M, has the same diameter as cylinder I, but is 24 inches long.
Cylinder H is in unison with the G Re Sol of the harpsichord's fundamental octave, while L is in unison with G Re Sol one octave above. H and L therefore sound in the octave, and since the ratio of their volumes is 8 : 1, they satisfy the proposition. K is in unison with D La Re, and forms a just fifth (minus a comma) with H. The ratio of the volumes of the two cylinders is 27 : 8, so the proposition is fulfilled. Cylinder M forms a fourth with cylinder I, while the volume ratio is 2 : 1, so we find a result similar to the first set of experiments, i.e. the hypothesis is not satisfied

IV A final set of experiments involves four wooden parallelepipeds N, O, P, Q. The conclusion is that the arrangements of the parallelepipeds follow no rule unless they are similar, as for the cylinders.

Carré ended his experiments by expressing his desire to repeat them with cylinders of different metals, even hollow ones, to test whether his proposition is always true, provided that similar solids are considered. The memoir concludes with some geometrical problems on solid and hollow cylinders, which could be the starting point for new experiments on metal cylinders.

4.4 Sauveur as a Mathematician

There are few biographical records of Sauveur's life; according to [51][118] it is most likely that all biographers drown from the obituary of the permanent secretary of the Académie des sciences of the time, Bernard Le Bovier de Fontenelle [39]. The obituary was published in the *Historie de l'Académie* in 1716, the same year as Saveur's death, as the *Éloge de M. Sauveur* [39]. If there are discrepancies with Fontenelle's *Éloge*, they are due to misunderstandings; this is the case, for example, of the tale that Sauveur was born deaf and dumb [82],[119] while Fontenelle refers only to his hardness of hearing and his difficulty in speaking fluently. Assuming the uniqueness of the source of Sauveur's biography, we decide to report Fontenelle's *Éloge*, almost verbatim, supplemented by a few notes coming from Sauveur's writings, from the procès-verbaux, and from some French archives [67]. Interesting comments on Sauveur can be found in [7, 24, 32, 50, 53, 82, 85].

4.4.1 Fontenelle's Éloge

Joseph Sauveur (1653–1716), a contemporary of Newton and Leibniz, was born in La Flèche, the son of a notary. Until the age of seven, he remained completely silent due to a defect of the vocal organs, which slowly, and never completely, began to free. This difficulty in speaking spared him all the useless talk of childhood and forced him to think more. He was already the engineer of the other children; he built small mills, made siphons with straw torches, water jets.

Admitted to the Jesuit college, he did not shine, spoke only with great difficulty, and had problems in memorizing. His mind rejected everything that was mere notional and did nothing except with the help of reason. The *Orations* of Cicero, the poems of Virgil did not touch him. By chance he read the *Arithmetic* presented by Jacques Pelletier du Mans (1517–1582). He was fascinated by it and learned it himself. His growing passion for science gave birth to a strong desire to go to Paris, for he felt all too well what was lacking in La Flèche. Having an important uncle, he decided to seek him out in order to obtain a pension to survive in Paris. He made the voyage in 1670 with M. Coubard, his friend, at that time the King's hydrographer in Brest; it was a very philosophical voyage, not only in the intention but also in the approach. They wrote down everything they could see along the way, even more than they should have. In Lyon, Sauveur, listening to the city's famous clock, which did much more than strike the hour, guessed the composition and the mystery of the machine.

His family had destined him for the Church, and in this perspective his uncle granted a pension to study philosophy and theology in Paris. While studying philosophy, he learned the first six books of Euclid in one month, without a teacher. This

[118] pp. 1–2.
[119] p. 15.

success stimulated his interest in mathematics, something that the philosophy of the schools could not satisfy, so he abandoned philosophy and began to study medicine after courses in anatomy and botany. At the same time, he eagerly attended the classes of the Cartesian philosopher of nature Jacques Rouault (1620–1675). Sauveur then met M. Louis Gérard Cordemoy (1620–1684), reader to the Dauphin—the son of Louis XIV—and an able philosopher, who spoke of him to Jacques Benigne Bousset (1627–1704), bishop of Condom, later bishop of Meaux, the young prince's tutor. This prelate wanted to meet Sauveur, suggested various topics in physics, examined him and came to appreciate him. He suggested that Sauveur give up medicine, because he was too direct and did not make turns of phrase, but correct and dry reasoning. In fact, a doctor had to deal with the imagination of his patients almost more often than with their physique; and it was necessary to know how to deal with this imagination, which required special peculiarities.

There was another reason for Sauveur to follow the wise advice of the bishop of Condom. His uncle, seeing that he no longer thought of the ecclesiastical state, stopped paying him his pension, and the young medical student still far from having any income, turned entirely to mathematics and its teaching. Some ladies helped him to get a reputation. He was only 23 when he had a student of very high birth, Prince Eugene of Savoy. A foreigner of the first rank who wanted to learn Descartes' geometry from him. But Sauveur did not know it, so he asked for a week, quickly looked up the text and began to study it. More out of pleasure than necessity. He spent whole nights reading, tired. Sometimes he let the fire go out, and since it was winter, he found himself shivering in the morning without realizing it. When the Ramus Chair of Mathematics fell vacant, which was awarded by competition to the Royal college, he prepared to take it, but having learned that he had to open the competition with a speech, he abandoned the enterprise.

He didn't read much for lack of time time, but he meditated a lot. It kept his attention away from useless conversations and made good use of the time he spent traveling. He would guess what he would find in books when he needed to, to save himself the trouble of searching and studying.

A geometer devoted entirely to his subject would have had no luck at card games, but the *bassette* (a game of chance) was easier for Sauveur than for most of those who played it with fury. Philippe de Coucillon (1638–1720), Marquis de Dangeau, asked him in 1678 to calculate the advantages of the *Banquier* [the holder of the deck of cards] over the *Pontes* [the players]; he did so to the astonishment of many people who saw what they had just guessed clearly expressed in precise numbers. As the bassette was very fashionable at the court, this helped him to be admitted here and had the honor of explaining his calculations to the king and queen. He was then asked for the calculations of other games that he didn't know and whose rules he learned, transforming them into algebraic equations that the players no longer recognized.[120]

---

[120] Here Fontenelle forgot to mention Blaise Pascal (1623–1662) and Pierre Fermat (1601–1665), who founded the theory of probability, and Christiaan Huygens' gambling calculus with his *De ratiociniis in ludo aleae* of 1657.

In 1680 Sauveur was chosen master of mathematics of the pageboys of Marie Anne Christine Victoire of Bavaria (1660–1690), the dauphine.[121] In the following years, he is signaled as mathematical teacher of many noble men. In a letter dated Paris, October 9, 1703, and addressed to Michel de Chamillart (1651–1721), then Minister of War, Sauveur offered to teach mathematics to his son. He disputed the reputation that he would only teach princes [67].[122]

During a trip to Fontainebleau, Marshal Bellefonds commissioned him to give a small course in anatomy to the courtiers. He was out of his usual sphere, but not out of his knowledge. It is said that the whole court listened to him.

In 1681 Sauveur went to Chantilly with Edme Mariotte to experiment with water, a challenging undertaking for a mathematician. He was known to Louis II prince de Condé, a man of genius and lively curiosity, who developed a great affection for him. One day, when Sauveur was discussing some scientific subject with the prince, in the presence of two other scholars, or who professed to be such, they interrupted him, which was never difficult, and began to explain what he had been talking about. When they had finished, the prince said to them: "You thought Sauveur didn't quite understand, because he speaks with difficulty, but I followed him and understood him perfectly. You spoke to me much more eloquently than he did, but I didn't understand you, and perhaps you didn't understand yourself".

He used the time of his trips to Chantilly to work on a treatise on fortification. But a few years later, distrusting mere speculation, he wanted to add practice, even the most dangerous. He took part in the siege of Mons in 1691 and went to the trenches every day. He risked his life so as not to neglect his education, and love of science had made him a brave warrior. After the siege, he visited all the places in Flanders. He learned the details of military developments, camps, army marches, in short, everything that belonged to the art of healing, where intelligence is more important than bravery.

He was the only mathematician at the court, and mathematics was little known there; and as in this country the fashion is more universal than elsewhere, and fortunately in this century there is no longer any education in mathematics, he had the honor of teaching it to all the young princes and sons of France.It would also be pointless to describe in detail his various works. Abbreviated methods for large calculations, tables for the flow rate of water jets, maps of the coasts of France, ordered by M. de Seignelai, the ratio of weights and measures of different countries, a method of measuring all kinds of barrels with great ease and precision, a universal and perpetual calendar. The maps were published as *Neptune François* in 1693 [81]. Other works on maritime subjects were found in his edition of Bernard Renau d'Elizagaray *De la théorie de la manoeuvre de vaiseaux* of 1699 [33] and his *Explications des echelles pour les calculs de la marine* of 1692 [75].

---

[121] It was in the same year that Sauveur was first mentioned in the *Histories de l'académie*, a nomination related to his first interest in music [51], vol. 1, pp. 7–8. The other time he received this honor was in 1681, 1685 and 1696, in correspondence of advancement as a mathematician: maitre of mathematics, professor at the Royal college, member of the Académie.

[122] p. 9.

Only with too long a discourse could one try to make people understand the difficulty and the value of this kind of work, which perhaps those who love only the highest peak of theory do not consider enough. Sauveur was concerned only with useful mathematics, the effect of his natural soundness of mind, and perhaps also of the habit of teaching. It must also be said that he was not too prejudiced against the new geometers of infinity, whom he called *infinitari*, as were those who did not want to accept them. It's not that he didn't understand their methods well but used them only when strictly necessary.

In 1686 Sauveur was appointed to a chair of mathematics at the Royal college. The stutter did not hold him back, for he was now a famous man. He did not write any of the papers he dictated as things that involved reasoning and did not require memory were so present in his mind and so well understood that he simply had to get them out. There were copyists who wrote for him to sell his treatises; he himself bought a copy at the end of each year. Sometimes, when he found attentive and intelligent listeners, he was carried away by the pleasure of instructing them, and would have devoted the whole day to them without realizing it, had not a servant accustomed to correct his distractions warned him that he was otherwise occupied.

Sauveur had been active in the Académie since 1680, although he was not made a full member until 1696, on the recommendation of Louis-Phélypeaux de Pontchartrain (1643–1727), chancellor of France. The minutes of the Académie mention 10 lectures by Sauveur on mathematical subjects for the period 1680 to 1694. It cannot be determined how often he attended the meetings when he was not lecturing. The names of those present at the academic meetings are recorded from the beginning of 1695. Sauveur's name, however, does not appear until his official appointment as academician geometer on February 29, 1696.

In early 1699, the Académie was reorganized and by then consisted of ten honorary members, twenty pensionnaires, twenty associates (some foreign), and twenty pupils. Sauveur was appointed one of the three associate mecaniciens on January 28, 1699. From March 13, 1700, to July 1, 1716 (a few days before his death), he was present at most of the meetings [67].[123]

Having just entered the Académie in 1696, he proposed a great project, on which he had been meditating for some time, to lay the foundations of an almost entirely new science that was to be related to optics: acoustics.

When Sauveur began his scientific studies of music and acoustics, he was already forty years old. Having neither voice nor ear he had to borrow the voice or the ear of others. In all parts of his system he often and usefully consulted the duke of Orléans, who had learned mathematics from him and knew music perfectly.

A new musical language, more comfortable and complete, a new system of sounds, a unique monochord, an echometer, the fixed sound, the nodes of the waves, were the fruits of Sauveur's research. He had gone as far as the music of the ancients. Greeks and Romans, Arabs, Turks and Persians; he was so jealous that nothing escaped him of this science of sounds. We have spoken too much of his discoveries in our *Histoires de l'Académie* to repeat anything here. The death of a scientist never hurts science so

[123] pp. 9–10.

much as when it interrupts long-term projects. A great number of discoveries and a certain precious and sometimes unique thread of ideas perish with the first inventor.

M. de Vauban, who, after being appointed Marshal of France in 1705, was in charge of examining engineers, proposed Sauveur to the king for this post, which no longer corresponded to his dignity. Sauveur was approved by the king and honored with a pension. He removed from his position as inspector all the useless or even harmful things that others might have put there, and he maintained only a gentle but fine and penetrating attention. Sometimes the engineers were hanging out thinking of a simple conversation, tested without believing they were.

Although Sauveur had always been in good health and seemed to have a robust constitution, he was carried away in two days by an inflammation of the chest; he died at the age of 64. The Académie was informed of Sauveur's death on July 11, 1716, one day after it occurred on July 10, 1716. Shortly thereafter, Fontenelle wrote his éloge. Unfortunately, he gave July 9, 1716 as the date of death, which has been copied in the numerous biographical notes on Sauveur based on his text.

Sauveur had no presumptions; he was heard to say that what one man can do in mathematics, another man can also do. The statement may not be true, but it is a sign of modesty. He found it very difficult to be satisfied with his works, and he had to take them out of his eyes and tear them out himself to stop retouching them; he was thoughtful, kind, and without harshness, even in his intimacy. Although he was widely known, his natural simplicity and ingenuity had not been affected, and the mathematical character had always prevailed.

Sauveur married Jeanne Haffrey for the first time in 1686. According to Fontenelle, before marrying, Sauveur first took a precaution. He did not want to see the one he was going to marry until a notary had been sent to write down the conditions he wanted. His wife bore him two sons, Jean-Urbain and Alexandre-Simon. After her death, he married Catherine-Corentine Leclerc. His second wife again bore him two children, Jeanne-Catherine-Renee (1700) and Jean-Baptiste-Joseph (1702), the son was mute until the age of 7 like his father.

4.4.2 Writings Not Acoustical

Before devoting almost entirely to the study of music, Sauveur had dealt with other subjects in mathematics and mathematical physics, of which, despite his lack of interest in publication, a few manuscripts remain, some of which have been edited and printed. Among the most interesting manuscripts are those on arithmetic and geometry, found in various libraries, including the Diderot library in Lyon; a good part of them were collected in the *Geometrie élémentaire et pratique* of 1754 [79]. An *Abregé de mécanique* [80], is also kept in the Diderot library. Finally, there are a manuscript on fortifications written on the occasion of the siege of Mons in 1691 [76] and a small treatise on *Echelles* [75]. There are no traces of writings on game theory that occupied the young Sauveur. Table 4.10 refers to non-musical lectures delivered to the Académie.

Table 4.10 Lectures given by Sauveur for the Académie Royal des sciences about non-musical subjects. Only for full texts of lectures the pages in the procès-verbaux (PV) of the Académie are given [67], p. 11

Date(s)	Subject
1680, July 13	Dioptrics, and cogwheels
1681, May 3	The measurement of water flows
1681, May 31	A mathematical problem (PV 9, 100v-101v); solution by De La Hire of the same problem: fol. 104r
1681, July 26	Experiments, begun with Mariotte
1683, June 14	Tools for designing water works
1684, November 18	Calculating water flows
1685, January 13	Air resistance
1685, February 13	Calendar
1690, June 17	Calendar
1694, August 18	Water flows
1696, March 24	The weighting of air. The text was handed in one week later, on 31 March 1696, but is not included in the procès-verbaux
1696, May 5	The logarithmic curve (PV 15, 47r-49r)
1696, June 23	Differential calculus
1696, June 30	Algebraic powers (PV 15, 111v-114v)
1696, July 17	The power series
1697, January 5	A problem of Bernoulli
1697, March 30	The tangent of a logarithmic curve (PV 16, 76r-78r)
1697, April 27	Chance calculations for the game of quinquenove, in the form of a letter to the prince of Condé (PV 16, 107v-114v)
1697, July 20, 27	The measurement of barrels (PV 16, 203r-206v)
1698, January 18,	The measurement of the speed of the water flow of a river
1698, November 15	Curves described by two connected objects
1702, July 12, 15	A letter about a Siamese twin, followed by a lecture on thermometers and barometers (PV 21, 277r-279r and 281r-283v)
1702, November 25	Gauging
1703, July 14	Friction (PV 22, 247v-249v)
1707, February 19, 26	Chronology
March 23	
1709, December 7, 20	Magic squares
1710, December 3	Magic cubes (PV 29, 419r-420v)
1711, January 14	A geodesic problem (PV 30, 31r-33r)
1712, March 4	The speed of the water flow in the River Seine on 2 March 1711 (PV 30, 10lr-103v)
1716, March 11, 18	A problem by Lagny (PV 34, 97r-98v)

4.4.2.1 Geometry

Below we will mention a few of the writings on geometry as they are those most directly related to acoustical studies. Sauveur's geometry is largely based on the geometry of Port royale, which is well represented in the text *Nouveaux élémens de géométrie* published in organic form by Antoine Arnauld (1612–1694) in 1667 [6]. Arnauld's goal was to revise Euclid's *Elements* so that evidence was not sacrificed to the demands of rigor and exactness as it was in the original treatise. The preface, written by Pierre Nicole (1625–1695), reads:

> He [Arnauld] did not consider geometry per se, but for the use that can be made of it. And he believed that by avoiding those errors that can be avoided, it can be used to train young people, not only for the rightness of mind, but in some way for the regulation of ethics [6].[124]

Port royale's geometry, because of its undoubtedly greater simplicity of presentation, largely influenced the teaching of mathematics at the time, especially in France. This was despite the not always positive opinions of French mathematicians such as Ignace Gaston Pardies (1636–1673) and Jean-Étienne Montucla (1725–1799) [44].[125]

Sauveur stood in the wake of Arnauld with original contributions, extending geometry to solids and also introducing a nod to Calculus. His *géométrie* was widely read in the first half of the 1700s and inspired, for example, Dominique François Rivard (1697–1778) in the composition of his highly regarded *Élemens de geometrie avec un abregé d'arithmetique et d'algebre* of 1732 [69].

Le Blond's edition is divided into two parts, elementary geometry (six books) and practical geometry (seven books). In the following, we will focus on only two subjects, one taken from elementary geometry, concerning elements and geometric proportions, and the other taken from practical geometry, concerning the use of logarithms, as these topics are of some interest for Sauveur's use in his writings on acoustics.

Straight or Curved Lines

> The straight line is the one that goes directly from one point to another, like AB. The curved line goes from one point to another line with some detour, like CD [79].[126]

Below Sauveur's definition of parallel lines:

> A line CD is parallel to another AB when all the points of one are equally distant from those of the other, that is, when all the perpendiculars CA, DB drawn from one and the other are equal to each other [79].[127]

to compare with Euclid's definition:

---

[124] Preface, p. 3, not numbered pages.

[125] p. 37.

[126] p. 58.

[127] p. 74.

> Parallel lines are those which, being in the same plane and being produced ad infinitum in both directions, do not meet each other in either direction [55].[128]

The definition is completed by a theorem:

> Théoreme I. 203. Two parallel lines continue to infinity, do not meet; but if two lines are not parallel, they meet on the side which they approach. This theorem follows from the definition of parallel lines [79].[129]

Sauveur did not prove the theorem of parallels as stated in Euclid's Proposition 27, for which if a straight line makes the alternate angles equal to each other, the straight lines will be parallel [55],[130] but only the converse, that is, Euclid's Proposition 29, for which the straight line falling on parallel lines makes the alternate angles equal to each other [55].[131] But while in Euclid the proof uses the fifth postulate, this is not the case here, simply because the fifth postulate of Euclid is not a postulate of Sauveur's geometry (note that Arnauld replaced the fifth postulate of Euclid with the following one: Two straight lines which, when extended, gradually approach each other, they finally meet [6].[132]

Geometric Proportions

Sauveur gave great relevance to the theory of ratios and geometric proportions to the point of devoting to it a small treatise, *Traité des proportions*, after a long arithmetic premise on the practical numerical calculation of square and cubic roots, which prefaces the expositions of elementary geometry.

This fact alone proves that his geometry follows a traditional approach in the wake of the classical Euclidean treatment. For him, only (?) in his treatise of geometry, the *ratio* is a relation between two homogeneous quantities, the first of which is called the antecedent or first term, the second the consequent or second term. The ratio or reason is the way in which the antecedent contains its consequent or some of its rates" [29].

Two ratios are equal if the antecedents contain the consequents in the same way. The equality of two ratios is called *proportion*, and the quantities between which there are equal ratios are called *proportional*. If the middle terms of a proportion are equal, the proportion is said to be continuous as 4, 6 :: 6, 9; in the continuous proportion, the middle term is called *proportional mean*. If the continuous proportion has more than three terms, it is called *progression*.

Sauveur, as a mathematician with strong practical tendencies, realized, or should have realized, that ratios could be treated as fractions. A modern reader of the *géométrie* is tempted to make this identification because Sauveur's proportions are always expressed by numbers and never by letters, as is the case in the geometry of

---

128 p. 318.

129 p. 574

130 p. 321.

131 p. 321.

132 Preface, p. 83.

Table 4.11 Transformation of proportions. Adapted from [79], p. 38

original proportions	6	·	8	::	15	·	20
	3, 2	·	4, 2	::	3, 5	·	4, 5
invertion	8	·	6	::	20	·	15
	4, 2	·	3, 2	::	4, 5	·	3, 5
permutation	6	·	15	::	8	·	20
	3, 2	·	3, 5	::	4, 2	·	4, 5
composition	14	·	6	::	35	·	15
	7, 2	·	3, 2	::	7, 5	·	3, 5
or	14	·	8	::	35	·	20
	7, 2	:	4, 2	::	7, 5	·	4, 5
division	2	·	6	::	5	·	15
	1, 2	:	3, 2		1, 5	·	3, 5
or	2	·	8	::	5	·	20
	1, 2	:	4, 2	::	1, 5	·	4, 5
composition + division	14	·	2	::	35	·	5
	7, 2	:	1, 2	::	7, 5	·	1, 5

Port royale, for example, which allowed for a more abstract treatment. At some point, the fraction sign also appears, for example, when comparing the ratios 12, 3 :: 20, 5, the diagram he gave is as follows [79]:[133]

$$\frac{12}{3}\ \{4 \;\Big|\; \frac{20}{5}\ \{4$$

But the time was not yet ripe; d'Alembert still had to justify, a few generations later, the equation of ratios with fractions, to extend the idea of proportionality to heterogeneous quantities by considering the relations between the numbers expressing their measures [26].[134]

The traditional treatment of proportions followed by Sauveur leads to complex rules, which become theorems; the following table illustrates how proportions can be transformed by following some basic rules, classified as: inversion, permutation, composition and division.

In Table 4.11 Sauveur referred to numbers—considered more appropriate for a didactic text—but, he later, also explained the operations of proportions in terms of letters. Assuming modern notation for simplicity, the operations on the proportion $a : b = c : d$ can be written as:

1. Inversion, from $a : b = c : d$ to $b : a = d : c$.
2. Permutation, from $a : b = c : d$ to $a : c = b : d$

---

[133] p. 33.

[134] p. 16, footnote 1.

3. Composition, from $a : b = c : d$ to $(a + b) : b = (c + d) : d$
4. Division, from $a : b = c : d$ to $(a - b) : b = (c - d) : d$
5. Composition and division, from $a : b = c : d$ to $(a + b) : (a - b) = (c + d) : (c - d)$

which for a modern reader are trivial results.

Calculating the fourth term of a proportion, given the other three terms, also becomes a small theorem: the *rule of three simple* from the medieval abacus schools (if proportions had been equated with equality of fractions, it would have been a matter of solving a trivial linear equation, and no theorem would have been needed).

The reluctance to treat ratios as fractions and proportions as equality of fractions was also evident when Sauveur introduced logarithms into proportions, where this ratio-fraction equality seems inevitable. Sauveur got away with treating the relationship between the logarithms of a geometric ratio as an arithmetic ratio in a section of his book on logarithms entitled *The rule of three or the proportion for logarithms*, where he argued as follows: given three numbers, to find the fourth proportional one must add the logarithms of the last two numbers and subtract from their sum the logarithm of the first; the remainder will be the logarithm of the fourth proportional.

Logarithms

Calculus with logarithms is introduced at the beginning of practical geometry. Sauveur did not give an explicit definition of the logarithm, for which the logarithm of n—base 10—is the solution of the equation $n = 10^x$ in the unknown x, but he did define its properties, such as those relating to multiplication:

> To multiply two numbers, one must add their logarithms, and the sum of these logarithms will be the logarithm of the product [79].[135]

In a similar informal way, division, power, and roots are defined.

From this 'theoretical' premise he proceeded to the calculation of logarithms of assigned numbers; first he considered the powers of ten, 1, 10, 100, 1000, 10, 000, to which the logarithms 0, 1, 2, 3, 4 are assigned without comment. From the association between numbers and logarithms, it is clear that Sauveur used base 10 logarithms, although he did not say so explicitly.

Later, Sauveur proposed to calculate the logarithm of 2 (the choice is not accidental, since the interval of an octave is exactly 2 : 1, and Sauveur had just become interested in music). The calculation of the logarithm of this number through the dichotomous method, based on the geometric mean, is given in Table 4.12.

The two starting numbers, whose logarithms define the interval containing the logarithm of 2, are 1 and 10; their geometric mean is $\sqrt{10} = 3.1623$, whose logarithm is given by $\log_{10} \sqrt{10} = 1/2 \log_{10} 10 = 0.5$. The second step is to make the geometric mean between 1 and 3.1623, which is worth $\sqrt{3.1623} = 1.7783$, which corresponds to the logarithm of 0.25. In the third step, he took the geometric mean between 1.7783 and 3.1623, the two numbers containing 2, which is worth $\sqrt{1.7783 \times 3.1623} =$

[135] p. 182.

Table 4.12 Logarithm of 2. Adapted from [79], p. 183

Numbers		Logarithms	Numbers		Logarithms
A	1.0000	0.00000	D	1.7783	0.25000
C	3.1623	0.50000	G	1.9110	0.28125
B	10.000	1.00000	F	2.0535	0.31250
A	1.0000	0.00000	G	1.9110	0.28125
D	1.7783	0.25000	H	1.9809	0.29687
C	3.1623	0.50000	F	2.0535	0.31250
D	1.7783	0.25000	H	1.9809	0.29687
E	2.3714	0.37500	K	2.0169	0.30469
C	3.1623	0.50000	F	2.0535	0.31250
D	1.7783	0.25000	Z	2.0000	0.30130
F	2.0535	0.31250			
E	2.2314	0.37500			

2.3714, whose logarithm is 0.375. Continuing, he obtained a five-digit approximation of the logarithm of 2 given by 0.30130.

Sauveur distinguished two parts of the logarithm, the part before the decimal point, which can be positive or negative and which he called *characteristic* or *figurative*, and the part after the decimal point, which he called *proper logarithm* (the modern terms are characteristic and mantissa, respectively). Take the two numbers $31.42 = 3.142 \times 10$ and $0.03142 = 3.142 \times 10^{-2}$ for the properties of the logarithms one has, respectively: $\log_{10} 31.42 = \log_{10} 3.142 + \log_{10} 10 = \log_{10} 3.142 + 1 = 1.4972062$; $\log_{10} 0.0342 = \log_{10} 3.142 + \log_{10} 10^{-2} = \log_{10} 3.142 - 2 = \bar{2}.4972062$. From this we get the rules: two numbers differing only by a power of 10 have the same mantissa and different characteristic.

Of course, the use of logarithms is of some help only with numerical tables. For what Sauveur said, the most common tables that circulated were for numbers up to 10,000 with 7 places.[136] According to him, these tables with a simple process of linear interpolation [77][137]—the same one used until a few years ago, before the spread of computers—allow to find with good accuracy the logarithms of numbers up to $1\ 10^7$. For larger numbers, one must use *les grandes tables* [79],[138] which give the logarithms of numbers up to 100,000 with 10 decimal places. However, according to Sauveur, for ordinary calculations it is sufficient to work with logarithms to 5 decimal places, and this is what he did in the acoustics papers.

[136] Because by multiplying or dividing a number n by a power of 10 its mantissa remains the same, the tables of numbers from 1 to 10,000 are useful to find the mantissa of all numbers with 4 significant digits. The numbers 0.00003142, 3.142, 3,142,000,000 have the same mantissa, the number to enter in the table is 3142 (a number between 1 and 10,000). From the table, we get the mantissa, which is 0.4972062, the characteristics for the three numbers are −5, 0, 9, respectively.

[137] p.189, footnote a.

[138] p. 183.

The tables Sauveur was to use and recommend were those of the Dutch publisher Adriaan Vlacq (1600–1666), who is explicitly mentioned in Sauveur's 1707 memoir *Méthode générale pour les systèmes tempérés de musique, et du choix de celui qu'on doit suivre*: "we have used the little tables of Vlacq, which are very common" [77].[139]

4.4.2.2 Other Mathematical Writings

If one had to give a single answer to the question of what Sauveur's profession was, the answer would undoubtedly be: that of a mathematician. A good mathematician, but not a great mathematician, partly because he had little interest in mathematics itself.

He was involved in mathematics all his life and taught it to many French nobles as a tutor. Unfortunately, we have almost nothing left of his mathematical output, Apart from the *Geometrie élémentaire et pratique*, which in any case had remained manuscript, Sauveur's only publication on matters of theoretical mathematics is the short paper *Solution du problême proposé par M. De Lagny*, which appeared in the Académie's memoirs in 1716, the year of his death [78].

This is an answer to a problem associated with the French mathematician Thomas Fantet de Lagny (1660–1734), which Sauveur formulated as follows:

> Find the ratio of two quantities x and y such that the greater x divided by the lesser y gives the quotient n and a first remainder; that the lesser y divided by the first remainder gives q and a second remainder; that the first remainder divided by the second gives the quotient q with a third remainder, which divided by the second remainder gives the quotient r and another remainder [78].[140]

According to Sauveur, the problem was very interesting, perhaps because it is in some way related to musical proportions, to the point that he stated: "I had decided to devote myself entirely to acoustics, but, attracted by the usefulness of the researches made by M. Lagny on the divisions of two numbers, I sought a general solution [of the problem] " [78].[141] The problem has two solutions; the first when we end up with a division without remainder; in this case the ratio of x to y is rational. The second, when the divisions go to infinity, and then x and y are incommensurable [78].[142]

In his *Éloge* Fontenelle said that Sauveur was quite a traditionalist in mathematics. This is not entirely true. In fact, Sauveur visited the so-called Malebranchians or the members of Malebranche's circle, the reference for many French mathematicians; among them: de l'Hôpital, Reyneau, Jaquemet, Byzance, Bernard Lamy, Varignon, Carré, Rémond de Montmort, Saurin, Guisnée, Renau d'Élizagaray, Fontenelle and

---

[139] p. 209. In 1628 Vlacq published his *Arithmetica logarithmica*, containing tables of logarithms from 1 to 100,000 to 10 places. Vlacq also published smaller tables of logarithms, those named by Sauveur, from 1 to 10,000 to 7 places, which remained in print well into the 19th century [72].

[140] p. 26.

[141] p. 26.

[142] pp. 26–27.

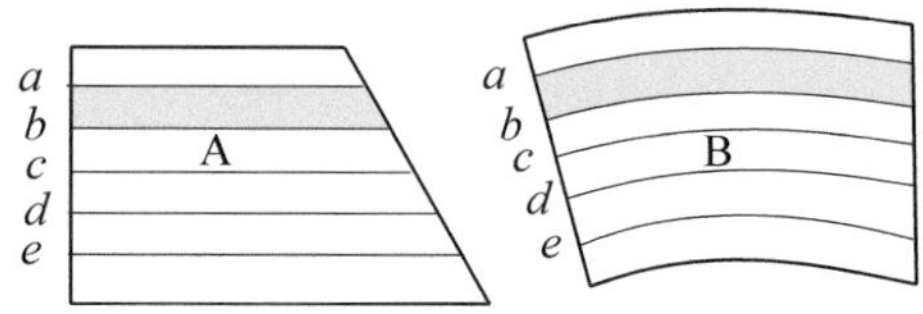

Fig. 4.27 Indivisibles for plane figures

Polignac, Nicole, Privât de Molières, etc.. They united their efforts against the traditional Cartesians, referring to Leibniz and and Johann Bernoulli "àleur service" to teach Calculus [71].[143] Dating from 1696 is the treatise *Analyse des infiniment petits* by Guillaume François Antoine de L'Hôpital, (1661–1704) [4], with which Sauveur could not have been unfamiliar.

Already in the *Geometrie élémentaire et pratique* there are signs that Sauveur did not despise Calculus. Though he only used Cavalieri's method of indivisibles to the calculation of areas and volumes of the entities of Euclidean geometry, his language was already that of the Calculus of his time.

Here is how he introduced indivisibles for the calculation of areas: consider two plane figures such as A or B, as shown in Fig. 4.27, bounded by straight or curved parallel lines which are infinitesimally small and equidistant from each other and bounded at the extremity of the figure by a straight segment; these lines form rectangular parallelograms whose lengths occupy one of the dimensions of the figure and whose infinitesimal widths are equal to each other, these types of parallelograms are called indivisibles, The following theorem holds:

Theorem I. 399. Every figure can be considered as filled with its indivisibles [79].[144]

A paper by Sauveur on infinitesimal calculus is documented in the procès-verbaux of 1696 (see Table 4.10). But there is a writing in which Sauveur made extensive use of the new differential calculus, showing that he could handle it perfectly. This is the section of his 1713 work on vibrating strings, when he calculated the length of the simple pendulum equivalent to the compound pendulum formed by a string deformed by its own weight rotating about the horizontal axis connecting the bridges that support the string (see Sect. 5.7.3). Here he carried out the differential of products of functions, using infinity either as a finite, though very large, quantity or as infinite in action, depending on the circumstances.

References

1. Adkins C (2001) Trumpet marine. Grove music on line. https://doi.org/10.1093/gmo/9781561592630.article.28494. Accessed 25 July 2022
2. Anderson GH (1979) La gamme du Si: a chapter in the history of solmization. Indiana Theory Rev 3(1):40–47

[143] p. 287.

[144] p. 123.

3. Anonymous (1666) Méthode facile pour apprendre a chanter la musique. Ballard, Paris
4. Anonymous (1696) Analyse des infiniment petits. Imprimerie Royale, Paris
5. Aristoxenus (1902) The harmonics of Aristoxenus. Clarendon Press, Oxford, Edited and translated into English by Macran HS
6. Arnauld A (1667) Nouveaux élémens de géométrie; contenant, outre un ordre tout nouveau, & de nouvelles demonstrations des propositions. Savreux, Paris
7. Auger L (1948) Les apports de J. Sauveur (1653-1716) à la création de l'acoustique. Revue d'Histoire des Sciences et de leurs Applications 1(4):323–336
8. Barbieri P (1980) Il corista bolognese secondo il rilevamento di V.F. Stancari L'Organo 18:15–29
9. Barnett G (2008) Tonal organization in seventeenth-century music theory. In: Christensen T (ed) The Cambridge history of Western music theory. Cambridge University Press, Cambridge, pp 407–455
10. Baskevitch F (2005) Notion de vibration sonore au XVIIème siècle. https://www.academia.edu/7180028/. Accessed 25 July 2023
11. Beeckman I (1939–1953) Journal tenu par Isaac Beeckman de 1604 à 1634 (4 vols). Nijhoff, The Hague
12. Bennett M, Schatz MF, Rockwood H, Wiesenfeld K (2002) Huygens's clocks. Proc R Soc 458:563–579
13. Bernoulli D (1753a) Réfléxions et eclaircissemens sur les nouvelles vibrations des cordes exposées dans les Mémoires de l'Académie de 1747 & 1748. Mémoires de l'Académie Royale des Sciences et Belles Lettres de Berlin 9:147–172
14. Bernoulli D (1753b) Sur le mélange de plusieurs especes de vibrations simples isochrones, qui peuvent coexister dans un meme système des corps. Mémoires de l'Académie Royale des Sciences et Belles Lettres de Berlin 9:173–195
15. Bernoulli Johann (1728) Meditationes de chordis vibrantibus, cum pondusculis aequali intervallo a se invicem dissitis, ubi nimirum ex principio virium vivarum quaeritur numerus vibrationum chordae pro una oscillatione penduli datae longitudinis D. Commentarii Academiae Scientiarum Imperialis Petropolitanae 3:13–28
16. Bernoulli Johann (1742) Theoremata selecta pro conservatione virium vivarum demonstranda et esperimenta confirmanda. In: Bernoulli J (ed) Opera omnia, vol 3, Bousquet, Lausanne and Geneva, pp 124–130
17. Biasiol P (1981) Christophe de Villiers: un "savant" di provincia nella corrispondenza di M. Mersenne. Rivista Critica di Storia della Filosofia 36(4):379–400
18. Bonanni F (1722) Gabinetto armonico pieno d'istromenti sonori indicati, e spiegati dal padre Filippo Bonanni della Compagnia di Gies offerto al Santo Re David. Plachor, Rome
19. Bühler W (2013) Musikalische Skalen bei Naturwissenschaftlern der frühen Neuzeit. Internationaler Verlag der Wissemschaften, Frankfurt
20. Busse-Berger AM (2008) The evolution of rhythmic notation. In: Christensen T (ed) The Cambridge history of Western music theory. Cambridge University Press, Cambridge, pp 628–656
21. Cannon J, Dostrovsky S (1981) The evolution of dynamics: vibration theory from 1687 to 1742. Springer, Dordrecht
22. Carré L (1707) Sur la theorie generale du son. Le Journal des Sçavants Supplement pp 115–125
23. Carré L (1709) De la proportion que doivent avoir les cilindres pour former par leurs sons les accords de la musique. Mémoires de l'Académie Royale des Sciences de Paris pp 47–62
24. Cohen A (1981) Music in the French Royal academy of sciences. A study in the evolution of musical thought. Princeton University Press, Princeton
25. Cohen H (1984) Quantifying music. Reidel, Dordrecht
26. d'Alembert J (1758) Traité de dynamique (second edition). David, Paris
27. Darrigol O (2007) The acoustic origins of harmonic analysis. Arch Hist Exact Sci 61:343–424
28. Descartes R (1653) Excellent compendium of musik with necessary and judicious animadversions. Harper, London
29. Descartes R (1656) Musicae compendium. Janssonium, Amsterdam

30. Descartes R (1668) Traité de la méchanique. Abrégé de musique. Astragal Press, Angot
31. Descartes R (1897–1913) Oeuvres de Descartes, vols 13. In: Adam C, Tannery P (eds) Cerf, Paris
32. Dostrovsky S (2007) Sauveur, Joseph. In: Gillispie CC (ed) Complete dictionary of scientific biography. Scribner, New York
33. Elizagaray RBd (1689) De la theorie de la manoeuvre des vaisseaux. Sauveur J (ed) Michallet, Paris
34. Ellis A (1885) On the musical scales of various nations. J Soc Arts 33:485–527
35. Euler L (1748) De propagatione pulsuum per medium elasticum. Novi Commentarii Academiae Scientiarum Petropolitanae 1:67–105
36. Fabbri N (2007) Genesis of Marin Mersenne's "Harmonie universelle": the manuscript 'Livre de la nature des sons. Nuncius 22:287–308
37. Fontenelle de BB (1707) Eloge de M. l'Abbé Gallois. Histoire de l'Académie Royal des Sciences de Paris pp 176–181
38. Fontenelle de BB (1711) Éloge de M. Carré. Histoire de l'Académie Royale des Sciences pp 102–107
39. Fontenelle de BB (1716) Éloge de M. Sauveur. Histoire de l'Académie Royale des Sciences, pp 97–107
40. Fourier JBJ (1822) Théorie analytique de la caleur. Didot, Paris
41. Gruber A (1970) Mersenne and evolving tonal theory. Journal of Music Theory 14(1):36–67
42. Huygens C (1673) Horologium oscillatorium. In: Societé Hollandaise des Sciences (ed) Oeuvres complètes de Christiaan Huygens (22 vols), vol 18, Nijhoff, The Hague, pp 27–438
43. Huygens C (1888–1950) Oeuvres complètes de Christiaan Huygens, vols 22. Nijhoff, The Hague
44. Itard J (1940) La géométrie de Port-Royal. Enseign Math 2(38):27–58
45. Kromhout MJ (2020) George Simon Ohm and the mathematical analysis of sound. Isis 111(3):471–492
46. La Hire Pd (1695) Traité de mécanique. Imprimerie Royale, Paris
47. La Hire Pd (1716a) Continuation d'expériences sur le son. Mémoires de l'Académie Royal des Sciences de Paris pp 264–268
48. La Hire Pd (1716b) Expériences sur le son. Mémoires de l'Académie Royal des Sciences de Paris pp 262–264
49. La Hire Pd (1730) Explication des differences des sons de la corde tendue sur la trompette marine. In: l'Académie Royale des Sciences (ed) Memoires de l'Académie Royale des Sciences. Depuis 1666, jusqu'à 169, vol 9, Compagnie des libraries, Paris, pp 500–529
50. Lindsay BR (ed) (1973) Acoustics: Historical and philosophical development. Dowden, Hutchinson & Ross, Stroudsburg, Pennsylvania
51. Maxham RE (1976) The contributions of Joseph Sauveur (1653-1716) to acoustics (2 vols). PhD thesis, University of Rochester
52. Mersenne M (1636–1637) Harmonie universelle, vols 2. Cramoisy, Paris
53. Miller DC (1935) Anecdotal history of the science of sound to the beginning of the 20th century. Macmillan, New York
54. Miller LE (1990) John Birchensha and the early Royal society: Grand scales and scientific composition. J R Music Assoc 115(1):63–79
55. Muller I (1981) Philosophy of mathematics and deductive structure in Euclid's Elements. Dover, New York
56. Muzzulini D (2015) The geometry of musical logarithms. Acta Musicologica pp 193–216
57. Muzzulini D (2020) Isaac Newton's microtonal approach to just intonation. Empir Musicol Rev 35(3–4):223–248
58. Newton I (1664–1666) Musical notes and of musik. https://www.newtonproject.ox.ac.uk/view/texts/diplomatic/NATP00128 (MS Add. 4000, ff. 104v-113r). The Newton project. Accessed 25 March 2025
59. Newton I (2013) Musical notes and Of musik. In: Wardhaugh B (ed) The Compendium musicae of René Descartes: Early English responses. Brepolis, Turnhout, pp 85–128

60. Norton F (1677) A philosophical essay of musick directed to a friend. Martyn, London
61. Perrault C (1680) Essais de physique, ou Recueil de plusieurs traités touchant les choses naturelles, vol 4. Coignard, Paris
62. Pesic P (2006) Isaac Newton and the mystery of the major sixth: a transcription of his manuscript 'Of musick' with commentary. Interdisc Sci Rev 31(4):291–306
63. Pirro A (1907) Descartes et la musique. Fischbacher, Paris
64. Plot R (1677) The natural history of Oxford-shine. Millers, Oxford
65. Ramos de Pareja B (1482) Musica practica. Baltasar de Hiriberia, Bologna?
66. Ramos de Pareja B (1993) Musica practica. Translation into English by Miller C A. Hänsler, Stutgart
67. Rasch R (ed) (1984) Joseph Sauveur. Collected writings on musical acoustics (Paris 1700–1713). The Diapason Press, Utrecht
68. Rasch R (2008) Tuning and temperament. In: Christensen T (ed) The Cambridge history of Western music theory. Cambridge University Press, Cambridge, pp 193–222
69. Rivard F (1732) Elemens de geometrie avec un abregé d'arithmetique et d'algebre. Henry, Paris
70. Robartes F (1692) A discourse concerning the musical notes of the trumpet, and trumpet-marine, and of the defects of the same. Philos Trans R Soc Lond 17(195):559–563
71. Robinet A (1960) Le groupe malebranchiste introducteur du calcul infinitésimal en France. Revue d'Histoire des Sciences 13(4):287–308
72. Roegel D (2010) A reconstruction of De Decker-Vlacq's tables in the Arithmetica logarithmica (1628). https://inria.hal.science/inria-00543941. Accessed 25 Aug 2023
73. Rousseau J (1710) Méthode claire, certain et facile pour apprendre à chanter la music (5th edition). Mortier, Amsterdam
74. Rousseau JJ (1768) Dictionnaire de musique. Duchesne, Paris
75. Sauveur J (1692) Explication des echelles pour les calculs de marine. Sevestre, Paris
76. Sauveur J (1701–1800) Pratique de la fortification des places. in: Recueil de quelques pièces par raport aux fortifications. Ark:/12148/btv1b9061377v. Bibliothèque Nationale de France. Département des Manuscrits, 12381
77. Sauveur J (1707) Methode generale pour les systêmes tempérés de musique, et du choix de celui qu'on doit suivre. Mémoires de l'Académie Royale des Sciences de Paris, pp 203–222
78. Sauveur J (1716) Solution du problême proposé par M. De Lagny. Par M. Sauveur. Mémoires de l'Académie Royal des Sciences de Paris, pp 26–29
79. Sauveur J (1754) Geometrie élémentaire et pratique. In: Le Blond (ed). Jombert, Paris
80. Sauveur J (c 1700) Abregé de mécanique. 1R 137225 and 1R 204413. Bibliothèque Diderot de Lyon
81. Sauveur J, al (1693) Le Neptune François ou recueil des cartes marines levées et gravées par ordre du Roy. Premier volume contenant les costes de l'Europe sur l'Ocean depuis Dronthem en Norvege jusques au détroit de Gilbraltar, avec la mer Baltique. Imprimerie Royale, Paris
82. Scherchen H (1972) The nature of music. Scholarly Press, Saint Clair Shores
83. Taylor B (1713) De motu nervi tensi. Proc R Soc Lond 28(317):26–32
84. Telonius CG (2018) Concert for trumpet marine, strings and timpani No. 13 in D major. https://www.youtube.com/watch?v=ShU_XXDtjX4. Accessed 12 Aug 2024
85. Truesdell CA (1960) The rational mechanics of flexible or elastic bodies. In: Leonhardi Euleri Opera omnia (1911, in progress), 2, vol 1, part 2, Teubneri GB; [then] Fussli Turici; [then] Birkhäuser, Basel
86. Wallis J (1671) Mechanica, sive de motu, tractatus geometricus. Godbib, London
87. Wallis J (1677) Dr. Wallis's letter to the publisher, concerning a new musical discovery. Philos Trans 12(134):839–842
88. Wallis J (1698a) A letter of Dr. John Wallis to Samuel Pepys Esquire, relating to some supposed imperfections in an Organ. Philos Trans 20(242):249–256
89. Wallis J (1698b) A question in musick lately proposed to Dr. Wallis, concerning the division of the monochord, or section of the musical canon with his answer to it. Philos Trans 20(238):80–84

90. Wallis J (2016) John Wallis: writings on music. In: Cram D, Wardhaugh B (eds) Routledge, New York
91. Wardhaugh B (2008) Musical logarithms in the seventeenth century: Descartes, Mercator, Newton. Hist Math 35:19–36
92. Wardhaugh B (2013) The Compendium musicae of René Descartes: early English responses. Brepolis, Turnhout
93. Wardhaugh B (2016) Music, experiment and mathematics in England 1653–1705. Routledge, New York
94. Zarlino G (1588) Sopplimenti musicali. Senese, Venice

Chapter 5
Joseph Sauveur and Acoustics

Abstract This chapter is entirely devoted to Joseph Sauveur. First, the *Traité de la théorie de musique* is presented, which probably dates back to 1697, but was not published until the 20th century. The *Traité* contains a comprehensive analysis of most aspects of music theory of the time and is the basis for all the memoirs that Sauveur would publish for the Académie des sciences of Paris from 1701 to 1713. The text is fully analyzed in the first part of the chapter. Sauveur's memoirs are then considered, including the 1700 account published in the *Histoire de l'Académie*. First, the 1701 memoir is presented in detail, which is certainly the most important, since it takes up in a concise but more mature way all the aspects reported in the 1697 *Traité*. In this memoir the division of the octave, the considerations on harmonic sounds and finally the method of determining the standard sound (fixed sound) are given. Three very interesting memoirs follow: one that applies the theory of harmonic sounds to the classification of organ stops, and two that give general systems for dividing the octave and for obtaining tempered systems. The last work, of 1713, completes the first memoir somewhat; the main result is the determination of an analytical relation of the absolute value of the frequency of a vibrating string as a function of the parameters on which it depends. In the last part of the chapter, we look at the minutes of the procès-verbaux of the Académie, in which some interesting results are given that shed light on Sauveur's approaches and that have not yet been considered in the literature.

5.1 Sauveur's Writings on Music

Sauveur's interest in music manifested quite late, in the second half of the 1690s, at least according to the available written evidence, when he was well into his 40s. Aside from a few entries in the minutes of the procès-verbaux of the Académie des sciences, the earliest written record of Sauveur's musical interests is a treatise known as *Traité de la théorie de musique* (herein after simply *Traité*) [70].

Sauveur wrote the *Traité* for his classes at the Royal college. Dated 1697, it has come down to us in manuscript form, preserved in the National library of France (Ms. fr n.a. 4574) of about two hundred pages divided into four books [70]. It was written

D. Capecchi and G. Capecchi, *Joseph Sauveur and His Time: Music and Mathematicians*,
https://doi.org/10.1007/978-3-032-04707-6_5

Table 5.1 Sauveur's *Memoires* de l'Académie des sciences of Paris

1701	Systême general des intervalles des sons, et son application à tous les systêmes et à tous les instrumens de musique	pp. 299–366
1702	Application des sons harmoniques à la composition des jeux d'orgues	pp. 121–137
1707	Methode generale pour former les systêmes temperés de musique, et du choix de celui qu'on doit suivre	pp. 203–222
1711	Table generale des systêmes temperés de musique	pp. 307–315
1713	Rapport des sons des cordes d'instruments de musique, aux flêches des cordes; et nouvelle détermination des sons fixes	pp. 324–350

by one of the copyists mentioned by Fontenelle in his *Éloge*, shortly after the author had been appointed a geometer at the Académie. According to his witnesses: "Then I gave an essay on acoustics in a *Traité de musique spéculative* which I discussed at the Royal college in 1697. I should have liked to print it; but the following reasons prevented me from doing so" [73].[1] The reasons added by Sauveur are essentially that he had not yet clarified certain aspects, and that his treatise was not suitable for reading by professional musicians, who generally had little interest in speculative music. The manuscript has been transcribed and translated into English by Richard Semmens in 1987, who also included a comprehensive commentary [81].

However, the writings that made Sauveur famous in the history of music are the memoirs he published in the journal of the Académie des sciences of Paris from 1701 to 1713. They can be counted exactly on one hand, being only the following five listed in Table 5.1.

The first memoir, dated 1701, consists of condensed expository material, omitting most of the demonstrations that would have been appropriate for a complete treatise on acoustics. The result is an article that could have been read with equal interest by practical and theoretical musicians, the latter providing with their own ingenuity those proofs and explanations that the former would have considered superfluous. The terminology is defined with great rigor and thoroughness, using various neologisms, including the term *acoustics*.

The second memoir was published the following year and, after an introduction, deals with the construction of organs, the characteristics of the different registers and sizes of longer and shorter organ pipes, and the theory of harmonic sounds. The third memoir, published in 1707, presents a general method of forming tempered

[1] p. 300.

Table 5.2 Sauveur's *Histoires* de l'Académie des sciences of Paris

1700	Sur la determination du son fixe	pp. 134–143
1701	Sur un nouveau systeme de musique	pp. 121–137
1702	Sur l'application des sons harmoniques aux jeux d'orgues	pp. 90–92
1707	Sur les systemes temperés de musique	pp. 117–120
1711	Sur les systemes temperés de musique	pp. 80–81
1713	Sur les cordes sonores et sur une nouvelle détermination des sons fixes	pp. 68–75

musical systems and lays down rules for choosing among them. The fourth memoir, published in 1711, is a response to a publication by Conrad Henfling (1648–1716), a scholar from Ansbach, who proposed a new temperament of 50 small parts in the octave. The fifth and final memoir on acoustics of 1713, with two sections devoted to infinitesimals, was published in 1716 without the benefit of final corrections due to the author's death. Sauveur returned here to many of the problems he had addressed in the 1701 memoirs.

Each memoir is preceded, according to the tradition of the Académie, by a commentary by the perpetual secretary, at the time Fontenelle, given in the same volume, in the part devoted to the *Histoires*. The commentaries are usually not simple summaries, but rather critical writings, sometimes very interesting; usually a few pages long, with a title reminiscent of that of the memoirs; they are listed in Table 5.2.

Sauveur's manuscripts are preserved in digital form and are easily accessible at the French national library. The *Memoires* and the *Histories*, and the *Traité*, have recently been reprinted and/or edited, as reported in the list:

1976 Maxham. *The contributions of Joseph Sauveur (1653–1716) to acoustics* [55]. The work, a PhD thesis, is divided into two volumes; the first is a fairly extensive introduction, but very much focused on the division of the octave and temperament. The second volume contains an English translations of Sauveur's five memoirs; Fontenelle's writings on the *Histoire* are omitted. The translation, in our opinion, should be revised, for although it is well done on the whole, it contains many passages that do not render the original well.

1984 Rasch. *Joseph Sauveur. Collected writings on musical acoustics (Paris 1700–1713)* [60]. It contains all of Sauveur's original-language texts, *Mémories* and *Histoires*. It is preceded by a brief but well done introduction.

1987 Semmens. *Joseph Sauveur "Treatise on the theory of music". A study, diplomatic transcription, and annotated translation* [81]. It contains transcriptions and a convincing translation of the *Traité*. It is preceded by a sufficiently comprehensive introduction.

2021 Jedrzejewsk and Papadoulos. *Joseph Sauveur. Écrits sur la musique et l'acoustique* [51]. It contains, in a fairly good editorial form, transcriptions of all of Sauveur's papers on acoustics in the original language, including the *Histories* and the *Traité*. The introduction is a bit general, but interesting.

In the following, the works mentioned above have been taken into account. Compared to them, apart from new translations, more space has been given to comments and critical discussions. In addition, English translations and comments of Sauveur's most important writings found in the minute of the procès-verbaux have been reported.

5.2 The Beginning of a Music Career: The *Traité de la Théorie de Musique*

The *Traité de la théorie de musique* is a remarkable text; it contains (almost) all the topics that Sauveur would cover in later years. Some of them, such as the division of the octave and temperament, are extensively developed; others, such as the theory of overtones or the determination of a standard sound (the fixed sound), are only hinted at.

Here is how Semmens—the editor of the transcription and translation of the original manuscript [81]—expresses his opinion on the value of the *Traité*, at the end of his introduction:

> His *Traité de la theorie de la musique* is indeed both sophisticated in its detail and logical in its presentation. Viewed in the context of his major contributions to acoustics, preserved in the *Histoires* and *Memoires* of the *Académie royal des sciences*, Sauveur's *Traité* provides a comprehensive background that allows insight into the sources of his ideas and discoveries, and provides an opportunity to trace their development and transformation. But the *Traité* also makes important contributions to our understanding of the history of musical acoustics in its own right. It is a compendium of the many crucial scientific, philosophical, and musical ideas that shaped the study of acoustic phenomena in the seventeenth century [81].[2]

In the following, we will give a fairly detailed account of the *Traité*, although it has already been effectively commented on by Semmens trying to present a different view, more focused on ideas than on editorial aspects.

It is very difficult to say what Sauveur's musical background was when he began to write his *Traité*, as there are few references to musicians or mathematicians in it: Mersenne, Guido of Arezzo, Étienne Loulié, Descartes, Huygens, Ole Rømer. Possibly it is true, as Fontenelle pointed out, that Sauveur was not an avid reader, but it is in no way conceivable that he could write such an accurate textbook as the *Traité* relying only on his acute intelligence. As a solitary scholar and a non-practicing musician, he could not have been able to correctly identify the problems addressed by past and contemporary music theorists.

[2] p. 85.

A part some reading, not cited by him, he could have received information about music development by the community of Académie, that at the turn of the 18th century were involved in music matter, the like of Claude Perrault (1614–1688), Christiaan Huygens (1629–1695), Philippe de La Hire (1640–1718), Joseph Guichard Duverney (1648–1730) [19].

The object of the *Traité* are expressed very clearly at its very beginning:

> Music is of two sorts, one speculative and the other practical. The speculative variety examines the properties of sound, and the manner in which it may be agreeable to the ear, and it is of this part that we shall speak in this treatise. Practical music seeks out ways to render sound attractive, and it is of this latter that musicians attempt to speak [70].[3]

In the quoted passage Sauveur restored the term *speculative* to its original meaning; that for example assigned to it by Boethius. It will be clear from the text that in any case Sauveur, unlike Boethius, rejected all metaphysical suggestions. In fact, his is a treatise on mathematical physics, based on reasoning and experiments. Using modern terminology it is a treatise on musical acoustics, although its field of interest is not as broad as that of a modern treatise on acoustics, and the purely mathematical aspects largely prevail on the physical ones.

The articulation of the treatise is summarized at the beginning of Chap. 1, which can be considered as a short preface:

1. Sound in general.
2. Relationships of sounds, that is intervals.
3. General system to measure intervals.
4. Application of the general system to all particular systems of music.
5. Application of the general system to musical instruments.

The preface closes cryptically with the statement: "After that, only the rules of composition remain" [70].[4]

In fact, it is not clear what this last sentence means, i.e., either the rules of composition are part of speculative music and are not included in the treatise for the sake of space—as was the case in contemporary music speculative treatises—or they are not part of speculative music.

The four books of the *Traité* are divided into chapters according to the list:
Book I
Of sound considered physically and mathematically
Chapter 1–Of sound in general
Chapter 2–Of sound considered in sonorous bodies
Chapter 3–Of sound considered in its media
Chapter 4–On the organ of hearing
Chapter 5–On the sensation of hearing
Chapter 6–Of sound considered mathematically
Book II

---

[3] p. 3. Translation into English in [81].

[4] p. 4.

Of intervals and systems in general
Chapter 1–Of intervals in general
Chapter 2–Of consonances and dissonances
Chapter 3–Of systems of music in general
Chapter 4–Of the diatonic system
Chapter 5–Properties of the diatonic system
Chapter 6–Of tempered systems in general
Book III
Explanation of the new system of music
Chapter 1–Principles of this new system
Chapter 2–Of the division of the octave
Chapter 3–Of the gamut or of the scale of the new system
Chapter 4–The musical characters for expressing the sounds
Chapter 5–The manner of marking notes in ordinary practice
Chapter 6–Of tempos and measures
Book IV
Application of the new system to the particular systems of music and to the instruments
Chapter 1–Of the diatonic system
Chapter 2–Of the chromatic and enharmonic systems
Chapter 3–Of the system of mean semitones
Chapter 4–Of the ordinary chromatic system
Chapter 5–Of the trumpet
Chapter 6–Of the bass viol
Chapter 7–Of the teorbo
Chapter 8–Of the violins
Chapter 9–Of the flutes

In the following quite extended account of the *Traité* we highlight the differences in approach between a professional mathematician and a musician who had a good mathematical background, such as the 16th-century treatise writers, or even such as Mersenne.

5.2.1 Book I

Chapter 1. *Of Sound in General*
Sound is not introduced intensively with an abstract definition, as the ancients, and to some extent, Mersenne also did, but rather extensively, reporting its properties, as Descartes did. Sound is considered from three different perspectives: from the point of view of the object that produces it, from the point of view of the medium that transmits it, and from the point of view of the organ of hearing. Sauveur never defined the concept of note in his treatise, and used the term *note* only on a few occasions, preferring the more general term *sound*.

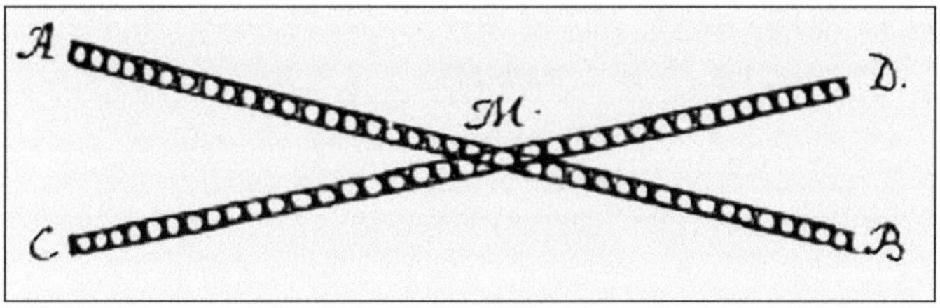

Fig. 5.1 Mechanism of transmission of sound; from [70], p. 17

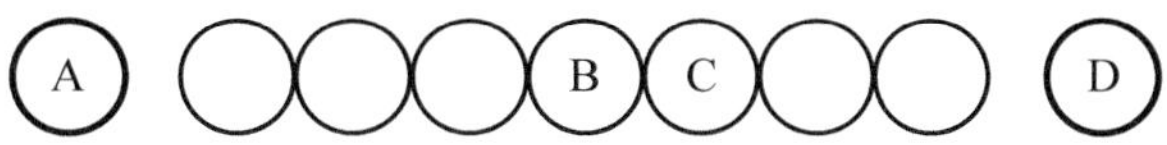

Fig. 5.2 Light propagation in a line. Redrawn from [48], p. 16

Chapter 2. *Of the sounds that are considered in the Corps Sonorisques*
In order to produce sound, a body must move rapidly and with a certain amplitude, either in its entirety or in some of its parts, in order to make its impression on the surrounding air. Different sonorous bodies produce different sounds. A string vibrates faster the more it is stretched and the thinner it is, and slower in the opposite case. In a solid body, the sound is due to the repeated contractions and expansions, called vibrations, of its tiny pores. Hollow bodies produce sound due to the movement of the air inside [70].[5] Sauveur claimed that if the motion is too slow, there is no sound, i.e. there is a lower limit in the range of vibration speed for the sound to be heard by humans. A sound is louder the larger its vibrations are; if they are too small, no sound is heard.

Chapter 3. *Of Sound Considered in the Media*
The vibration of sonorous bodies are communicated to the air, which transmits them to the ear. The mechanism of transmission presented by Sauveur is analogous to the mechanist one that Huygens assumed for light. Without engaging too much in ontological aspects, particularly whether air has a continuous or discrete nature, he proposed an analogy with a corpuscular model, such as the one shown in Fig. 5.1, which is also used to explain the relative lack of interaction of sounds from different sources. Tubes AB and CD are filled with elastic balloons that represent the elasticity of air. When the first balloon, located at A in tube AB, is compressed by the sonorous body, it gradually transmits its compression to the adjacent balloons, and similarly to the others. After a finite time, a certain amount of compression reaches the last balloon at B. When there are two sonorous bodies, for example at A and C, it is clear that the two sounds propagating in the two balloon tubes, although interacting at the point of intersection M of the tubes, can travel independently of each other.

Figure 5.1 has strict analogies with the following Fig. 5.2, which Huygens used in his *Traité de la lumiere* to explain the transmission of light. Sauveur compared his model with that:

[5] pp. 8–9.

> From this it follows that if the globules of Mr Descartes are perfectly solid the impression of the light will be made in an instant from one place to another, and that if these globules have pores and are capable of elasticity, or if the light is communicated as well by the intervention of some other elastic body, this communication will not be made instantaneously, as is claimed by Rømer and Huygens in their description of the phenomena of the satellites of Saturn. Finally, sound, since it is communicated through the intervention of the medium of air, which is capable of great elasticity, requires a certain amount of time for this communication [70].[6]

Huygens was named by Sauveur because he had argued for the finite speed of light based on astronomical observations, particularly those of the Danish astronomer Ole Rømer (1644–1710), who made use of the eclipses suffered by the 'small planets' that orbit Jupiter and often enter its shadow. Rømer found for the speed of light the value of 11 hundred times a hundred thousand toises (i.e. 110,000,000 toises) in one second.[7]

Sauveur ended this chapter by comparing the speed of light and that of sound: "Experience shows that sound takes one second more than light to travel 180 toises" [70].[8]

This is a somewhat ambiguous sentence. It can be read as a description of an actual experiment in which an observer measures the time it takes to hear a sound after seeing a flash of light is produced simultaneously with the sound. Sauveur had a Cartesian education in natural philosophy, he attended Rohault's classes, and for Cartesians light had an infinite speed. If he had been fully convinced of this, he would have simply written: "Experience shows that sound takes one second to travel 180 toises." But he was more nuanced because he knew of Rømer's experiment and should take it into account.[9]

During the 17th century there were many experimental calculations of the speed of sound; see, for example [19].[10] The value of 180 toises per second is the value measured by Cassini, Picard, and Rømer in 1677, a fact that Sauveur probably knew: "June 23 [1677] Cassini, Picard, and Rømer measured the time elapsed for the perception of sound and light. For the space [...] of 1280 toises [hexapedes] an interval of 7 minutes seconds [that is 180 toises/s] elapsed between the blow of a bombard

---

[6] pp. 15–16. Translation into English in [81].

[7] The speed obtained is 214,000 km/s; less than the modern estimate of 300,000 km/s.

[8] p. 60. A toise is exactly six feet. But there were several different measures of a foot in use in France, also quite different from each other. One kind of foot Sauveur quoted in the *Traité*, is the king's (or Paris) foot, which is about 32.5 cm. Or at least this is the value adopted by the French Republic in the law of December 10, 1799, which states that a decimal meter is exactly 3 feet and 11,296 lines (one line is 1/144 of a foot). Another kind of foot Sauveur used was the *astronomical foot*, defined so that 3 units of it gave the length of a simple pendulum beating to the second (actually 2 seconds), equal to 33.15 cm. See footnote 196.

[9] Assuming a foot equal to 32.5 cm, 180 toises per second gives a speed for sound equal to $6 \times 180 \times 32.5 \approx 351$m/s, a value quite higher than the 340 m/s (at 20 C) accepted today for the speed of sound. A better result than that reported by Mersenne, which is equal to 161 toises [56], vol. 1, pp. 213–214.

[10] pp. 8–9.

and the sight of the flame" [29].[11]

Chapter 4. On the Organ of Hearing

In this chapter, Sauveur briefly described the structure of the ear, for which he may have been strongly influenced by his colleague at the Académie, the anatomist Joseph Guichard Duverney (1648–1730). Based on Duverney's work, Sauveur considered the process of hearing, which begins with the ear. After being excited by the air, its vibrations are carried by the auditory nerves to the center of the brain, which gives rise to hearing by the soul. If the animal spirits in the brain are distracted by some other object, the soul is unable to receive a correct impression.

Chapter 5. On the Sensation of Hearing

At the beginning of the chapter, Sauveur discussed the phenomenon of resonance. A sonorous body transmits its vibrations, through the intervention of the air, to surrounding bodies, and if the body receiving the vibration is capable of "isochronous vibrations with those of the resonant [i.e. exciting] body" [70],[12] it makes a "going and coming" in the same time as the resonant body. Here Sauveur probably meant, although the statement is not entirely clear, vibrating strings, for which the concept of isochronous vibrations is easy to grasp.

In such a situation, the two bodies are said to be in unison. Resonance can occur even when the natural frequency [modern terminology] of the second body is twice that of the first. However, Sauveur continued, if the ratio of the natural frequencies is expressed in terms of large numbers, the first body produces no effect on the second.

Sauveur, although assuming the wave nature of sound, did not have our images of sound as sinusoidal waves, but rather as pulses (see Sect. 4.2); he could conceive of them as in phase or out of phase, as did Wallis Sect. 4.3.1.1. This is one possible interpretation of the following quotation:

> If each impression is alive, I mean, if it has completed its action, if the cessation occurs just at the beginning of the next vibration, then the sound will be pure. It will be even more pure if several unisons make their impressions and their cessations at the same time. And the sound is not pure or distinct if the impression continues its action until the following impression, so that the cessation is hardly marked [70].[13]

After giving some examples of the resonance of bodies whose natural frequencies can be made to coincide with the frequency of the exciting body, Sauveur began to talk about the way the ear perceives sound. Like Duverney, he maintained that the ear could be stretched and relaxed by the tympanic muscles so that it could vibrate in sympathy with the received sound [30]:[14]

> Thus, when the eardrum is in tune with the bodies producing a sound, then the eardrum moves that part of the membrane or the cochlea that is in harmony with the current situation

[11] p. 161.

[12] p. 22.

[13] p. 29. Translation into English in [81].

[14] pp. 80–81.

> of the eardrum, that is [in harmony] with the body that produces the sound. And then this movement is carried distinctly by the auditory nerves to the center of the brain, which gives rise to the soul's hearing it distinctly [70].[15]

Duverney's theory, no longer maintained today, works only when the sonorous body produces sound characterized by a well-defined frequency, such as notes or a group of them. Otherwise, according to the theory, the eardrum cannot be put in tune with the sonorous body; it will be imperfectly agitated and will receive the sensation of a confused sound. Therefore, when several people speak together, only the voice with which the eardrum is in unison can be heard.

At the end of the chapter, Sauveur introduced the concept of musical interval:

> This union or impressions is called a chord [interval] which is expressed by the ratio of the number of vibrations between these unions or impression of one to the number of vibrations or the other. If two sounds make their simultaneously impression in a certain time, after, five vibrations or one, and six of the other. their impressions will reunite once more. This ratio, 5 to 6, of the vibrations or these two sonorous bodies will be called a chord [70].[16]

The interval, is related to the sensation and for this reason it is called *chord*. This is the term used by Sauveur throughout his treatise. In the following, we have decided to translate *chord* with *interval*, following in this the editor of the *Traité*.

Note that Sauveur did not initially assign names to intervals; for example, when he considered the interval $(3 : 2)$, he did not call it a fifth or a diapente. He did not mention the theory of coincidence, only emphasized the fact that if an interval between two sound is expressed by the ratio $(m : n)$, for example $(6 : 5)$, then it happens that after six cycles of the first and five of the second "their impressions will reunite one more".

Intervals are classified as consonant if they are expressed by ratios of simple integers, such as 3 and 2, and non-consonant if they are expressed by ratios of large numbers, such as 16 and 15.[17] Sauveur did not give examples of other intervals or even comment on their hierarchy. Finally:

> Intervals expressed by incommensurable numbers will not be discussed at all, because in physics we do not recognize them unless they approximate to commensurable ratios, when the senses cannot detect a difference [70].[18]

Implicit in this quotation is the assertion that it is not possible by hearing to distinguish an interval expressed by the ratio of two simple integers from an irrational ratio close to it, and also that irrational intervals close to consonant ones may appear consonant, opening up the possibility of a temperament.

Chapter 5 ends with the interesting consideration that underlies Sauveur's entire conception of the theory of sound:

---

[15] p. 20. Translation into English in [81].

[16] pp. 30–31. Translation into English in [81].

[17] In the *Traité* Sauveur indicated the value of intervals with the symbol of fraction, and as a matter of fact in many occasions considered them as fractions.

[18] p. 32. Translation into English in [81].

> By paying more attention to the explanations we have just given, one could write a treatise on hearing that would be as precise as the one we have on sight, and that would contain phenomena that are just as curious [70].[19]

Chapter 6. *On Sound Considered Mathematically*
In this chapter Sauveur introduced the sum and difference of two intervals and showed that it is necessary to define them by the product or division of its ratios (note that in the following he used fractions instead of ratios, so that $(m : n)$ is written more simply as m/n). The demonstration is analogous to that given in Sect. 1.3.2. The multiple of an interval is the sum of n equal intervals; its ratio is given by the n-th power of the fraction representing the interval. So the interval corresponding to double of $(3 : 2)$ is $(3 : 2)^2 = 9/4$, the triple is $(3 : 2)^3 = 27/8$. In contrast, the division of an interval into n equal parts is expressed by the n-th root of the fraction representing the interval [70].[20]

According to Sauveur, the best way to work with intervals is to consider the logarithms (base 10) of the fractions representing them. By the time the use of logarithms had become commonplace, Sauveur himself devoted an extensive section to them in his *Geometrie élémentaire et pratique* [78].

An example of the use of logarithms given by Sauveur concerns the division of the octave into 12 equal parts; each division is defined by an interval of length $\sqrt[12]{2}$. Using logarithms, one must calculate $^1/_{12} \log 2 = 0.3010/12 = 0.0251$. The truncation—or approximation—to only four decimals is no longer followed in the remainder of the *Traité*; it is replaced by one with five decimals.

An interesting simple and well approximated way to divide an interval into n equal parts is presented at the end of the chapter, analogous to the way used by Cardano to divide the major tone into two equal parts (see Sect. 3.2.2.2). Given the interval p/q, first take it back to the equivalent interval np/nq, then write the arithmetic series $nq, nq + 1, nq + 2, \ldots, np$ and take the ratio of the two middle numbers. For example, divide the interval $16/15$, a major semitone, into four equal parts. The given interval is equivalent to the interval $64/60$, which gives the series $60, 61, 62, 63, 64$, whose center is 62. The candidates for the ratio equal to a quarter of the interval $16/15$ are the two ratios $62/61$ and $63/62$. Sauveur, without any justification, chose the second, which is the smaller one. The ratio $63/62 = 1.01613$ is very close to the value 1.01626 obtained by taking the fourth root of $16/15$.

5.2.2 Book II

Chapter 1. *On Intervals in General*
Book II is essentially mathematical in nature and deals with the division of the octave into an appropriate number of intervals of various sizes, taking up in greater detail what was discussed in the last chapter of Book I. Intervals are classified as simple

[19] pp. 35–36. Translation into English in [81].
[20] p. 42.

Table 5.3 Division of the octave according to the diatonic system

Ut	Re	Mi	Fa	Sol	La	Si	Ut
C	D	E	F	G	A	B	c
t	T	S	T	t	T	S	

if they are formed by two numbers in the range 1–6, otherwise they are complex. Simple intervals are consonant, complex intervals are dissonant. Here we are in the spirit of Zarlino's senario, which omits the minor sixth interval $8/5$. While Zarlino tried to justify his consonances, Sauveur said nothing.

As will become clear later, Sauveur divided the octave into seven intervals according to the diatonic system, in the form: tTSTtTS, where t is the minor tone, T the major tone, and S the major semitone. Starting from C, the previous sequence can be rewritten as in Table 5.3. Sauveur followed the French tradition in choosing the order of major and minor tones at the beginning as tTS. But he was not consistent; for example, on pp. 102 and 106 of the *Traité* the initial sequence is Tt.

In Italy, as in Europe, the sequence was TtS instead, following Zarlino [92][21] and Ptolemy.[22] In French music theory, the placement of a minor tone between Ut and Re was of some longevity; and as late as 1722, Jean Philippe Rameau used the minor tone [81].[23] In any case, in the C major scale today it is generally accepted that the sequence begins with a major tone, i.e., TtS (if the difference in tone is made).

One of the reasons why in France at the turn of the 18th century there was the habit of considering a minor tone between Ut-Re, is most likely due to the influence of Descartes, who wrote in his *Musicae compendium*:

> It is further clear that Ut to Re must always be a minor whole tone, Re to Mi a major whole tone, Mi to Fa a major semitone, Fa to Sol a major whole tone, and finally Sol to La a minor whole tone [24].[24]

Mersenne also followed Descartes, but not very consistently, as he alternated the necessity of a minor tone with the possibility of a major tone. Indeed, at the end of his reflections on the subject he stated: "It may then be concluded from what has been said, that it is indifferent to suppose the major or minor tone between Ut and Re" [56].[25]

According to Mersenne, the difficulty in deciding the correct position of the major and minor tones is due simply to the difficulty in deciding whether to represent the interval as a ratio of lengths or of frequencies, in the first case one gets a minor tone,

---

[21] p. 125.

[22] That Ptolemy's syntonic diatonic tetrachord was defined by the sequence STtSTt was first pointed out by Girolamo Mei, correcting Vincenzo Galilei, who assumed that in Ptolemy's tetrachord the minor tone would precede the major tone [58], p. 68. Letter to Vincenzo Galilei, May 8, 1572.

[23] p. 40.

[24] p. 20.

[25] Part. 1, Livre troisieme Des genres de la musique, p. 152.

in the second a major tone. As explained in [81],[26] this depends on the ease with which a sequence of integers can be found to express the sequence of ratios.

Based on his division of the octave, Sauveur was able to examine how many intervals there are of each type, where the type of interval of two notes is defined by the number of steps (lines and spaces) there are in the staff between the first and last notes of the interval.

1. Intervals of second: between two consecutive steps. There are three types, T, t, S, respectively major tone (9/8), minor tone (10/9), major semitone (16/15).
2. Intervals of third: between three consecutive steps. There are two types: tT, ST; major third (5/4)—such as Ut-Mi—and minor third (6/5)—such as Mi-Sol.
3. Intervals of the fourth: between four successive steps. There are two kinds tTS, TtT, respectively, just fourth (4/3)—such as Ut-Fa—and tritone or false fourth (45/32)—such as Fa-Si.
4. Intervals of the fifth: between five successive steps. There are two kinds, tTST, STtS, respectively, just fifth (3/2)—such as UT-Sol—and false fifth (64/45)—such as Si-Fa (with Fa in the higher octave).
5. Intervals of sixth: between six successive steps. There are two types, tTSTt, STtTS, major sixth (5/3)—such as Ut-La—and minor sixth (8/5)—such as Mi-Ut.
6. Intervals of seventh: between seven successive steps. There are two kinds, tTSTtT, TSTtTS, respectively, major seventh (15/8)—such as Ut-Si—and minor seventh (9/5)—such as Re-Ut.

Complement. The complement of an interval is its difference from the octave. Thus the fourth (4/3) is the complement of the fifth (3/2), similarly the complement of the minor third (6/5) is the major sixth (5/3).

Replica. The replica of an interval is the same interval one octave higher. So the replica of the minor third 6/5 is given by 12/5. The name of a replica is obtained by adding seven to the name of the original interval. So the replica of the sixth is the thirteenth, the replica of a fifth is the twelfth. The replica of a replica is called a triplicate; the triplicate of the fifth (3/2) is 12/2 = 6/1.

After the examination of the types of intervals, Sauveur returned to consonances: the minor and major third, the fourth, the fifth, the minor and major sixth, and the octave, following the ascending order derived from the coincidence theory (not named), without pronouncing on the order of agreeableness. Note that the minor sixth (8/5) appears in the list of consonances. For Sauveur, replicas of consonances were also consonant. Under dissonances, Sauveur listed the second, the tritone and its replicas. Nothing is said about the interval of seventh.

The just intervals mentioned by Sauveur are the most common. To them he added other intervals which are of some interest in music theory, given in the following list [70],[27] which he took point by point from Mersenne [56].[28]

---

[26] pp. 34–38.

[27] p. 57.

[28] Part 1, Livre second des consonances, p. 116.

1. Maximum semitone 27/15, or nearly 13/12 or 14/13
2. apotome 2187/2048
3. major semitone 16/15
4. medium semitone 135/128
5. limma 256/243
6. minor semitone 25/24
7. minimum semitone 648/625
8. subminimum semitone 250/243
9. enharmonic diesis 128/125
10. major comma 81/80
11. minor comma 2048/2025
12. difference between the two commas 10936/10935.

Note that these intervals are not listed in descending order of coincidence, and that some of them come from Pythagorean tuning.

Sauveur proposed a simple and practical method for identifying the simple intervals that make up a complex interval. It involves decomposing the numerator and denominator of the interval into (prime) factors and then examining the ratios between these factors. For example, the minor semitone interval (25/24) is written as:

$$\frac{5 \cdot 5 \cdot 1 \cdot 1}{3 \cdot 2 \cdot 2 \cdot 2}$$

where Sauveur added two times 1 to the numerator to have the same number of digits as in the denominator. The simple intervals are easily identified as $5/3$, $5/4$ and $1/2$. The first two intervals are normalized in the standard way, since they are greater than 1 and less than 2. The third interval is less than 1, so we must take its inverse. Essentially, we have the following decomposition:

$$\frac{5}{3} + \frac{5}{4} - \frac{2}{1}$$

In the case of the octave, an indecomposable interval, Sauveur proposed the equivalent ratio $(12 : 6)$, which can be rewritten as $(3 : 2) + (4 : 3)$, i.e., as we know, the octave can be decomposed into a fifth and a fourth.

This procedure can be generalized "since a simple interval like 3/2 can be expressed in several ways, 18/12, 30/20, 120/80" [70].[29]

Chapter 2. *Of Consonances and Dissonances*
Sauveur took up the subject of consonance and dissonance here in a systematic way; he was certainly familiar with the two alternative theories of consonance that were being considered at the time—the numerological theory of the Pythagorean type and the physical theory based on the idea of coincidence. However, he was inspired by the aesthetic theories of Descartes, especially those reported in the *Musicae compendium*. They belong to a category not very different from the Pythagorean theories, but

[29] p. 60.

whereas the latter refer to a universal harmony, the former refer to human harmony, to a metaphysics with a psychological background instead of a metaphysics with a cosmological or, if you will, a theological background.

According to Descartes and Sauveur, the pleasure caused by the sense of hearing is analogous to that caused by the other senses. The sonorous message is carried by certain nerves to the brain, which, by receiving an impression, provides the *occasion* for the soul to form an idea, which is always the same if the impression received by the brain is always the same. The pleasure and displeasure of the soul result from the combinations of the various ideas.

To better explain his theory, Sauveur imagined an adult man in perfect health coming into contact with the sensation of sound for the first time. It has been suggested that this image is autobiographical in that Sauveur had difficulty hearing and speaking for several years after his birth [81].[30]

This man will be able to recognize and appreciate only the simplest of the sensations he perceives, such as those associated with unison. Then, however, he will become accustomed to this sensation and bored with hearing it, so that he will gradually seek out more complex sensations, beginning with the octave, then the fifth and fourth, then the thirds and sixths. After he gets used to hearing the sounds we call consonants, he will seek more complex sensations by seeking dissonances as well. Then he will move from the isolated sound to the appreciation of arias, first as simple as simple songs, and then more complex as concertos, which should not be too demanding at first.

However, the pleasantness perceived by the senses depends not only on the sounds, but also on the circumstances surrounding them. For example, celebrations or exhibitions of esteemed people; on these occasions our man will get accustomed to hearing and appreciating some dissonances in addition to consonances.

Here Sauveur introduced an idea that was not new anymore, and which we have also seen, for example, in Huygens, that the appreciation of sounds has no absolute value, but varies from culture to culture: "So in France he will be accustomed to certain dissonances; therefore he will become more familiar with the airs of one country, and (these) will seem more attractive to him than those of another, to which he is not accustomed" [70].[31]

After discussing the cultural influence on the perception of sounds, with a clear logical leap, he went on to present a theory of consonances by presenting, what he had not yet done, a hierarchy of them:

> We have said that the intervals that were expressed between the numbers 1, 2, 3, 4, 5, 6 were attractive, and it is true likewise that the interval of 2 to 3 is more attractive than that of 3 to 4, that the interval of 4 to 5 is even less attractive than it, and still less attractive is that of 5 to 6, from which it is concluded that 6 to 7 will still be less attractive as will be those of 7 to 8 and of 8 to 9 etc. And thus there is no reason to end the consonances at 6 than at 7 or 8, since it is the attractiveness which diminishes by degree [70].[32]

---

[30] p. 63.

[31] pp. 69–70.

[32] pp. 79–71. Translation into English in [81].

Here Sauveur argued for a fairly shared hierarchy among coincidence theorists, with intervals expressed by ratios of smaller numbers being the most consonant. And he came to the most natural conclusion that there is no clear cut between consonances and dissonances. So there is no reason to end consonances at number 6 rather than 7 or 8.

After this consideration, however, Sauveur resumed the classical criticism of intervals containing the number 7. For example, he tried to explain why the interval 9/8 is more attractive than 8/7 or 7/6. The explanation requires the assumption that intervals which are the difference of consonant intervals are also attractive. Thus 4/3 (the fourth interval) is attractive because it is the complement of 3/2, the fifth (the difference between the octave and the fifth), but less attractive because it is only a shadow of it; so sixths are less attractive than thirds because they are their shadow. It is clear that we are dealing with sophistry here; one could just as well say that the fifth, being the complement of the fourth, is its shadow, and the same for sixths.

The major tone, 9/8, is somewhat attractive because it is the difference between two consonances, that of the fifth and that of the fourth; the same is true of the minor tone, 10/9, the difference between the fourth and the minor third, and the major semitone, 16/15, the difference between the fourth and the major third. The interval 8/7, on the other hand, can in no way be associated with differences between consonances. Furthermore, the intervals formed with the number 6 are 6/5, 6/4, 6/3, and their complements are 10/6, 8/6, 6/6. Of these, only 6/5 cannot be reduced to simpler and therefore more pleasing intervals. The intervals formed by the number 7, on the other hand, are 7/6, 7/5, 7/4, and their complements 12/7, 10/7, 8/7, and none of them can be simplified, so that Sauveur felt authorized to say "the number 7 cannot be counted among the intervals which give pleasure to the ear" [70].[33]

Chapter 3. *Of systems of Music in General and the Elements of Intervals*
The different divisions of the octave are called *systems* in music, whose *elements* are the smaller intervals that make up the octave.

> To divide an octave into its elements it is necessary first that these elements be perceptible, because if they were imperceptible the ear would not be able to appreciate them, nor the voice execute them. Thus they would become useless to the ordinary man. That is why in the ordinary diatonic system the smallest element is the minor semitone,[34] for the comma can be appreciated only by a sensitive ear. Second, these elements must be more or less equal, in order that to begin with whatever division of the system is desired, all sorts of consonances may be formed by taking in sequence the number of elements required. Third, since in different countries and at different times men are accustomed to certain intervals, and different orders of interval, and since from time to time men have become more or less accustomed to music, this has given rise to different divisions of the octave, or to different systems which we will explain in the following [70].[35]

Chapter 4. *Of the Diatonic System*

---

[33] p. 74.

[34] The minor semitone is not actually part of the ordinary diatonic system.

[35] pp. 79–80. Translation into English in [81].

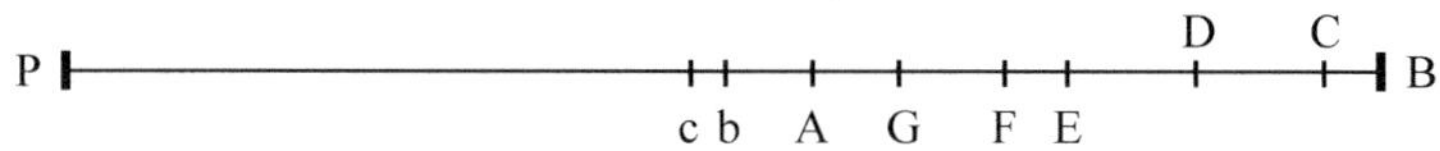

Fig. 5.3 Partition of the octave; from [70], p. 82

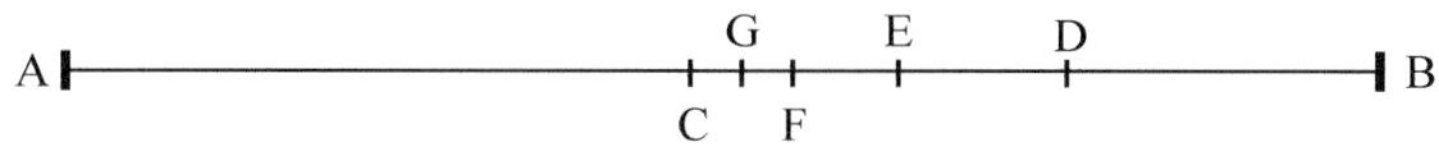

Fig. 5.4 Descartes' dichotomic partition of the octave; from [11], vol. 3, p. 136

> The diatonic system is that in which the octave is divided into tone and semitone. It is regarded as the principal system because it is the simplest and the most natural, and because from the combination of its simple intervals or of its elements, all kinds of consonances are made [70].[36]

Sauveur illustrated a geometric division of the monochord leading to a diatonic system through a series of divisions, partly following Descartes' approach. With reference to Fig. 5.3 the points are considered:

Point b.	Such that	bB = 1/2PB.
Point E.	Such that	bE = 1/2 bB.
Point G.	Such that	bG = 1/2 bE.
Point A.	Such that	bA = 1/2 bG.
Point F.	Such that	GF = 1/8PG.
Point C.	Such that	EC = 1/4PE.
Point D.	Such that	ED = 1/2EC.
Point c.	Such that	Pc = 1/2PC.

The octave bB is divided at the points C, D, E, F, G, A, as shown in Fig. 5.3, into seven intervals, all of which are major, minor tones, or semitones, as can be easily verified. But they did not form the "ordinary" diatonic system. For this purpose it is necessary to consider the octave cC so that cC = 1/2 PC and to start the octave from C and end it in c.

As shown in Chap. 4, Descartes' in his *Musicae compendium*, followed an apparently similar approach, as shown in Fig. 5.4. He adopted a sequence of dichotomous divisions of the right part of the monochord AB such that BD = 1/2 BC, CE = 1/2 CD, CF = 1/2CE, CG = 1/2 CF. This division gives correct values for major and minor tones, but semitones different from those of the diatonic scale. In fact, it is AF/AC = 9/8; AG/AC = 17/16; AF/AG = 18/19. Descartes' approach was more elegant, but Sauveur's is smarter.

The geometric construction of Fig. 5.3 is mirrored by Sauveur with an arithmetic one, assuming PB = 64, PC = 60, and arriving at the sequence of Table 5.4.

Chapter 5. Property of the Diatonic System

---

[36] p. 80.

Table 5.4 Aritmetic division of the octave

c		b		A		G		F		E		D		C
30		32		36		40		45		48		54		60
	$\frac{16}{15}$		$\frac{9}{8}$		$\frac{10}{9}$		$\frac{9}{8}$		$\frac{16}{15}$		$\frac{9}{8}$		$\frac{10}{9}$	

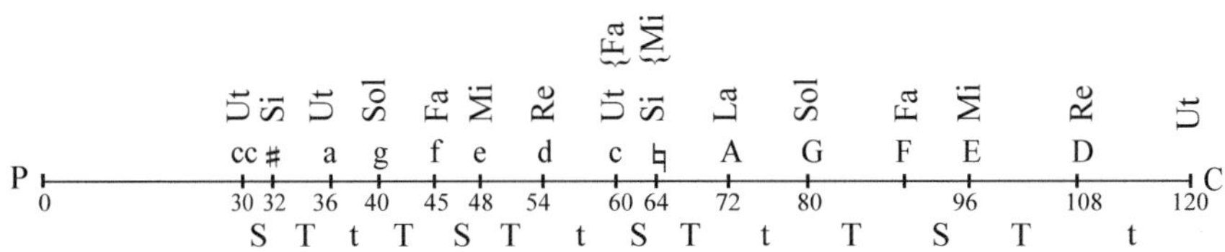

Fig. 5.5 Double octave of the diatonic system; from [70], p. 87

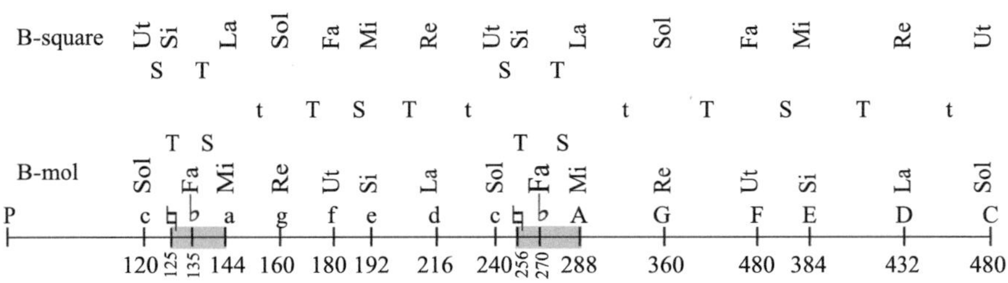

Fig. 5.6 Two different solmizations of the diatonic system; from [70], p. 87

He had previously avoided naming, at least explicitly, the notes that define intervals; now he did so explicitly, first by introducing literal ancient notation:

> First, since a string placed under these various divisions produces different sounds, the sounds were given different names. The ancients, and at present the Germans, give them the names or letters. Thus the sound which the entire string PC produces being called C, the sound that the portion PO produces will be called D, and similarly the other portions of the string will be called E, F, G, A, B. And to indicate a second octave they made use of lower case letters, while for a third octave they make use of double lower case ones, taking care to use a b natural, marked thus ♮, which they call b square for reasons we will indicate later [70].[37]

Then he referred to 'modern' notation, recalling Guido of Arezzo and his solmization: Ut, Re, Mi, Fa, Sol, La. Finally, he noticed that the note Si had been introduced at some point to simplify the notation and also to give a name to the note B square (♮).

Figure 5.5 shows 'ancient' and 'modern' notation systems side by side. A further step is taken by introducing the soft note ♭,[38] as shown in Fig. 5.6, where an octave defined between 120 and 480 is considered, so that the new note ♭ is represented by

[37] pp. 87–88.

[38] The possibility of introducing the note ♭ was commented on in the section devoted to Zarlino, in Chap. 2.

an integer (270). Indeed, the note ♭ cannot be represented as an integer in the octave 30–120. The note indicated as ♭ is sometimes denoted by the symbol Sa, similar to Si (Huygens did this, for example). Sauveur,however, preferred not to introduce the note explicitly, but instead to rename all the notes of the octave, arriving at the scale labeled B-mol, in Fig. 5.6. In this solmization, there is no need to introduce the symbol ♭, since the interval A–♭ becomes the interval Mi-Fa, which is a semitone.

Having introduced the diatonic system on two octaves defined by the sequence of tones and semitones: tTSTtTStTSTtTS, Sauveur repeated the interval analysis he had already done in Chap. 3, limited to a single octave, starting from Ut, a major scale. Considering two octaves instead of just one, there is a greater variety of intervals, as the list shows:

1. Intervals of seconds, between two successive steps. There are three types, T, t, S, respectively major (9/8), minor (10/9), major semitone (16/15).
2. Intervals of a third, between three successive steps. There are five potential types: SS, St, ST, tT, TT. SS is not used; St is the diminished third (32/27), such as Si-Re; ST is the minor third, such as Re-Fa; Tt is the major third, Ut-Mi. TT is not used.
3. Intervals of the fourth, between four successive steps. There are three types: StT, STT, TtT. The just fourth StT, Ut-Fa; the augmented minor fourth STT (27/30), Re-Sol; the tritone TtT, Fa-Si.
4. Intervals of the fifth, between five successive steps. There are three types, the fifths being the complement of the fourths. The false fifth, StTs (64/45) Si-Fa; the diminished fifth tTSt (40/27),[39] Sol-Re; the just fifth tTST (3/2), Ut-Sol.
5. Intervals of the sixth, between six successive steps. There are five kinds that are the complement of thirds, three kinds in practice. The minor sixth, TStTS (8/5), La-Fa, the major sixth, tTSTt (5/3), Ut-La, the augmented sixth, TSTtT (27/16), Re-Si.
6. Intervals of the seventh, between seven successive steps. Being complementary to the second, there are three types. The diminished minor seventh (16/9) Mi-Re; the minor seventh (9/5) Re-Ut, the major seventh (15/8) Ut-Si.

The chapter ends with an introduction to the next chapter, devoted to tempered systems, by noting that the variety of intervals depends on the presence of two tones of different sizes, and by advocating the needs of temperament. In fact, Sauveur noted, each small interval, i.e., the second, the third and the fourth, is major when it contains only tones, and minor when it contains a semitone. And in the large intervals, the fifth, sixth and seventh, the intervals are major when they contain only one semitone, and minor when they contain two. And since there are two tones, a major and a minor, their different combinations make the interval perfect or imperfect. But the imperfect intervals are unattractive to the ear, and musicians have preferred to take a mean between the perfect and the imperfect in order to reduce the unattractiveness of the imperfect rather than compromise the overall attractiveness. This is the origin

[39] The definition of false and diminished is proper of Sauveur and does not correspond at the modern usage.

of what are called tempered intervals, or mean tones, and the system that contains these types of intervals is called the tempered system [70].[40]

Chapter 6. Of Tempered Systems in General

A temperament, in general, is a modification of a tuning to avoid some hardness. Sauveur gave a more technical and more limited definition.

> A division of the octave into equal intervals is called a *tempered system* [emphasis added] which makes all the intervals perfect, either as exactly as could be desired, or with some discrepancy small enough so that the intervals are not unattractive [70].[41]

He was not the first to propose the division of the octave into $n \neq 12$ equal small parts. In Chapter 3 we saw that Mercator (1646), Huygens (1661) and Newton (1664) proposed different kinds of division, but they did not consider them as a way of tempering. If we want an antecedent of Sauveur's position, we have to look at the *Fabrica, y uso de varios instrumentos mathematicos* by José Zaragoza (1627–1678) of 1675 [91]. However, it is most likely that Sauveur did not know of Zaragoza.

Sauveur presented three different methods of temperament. The first method is partially independent of his definition of temperament in that it does not exactly consist in dividing the octave into n equal small parts. It is obtained by equating the minor and major tones, leaving the semitone unchanged. This is done by using logarithms. There are three major tones $(9 : 8)$ and two minor tones $(10 : 9)$ in an octave, to which the logarithms correspond respectively to 0.05115 and 0.04575.[42] In total, they cover the interval $3 \times 0.05115 + 2 \times 0.04575 = 0.24495$, which can be divided into five equal intervals, giving a middle tone T whose logarithm is $0.24495/5 = 0.04899$.

Table 5.5 compares the logarithm of just intonation intervals with tempered intervals, obtained by tone-semitone grouping as given in the second column, with T=t=T. The last column shows the differences. The maximum difference in the consonances is found in the minor third and major sixth, and is 0.00216, or less than half the syntonic comma (0.00539), or about one in 200 vibrations less than the just interval.[43]

The second method of temperament is that now known as *equal temperament*, which Sauveur introduced in the following way: considering that a tone is made up of two semitones, so that five tones of the octave contain ten semitones, and if the two other semitones are added, this makes 12 semitones. To create the temperaments, it is therefore necessary to divide the logarithm of the interval of the octave, 0.30103, into 12 parts, which gives 0.02509. This temperament can be called of equal semitones [70].[44] Table 5.6, analogous to what was done in Table 5.5, compares it with the just intonation.

---

[40] pp. 98–99.

[41] p. 100.

[42] Sauveur wrote 05115 and 04575; that is, he considered only the mantissa (with 5 digits).

[43] The interval corresponding to 0.00216, i.e. its antilogarithm, is 1.0049, or 49 units more than 10,000, and $10,000/49 \approx 200$.

[44] p. 103.

Table 5.5 First method of temperament versus just intonation; from [70], p. 102

		Accords		Just	Temp.	Diff.
Ut		unison	1/1	00000	00000	00000
	S	semiton maj.	16/15	02803	02803	00000
	t	ton min.	10/9	04575		+00324
		ton moy.			04899	
Re	T	ton maj	9/8	05115		−00216
	ST	tierce min.	6/5	07918	07702	−00216
Mi	tT	tierce maj.	5/4	09691	09798	+00107
Fa	StT	quarte	4/3	12494	12601	+00107
	tTT	triton	45/32	14806	14697	−00109
	SStT	fausse quinte	64/45	15297	15406	+00109
Sol	StTT	quinte	3/2	17609	17502	−00107
	SStTT	sexe min.	8/5	20412	20305	−00107
La	SttTT	sexe maj.	5/3	22185	22401	+00216
		sept. min-min.	16/9	24988		+00216
	SSttTT	sept. min-moy			25204	
	SStTTT	sept. min-maj.	9/5	25528		-00324
Si	SttTTT	sept. maj.	15/8	27300	27300	00000
Ut	SSttTTT	octave	2/1	30103	30103	00000

Sauveur noted that the equal semitone temperament has the advantage that, starting from any desired note, all intervals of the same type have the same ratio, since they are made up only of equal semitones. But the differences are considerable for the two thirds, almost a syntonic comma. The usefulness of this tempered semitone system derives from the fact that most fretted instruments follow this system, but Sauveur claimed that the large differences in thirds make it insupportable for sensible contemporary musicians [70].[45]

The third method of making tempered systems presented by Sauveur, which he will consider in detail in his memoirs of the Académie, is based on the division of the octave into n equal small parts starting from the observation that the major semitone (which for Sauveur is associated to the syntonic semitone) is larger than the half of either of the two major tones. When the tone is divided into an odd number of equal parts, it is necessary to assign the larger portion of the tone to the major semitone. And since the division can be done in several ways, the one that gives the tempered

[45] pp. 104–105.

Table 5.6 Equal semitone temperament versus just intonation; from [70], p. 104

Petit	Accords		Just		Temp.	Diff.
	unison	1/1	00000		00000	00000
S	semiton	16/15	02803	1	02509	−00294
t	ton mineur	10/9	04575			+00442
					2	05017
T	ton majeur	9/8	05115			−00098
ST	tierce min.	6/5	07918	3	07526	−00392
tT	tierce maj.	5/4	09691	4	10035	+ 00344
StT	quarte	4/3	12494	5	12545	+00052
tTT	triton	45/32	14806	6	15052	+00246
Grand	Accords		Just		Temp.	Diff.
SStT	fausse quinte	64/4	15297	6	15052	−00246
StTT	quinte	3/2	17609	7	17558	−00051
SStTT	sexte min.	8/5	20412	8	20068	−00344
StTT	sexte maj.	5/3	22185	9	22577	+00392
SSttTT	sept. min.	16/9	24988			
SStTTT	sept. min. maj.	9/5	25527	10	25086	−00442
SttTTT	sept. majeure	15/8	27300	11	27594	+00294
SSttTTT	octave	2/1	30103	12	30103	00000

consonances with the closest approximations to the just intonation should be chosen [70].[46]

Sauveur proposed to divide the tone into 3, 5, 7 or 9 parts and the major semitones into 2, 3, 4 or 5 parts respectively. In the first case, the octave being of 5 tones and 2 major semitones is divided into $5 \times 3 + 2 \times 2 = 17$ parts;[47] in the second case, the octave is divided into $5 \times 5 + 2 \times 3 = 31$ parts; in the third case, the octave is divided into $5 \times 7 + 2 \times 4 = 43$ parts. In the last case, the octave is divided into $5 \times 9 + 2 \times 5 = 55$ parts (Table 5.7).

Comparing the differences in the just consonances of these systems, it is interesting to observe the trend of the difference (as a logarithm) of the fourth; in the 19-part system it is +0.00138, in the 31-part system +0.00129, in the 43-part system +0.00106, and in the 55-part system +0.00087. The difference of the major third, on the other hand, is −0.00217 in the 19-part system; +0.00019 in the 31-part system; +0.00109 in the 43-part system; and +0.00155 in the 55-part system. Similar differences occur in the minor third.

[46] pp. 105–106.

[47] Each tone is divided into 3 parts, each semitone into 2 parts.

Table 5.7 Temperaments of the third kind versus just intonation; from [70], p. 106

Noms des accords			accords justes	parties	Sisteme de 19	erreurs	parties	Sisteme de 31	erreurs	parties	Sisteme de 43	erreurs	parties	Sisteme de 55	erreurs
1. partie	–	—	—	1	01579	—	1	00971	—	1	00700	—	1	00547	—
unisson	ut		00000												
		S	02803	2	03158	+355	3	02913	+110	4	02800	−3	5	02735	−68
		t	04575			+162			+280			+325			+348
				3	04737		5	04855		7	04900		9	04923	
ton	re	T	05115			−378			−260			−215			−192
tierce min.		ST	07918	5	07895	−023	8	07768	−150	11	07700	−218	14	07658	−260
tierce maj.	mi	tT	09601	6	09474	−217	10	09710	+19	14	09800	+109	18	09846	+155
quarte	fa	StT	12494	8	12632	+138	13	12693	+129	18	12600	+106	23	12581	+87
triton		tTT	14806	9	14211	−595	15	14565	−241	21	14700	−106	27	14769	−37
	sol[sic]	SSTT	15297	10	15892	+595	16	15538	+241	23	15403	+106	28	15334	+37
quinte		StTT	17609	11	17471	−138	18	17480	−129	25	17503	−106	32	12522[sic]	−87
sexte min.		SStTT	20412	13	20629	+217	21	20393	−19	29	20303	−109	37	20257	−155
sexte maj.	la	SttTT	22185	14	22208	+023	23	22335	+150	32	22403	+218	31	22445	+260
		SSttTT	24988			+378			+260			+215	sic		+192
				16	25366		26	24248		36	25203		46	25180	
		SStTTT	25527			−162			−280			−325			−348
	si	SttTTT	27300	17	26945	−355	28	27190	−110	39	27303	+3	50	27368	+68
octave	ut	SSttTTT	30103	19	30103		31	30103		43	30103		55	30103	

It follows that the greater the number of parts, the more just is the fourth, but on the other hand, the result is not so good for the thirds, and therefore it seems useless to go further than 55 parts [70].[48]

Sauveur also considered a fourth method, which is a variation of the third. Whereas in the third method the major and minor semitones (i.e. the difference between a tone and the major semitone) differ only by one unit, i.e. if n is the number of parts of the minor semitone, $n+1$ is that of the major semitone, in the fourth method the difference between the minor and major semitones is n to $n+j$, with $j>1$. For example, for $j=2$ and $n=3$, one has 3 parts for the minor semitone and 5 parts for the major semitone—and thus 8 parts for the whole tone—and the octave is divided into $5\times 8+2\times 5=50$ parts for the octave. Sauveur found that for this division the difference with the just values is the mean between those of the 19 and 31 systems.

If $j=2$ and $n=5$, you have 74 parts for the octave; if $j=2$ and $n=7$, you have 98 for the octave; if $j=2$ and $n=9$, you have 122 for the octave. Sauveur suggested considering the case of $j=3$ and more, but as with the first method, as the octave is divided into too many parts, these systems become too complex. Sauveur concluded that the third method is to be preferred, and that dividing into 43 is also to be preferred [70].[49]

48 pp. 110–111.

49 pp. 111–114.

5.2.3 Book III

Book III is largely an anticipation of part of the 1701 memoir; despite the overlap in the sequel, we will report it in fairly extensive form, as we did for the first two books, so that we could appreciate the historical evolution of Sauveur's thought.

Chapter 1. *Principle of the New System*
The starting point is the tempered system obtained by dividing the octave into 43 parts, which was considered at the end of Book 2 to be the most suitable. This system is considered from a different point of view, as a variation of the first method of temperament. It consists of taking a meantone among the three major and two minor tones that exist in the octave; the logarithms of the major tone (9/8) and the minor tone (10/9) are 05115 and 04576, respectively. Taking the weighted mean of the two, i.e. multiplying the major by three and the minor by two, gives 24500 (the correct result is actually 24 495, but the approximation is minimal). This number can be divided by five without further approximation, giving the middle tone 04900. The logarithm of the major semitone is 02803, which approximates 00280. Both the tone and the semitone logarithms can be divided by 7, giving 00700 and 00400 respectively, i.e. a ratio of 7 over 4. Thus the octave is divided—more or less—into $5 \times 7 + 2 \times 4 = 43$ parts "and we will call each of these parts a *meride*".[50]

Multiplying 43 by 7 gives 301, which is roughly the logarithm of the octave, $\log 2 = 30103 \approx 30100$; the octave is thus divided—more or less—into 301 equal parts called *heptamerides*.

The introduction of heptamerides can be done in a more direct way. If i represents a musical interval, its measure in heptamerides can be defined as:

$$E = 1000 \log i$$

For the interval of an octave, $i = 2$ and $E = 301.03$, Sauveur instead assumed $E = 301$, which implies that a heptameride is almost exactly the three-hundred and first part of the octave.[51]

Chapter 2. *Of the Division of the Octave*
Tables 5.8 and 5.9 show the intervals of the tempered diatonic system expressed in heptamerides. The fifth column shows the ratio of the number of oscillations to those of the lowest value, assumed to be 1000. These oscillations are obtained very simply by calculating the anti-logarithm of the heptamerides, since these give exactly the

[50] In fact, in the *Traité* Sauveur used the term *meries*, here we adopt merides for the sake of uniformity, since this term will be used in all of Sauveur's later work.

[51] The precise definition of the heptameride is unclear in the current literature. What Sauveur defined as a heptameride is often called a *savart*, after the French physicist Flix Savart (1798–1841), while the name heptameride refers to the 301-st part of the octave. Today, a savart is also intended as the 300-th part of an octave, so 25 savarts exactly cover a semitone [50].

intervals with respect to the reference value. The last column gives the increase in the number of oscillations with respect to the value of the previous row (Sauveur did not comment on this column).

Instead of using heptamerides as in the table, Sauveur added, one can use merides followed by dashes to indicate the heptamerides to be subtracted if the dashes are at the bottom, or to be added if they are at the top. For example, $25^{|}$, $25^{||}$, $25^{|||}$ and $26_{|}26_{||}$, $26_{|||}$ represent 25 merides plus 1, 2, 3 heptamerides and 26 merides minus 1, 2, 3 heptamerides, respectively. Since there are seven heptamerides between two consecutive merides, $25^{|||}$ and $26_{|||}$ represent the same sound.

Adding a dot next to the dashes denotes half a heptameride, so $32^{|||\cdot}$ denotes 32 merides plus $3\frac{1}{2}$ heptamerides, $18_{||\cdot}$ denotes 18 merides minus $2\frac{1}{2}$. Note that in later writings by Sauveur, the division into half heptamerides is placed side by side with the division into $\frac{1}{10}$ heptamerides, called *decamerides*.

These notations make it easy for Sauveur to distinguish between just and tempered intervals. For example, the tempered fifth is 25 merides, the just is $25^{|}$, which makes

Table 5.8 Merides and heptamerides of the notes of the diatonic tempered system for small intervals; from [70], p. 120

Accords		Merid.	Heptam.	Vibr.	Diff.
unison	Ut	0	0	1000	0
		1	7	1016	16
		2	14	1033	17
		3	21	1050	17
semit. min		4	28	1067	17
semit. maj		5	35	1084	17
		6	42	1102	18
ton	Re	7	49	1120	18
		8	56	1138	18
		9	63	1156	18
		10	70	1175	19
tierce min.		11	77	1194	19
		12	84	1213	19
		13	91	1233	20
tierce maj.	Mi	14	98	1253	20
		15	105	1273	20
		16	112	1294	21
		17	119	1315	21
quarte	Fa	18	126	1336	21
		19	133	1358	22
		20	140	1380	22
		21	147	1403	23
triton		22	154	1426	23

Table 5.9 Merides and heptamerides of the notes of the diatonic tempered system for large intervals; from [70], p. 120

Accords		Merid.	Heptam.	Vibr.	Diff.
fausse quinte		23	161	1449	23
quinte	Sol	24	168	1472	24
		25	175	1496	24
		26	182	1520	25
		27	189	1545	25
		28	196	1570	25
sexte min.		29	203	1596	26
		30	210	1622	26
		31	217	1648	26
sexte maj.	La	32	224	1675	27
		33	231	1702	27
		34	238	1730	28
		35	245	1758	29
sept. min.		36	252	1786	28
		37	259	1815	30
sept. maj.	Si	38	266	1845	31
		39	273	1875	31
		40	280	1906	31

the difference clear.

Chapter 3. *Scales in the New System*
In this chapter Sauveur gave the names of all 301 sounds expressed in heptamerides. For the sake of simplicity, we will not follow his somewhat cumbersome exposition exactly, but will use Table 5.10, which shows all the names and all the symbols.

The names of the new notes are given in the middle column of the second subtable; for clarity, we list them below.[52]

Ut	Re	Mi	FA	Sol	La	Si	Ut
PA	RA	GA	SO	BO	LO	DO	PA

The above notes are tempered; the just notes, which are one or two heptamerides apart, are indicated in Table 5.10 in round () or square [] brackets. According to Sauveur, the initials of the new names are not chosen at random, but in such a way as to be somewhat similar to the symbols he will introduce for the 301 sounds; the second letter plays a specific role; two successive sounds of the same letter (A or O)

[52] Note: in the *Traité*, the note Fa is denoted FO; in the later works, it is SO instead; we have unified the symbology with that of the later works by using SO.

Table 5.10 New names and symbols of the notes of the octave. Drawn from [73], Planche I

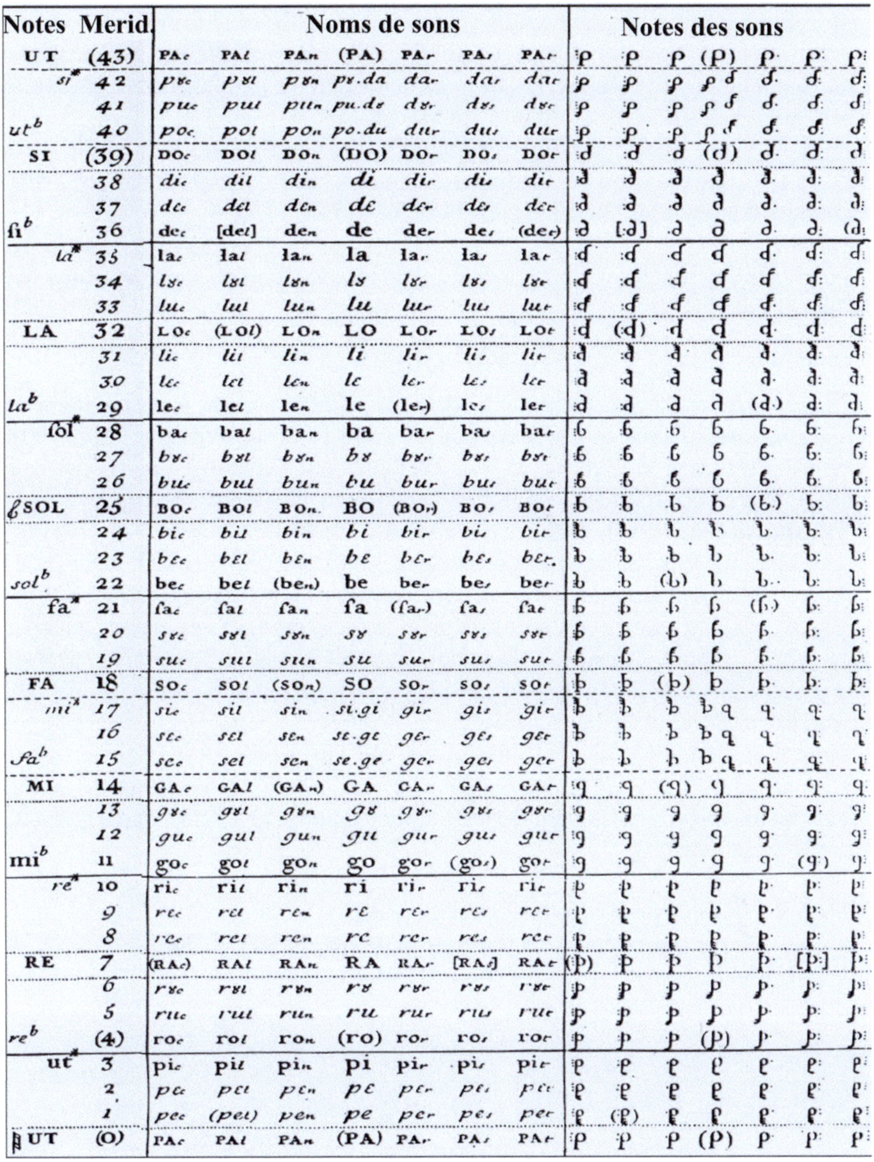

Notes	Merid.	Noms de sons						
UT	(43)	PAc	PAl	PAn	(PA)	PAr	PAs	PAt
si*	42	pɤc	pɤl	pɤn	pɤ.da	dar	das	dat
	41	puc	pul	pun	pu.dɤ	dɤr	dɤs	dɤt
ut b	40	poc	pol	pon	po.du	dur	dus	dut
SI	(39)	DOc	DOl	DOn	(DO)	DOr	DOs	DOt
	38	dic	dil	din	di	dir	dis	dit
	37	dec	del	den	de	der	des	det
si b	36	dec	[del]	den	de	der	des	(det)
la*	35	lac	lal	lan	la	lar	las	lat
	34	lɤc	lɤl	lɤn	lɤ	lɤr	lɤs	lɤt
	33	luc	lul	lun	lu	lur	lus	lut
LA	32	LOc	(LOl)	LOn	LO	LOr	LOs	LOt
	31	lic	lil	lin	li	lir	lis	lit
	30	lec	lel	len	le	ler	les	let
la b	29	lec	lel	len	le	(ler)	les	let
sol*	28	bac	bal	ban	ba	bar	bas	bat
	27	bɤc	bɤl	bɤn	bɤ	bɤr	bɤs	bɤt
	26	buc	bul	bun	bu	bur	bus	but
SOL	25	BOc	BOl	BOn	BO	(BOr)	BOs	BOt
	24	bic	bil	bin	bi	bir	bis	bit
	23	bec	bel	ben	be	ber	bes	bet
sol b	22	bec	bel	(ben)	be	ber	bes	bet
fa*	21	fac	fal	fan	fa	(far)	fas	fat
	20	sɤc	sɤl	sɤn	sɤ	sɤr	sɤs	sɤt
	19	suc	sul	sun	su	sur	sus	sut
FA	18	SOc	SOl	(SOn)	SO	SOr	SOs	SOt
mi*	17	sic	sil	sin	si.gi	gir	gis	git
	16	sec	sel	sen	se.ge	ger	ges	get
fa b	15	sec	sel	sen	se.ge	ger	ges	get
MI	14	GAc	GAl	(GAn)	GA	GAr	GAs	GAt
	13	gɤc	gɤl	gɤn	gɤ	gɤr	gɤs	gɤt
	12	guc	gul	gun	gu	gur	gus	gut
mi b	11	goc	gol	gon	go	gor	(gos)	got
re*	10	ric	ril	rin	ri	rir	ris	rit
	9	rec	rel	ren	re	rer	res	ret
	8	rec	rel	ren	re	rer	res	ret
RE	7	(RAc)	RAl	RAn	RA	RAr	[RAs]	RAt
	6	rɤc	rɤl	rɤn	rɤ	rɤr	rɤs	rɤt
	5	ruc	rul	run	ru	rur	rus	rut
re b	(4)	roc	rol	ron	(ro)	ror	ros	rot
ut*	3	pic	pil	pin	pi	pir	pis	pit
	2	pec	pel	pen	pe	per	pes	pet
	1	pec	(pel)	pen	pe	per	pes	pet
UT	(O)	PAc	PAl	PAn	(PA)	PAr	PAs	PAt

are separated by one tone; on the other hand, if the letters are different, for example one is A and the other is O, they are separated by a semitone.

In Table 5.10 there are 43 rows, one for each meride, starting from 0 at the bottom to 43 at the top. The rows in the second sub-table represent the 301 new names of notes, separated by a heptameride. They are indicated by six lowercase consonants:

c l n r s t

For example, in the row corresponding to Mi, new name GA, to the left of the central column marked by GA, are the sounds GAc, GAl, GAn, and on the right GAr, GAs and GAt.

Naming notes that move along a column is a bit more complicated; it is done with the use of vowels[53]:

a e ϵ i o u γ

Let's see what happens when moving along the middle column; for example, when going above and below RA of a meride; starting from the bottom, you have the symbols: ro, ru, rγ, RA, re, rϵ, ri. For example, since RA is 7 merides, ri is 10 merides, and rγ is 6 merides.

The order of the vowels changes if the second letter of the note is O instead of A; for example, for LO, starting from the bottom, you have: le, lϵ, li, LO, lu, lγ, la.

There is only one semitone between the notes GA-SO (Mi-Fa) and DO-PA (Si-Ut), so instead of six heptamerides there are three. In these cases there are two equivalent names for each note. Consider for example GA-SO, going up from GA you would have ge, gϵ, gi while going down from SO you would have, si, sϵ, se. So we have the equivalences se $\equiv$ ge, sϵ $\equiv$ gϵ, si $\equiv$ gi.

Note that Table 5.10 has both uppercase and lowercase letters. Sauveur is not consistent in his work, and uppercase letters should be reserved only for the sounds that are on the rows corresponding to the traditional notes, in all other cases he should use lowercase letters.

Chapter 4. *Symbols for Expressing Octave Sounds*

The symbols for the notes of the natural scale are shown in the last row of the following table:

Ut	Re	Mi	FA	Sol	La	Si	Ut
PA	RA	GA	SO	BO	LO	DO	PA
ρ	þ	9	ƀ	b	đ	d	ρ

---

[53] This is actually the notation of the 1701 memoir. In the *Traité* Sauveur used different names for the vowels: a, e, ê, i, o, u, û, and arranged them in reverse order [81], pp. 53–54.

They have four main characteristics:

1. The central body, an O.
2. The vertical position of a stem, which may be above or below the body.
3. The horizontal position of the stem, which may be to the left or to the right.
4. The head which is the continuation of the stem, high at most $^1/_3$ of it.

The basic figures are the two P and P the others are obtained by reflection around horizontal or vertical axes.

Sounds that are on the same line are indicated by dots on the left, to remove as many heptamerides as there are dots, and on the right, to add as many heptamerides as there are dots, as shown below:

-3 -2 -1 0 +1 +2 +3

Moving vertically adds one to three tails (or hooks); each tail adds or subtracts one meride. The tails look to the left to decrease merides, to the right to increase merides, as shown below:

-3 -2 -1 0 +1 +2 +3

Note that the more tails there are, the less merides are added or subtracted; this is probably because the most common alteration is that of 3 merides, and so the simplest notation should be used for it. Table 5.11 shows the different ways of representing sounds for the traditional and new notations; the greater simplicity of the new notation is evident.

After introducing the symbols for the notes, Sauveur went on to describe the representation of them as an alternative to the traditional arrangement on the staff (Sauveur did not use this term, he rather spoke of *tablature* or parts).

Sauveur first introduced new symbols for the clefs, for which he needed a reference tone. In the work of 1701 he will develop for this purpose his idea of the *fixed sound*; here he was content to take as reference sound the *ton de Chapel* or *ton d'Opéra*, which he understood to correspond to C Sol Ut, taken as PA; this sound is that of the middle octave in harpsichord, organ and all the instruments playing in an orchestra. The octave beginning with this PA is represented by ⊙. The upper octaves are denoted by ① ②, ③, and so on. The lower octaves are denoted by the symbols D D D etc. [70].[54]

Chapter 5. Marking the Notes of Music According to Ordinary Practice

Since in music the compass of a normal voice is 11 notes, to represent all the notes it is sufficient to have five parallel lines (the staff) which, with the six spaces (considering as such also those above and below the five lines, form the 11 steps). But, continued Sauveur, if one wants to indicate all the notes of a concerto, one must have at least 13 parallel lines; this creates problems of representation. Since five are usually sufficient

[54] p. 140–141.

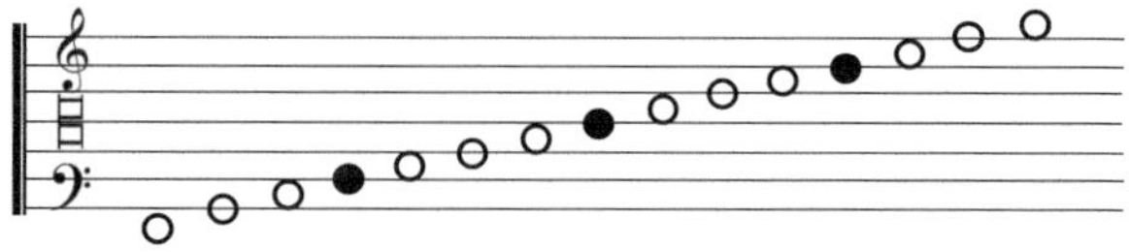

Fig. 5.7 All three clefs together. Redrawn from [70], p. 147

for normal instruments, one proceeds differently. One chooses one of the notes to be indicated on one of the lines or spaces; the other notes follow, going up or down [70].[55]

The sign that tells us the name of the reference note is called a clef. In ordinary music, there are three clefs: the clef of Ut, indicated by 𝄡, usually called the clef of C Sol Ut; the clef of Sol, indicated by 𝄞, usually called the clef of G Re Sol; and finally the clef of Fa, indicated by 𝄢, usually called the clef of F Ut Fa.

In normal writing, it is sufficient to place a clef on a line to identify the notes above and below it, as can be seen in Fig. 5.7, which gives an overview.

But since there are usually only five lines, only one clef is used.

If you wanted to lower or raise a note by a semitone, you would use two characters placed before the note. To lower or raise all the notes in a line or space, the two characters are placed at the beginning of the line or space in tablature. These are the ♭ (flat) for lowering and the ♯ (sharp) for raising. Referring to the staff of Fig. 5.8, the first Mi in the first line is lowered by a semitone, but not the second, while Fa in the first space is raised by a semitone; on the third line, both Si are lowered.

The symbols ♯ and ♭ that precede a note are called alterations. They can be avoided where they appear at the beginning of a line by changing the names of the notes so that any semitone due to an altered note coincides with a semitone of the diatonic system. Unlike Guido d'Arezzo, Sauveur uses a heptachord and thus his solmization can cover two semitones together, (Fa-Mi) and (Si-Do), of the diatonic scale. Figure 5.9 shows the examples of some solmizations; above those concerning flats, below those concerning sharps.

At the end of the chapter, Sauveur took up the topics he had covered with the traditional notation, using the new notation. Figure 5.10 is the equivalent of Fig. 5.7. Apart from the names of the notes, the main difference is in the use of clefs. The new clefs are not specialized for particular instruments, but, perhaps more rationally, divide the tablature into octaves.

The mutation problem does not exist, simply because there are no sharps and flats, at least in music notation. If a sharp occurs in the normal notation, in the new notation it is sufficient to change A to I and O to E. In the case of flat, A changes to O and O to A (see Table 5.11).

[55] p. 145.

Fig. 5.8 Sharps and flats; from [70], p. 149

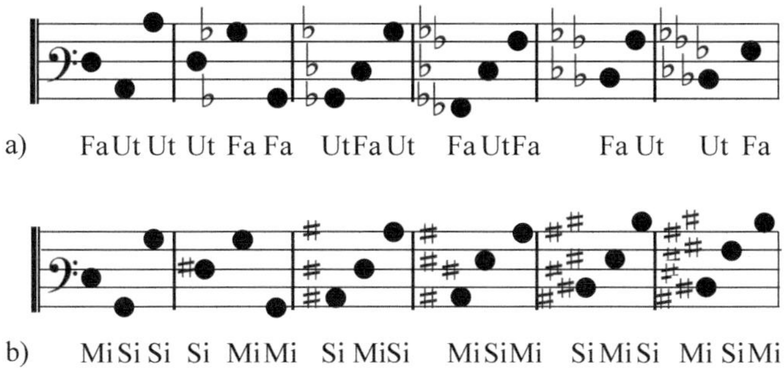

Fig. 5.9 Solmizations; from [70], p. 153 (**a**), p. 154 (**b**)

Fig. 5.10 All three new clefs together. Redrawn from [70], p. 155

Chapter 6. *Of Tempos and Bars.*
In this chapter, Sauveur set out to measure absolute and relative tempos or durations of sounds. Absolute tempos, such as the duration of a bar, can be measured in seconds with any clock, such as a pendulum. For this purpose, Sauveur presented Étienne Loulié's chronometer, criticizing it perhaps unfairly, insisting more on its defects than its merits; since it will be taken up in much the same way in the memoir of 1701, we will avoid any comment here.

Relative tempo concerns the relative duration of individual notes within a bar; the idea is well illustrated by Fig. 5.11a. Here the reference note is PA (Ut), but the concept is the same for all notes. The unit of measure, the conventionally assigned duration 1, is represented by a black note. Notes whose duration is less than the reference duration are also black; the dashes above these notes indicate the number of times the duration of the note should be halved. If the duration of the note is to be greater than the reference duration, the note is white; the dashes above it indicate the number of times the duration is to be doubled from the reference duration.

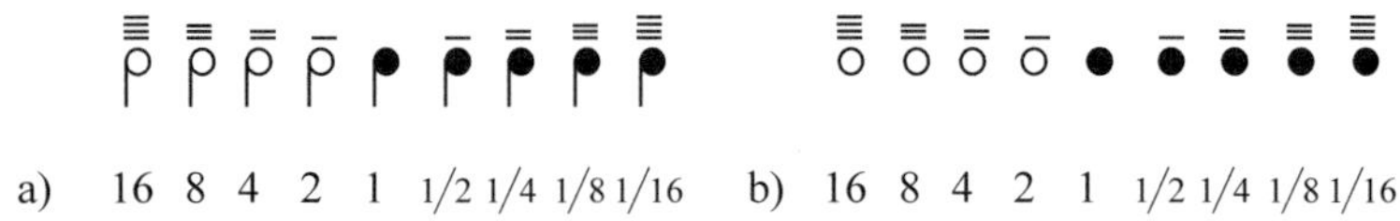

Fig. 5.11 Duration of notes and rests; from [70], p. 164

Table 5.11 Scales with new and traditional notation; from [70], p. 130

The three scales of the ordinary system

Scale of sharps	Ut#	Re#	Mi#	Fa#	Sol#	La#	Si#	Ut#
Scale natural	Ut	Re	Mi	Fa	Sol	La	Si	Ut
Scale of flats	Ut♭	Re♭	Mi♭	Fa♭	Sol♭	La♭	Si♭	Ut♭

The three scales of the new system

Scale of sharps	pi	ri	gi	se	be	le	de	pi
Scale natural	PA	RA	GA	SO	BO	LO	DO	PA
Scale of flats	po	ro	go	sa	ba	la	da	po

The three scales of the new system with merides

Scale of sharps	3	10	17	21	28	35	42	3
Scale natural	0	7	14	18	25	32	39	0
Scale of flats	40	4	11	15	22	29	36	40

The three scales of the new system with symbols

Scale of sharps
Scale natural
Scale of flats

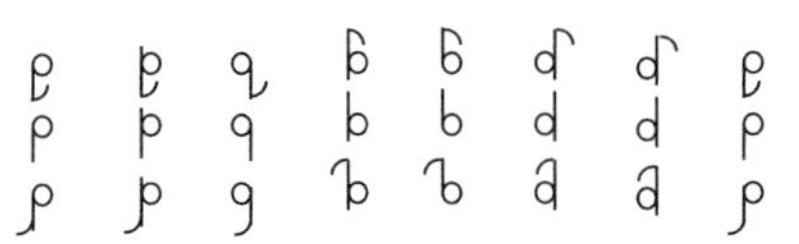

Rests in the new notation system are indicated by the O without a stem. The reference value 1 is indicated by a black O. The durations of the rests follow the same conventions as those of the notes, as shown in Fig. 5.11b.

5.2.4 *Book IV*

Book IV is less general in nature than the previous ones; it deals with the musical systems in use and how they fit the musical instruments of the time, using both traditional and new notation. Sauveur used merides and heptamerides as a convenient means of comparing the different intervals of the various systems.

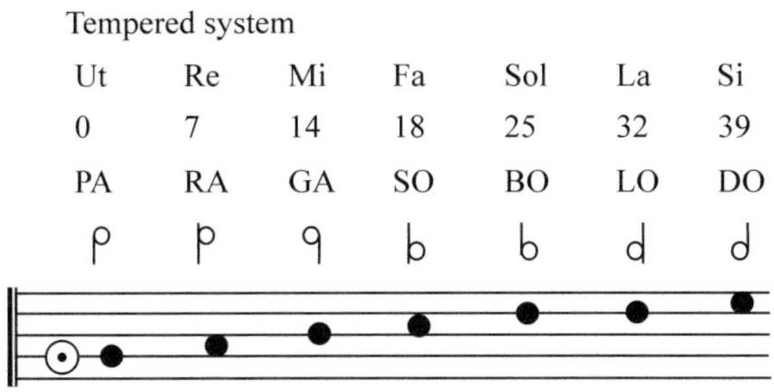

Fig. 5.12 Tempered diatonic system; from [70], p. 170

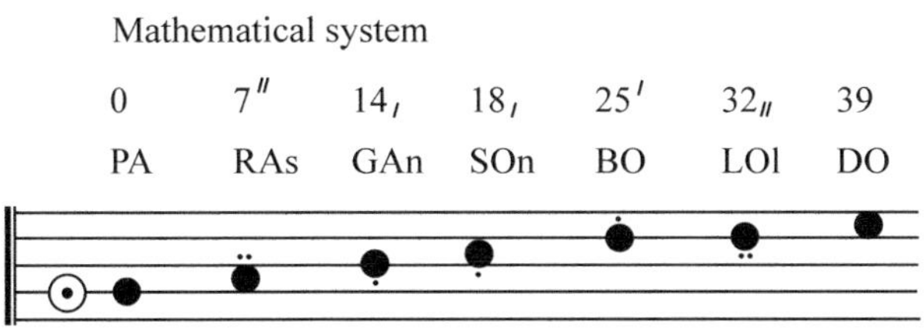

Fig. 5.13 Just diatonic system; from [70], p. 171

Chapter 1. *Of the Diatonic Systems*
The first system to be discussed is the diatonic system, characterized by the presence of five tones and two semitones. This system is presented with the use of merides in Figs. 5.12 and 5.13, where the tempered diatonic form, resulting from the division of the octave into 43 merides, is compared with the just form, also called the mathematical system. In Fig. 5.13 dashes have been used to raise, up, or lower, down, the value of the notes of as many heptamerides as there are dashes. In the staff, the raising or lowering of notes is represented by dots, above or below, respectively.

Chapter 2. *Chromatic and Enharmonic Systems.*
Sauveur began this chapter by pointing out that chromatic and enharmonic systems can be considered from two points of view, that of the Greeks and that of modern musicians; he first treated the Greek view by referring to Figs. 5.14 and 5.15, where the tetrachords form the unmodulating system of the Greeks, that is, the union of the greater perfect system and the lesser perfect system. The greater perfect system is formed by the tetrachords numbered I, II, IV, V (see Fig. 5.15)—corresponding in order to the tetrachords hypatōn, mesōn, diegzeumenōn and hyperbolaiōn—to which is added, in the lower part, the note proslambanomenos. The lesser perfect system is formed by the proslambanomenos and the tetrachords I, II, III, the last of which is the synēmmenōn.

The chromatic and enharmonic genera are shown in Fig. 5.14, together with the diatonic genus, in their tempered forms. Sauveur, following the tradition of identifying the lowest note, proslambanomenos, of the greater perfect system with La,

		hypatōn			mesōn			synēmmenōn				diezeugmenōn			hyperbolaiōn		
dia	32	39	0	7	14	18	25	32	36	0	7	39	0	7	14	18	25	32
chr	32	39	0	3	14	18	21	32	36	39	7	39	0	3	14	18	21	32
enh	32	39	41	0	14	16	18	32	34	36	7	39	41	0	14	16	18	32

Fig. 5.14 Merides of the Greek systems. Redrawn from [70], p. 174

	proslm		hypatōn I			mesōn II			synēmmenōn III				diezeugmenōn IV			hyperbolaiōn V		
dia	LO	DO	PA	RA	GA	SO	BO	LO	de	PA	RA	DO	PA	RA	GA	SO	BO	LO
	La	Si	Ut	Re	Mi	Fa	Sol	La	Si♭	Ut	Re	Si	Ut	Re	Mi	Fa	Sol	La
chr	LO	DO	PA	pi	GA	SO	sa	LO	de	DO	RA	DO	PA	po	GA	SO	sa	LO
	La	Si	Ut	Ut♯	Mi	Fa	Fa♯	La	Si♭	Si	Re	Si	Ut	Ut♯	Mi	Fa	Fa♯	La
enh	LO	DO	dγ	PA	GA	gu	SO	LO	lγ	de	RA	DO	dγ	PA	GA	gu	SO	LO

Fig. 5.15 New names and symbols of the Greek systems. Adapted from [70], p. 174

assigned it a value 32, which leads to a value 0 for the second string of the first tetrachord, corresponding to Ut. In this way, Sauveur can relate the Greek system beginning with La to his own system beginning with Ut.

The second note, the first of the first tetrachord, is one major tone away from the proslambanomenos, i.e. it is 7 merides away, giving a value of 39. The merides of the first note of the second tetrachord are obtained by adding a tempered fourth (18 merides) to the first note, i.e. $39 + 18 = 57$, which, reduced to the octave, i.e. subtracting 43, gives 14 merides. The first note of the third tetrachord (conjoint with the second) is obtained by adding a tempered fourth, which gives 32. The first note of the fourth tetrachord, disjoint with the second, is obtained by adding a fourth and a tone to the first string of the second tetrachord, giving $14 + 18 + 7 = 39$. The first note of the last tetrachord is obtained by adding a fourth to the first note of the fourth tetrachord, giving $39 + 18 - 43 = 14$. The merides of the middle notes are obtained by adding 4 to the first note and 7 to the second note, for the diatonic genus. By adding 4 and 3 for the chromatic genus; 2 and 2, for the enharmonic genus.

Figure 5.15 refers to the notes for all three Greek systems, with both new and old notation.

Chapter 3. *Of the Systems of Mean Semitones.*
In the system of mean semitone shown in Fig. 5.16, the octave is divided into 12 equal parts (semitones); and to understand the value of each in relation to the general system, it is necessary to divide the 311 heptamerides, or the octave, into twelve equal parts. Each semitone will contain 25 1/12 heptamerides. Since this system is determined numerically, the name corresponding to each sound with the new sym-

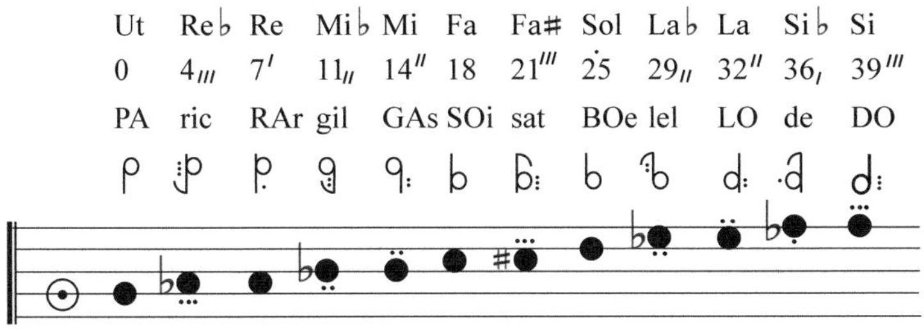

Fig. 5.16 Equal temperament chromatic system; from [70], p. 179

bols and the old symbols can be found by consulting Table 5.10.

Chapter 4 . The Ordinary Chromatic Systems.
Musicians' chromatic systems are nothing but diatonic systems in which tones are divided into semitones; one example is given by the system of equal semitones referred to in the previous chapter. But since this temperament, according to Sauveur, is disliked by musicians because of its excessive differences from the just temperament system, they use systems that can have three types:

1. The notes Ut, Re, Mi, Fa, Sol, La, Si are assigned their tempered values, PA, RA, GA, SO, BO, LO, DO (see Fig. 5.17). The notes Ut, Fa, Sol have sharp (♯), while Mi and Si have flat (♭), and Re and La are not altered. To have this system we need to add three merides for the sharp and remove three merides for the flat.
2. The second chromatic system is suitable for the harpsichord, in which all the chromatic keys are divided into two parts, in which all the notes, except those between Mi-Fa and Si-Do, have a sharp and a flat. Thus, between the sharp of the lowest note and the flat of the highest note there is a meride, as shown in Fig. 5.18. Thus the octave is divided by 17 keys.
3. To avoid such a large number of notes, the sharp of a note can be made to be the same as the flat of the adjacent note, as shown in Fig. 5.19. For this purpose it is sufficient that sharp and flat have the intermediate value of 3.5 merides. This system, said Sauveur, is used in fretted instruments but only when the compass has been extended by transposition, otherwise the system of equal semitones is preferred.

The chromatic systems presented above, according to Sauveur, are of the ordinary type, i.e., they are those commonly used by musicians of his time. But he illustrated them with his tempered system of 43 merides, which was unknown to ordinary musicians. Therefore, it is not clear to which musicians he is referring. His system of 43 merides corresponds to 1/5 of a comma temperament, but French musicians, according to Sauveur, preferred a 1/6 of a comma temperament, associated with the division into 55 parts of the octave (see Sect. 3.2.2.2); this however is not so certain [25].[56]

[56] p. 71.

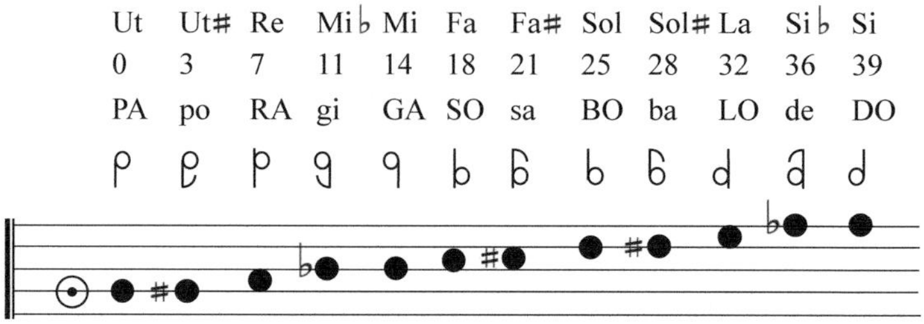

Fig. 5.17 First chromatic system for unequal temperament; from [70], p. 181

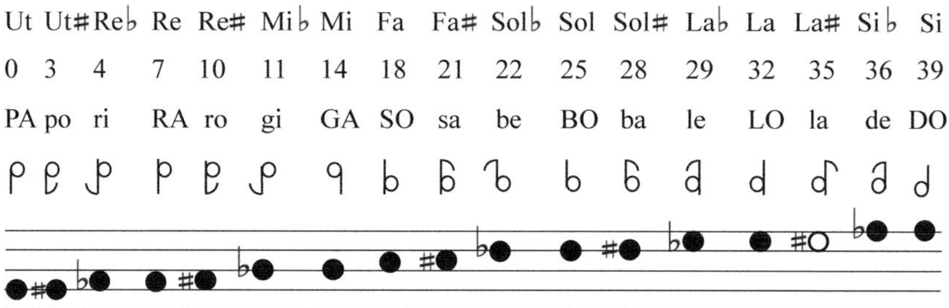

Fig. 5.18 Second chromatic system for unequal temperament; from [70], p. 182

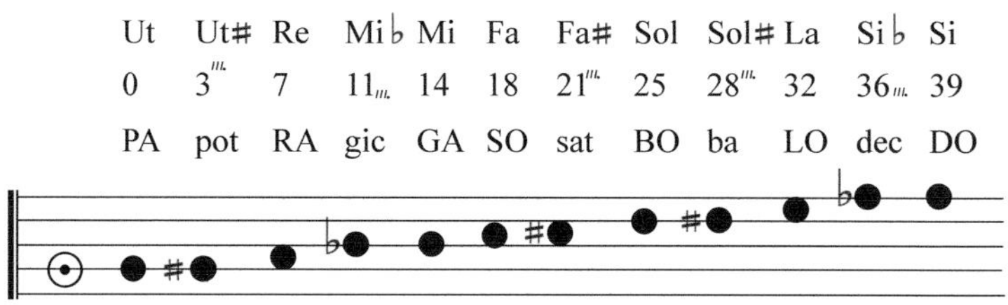

Fig. 5.19 Third chromatic system for unequal temperament; from [70], p. 182

Chapter 5. *Of the Trumpet*
Sauveur began this chapter by saying that there were two types of trumpets, the trumpet marine and the wind trumpet, both characterized by the fact that their sound make jumps. The trumpet marine was the first to be discussed in detail, even though it was less used—it would become obsolete in the 1750s—but its operation is easier to describe and this justifies, from a didactic point of view, devoting more space to it.

Sauveur, in his memoirs of 1701, will take up the explanation of the trumpet marine on the strength of his mastery of the theory of harmonic sounds and thus devote only a few lines to it. In this chapter the subject is treated in more detail precisely because it lacks an established reference theory, not even the one developed by Wallis and company.

Table 5.12 Sounds produced by a trumpet marine; from [70], p. 185. Stars are for not just notes

Aliquotes	Heptamerides	Merides		
1	000	0	PA	Ut
2	000	0	PA	Ut
3	176	25	BOr	Sol
4	000	0	PA	Ut
5	097	14	GAn	Mi
6	176	25	BOr	Sol
7	243	35	lal	La*
8	000	0	PA	Ut
9	051	7	RAs	Re
10	097	14	GAn	Mi
11	138	20	BOr	Sol
12	176	25	ler	Fa*
13	211	30	ler	La*
14	243	35	lal	La*
15	273	39	DO	Si
16	000	0	PA	Ut

The description of the instrument is similar to that of La Hire, and therefore reference is made to it (see Sect. 4.3.2.1). Sauveur was primarily interested in the sounds produced by the instrument. He assumed that the string of the trumpet marine is divided into aliquots, i.e. that the smaller part is an integral sub-multiple of the whole. The length of the smallest part of a string long 32 is thus given by l/n for $n = 1, 2, 3, \cdots, 32$. Table 5.12 summarizes the tones emitted by the trumpet, divided into octaves and limited to the first 16 harmonic sound, which represent the limit that can be reached in practice, although harmonic sounds above the thirteenth are rare [1].

The first column of the table is for the value of n, the second is for the heptamerides, where 301 heptamerides are set to zero; the third column is for the rounded values of the merides, the fourth column is for the new names of the notes, the last column is for the old names, where an asterisk means a not just interval.

As you can see from the table, the first octave $(1 - 2)$ contains no intermediate notes, the second octave $(2 - 3 - 4)$ contains an intermediate note given by a just fifth, the third octave $(4 - 5 - 6 - 7 - 8)$ contains a just major third and a just fifth. The fourth octave, contains various possible intervals.[57]

---

[57] The scale formed by the harmonic sounds, now often called the natural scale—not to be confused with the just scale—if we limit ourselves to the lowest harmonic sounds, cannot provide the notes of the diatonic or chromatic scale. It can only provide the notes that correspond to intervals with an even denominator (a power of 2 indeed). It gives the just values of the notes Ut, Re, Mi, Sol, Si. But for example, to get a good approximation for La (a major sixth), you have to go at least as far

Sauveur pointed out that the only just consonant intervals are the fifth and the major third. This means that in compositions for the trumpet marine it is necessary to indicate only the intervals that can be reproduced, instead of the just intervals that are missing.

With regard to the common trumpet, Sauveur gave only an acoustic explanation for the fact that it produces only harmonic sounds. The explanation, which is interesting because it shows how far Sauveur had already come in his knowledge of harmonic sounds, is given in full for its brevity:

> The ordinary trumpet has the same intervals as the trumpet marine, and the reason for this can be explained in this way. The molecular parts of a trumpet are capable of a certain vibration that amplifies the vibrations of the air exiting from the mouth, whenever the lips that excite the air in the trumpet have impressed on this air a vibration equal to or multiple of that of the trumpet, because the vibrations of the air exiting from the mouth then help to transmit the vibrations to the trumpet, which in turn amplifies those of the air giving it the sound or the trumpet. Conversely, if the number of vibrations of the air leaving the mouth is not a multiple of the vibrations or the trumpet, the one set of vibrations opposing the other hinders the sound of the trumpet. That is why blowing softly into a trumpet produces no sound unless the vibrations of the lips are an aliquot of the vibrations of the trumpet. Next, if the air is forced, it still will make no sound until the vibrations of the lips are double those of the trumpet, so that the fundamental sound of the trumpet will suddenly rise one octave. If the air is forced more, it will make no sound until the vibrations of the lips are triple those of the trumpet, and then quadruple, quintuple, and so on. That is to say, the sound will rise to the fifth, to the fourth, to the major third, the minor third, etc., as we have shown with the trumpet marine [70].[58]

As can be seen, Sauveur believed that the sound of the trumpet came from the interaction of the natural vibrations of its metal body with the air.

5.3 The First Memoir on Acoustics for the Académie des Sciences

The memoir of 1701, *Systême general des intervalles des sons, et son application à tous les systêmes & à tous les instruments de musique*, is Sauveur's most mature and perhaps most important writing, reporting the results of his reflections, studies, and experiments matured over much of the previous decade.

Before Sauveur published his memoir, he had presented his ideas to the Académie in 1700, as reported in the memoir *Sur la determination d'un son fixe* of the *Histories* of the same year [35], and for this reason it is referred to below, almost verbatim. Comments are generally avoided, mainly because they are given directly in the published memoir.

---

as harmonic 27, with a ratio of $27/16 = 1.683$, which is quite close to the just value of the major sixth $5/3 = 1.6666$. A similar argument applies to Fa.

[58] p. 186–188. Translation into English in [81].

5.3.1 The Histoire of 1700. Sur la Determination d'un Son Fixe

The memoir *Sur la determination d'un son fixe*, published for the *Histories* of 1700, is due to Fontenelle, the perpetual secretary of the Académie since 1697. Part of the content of the memoir is drawn from the procès-verbaux of 1700 relating to the determination of what Sauveur called the *son fixe*, although Fontenelle's account covers much more aspects of music theory.

The memoir begins by illustrating Sauveur's motivations:

> The science of hearing cannot be less extensive than that of sight, but it has hitherto been less thorough. The need that philosophers have had for telescopes and microscopes has forced them to study with extreme care the various paths and properties of light, but since they have not had the same need to know exactly everything that belongs to sound, and since they have often treated music as a matter of taste for which one does not have to look too hard for rules in philosophy, they have not directed their speculations to this part. So Sauveur thought that there was still a little known country to explore [35].[59]

Sauveur, said Fontenelle, found that the science of sound was becoming broader as it progressed and believed that, like optics, it deserved a special name and called it acoustics. The number and importance of the new discoveries justify this new name. One could already take for a piece of acoustics what we see from Dodart's writings in the memoirs of the *Académie* on the formation of the voice [27].

No one is unaware that to have all the chords[60] of music on two strings of an instrument, equally thick and equally tensioned of the same material—as we will always assume in the following—all that is necessary is that their lengths are relative to each other in certain numerical ratios. For example, if such strings are equal in length, they are in unison, if their strings are in the ratio of 1 to 2, they give the octave, if they are in the ratio of 2 to 3, we have the fifth, if 3 to 4, the fourth, if 4 to 5, the major third, and so on.

Since these main chords are formed by the ratios of the lengths of two strings, it is clear that other ratios, such as 8 to 9 or 15 to 16, will be other chords. And since the ratios of numbers are infinite, it seems that the chords of music must also be infinite, and indeed they would be if voices and instruments were not necessarily limited to a certain range, and if the ear were not so made that it cannot distinguish sounds below a certain tone or above a certain other tone.

It is true that in the extension which the voice and the instruments can reach, and in which the ear can distinguish the sounds, an infinite number of chords can still be made. For example, that of 929 to 930, or 1000 to 1001, and so on. But the difference between the numerator and denominator in these ratios is so small that they would

---

[59] p. 134.

[60] As already noted, for Sauveur a chord is more or less the same as an interval; and in many English translations *chord* is rendered as *interval* so as not to confuse the concept expressed by this term with the modern one of a chord, which generally concerns three and not two sounds played simultaneously. We accepted this choice while commenting on the *Traité*. From now on we have preferred to keep the term because, Sauveur from the 1701 paper will use both chord and interval.

produce the same effect as if they were equal, i.e. , if they were in unison, because it is always a question of the ear, which, being a material organ, cannot distinguish differences beyond a certain fineness.

Before a string of length 2 is shortened to 1, that is to say, to be one octave higher than the sound it previously produced, it can pass through as many divisions as one wishes. Sauveur fixed this number of divisions in 43 parts, and these 43 parts, which he called merides, fill the entire extension of the octave and give the most sensitive and ordinary sounds contained in the octave. But if one wants to obtain much more delicate differences in sounds, one must further divide each meride into 7 parts, which he called heptamerides, and consequently in an octave there will be 301 heptamerides. These numbers 43 and 301 have conveniences and uses that make Sauveur prefer them to all the others that can be chosen, but that is not the point now.

A shorter string produces a greater number of vibrations in the same amount of time than a longer one, and therefore produces a higher sound. The number of vibrations of two different strings are in an inverse length ratio, i.e. if the two strings are like 2 to 3, the shorter of the two makes 3 vibrations while the other makes 2.

Therefore, it is clear that the vibrations of two equal strings must always go together, begin, end, begin again at the same moment, but those of two different strings must sometimes be reunited and even separated for a longer time when the numbers expressing the difference of the strings are greater. If two strings are equal to 1 and 2, and begin to vibrate at the same time, it is certain that after two vibrations of the shorter one and one of the other, they will begin to vibrate together again, and every two vibrations of the shorter one there will always be a concurrence of the vibrations of both. If they were like 24 to 25, there would only be a meeting of the vibrations every twenty-fifth vibration of the shortest. It is clear that for larger numbers the meetings will be even rarer.

Up to this point we have been dealing only with ratios of sounds, which says nothing about their absolute values. It is necessary to set a limit above the high sounds and below the low sounds. It is easy to choose this limit arbitrarily, the difficulty is to find it safely when you need it.

Musicians use a kind of wooden or metal whistle of a certain length to determine the pitch to which the voices and instruments must agree in a concert, and since they want these pitches to be always the same, they assume that this whistle will always do the job. But this assumption is not entirely true.

1. The material of the whistle is very susceptible to alteration. Simply because of the use made of it, the times, the accidents, the sound changes considerably after a few years.
2. It is certain that by blowing more or less strongly on the whistle, the pitch rises or falls, and one cannot be sure of always giving the same blow.
3. Finally, if the whistle is lost, it is impossible to find the sound that has been determined.
4. Even an organ pipe, which is much more reliable than a whistle, does not always produce exactly the same sound.

These inconveniences made Sauveur want to be able to determine a standard for sound, which he called *fixed sound*, more easily and more surely, and this was his idea.

When one tries to tune an organ, and two pipes approaching unison play together, there is a moment when the common sound they make is strongest, and this moment seems to return at equal intervals of time. Sauveur, after searching for the cause of this phenomenon, imagined with extreme probability that the sound of the two pipes together must be more powerful when their vibrations, after having been separated for some time, come together and coincide to strike the ear with the same shot. It also seems that the common expression of musicians who say that the pipes beat when their sounds meet again has its origin in this idea. It is important that the pipes come close to being in unison, because the closer the numbers that express their lengths, the fewer the vibrations that meet and beat, and therefore the easier it is for the ear to perceive and distinguish the beats.

If two pipes were taken in such a way that the interval between their beats was large enough to be measured by the oscillations of a pendulum, one would know exactly, from the length of the pendulum, the interval between the beats, One would also know from the nature of the chord of the pipes how many vibrations one would make if the other made a certain number, and as these two numbers would be included in the interval of two beats whose duration one knows, one would know exactly how many vibrations each pipe makes in a given time. Now it is only a certain number of vibrations made in a given time that makes a certain sound, and since the ratios of the vibrations of all the sounds are known, we would know how many vibrations each sound makes in that given time.

This is how far human industry has come. Sauveur took as a fixed sound that which makes 100 vibrations in one second and called a fixed octave high the one that is above and a fixed octave low the one that is below.

This fixed sound can be easily found at any time and in any place by experiencing the beats of two pipes, which will certainly produce a sound that has 100 vibrations per second. Thus, the tone of the Paris Opera will be determined by the relationship to the fixed sound, and will certainly be the same in China. And it does not matter that the two pipes on which the fundamental beat experiment is carried out in China are of the same length or the same thickness or the same material as those in Paris. All you have to do is find a sound for two pipes, whatever they may be, that makes 100 vibrations per second. This sound is always the same, regardless of the instruments that produce it, and its whole nature consists in this specific number of vibrations per second.

When Sauveur communicated all these points of view to the Académie, it immediately went to secure the experience of the beats and appointed some academicians to assist him. The experiment failed, as Sauveur admitted, but he assured that on other occasions and in the presence of more skilled musicians from Paris, it had given very correct and very precise results. The difficulty of starting from scratch and the limited equipment he had, were the cause of the failure. But we know that we shouldn't get discouraged so easily with experiments, and they have their quirks, so to speak, which are overcome with time. While we are waiting for the Académie to take it up again, it is proposed to the musicians as an experience to be verified.

In order to better illustrate how much it deserves to be followed, we will here set out some of the consequences that will be drawn from the fixed sound once established, both in relation to music theory and in relation to physics in general.

Sauveur had found in his experiments that an organ pipe of about five feet in length produced the fixed sound, while a pipe of forty feet in length produced a sound so low as to be indistinguishable. This forty-foot pipe, being eight times longer than the five-foot one, has eight times fewer vibrations in a second, and as the five-foot one, according to the hypothesis of the fixed sound, makes 100 vibrations, it follows that the lower tone which the ear can hear is that of 12$^1/_2$ vibrations in a second. Furthermore, if the shortest tube from which the sound can be distinguished has a length of 1 inch less $^1/_{16}$ and thus a ratio of 1 to 64 to the tube of 5 feet long, the highest sound will make 6400 vibrations in one second.[61]

From this we see the limits of the ear's ability to discriminate tones, and we see that from the point where this organ begins to be sensitive to the point where it ceases to be so, is like the ratio of 12$^1/_2$ to 6400, which is more or less than that of 1 to 512, it can be concluded that the ear is capable of producing the sensation after a certain degree up to another which is 512 higher.

What has been said about organ pipes can be applied to the strings of instruments, for the strings can be put in unison with such pipes at will.

The vibrations of the strings are visible, while those of the pipes are not. It is possible, especially with a microscope, to observe the path that a plucked string makes in its center, that is, the extension of its vibration. When this string produces the fixed sound, its path multiplied by 100 is the entire space covered by the central point of the string in one second.

The first vibrations of a string that is plucked are much greater than the last, but they are always of the same duration, and this means that the sound fades from beginning to end without the note changing. Therefore, in the previous observations, it is necessary to try to have the path that the center of the string takes in its first vibrations, when its sound is strongest, and what it takes in the last vibrations, when the sound is weakest. Sauveur found that the center of a string which produces the fixed sound and which has its diameter of $^1/_6$ of a line, makes in the last observable vibrations $^1/_{18}$ of a line and therefore about 6 lines per second and during the first vibrations 72 times more, that is to say 3 feet per second.

The low sounds make fewer vibrations in the same time than the high sounds, but they are larger, and since, as is very likely, the amplitude of the vibrations exactly compensates the defect of the number, all the strings of a $^1/_6$ line diameter, whatever their tone, cover the same distance in one second as that covered for the fixed sounds.

Momentum is the product of the mass of an object and its velocity. The strings we are considering here have the same diameter and different lengths, they can be of different pitches, but they all have the same velocity because they travel the same distance at the same time. Therefore, their momentum is proportional to their length,

[61] The perception range accepted today of the ear is between 20 and 20,000 Hz.

and therefore the strings of a low note have more momentum and hit the ear more, but the hits they deliver are less tight. One can imagine a kind of extension that starts from the lowest sound, that is, from the vibration with greater amplitude and less liveliness. From this point, the amplitude decreases and the liveliness increases until it is the greatest and the amplitude the smallest that is possible. In the middle they are equal.

From the experiments made by Sauveur to determine the fixed sound, it should be noted that when two pipes are in such agreement that they beat 6 times, that is, their vibrations meet 6 times per second, their beats can still be distinguished, but not with ease [28].[62] Therefore, in those chords where the vibrations meet more than 6 times per second, the beats are not heard at all, and on the contrary, they are heard with much greater ease when the vibrations meet less than 6 times per second. Now for the fixed sound system, i.e. the system of octaves measured from the fixed sound, it is easy to determine how often the vibrations of any chord meet.

For example, if one string is one octave higher than the fixed sound, it will make 200 vibrations in one second, while the other will make 100, and their vibrations will meet 100 times, which is far from meeting six times. If this string is in the lower octave of the fixed sound, it will make 50 vibrations per second, and their vibrations will meet 50 times. If it is in the second sub-octave, the vibrations will meet 25 times. Thus, the fourth sub-octave must be taken into account, because the vibrations meet only six times and the beats can be heard.

It is therefore nearly impossible to ever hear the beats for two sounds that are one octave apart. If the octave is above the fixed sound, the impossibility is complete and absolute. If it is below it, an instrument producing a sound four octaves below the fixed sound would be needed, such as an 80-foot organ pipe, because a 5-foot pipe makes the fixed sound. Now, not counting the enormous size of the instrument, we have seen that the ear cannot distinguish tones lower than those of a 40-foot pipe.

The beats do not please the ear at all because of the unevenness of the sound, and it is very likely that what makes the octaves so pleasant is that one never hears the beats.

> Following these ideas, we find that the chords whose beats are not heard are what musicians call consonances, and those in which the beats are heard are dissonances, and that if a chord is dissonant in one octave and consonant in another, it is because it beats in one and does not beat in the other. The same is true of imperfect consonants [36].[63]

If this hypothesis is true, it will provide the true source of the rules of composition, hitherto unknown to philosophy, which leaves them almost entirely to the judgment of the ear. These kinds of natural judgments, even if they sometimes seem bizarre, are not at all, they are very real causes of consonance.

---

[62] p. 202.

[63] p. 143.

5.3.2 *1701.* Systême General des Intervalles des Sons

The main topics of the memoir *Systême general des intervalles des sons, et son application à tous les systêmes & à tous les instruments de musique* of 1701, are:

1. The determination of a fixed or standard sound, i.e. the formulation of a procedure for producing sounds, notes in fact, of a given and constant frequency, in this case 100 Hz, against which the frequencies of all the notes used in music can be measured.
2. The development of what Sauveur called the *general system of music*, i.e. the precise exposition of a musical system in which the octave is divided into many equal parts. This division was already present in the *Traité*, where it was shown that the division of the octave into 43 parts represents a fair compromise between precision, i.e. the ability to best approximate the intervals of just tuning, and economy; the unit of this division is called *meride*. Each meride is divided into seven *heptamerides*, for a total of 301 units. A further refinement, which Sauveur proposed but hardly used, is the division of the heptamerides into 10 equal parts, called decamerides, which gives the division of the octave into 3010 equal parts.
3. The presentation of a new musical notation, which allows the immediate recognition of the frequency of a note expressed in heptamerides.
4. The study of overtones.

The memoir is divided into 12 sections, each of which is subdivided into a variable number of subsections, the contents of which are given below in full, with a more or less free translation of the parts considered most relevant. The memoir begins with a preface recalling the aims and objectives of the work:

> *Preface*. The occasion on which I found myself explaining the theory of music to very enlightened princes and people of deep understanding, gave me occasion to notice that those who are interested in speculative music have in mind only a few properties of sounds, especially of the practice of singing in use in their time. They have been content to bring back a few musical systems in this connection, which others have gradually changed as musical tastes have changed. No one that I know of has taken this question further, and considered it in a more important way, as the subject of a higher science than music, and then separated from it that part which was properly suited to it, and which has a natural and easy connection with the other parts of the science.
> I believed, therefore, that there was a science superior to music, which I call *acoustics*, and which had for its object sound in general, while music had for its object sound as long as it was pleasant to hear. In order to treat this science in the manner of the other [sciences], and especially of optics, with which it has many connections, it would have been necessary to explain the nature of sound, the organ of hearing, and all the properties of sound in detail; to understand the causes of the pleasantness and unpleasantness of the sounds which serve as the object of music, and of the harmony [sympathie] of sounds; and finally, the instruments not only for music in particular, but also for acoustics in general [73].[64]

After this premise, Sauveur presented his theory of sound with a rather traditional approach, but with the concise language of mathematicians; this makes the presentation somewhat original. First of all, he recalled that sound is caused by the vibrations

[64] p. 299.

of the parts of a sonorous body and that the main property of these vibrations consists in the ratio of the number of vibrations of one sound to that of another sound. Nothing is said about the propagation of sound in the air and its arrival at the ear, as was done instead in the *Traité* of 1697.

Sauveur paused to recall his discussion at the Académie in 1700, which dealt mainly with the appropriateness of using a fixed sound and the reasons why he had not decided to publish his *Traité*:

1. As the names and signs I gave to the sounds were new, I had no doubt that I would find people, especially among musicians, who would disagree with me; and I hope that their objections will be the occasion for some corrections. But since they made them only in relation to use, and since they considered the sounds only for their own needs, I was forced to make some minor changes of my own.
2. While I was working on the treatise on speculative music, I felt the necessity of having a fixed sound to serve as a standard by which to compare all other high and low sounds: in 1700 I gave the way I had imagined to find it, but as in the *Histoire* of the Académie I only showed the necessity and the advantages that would be derived from it, here I give the way to find it.
3. While meditating on the phenomena of sounds, it was pointed out to me that, especially at night, especially with the long strings, other sounds were heard in addition to the fundamental sound, the twelfth and seventeenth of this sound; that the trumpets, in addition to these sounds, has other sounds whose number of vibrations is a multiple of that of the fundamental sound. I found nothing in the explanations of the trumpet marine that satisfied me on this point. But searching for the cause of this phenomenon, I came to the conclusion that strings, in addition to the undulations they make along their length to form the fundamental sound, are divided into two, three, four, etc., equal undulations, which form the octave, twelfth, and fifteenth of this sound. Then I conclude the necessity of the presence of nodes and anti-nodes of these undulations, and the way of perceiving them by touch and sight, as I explain it in the harmonic sounds.
4. This phenomenon gave me the cue for researches about the sympathy of sounds, wind instruments, and acoustic instruments, which can be perfected to the same degree as those of optics; and I wait until these things are in their perfection to give rise at last to a perfect treatise on acoustics [73].[65]

5.3.2.1 On Ratios and Intervals of Sound

Sauveur continued his memoir by echoing what he had written in the *Traité* of 1697; however, he did so in a more compact manner, assuming here a more educated audience in musical matters than that for which the *Traité* was intended. He began with the definition of *ratio* and *interval*:

> Acoustics teaches us that when two bodies make the same vibrations at the same time, they are in unison; and that when one makes more than the other at the same time, the one that makes less makes a low sound, and the one that makes more makes a high sound; and that the ratio of the low and high sounds, therefore, consists in the ratio of the number of vibrations

[65] pp. 300–301.

> that one and the other make at the same time. The ratio of two sounds at the fifth is that of 2 to 3; for while the lower sound makes 2 vibrations, the one at the fifth makes 3 vibrations.
>
> We have just compared sounds using the ratio of the number of vibrations of one of these sounds to the number of vibrations of the other, which we will simply call the ratio of sounds. There is another way to compare sounds, and that is by their intervals.
>
> To understand what the interval of two sounds is, let us first imagine two sounds that are the same or in unison; then one of them goes up, getting higher and higher in pitch until it reaches infinity. It will move further and further away from the other; *and it is this separation or distance of these two sounds that we call interval* [emphasis added]. The same thing will happen as this sound goes down, becoming lower and lower [73].[66]

For Sauveur, ratios are relationships between two numerical values that represent the frequencies of the sounds under consideration and that can be interpreted as fractions of numbers when necessary. Intervals, on the other hand, are intended as relationships between pitches, useful for defining their distance on an intuitive level, and are named by ordinal numbers: first, second, third, etc.

Intervals are first divided into *octaves*, which occur when the highest tone makes two vibrations against one of the lowest. Thus a sound going up passes through the intervals of a first, second, third, fourth, etc. octave when it makes 2, 4, 8, 16, etc. vibrations against one of the first; and passes through similar octaves going down when it makes only $1/2$, $1/4$, $1/8$, $1/16$ of the vibrations of the first.

Sauveur pointed out that the pitches of sounds are currently undefined, and that one must begin with what is conventionally called the fundamental, to which both the highest and lowest sounds are compared. Musicians take as their fundamental the C Sol Ut, which is the ton de Chapel or ton d'Opéra: this sound is located in the center of the keyboard of a harpsichord and is that of a four-foot open organ pipe. But since this sound is not sufficiently determined, one should take for the fundamental sound a sound whose pitch is known and invariable; this is what Sauveur called *fixed sound*, and he decided that this sound should make 100 vibrations per second, or in modern term, it has the frequency of 100 Hz.

The reasons for choosing 100 Hz are not specified by Sauveur; they could be as follows:

1. For a mathematician, the choice of a numerical constant also has an aesthetic value. In Sauveur's time, the most natural choices would be a power of 10 or a power of 2. At this stage, Sauveur chose the power of 10, but in later work he will choose a power of 2.
2. Possible powers of 10 are 10, 100, 1000. Sauveur, at the time of the experiment, must have had at least a vague idea of the value of the frequencies of the various notes, and 10 Hz is too low a value to be heard, while 1000 is too high. This leaves 100 Hz [82].[67]

For the most widely used diatonic system, the seven intervals Sauveur assumed to characterize it are defined by the ratios of the numbers representing frequencies:

---

[66] p. 302.

[67] pp. 10–11.

24.27.30.32.36.40.45.48

Thus they are scaled by a factor of 1.25 from those he had considered in *Traité*, which were representing lengths:

30.32.36.40.45.48.54.60

even though the ratios remain the same.

The following definitions are introduced:

I The interval from one sound to the next is called a second; the interval from one sound to the third is called a third, and so on; finally, the interval from the first to the last is called an octave. If we continue, the numbers that define the diatonic system assumed above, beyond the octave, multiplying them by 2, then by 4, by 8, by 16, and so on, these intervals become the ninths, the tenths, and so on to infinity.

II Comparing one sound to the next, there are three types of seconds.

a. 24, 27 and 32, 36 and 40, 45 whose ratios are equal to that of 8 to 9; this interval is called the major tone.
b. 27, 30 and 36, 40 whose proportions are 9 to 10; this interval is called the minor tone.
c. 30, 32 and 45, 48 which are in the ratio of 15 to 16; this interval is called the major semitone.

The major and minor tones are called major seconds, and the semitone is called minor second. These three kinds of seconds are the *elements* of the diatonic system. In addition to these intervals, musicians recognize the minor semitone, which is the difference between the minor tone and the major semitones, and the comma, which is the difference between the major and the minor tones [73].[68]

According to different combinations of tones and semitones, there are major and minor thirds, fourths, fifths, etc.; and according to different combinations of major and minor tones, there are just or altered intervals. There is only one kind of octave; the major fourth is called tritone, and the minor fourth is simply called fourth. The major fifth is simply called the fifth; the minor, the false fifth.

Among the intervals of the diatonic system, the second, third, and fourth are called minor intervals; the fifth, sixth, and seventh are called major intervals. Moreover, one interval is the *complement* of the other to the octave when it completes the octave; so the fourth is the complement of the fifth to the octave; the third is the complement of the sixth; and the second is the complement of the seventh, and vice versa. The complements of major intervals are minor, and vice versa; so the major sixth is the complement of the minor third.

As in the *Traité*, Sauveur classified intervals as consonant and dissonant, without any reference to the acoustic meaning of the term consonance, as if it were a simple

[68] p. 304.

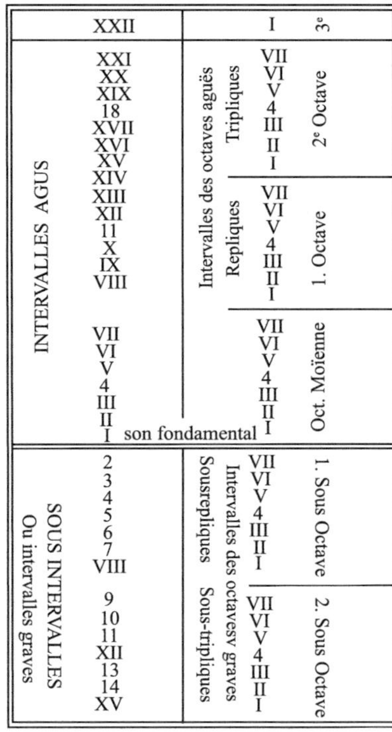

Fig. 5.20 Various types of octaves; from [73], p. 305

nominal definition: "Unison, octave, fifth, fourth, thirds, and major and minor sixths are called consonances, and other intervals are called dissonances" [73].[69]

Figure 5.20 shows Sauveur's classification of octaves and related intervals. The sounds between the fundamental and its acute octave are called sounds of the middle octave. Those that are from VIII to XV are sounds of the first acute octave, those that are from XV to XXII are sounds of the second acute octave, and so on going up. Similarly, the sounds between the fundamental and the first sub-octave will be called sounds of the first sub-octave; those between sub VIII and sub XV will be called sounds of the second sub-octave; and so on going down.

The intervals of the first, second, third octaves are called replicas, triplicas, quadruplicas of those of the middle octave, or their first, second, third octaves, etc.; thus the twelfth is the replica of the fifth, the seventeenth the triplicas of the third. Likewise, the intervals of the first, second, third, etc. sub-octaves are the sub-replicas, sub-triplicas, sub-quadruplicas of those of the middle octave, or their first, second, third, etc. sub-octaves.

Sauveur claimed that octaves or sub-octaves of consonances are still consonances, and octaves or sub-octaves of dissonances are still dissonances. It is not clear whether this is also a nominal definition, or whether he forgot that consonances are only nominally defined.

---

[69] p. 304.

The definition of intervals, although accepted by all contemporary musicians, has serious drawbacks, according to Sauveur.

> The intervals of which we have just spoken are accepted by those who follow the diatonic, chromatic, and enharmonic systems; and therefore we are obliged to accept their names: but they are not suitable for explaining all the properties of sounds and their intervals. For:
>
> 1. Every interval has a margin of ambiguity: a third, for example, can be major and minor, superfluous and diminished, maximum and minimum; just, altered, and tempered.
> 2. The double of an interval does not mark the double of the number it denotes; thus two thirds do not make a sixth, but a fifth, and the fifth is the sum of two different thirds.
> 3. The multiple of one element of the intervals, for example the major tone, does not form an accepted interval in the diatonic system [73].[70]

In order to know these intervals exactly, it is therefore necessary to find a common measure, at least of all the intervals that are the elements of the others; that is, the major tone, the minor tone and the major semitone. "We have therefore divided the octave into 43 equal intervals, which we call merides, and each meride into 7 equal heptamerides; with these parts we shall express not only the ordinary intervals of music, but also all the intervals that can be imagined in acoustics" [73].[71]

To present the intervals of the diatonic system and his system in a systematic and compact way, Sauveur reported them in a table of his memoirs (Planche I), part of which is reproduced here as Table 5.13; an overwhelming amount of data indeed.[72] Note that a modern reader cannot follow the arguments below unless he knows the *Traité* or other Sauveur writings.

Table 5.13 is divided into two parts; the first contains information on the diatonic system, the second contains data on the general system. The two systems, diatonic and general, are put together to show the relationship of one to the other.

In the first part of the table there are four columns. Col. 1 shows the intervals of the diatonic system in the middle octave. Consonances are written in upper case, dissonances and false consonances in lower case. The other intervals, i.e. the comma and the minor semitone, which are not part of the diatonic system, are shown in italics. On the left are either Roman or Arabic numbers associated with the intervals, whose use is clear from the table; for example, VI means major sixth; and 6, minor sixth. After these intervals are the signs ( ) and [ ], which serve as references in the parts of the general system tables to mark the just intervals.

Col. 2 of the table indicates the elements, i.e. the small intervals of which the other intervals are composed. T means major tone, t minor tone, S major semitone, s minor semitone, c comma. We can see that the fifth, 2T-t-S, is composed of 2 major tones, 1 minor tone and 1 major semitone.

To obtain the elements of the acute octaves, multiply the octave 3T-2t-2S, by 2, 3, 4, 5, 6 and add the elements of the intervals. Thus, to have the elements of the

[70] p. 307.

[71] pp. 306–307.

[72] The other part of Planche I is represented by Table 5.10 in Sect. 5.1.

Table 5.13 General system of intervals

SYSTEME GENERAL DES INTERVALLES ET DES RAPPORTS DES SONS

SYSTEME DIATONIQUE. I. Intervalles Diatoniques —	II. Elemens	III. Raport des Sons	IV. Noms des Sons	TABLES DU SYSTEME GENERAL. I TABLE. Des MERIDES avec les Eptamerides.							II. TABLE. Des EPTAMERIDES.							III. TABLE. Du RAPORT des Sons, avec le Son fondamental.						
VIII. OCTAVE ()	3T.2t.2S	1. 2	UT	43	43	43	(43)	43	43	43	298	299	300	(301)	302	303	304	19861	19907	19953	(20000)	20045	20091	20137
			si♯	42	42	42	42	42	42	42	291	292	293	294	295	296	297	19543	19588	19634	19679	19724	19770	19815
				41	41	41	41	41	41	41	284	285	286	287	288	289	290	19231	19275	19320	19364	19409	19454	19498
			ut♭	40	40	40	40	40	40	40	277	278	279	280	281	282	283	18923	18967	19011	19055	19099	19143	19187
VII. Septieme Majeure ()	3T.2t.S	8. 15	SI	39	39	39	(39)	39	39	39	270	271	272	(273)	274	275	276	18621	18664	18707	(18750)	18793	18836	18880
				38	38	38	38	38	38	38	263	264	265	266	267	268	269	18323	18365	18408	18450	18493	18535	18578
				37	37	37	37	37	37	37	256	257	258	259	260	261	262	18030	18072	18113	18155	18197	18239	18281
7 Septieme {Mineure () / Minime ()}	3T.t.2S / 2T.2t.2S	5. 9 / 9. 16	si♭	36	[36]	36	36	36	36	(36)	249	[250]	251	252	253	254	(255)	17742	[17783]	17824	17865	17906	17947	(17989)
			la♯	35	35	35	35	35	35	35	242	243	244	245	246	247	248	17458	17498	17539	17579	17620	17660	17701
				34	34	34	34	34	34	34	235	236	237	238	239	240	241	17179	17219	17258	17298	17338	17378	17418
				33	33	33	33	33	33	33	228	229	230	231	232	233	234	16904	16943	16982	17022	17061	17100	17140
VI. SIXTE Majeure ()	2T.2t.S	3. 5	LA	32	(32)	32	32	32	32	32	221	(222)	223	224	225	226	227	16634	(16672)	16711	16749	16788	16827	16866
				31	31	31	31	31	31	31	214	215	216	217	218	219	220	16368	16406	16444	16482	16520	16558	16596
				30	30	30	30	30	30	30	207	208	209	210	211	212	213	16106	16144	16181	16218	16255	16293	16331
6 SIXTE Mineure ()	2T.t.2S	5. 8	la♭	29	29	29	29	(29)	29	29	200	201	202	203	(204)	205	206	15849	15885	15922	15959	(15996)	16032	16069
			sol♯	28	28	28	28	28	28	28	193	194	195	196	197	198	199	15596	15631	15668	15704	15740	15776	15812
				27	27	27	27	27	27	27	186	187	188	189	190	191	192	15346	15382	15417	15453	15488	15524	15560
				26	26	26	26	26	26	26	179	180	181	182	183	184	185	15101	15136	15171	15205	15241	15276	15311
V. QUINTE ()	2T.t.S	2. 3	SOL	25	25	25	25	(25)	25	25	172	173	174	175	(176)	177	178	14859	14894	14928	14962	(14997)	15031	15066
				24	24	24	24	24	24	24	165	166	167	168	169	170	171	14622	14655	14689	14723	14757	14791	14825
				23	23	23	23	23	23	23	158	159	160	161	162	163	164	14388	14421	14454	14488	14521	14555	14588
5 Fausse Quinte ()	T.t.2S	45. 64	sol♭	22	22	(22)	22	22	22	22	151	152	(153)	154	155	156	157	14158	14191	(14223)	14256	14289	14322	14355
IV. Triton ()	2T.t.	32. 45	fa♯	21	21	21	21	(21)	21	21	144	145	146	147	(148)	149	150	13932	13964	13596	14028	(14060)	14093	14125
				20	20	20	20	20	20	20	137	138	139	140	141	142	143	13709	13740	13772	13804	13836	13868	13900
				19	19	19	19	19	19	19	130	131	132	133	134	135	136	13490	13521	13552	13583	13614	13646	13677
4. QUARTE ()	T.t.S	3. 4	FA	18	18	(18)	18	18	18	18	123	124	(125)	126	127	128	129	13274	13305	(13335)	13366	13397	13428	13459
			mi♯	17	17	17	17	17	17	17	116	117	118	119	120	121	122	13062	13092	13122	13152	13183	13213	13243
				16	16	16	16	16	16	16	109	110	111	112	113	114	115	12853	12882	12912	12942	12972	13002	13032
			fa♭	15	15	15	15	15	15	15	102	103	104	105	106	107	108	12647	12677	12706	12735	12764	12794	12823
III TIERCE Majeure ()	T.t.	4. 5	MI	14	14	(14)	14	14	14	14	95	96	(97)	98	99	100	101	12445	12474	(12503)	12531	12560	12589	12618
				13	13	13	13	13	13	13	88	89	90	91	92	93	94	12246	12274	12303	12331	12359	12388	12417
				12	12	12	12	12	12	12	81	82	83	84	85	86	87	12050	12078	12106	12134	12162	12190	12218
3 TIERCE Mineure ()	T.S	5. 6	mi♭	11	11	11	11	11	(11)	11	74	75	76	77	78	(79)	80	11858	11885	11912	11940	11967	(11995)	12023
			re♯	10	10	10	10	10	10	10	67	68	69	70	71	72	73	11668	11695	11722	11749	11776	11803	11830
				9	9	9	9	9	9	9	60	61	62	63	64	65	66	11482	11508	11535	11561	11588	11615	11641
				8	8	8	8	8	8	8	53	54	55	56	57	58	59	11298	11324	11350	11376	11402	11429	11455
II. Seconde Maj {Ton Maj () / Ton min ()}	T. / t.	8. 9 / 9. 10	RE	(7)	7	7	7	7	[7]	7	(46)	47	48	49	50	[51]	52	(11117)	11143	11169	11194	11220	[11246]	11272
				6	6	6	6	6	6	6	39	40	41	42	43	44	45	10940	10965	10990	11015	11041	11066	11092
				5	5	5	5	5	5	5	32	33	34	35	36	37	38	10765	10789	10814	10839	10864	10889	10914
2. Seconde min Semiton Maj	S	15. 16	re♭	4	4	4	(4)	4	4	4	25	26	27	(28)	29	30	31	10593	10617	10641	(10666)	10691	10715	10740
Semiton Mineur ()	s	24. 25	ut♯	(3)	3	3	3	3	3	3	(18)	19	20	21	22	23	24	10423	10447	10471	10495	10520	10544	10568
				2	2	2	2	2	2	2	11	12	13	14	15	16	17	10257	10280	10304	10328	10351	10375	10399
Comma ()	C	80. 81		1	(1)	1	1	1	1	1	4	(5)	6	7	8	9	10	10093	(10116)	10139	10162	10186	10210	10233
I. UNISSON fondamental		1. 1	UT	0	0	0	(0)	0	0	0				(0)	1	2	3	9931	9954	9977	(10000)	10023	10046	10069

octave of the fifth, that is, the elements of the twelfth, add 2T-t-S [the fifth] to 3T-2t-2S [the octave], which gives 5T-3t-3S; to have the elements of the twenty-fifth, or third octave of the fourth, add T-t-S to 9T- 6t-6S, which gives 10T-7t-7S. To have the elements of the sub-octaves or major octaves of the intervals, take away from the same products the elements of these intervals; thus, to have the elements of the sub-18th or third sub-octave of the fifth, subtract 2T-t-S from 9T-6t-6S, there will remain 7T-5t-5S, which are the elements of the sub-XVIII.

Col. 3 contains the ratios of the sounds of each interval to the fundamental marked *unison* at the bottom of Col. 1.[73] To find the ratios of the acute octaves of the intervals in this column, multiply the second number of their ratios by 2, 4, 8, 16, 32, 64. To find the grave octaves of the intervals, multiply the first number of their ratios by the same numbers. Thus, to obtain the second octave of the fifth, (2 : 3), multiply 3 by 4 and obtain (2 : 12), or (1 : 6); and to obtain its second sub-octave, multiply 2 by 4 and obtain (8 : 3).

Col. 4 shows what Sauveur called the ancient names, i.e. those that Guido of Arezzo gave to the sounds of the diatonic system, circa 1024, to which custom added Si, circa 1670. The principal names, which are the black keys of organs or

[73] Note that the second number of the ratios in Col. 3 is greater than the first. This would indicate a ratio $(m : n)$, with $n > m$, different from Sauveur's use in the *Traité*. However, nowhere in the paper did Sauveur explicitly state a ratio as $(m : n)$ or m/n); he always said something like "m is to n".

harpsichords, are in capital letters, and in small letters their flat and sharp, used in the chromatic system, which are the white keys of organs or harpsichords.[74] The symbol ♭ lowers the sound by a minor semitone, or 3 merides, and the symbol ♯ raises it by the same amount.

The sounds of the other octaves have the same names. But to distinguish the notes that have the same name, three clefs are used. The first, called C Sol Ut, marks the fundamental note Ut. The second, G Re Sol, marks its fifth or Sol. The third, F Ut Fa, marks the sub-fifth or Fa of the first sub-octave. From these three clefs we know the other notes that are at most 2 octaves away from the fundamental.

The second part of Table 5.13 shows in the first two sub-tables, called tables by Sauveur, the values of the division of the octave into merides and heptamerides. There are basic assumptions implicit in Sauveur.

1. First of all, he assumed his division of the octave to be known, even though there is no reference to it in the memoir. Moreover, there are inconsistencies in the exposition. Sometimes he said how to build the table, sometimes how to use it.
2. The division of the octave into 301 parts is considered sufficiently refined to not carry significant errors to represent values of any interval. In fact, the maximum error that can be made is at most $1/2$ of the heptamerides, corresponding to the interval $\sqrt[602]{2} = 1.0011$, i.e. the deviation of one cycle in 1000. In any case, Sauveur ventilated the possibility of using decamerides, which give a more precise representation.
3. The division into 43 merides of the octave is related to the $1/5$ of a comma temperament.

The merides are listed in the middle column of Table I, starting at the bottom with the fundamental or Ut, which contains 0 merides, and going up to the octave, which contains 43 merides. To the left and right of this middle column are three columns with dashes placed either at the bottom left or bottom right to indicate the heptamerides to be added or subtracted from the indicated meride.

In Table I and the following Table II, the merides and heptamerides of just intervals are enclosed between the signs ( ) or [ ]. To obtain the intervals of the diatonic system tempered by the division of the octave into 43 parts, it is necessary to take only the merides, neglecting the heptamerides; thus, for example, the tempered fifth is 25 merides.

To have the merides of the intervals of the acute octaves, add to 43, 86, 129, 172, 215 the merides of the proposed interval; thus the 19th or second octave of the fifth measures $86 + 25 = 111$ merides. To have the sub-octaves, from the numbers of the octave, remove the merides of the proposed interval; thus the second sub-octave of the fifth measures $86 - 25 = 61$ merides. To avoid the embarrassment of large numbers, the higher octaves are indicated by 1+, 2+, 3+, etc., and the lower by 1−, 2−, 3−, etc., so $2 + 25'$ will mark the acute double octave of the fifth, and $2 - 25'$ will mark the double sub-octave of the same fifth.

[74] Remember that the French keyboards had color reversed with respect to Italian keyboards.

To find the merides and heptamerides that each interval in the just diatonic system contains, one must take the elements from Col. 2, where the major tone T contains $7''$, that is, 7 merides plus 2 heptamerides; the minor tone t contains $7_{'''}$, or 7 merides minus 3 heptamerides, and the major semitone S contains 4 merides. For example, to have the merides of the fifth, whose elements are 2T-t-S, take their values $7''$, $7''$, $7_{'''}$, 4, whose sum is $25'$.

Table II is that of the heptamerides. The seven columns of this table correspond to the seven columns of the previous Table I.

Table III shows how many vibrations a sound makes, while the fundamental, i.e. Ut, makes 10,000 vibrations. To obtain the vibrations of the upper octaves we must multiply the values of Table III by 2. 4. 8. 16. 32. 64, while for the sub octaves, it is necessary to divide them by the same numbers.

After having presented Table 5.13, Sauveur explained how to construct it. Of course, the coefficients of Tables I and II do not require any calculation, since they form an arithmetic sequence with a unitary common difference. The case is different for the coefficients of Table III and for the assignment of the heptamerides to the ratios of the diatonic intervals as given in Col. 3 of the first part of Table 5.13. For this purpose, he proposed the following algorithm, without giving any explanation:

1. For the elements assume T of 51 heptamerides, t of 46 and S of 28.
2. For the ratios of the sounds of Col. 3 of the diatonic system, take a table of logarithms of Vlacq with 7 digits and look for the logarithms of the two numbers defining the ratios of the sounds, subtracting the greater logarithm from the lesser by removing the last 4 digits, the first 3 digits will mark the heptamerides of this interval. For example, to have the heptamerides of the fifth, whose ratio is 2:3, take their logarithms [base 10] 3010300 and 4771213, subtract from each other, the remainder will be 1760913. Removing 0913 leaves 176 heptamerides for the fifth (marked with round brackets). Note that if the first digit removed was 5 or greater, it would be necessary to add 1 to the heptamerides.
3. To find the heptamerides without the tables of logarithms:
    a. Take the sum of the numbers of Col. 3 of the diatonic system, which mark the relationship of the sounds.
    b. Take their difference (the difference must be less than 1/6 of the sum, otherwise the rule of point 4 applies).
    c. Multiply 875 by the difference (if the difference is 1, no multiplication is necessary).
    d. Divide 875, or the product of the previous multiplication, by the sum, the quotient gives the heptamerides. Example: the fourth has the ratio (3 : 4).
        i. The sum is 7.
        ii. The difference is 1.
        iii. The product of 875 and 1 is still 875.
        iv. 875 divided by 7 gives 125 heptamerides for the 4th.

4. If the difference is greater than 1/6 of the sum, divide the given interval into two or three or other intervals; look for the heptamerides of each interval, their sum will give the heptamerides of the proposed interval. For example, to obtain the heptamerides of

> a major sixth whose ratio of tones is (3 : 5), divide this ratio in (3 : 4) and (4 : 5). As above, look for the heptamerides of the ratio (3 : 4), which are 125, and those of (4 : 5), which are 97; the sum 222 will be the number of heptamerides of the major sixth.
>
> Alternatively, multiply the smallest number of the ratio for 2, 4, 8, 16, 32 and so on, so that the difference of this product from the largest is less than 1/6 of their sum, and look for heptamerides by the first rule, which you must subtract from 301. 602. 903. 1204, etc. The sum or the remainder gives the heptamerides of the interval of these two sounds. Example: let the ratio of the tones be (2 : 15). Multiplying the smaller 2 by 8 gives 16. Search by the first rule for the heptamerides of the ratio (15 : 16). This gives 28 heptamerides, which subtracted from 903 gives 875 heptamerides for the ratio (2 : 15) [73].[75]

In the following, we will justify Sauveur's move, a justification that he thought was hardly relevant; however, this is not our feeling, and we believe that a reconstruction of the algorithm presented in the above quotation is very useful because it shows the great ingenuity of Sauveur.

The part concerning the elements is not very easy to decipher, in the sense that it is difficult to understand why there should be a different approach to finding the merides and heptamerides for them and for the other intervals, such as the fifth, for example.

As for the intervals of Col. 3 (and of any intervals), two methods are proposed to find the heptamerides of an interval (and of course the merides), one that requires the use of a logarithmic table, the other that does not. The basic relation is the definition of heptamerides of an interval i, as given in the *Traité* (see Sect. 5.2.3), that is rewritten below:

$$E = 1000 \log i \approx \frac{301}{\log 2} \times \log i = 301 \log_2 i$$

As we can see, the heptamerides have a very close relationship with cents c, the unit of measure currently used for musical intervals, defined as $c = 1200 \log_2 i$, which are more interesting than the heptamerides for the equal temperament only, in which there are 12 equal semitones of 100 cents each.

Note that in the previous equations the sign $\approx$ is used instead of $=$, because a heptameride is not exactly 1/301 of the octave, since $\log 2 = 0.30103$ has been approximated by 0.30100. Thus 301 heptamerides gives the interval (1 : 1.99987) instead of (1 : 2).

The second method does not use tables of logarithms, but uses an approximate method to calculate them, based on the bimodulus, according to which:

> When the difference of two numbers is small, the difference of their logarithms is approximately equal to the bimodulus multiplied by the difference and divided by the sum of the numbers themselves [31].[76]

For a musical interval $i = m/n \geq 1$, the previous proposition implies the relation:

---

[75] pp. 311–312.

[76] p. 382.

$$E = 1000 \log i \approx 2000M\frac{m-n}{m+n}$$

where M is the modulus of the logarithm (and consequently $2M$ is the bimodulus), base 10, given by $M = 1/\ln 10 \approx 0.4343$. Thus the heptamerides E of an interval $i = m/n$ are given by:

$$E \approx 869\frac{m-n}{m+n}$$

which agrees with Sauveur's algorithm, except that he used 875 instead of 869. The reasons for this variation are empirical, since 869 is only approximate and 875 gives more precise values.[77]

To find the heptamerides of the higher and lower octaves of the intervals marked in the table, one must proceed as for the merides, adding or subtracting the heptamerides to/from the numbers 301. 602. 903. etc. To find the decamerides D of an interval, proceed as for the heptamerides above, but subtract only the last 3 digits. The error, using decamerides (1/2D), can be at most 1 vibration out of 10,000 vibrations.[78]

The elements of Table III, are the oscillations corresponding to the various intervals, taking as reference Ut at 10,000 Hz. They correspond to the antilogarithms of the columns of the previous Table II; more precisely, the antilogarithm of a heptameride divided by 1000 gives the numerical value, contained in $[1, 2]$, of the interval corresponding to it. This value multiplied by 10,000 gives the number of oscillations of Table III.

Antilogarithms can be found in three ways:

1. With Vlacq's tables of logarithms with 7 digits.
2. Using the bimodulus method in reverse. If the heptamerides do not exceed 150, or half an octave:

    a. The proposed heptamerides are subtracted from 875 to obtain the subtraction.
    b. The heptamerides are added to 875 to get the sum.
    c. 0000[79] is added to the sum, which, divided by the difference, gives the value of the proposed oscilations [73].[80]

    Sauveur's procedure can be justified by applying the bimodulus method in reverse. Starting with the relation giving E as a function of i and solving with respect to i, one has

$$i \approx \frac{875 + E}{875 - E}$$

[77] The bimodulus method can be proved quite easily by following [31]. If we rewrite the ratio $i = m/n$ as $1 + p$, we have $\ln i = \ln(1 + p) = p - p^2 + p^3 + \text{etc.}$ Assuming $q = p/(2 + p)$, we get $\ln i = 2\left(q + 1/3q^3 + \text{etc.}\right)$. Approximating to the first order, we get $\ln i \approx 2q = 2\,\frac{p}{2+p} = 2\,\frac{m-n}{m+n}$, which, if we switch to base 10 logarithms, gives the bimodulus rule.

[78] Sauveur suggested the value 8686 instead of 10,000, but he was probably wrong [73], p. 313.

[79] Sauveur erroneously said 000.

[80] pp. 313–314.

which justifies Sauveur's algorithm.
If the heptamerides are greater than 150 and less than 301, subtract these heptamerides from 301 to get the complement of the interval, then proceed as above. Dividing 2 by the ratio found gives the ratio corresponding to the proposed heptameride.

3. To reduce an interval of Col. 3 of the first part of Table 5.13, to that of Table III, the ratio of this interval n/m is changed in the ratio with respect to the fundamental sound, and then rounded to a value of Table III. For example, the ratio of the fifth (2 : 3) corresponds to the ratio 10,000:15,000, which is not in the tables. One then takes the value 14,997, which is very close to it; for the difference is only 3 in 15,000, or 1 in 5000 vibrations, or 1/12 of the heptameride, which is absolutely negligible. Reading the value corresponding to 14,997 in Table II gives 176 heptamerides for the fifth.

Division and use of the echometer in general.
In this section Sauveur took a graphical-analytical method for the practical execution of simple calculations, with the use of what he called *scales*. He was not new to this approach as he had already published a small treatise on the use of scales for solving simple problems of navigation [69]. An example is a scale for finding the day of the week by knowing the day of the month and the year. This strategy is due to the fact that Sauveur considered as possible readers also professional musicians with limited mathematical skills.

The instrument responsible for avoiding the 'boring' calculations that occur in music, is called *echometer* by Sauveur. It is a ruler with several divided lines that serve as scales for measuring the duration of sounds and finding their intervals. From the drawing Sauveur reported, it appears as a ruler at least five feet long and more than a foot wide, or about 150 × 40 cm. "Mr. Chapotot[81] one of the most skilled engineers of mathematical instruments in Paris, has echometers divided in all the ways indicated above. He made one in copper for his royal highness the duke of Orleans" [73].[82]

The first scale concerns Loulié's chronometer, already commented on in the *Traité*, which Sauveur described briefly, saying only that it is basically a scale divided into astronomical inches, from 1 to 64. It must be said that Sauveur somewhat downplayed Loulié's role; in fact, the latter had not only divided a ruler into 12 equal parts, but had also devised an interesting system for measuring the periodicity of a phenomenon.

The chronometer in consideration, shown in Figs. 5.21 (first scale AB) and 5.22, consists of a ruler graduated in astronomical inches from which projects a small shelf or cantilever AD supporting a pendulum formed by a wire and a mass point E. The length DE of this pendulum is adjusted by a pin F which fixes one end of the wire of the pendulum at a certain position on the graduated ruler. A musician, while playing a piece X with the duration of a bar that instinctively seems most appropriate, sets the pendulum in motion and moves the pin until the oscillations of the pendulum

[81] Louis Chapotot, French scientific instrument maker, active in Paris in the period 1670–1700.
[82] p. 322.

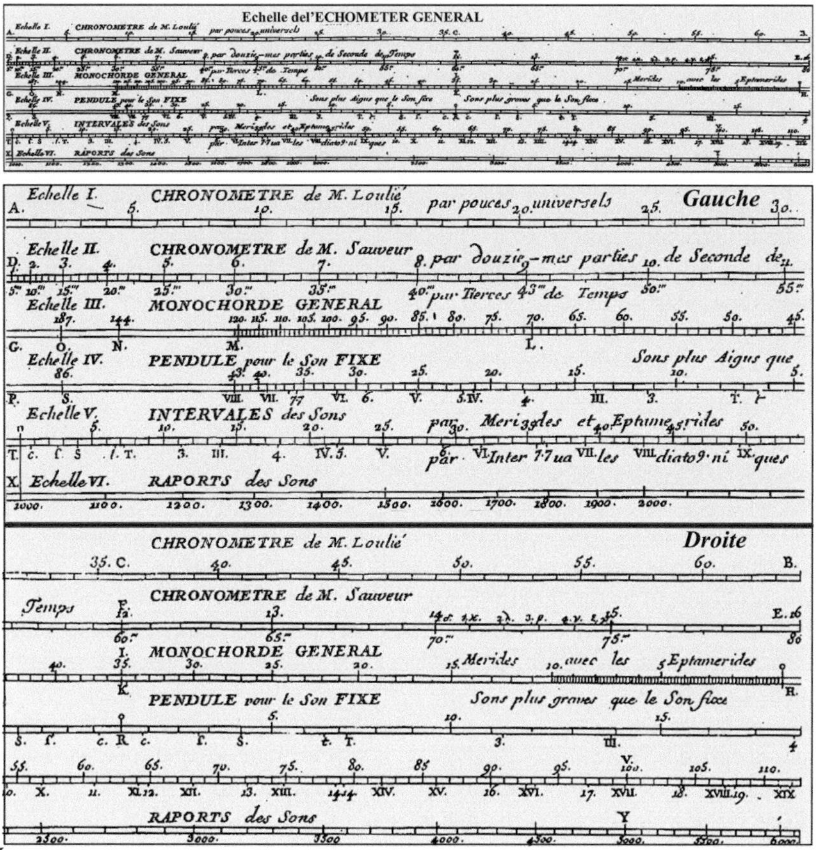

Fig. 5.21 Different views of the echometer: above an overview; below two enlarged halves. Drawn from [73], Planche 2

overlap at the beginning and end of the bar. At this point, the musician can read the length of the pendulum indicated by pin F and record it in the staff of composition X so that he or other musicians can perform the piece with the same duration of bars [53].[83]

The second scale DE of Fig. 5.21, is Sauveur's chronometer. In contrast to Loulié, Sauveur proposed to indicate in the scale the half-period of the oscillation.[84] Thus, if Loulié's chronometer shows astronomical inches, Sauveur's shows seconds; more precisely, twelfths of a second.

Here is Sauveur's comment on the use of the two scales.

---

[83] pp. 81–88.

[84] In fact, Sauveur wrote *period* because, as will be commented on in more detail below, in his time the period of a pendulum was counted as the time it took to pass from one position of rest to the next and not for a complete go and back.

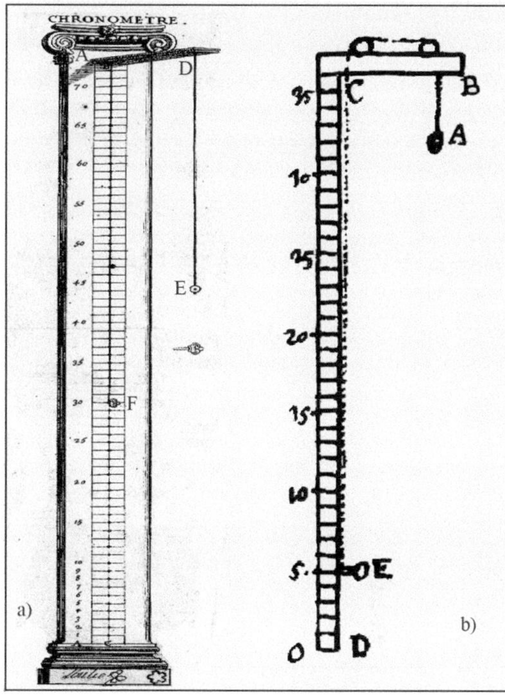

Fig. 5.22 Loulié's chronometer: **a** original drawing, from [53], pp. 82–83; **b** Sauveur's draft, from [70], p. 159

> In order to regulate the duration of a note or a measure in music, one uses the movement of the hand; but in order to mark exactly the time we take to make this movement, one must have a simple pendulum which one lengthens or shortens until each of its oscillations is isochronous, i.e., of the same duration, as the movement of the hand; then one measures the length of this pendulum from the point of suspension to the center of the sphere. If you use scale I [see Fig. 5.21], you will have the length of this pendulum in astronomical inches from Monsieur Loulié's chronometer. If you use scale II, the length is in twelfths of a second on one side and in *tierces* of time on the other [73].[85]

After Sauveur, there were a number of works on the measure of time for musicians, that did not, however, introduce any major innovations, but simply helped to maintain the original idea; among them for example *Description & usage d'un métrometre, ou machine pour battre les mesures & les temps de toutes sortes d'airs* of 1732 by Louis Léon Pajot [57], which introduced the term *metrometre*. The decisive breakthrough to the modern mechanical metronome was made only in the 19th century by the clockmaker Dietrich Nikolaus Winkler (1777–1826), who is the true inventor of the metronome, although it is generally associated with the name of Johann Nepomuk Mälzel (1772–1838), who patented it in 1816 and contributed to its widespread use.

The modern mechanical metronome is inspired by Sauveur's chronometer, in the sense that it retains the idea of a pendulum and a graduated scale. However, the graduation does not indicate the period in fractions of a second, but the frequency in

[85] pp. 319–320. The tierce is 1/5 of a twelfth, i.e. 1/60 of a second.

beats per minute (bmp). Then the pendulum is not simple but compound, consisting of two masses, a larger one placed at a fixed distance under the pivot, usually hidden from view, and the other smaller upper one sliding along a rod, suggesting the deceptive image of an inverted pendulum. By moving the upper mass, the frequency of the pendulum can be adjusted with much more compact dimensions than in the case of Sauveur's simple pendulum: the frequency decreases when the mass is moved upwards, and increases when it is moved downwards. A spring keeps the assembly in motion by compensating for the mechanical energy losses due to friction; in addition, a sound is heard on every beat, which is a great help to the musician.

To describe his chronometer, Sauveur, instead of developing simple mathematics based on the theory of the pendulum, proposed an algorithm to be executed blindly. It is as if he were giving instructions to a carpenter.

1. Take the first scale, AB (see scale I of Fig. 5.21), and mark divisions at the numerical values $1, 4, 9, 16, 25, 36, 49, 64$, from D of the ruler to E (of scale II). On these divisions, write the even numbers, 0, 2, 4, 6, 8, 10, 12, 14, 16, etc.
2. Divide the first inch of the line DE into quarters and mark the intervals $1/4$, $1 + 1/4$, $2 + 1/4$, $3 + 1/4$, $4 + 1/4$, $5 + 1/4$, $6 + 1/4$, $7 + 1/4$, etc. after the divisions of the even numbers starting with D and write on these new divisions the odd numbers 1, 3, 5, 7, 9, 11, 13, 15, etc. This gives the first chronometer divided into twelfths of a second.
3. In each of these preceding parts, take to the right $\epsilon\gamma$ of $1/4$ inch, and divide the remainder $\delta\epsilon$ into 5 equal parts at points 1, 2, 3, 4.[86] Then divide the $1/4$ inch portion into 25 small equal parts; take 1 small part and carry it to 1 and mark it χ; take 4 small parts and carry it to 2 and mark it λ; take 9 small parts and carry it to 3 and mark it μ; finally take 16 small parts and carry them to 4 and mark it ν. These four points χ, λ, μ, ν, divide $\delta\gamma$ into 5 parts. Doing the same for all the intervals, we have the divisions of our second chronometer into *tierces* of time [73].[87]

Sauveur explained the algorithm for dividing his chronometer in a later section as follows. The relation between the (half) period of the pendulum, expressed in seconds, and the length, expressed in astronomical inches, is given by $36T^2 = l$; or equivalently, $T = 1/6\sqrt{l}$. When T is measured in twelfths of a second, the relationship becomes $T = 2\sqrt{l}$, or equivalently $l = T^2/4$. Applying this law to the sequence of distances $1, 4, 9, 16, 25, 36, 49, 64$, we obtain the sequence of periods $2, 4, 6, 8, 10, 12, 14, 16$, etc. To the sequences of distances $1/4$, $1 + (1 + 1/4)$, $4 + (2 + 1/4)$, $9 + (3 + 1/4)$, $16 + (4 + 1/4)$, etc. $= 1/4, 9/4, 25/4, 49/4, 81/4$, correspond to periods $1, 3, 5, 7, 9$, etc.

Sauveur also considered a finer subdivision, in *tierces*, i.e. $1/60$ of a second. To explain the way to continue this subdivision, 5 parts for each interval of twelfths, he focused on the interval between 14 and 15 twelfths, as shown in Fig. 5.23. The distance between two successive divisions of time is given by $1/4[(T+1)^2 - T^2] = (T/2 + 1/2)^2 - T^2/4$. Denoting by a^2 the distance $T^2/4$ corresponding to the period

[86] To understand Sauveur's writing, we must refer to the interval of the second scale between 14 and 15 twelfths of a second.

[87] pp. 317–318.

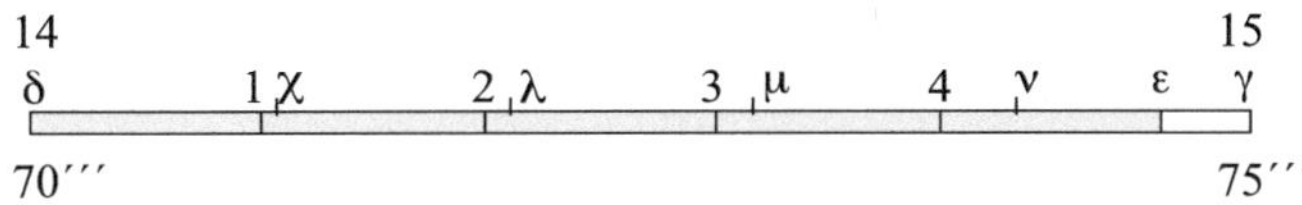

Fig. 5.23 Division of an interval into tiers

T of the initial value of the interval in twelfths, the distance in astronomical inches between the two successive divisions is given by $(a + 1/2)^2 - a^2 = a + 1/4$. Sauveur said that the part corresponding to 1/4 inch should be divided into 25 parts, while the aliquot part a should be divided into 5 parts. The points χ, λ, μ, ν are given respectively by the relations $a^2 + a/5 + 1/100$, $a^2 + 2a/5 + 4/100$, $a^2 + 3a/5 + 9/100$, $a^2 + 3a/5 + 9/100$, which correspond to the sequence $(a + n/10)^2$ for $n = 1, 2, 3, 4$ and thus to distances obeying the law of the pendulum.

The third scale in Fig. 5.21, GH, is the common monochord, which has a span of about 63 astronomical inches. It is a logarithmic scale adapted to merides and heptamerides. This type of scale had become sufficiently widespread in France after its first introduction by Edmund Gunter (1581–1626) in the 1620s, shortly after the 'invention' of logarithms. Sauveur must have been familiar with it, as well as with the slide ruler, since it is reported that "about 1700 Sauveur had slide rulers made by the craftsmen Gevin and Le Bas" [17][88]

Sauveur's instructions for its construction are unclear and do not seem to us to correspond to the illustration. Basically, the scale is divided into octaves, and each octave is divided into 43 merides by taking the distances between them in astronomical inches with the help of Table 5.13. The division can be pushed up to heptamerides.

The fourth scale relates to the pendulum used to find the fixed sound. To obtain it, take PR equal to GK of the third scale, but mark only one division of two and write 5, 10, 15 where in the previous scale there was 25, 15, 5. Then, on the left side, carry KO to RS, marking every second division and writing 5, 10, 15, 20, 25, etc., where in the previous scale there were 45, 55, 65, 75, 85, etc. Thus we have the pendulum divided by merides and heptamerides [73].[89]

Sauveur's instructions, though very precise, are rather cryptic. The scale is used to find the interval, expressed in merides M_0, between the frequency of the pendulum beating one second, of length $L_0 = 36$ astronomical inches, and the generic pendulum of length l. This value is given by:

$$M_0 = \frac{1000}{7} \log \sqrt{\frac{L_0}{l}} = \frac{500}{7} \log \frac{L_0}{l}$$

The merides M of the monochord of the third scale of total length L, which correspond to a fraction l are instead given, again apart from the same constant, by the relation:

[88] p. 26.

[89] p. 319.

$$M = \frac{1000}{7} \log \frac{L}{l}$$

With simple calculations, assuming $L \approx 63$ astronomical inches we get the relation:

$$M_0 = \frac{1}{2} (M - 35)$$

which justifies Sauveur's algorithm.

The fifth scale, TV, marks the intervals of the sounds in merides and heptamerides, corresponding to the intervals of the diatonic system. Its use is an alternative to the use of Table 5.13, and possibly easier. For its construction, TV is divided equally into 100 merides, and each meride into 7 heptamerides, which can be divided into 10 decamerides. Mark the values 1, 2, 3, 4, etc. above to indicate the division into merides; and mark the diatonic intervals s. S. T. 3. III. 4. IV. 5. V. 6. VI. 7. VII. VIII.

The sixth scale, XY, refers to the number of vibrations per second corresponding to the merides and heptamerides of the intervals of the fifth scale. Note that the vibration corresponding to 0 heptamerides is 1000, instead of 10000 as in Table 5.13; the vibration corresponding to 43 heptamerides is 2000.

Application of the General System and Echometer to Every Musical Systems
In this section Sauveur used his general system, either as given by Table 5.13 or by the echometer, to describe in detail any musical system. Here his division of the octave into small parts is used only with a classificatory purpose, different from what happened in the *Traité*, where this division was presented as a way of tempering the just system. Later, he was also tempted to use his division to build a new system of his own. A clear example of this are the comments on a keyboard with 31 keys, as reported in Sect. 5.8.4.

The section begins by stating that in each musical system the sounds are expressed either by the ratios of their vibrations or by the lengths of the string of a monochord, or finally by the ratios of the intervals of one sound to the others. It is necessary to know about all the systems:

1. Whether the intervals, according to the proposed system, form just consonances or dissonances, and if they do not form just consonances, to know their differences.
2. To find the ratio of each sound in this system to the fundamental.
3. Mark the names of the notes of both the ordinary diatonic system and the general system that correspond to the sounds of the proposed system.
4. Search for the reciprocal relations or intervals of a sound of this system with all other sounds of this system [73].[90]

I. The first system to be examined is the just diatonic system, expressed by the numbers in column A of Table 5.14. As usual, Sauveur provided rules to be followed without comment:

---

[90] p. 323.

Table 5.14 Analysis of a just diatonic system; from [73], p. 324

A	B	C	D	E	F	G	H	K	L	M	N
48	20,000	20,000	PA	Ut	VIII	43	301	28	S	43	PA
45	18,750	18,750	DO	Si	VII	39	273	51	T	39	DO
40	16,667	16,672	LOl	La	VI	$32_{''}$	222	46	t	32	LO
36	15,000	14,997	BOr	Sol	V	$25'$	176	51	T	25	BO
32	13,333	13,335	SOn	Fa	4	$18_{'}$	125	28	S	18	SO
30	12,500	12,503	GAn	Mi	III	$14_{'}$	97	46	t	14	GA
27	11,250	11,246	RAs	Re	II	$7_{'}$	51	51	T	7	RA
24	10,000	10,000	PA	Ut	I	0	0	0	0	0	PA

1. Add 0000 to each of these numbers and divide them by the smallest number, 24. The quotients will give the numbers in column B. That is, column B contains a set of numbers in the same ratio as column A, but normalized so that it starts at 10,000. For example, if you add 0000 (that is, multiply by 10,000) to 27, you will have 270,000, which you must divide by 24 to obtain a quotient of 11,250, which you must enter in column B, and so on with the others, all of which are in the same ratio as those in column A.
2. Look at Table III of the general system of Table 5.10 for the numbers of vibrations in column B, or those closest to them to have the numbers in column C. For example, take 11,246 instead of 11,250. Note that the differences between the numbers C and the numbers B are insensitive to sound; for the largest difference is 4 in 11,250, or 1 in 2812 vibrations, which is about 1/6 of a heptameride.
3. Look at Tables 5.10 and 5.13 to find the names D of our system, the old names E and the intervals F which correspond to the numbers C.
4. Similarly we find the merides G and the heptamerides H.

If we carefully examine all the properties of this diatonic system, we see:

1. That the elements or intervals from one sound to the next are T, t, S, marked in column L; this is seen by taking the difference of two successive numbers in column H to form K.
2. That the intervals of each tone with respect to the fundamental tone PA or Ut are just consonances or dissonances, because they are included in the markings ( ) [ ] of the general tables.
3. The just system can be reduced to a tempered one by removing in column M the heptamerides that mark the differences from the tempered intervals to the just intervals [73].[91]

II. The second system to be analyzed is that defined by the different string lengths of a monochord, for example a tempered system of the harpsichord, expressed by the numbers in column A of Table 5.15. Find the numbers B and C and the intervals

[91] pp. 323–324.

Table 5.15 The tempered system of a harpsichord; from [73], p. 325

A	B	C	G	H	g	h
810	10,000	10,000	0	0	43	301
863	10,657	10,666	4	28	39	273
904	11,160	11,169	7	48	36	253
967	11,938	11,940	11	77	32	224
1016	12,543	12,531	14	98	29	203
1083	13,370	13,366	18	126	25	175
1154	14,247	14,191	22	154	21	147
1212	14,963	14,962	25	175	18	126
1289	15,913	15,922	29	202	14	99
1353	16,704	16,711	32	223	11	78
1448	17,877	17,865	36	252	7	49
1542	19,037	19,055	40	280	3	21
1620	20,000	20,000	43	301	0	0

Table 5.16 A tempered chromatic system

A	B	C	D
VIII	55	301	43
VII	50	274	39
7	46	252	36
VI	41	224	32
6	37	203	29
V	32	175	25
IV ou 5	28	153	22
4	23	126	18
III	18	99	14
3	14	77	11
II	9	49	7
2	5	27	4
I	0	0	0

in heptamerides H or merides G, as before. Take their complements to the octave by removing the numbers H from 301 heptamerides,[92] giving the numbers h, or by removing the numbers G from 43 merides, giving the numbers g.

III. The third system to be considered is that of Table 5.16, defined by the ratios of the intervals of a tempered chromatic system in commas, which assumes that

[92] This is made to change the order from lengths to frequencies.

TETRACHORDES

		I		II	III		IV
	Ton ajoiuté	des graves		des moyennes	des separées		des aiguës
Merides	$-7''$	0	$18_{\prime}$	$36_{\prime\prime}$	43	$61_{\prime}$	$79_{\prime\prime}$
ou	$1-32_{\prime\prime}$	$1-39$	$14_{\prime}$	$32_{\prime\prime}$	39	$1+14_{\prime}$	$1+32_{\prime\prime}$
Noms nouveau	*sub*LO*l*	*sub*DO	GA*n*	LO*l*	DO	*sem*GA*n*	*sem*LO*l*
Noms anciens	La	Si	Mi	La	Si	Mi	La

Fig. 5.24 Tetrachords. Greater perfect system; from [73], p. 327

the octave is divided into 55 parts or commas, of which the major semitone has 5 commas, the minor semitone 4, and the tone 9.

1. Mark in column A of this table the consonances and dissonances of the diatonic system as in Table 5.13.
2. Look in Col. 2 of the first part of Table 5.13 and take 5 commas for S and 9 commas for T and t—we are considering a tempered interval where the major and minor tones are equal—and we will have in column B the number of a commas that make up each interval.
3. Dividing the 301 heptamerides that make up the octave into 55 equal parts, the quotient gives $5^{26}/_{55}$. Multiplying the numbers in B by this value gives the numbers C, which indicate the values of the intervals in the heptamerides, which are then reduced to the merides by dividing by 7 and ignoring the decimals, in column D.

By a similar method we shall apply our general system to the system of mean semitones, which assumes that the octave is divided into 12 equal semitones, of which 2 form the tone; to the system which divides the octave into 19 equal parts, of which 2 form the major semitone and 3 the tone; to the system which divides the 'octave into 31 parts, of which 3 form the major semitone and 5 the tone.

Sauveur analyzed the system of the Greeks, as in the *Traité*, limiting it to the greater perfect system; in it, a fourth contains 18 merides. The four tetrachords of the Greeks were represented by the intervals given in Fig. 5.24.

The first line marks the order of the tetrachords, the second their names, the third the intervals in merides and heptamerides of the fixed notes, beginning with the adjoint, the proslambanomenos, which is one major tone, below the first tetrachord. In the fourth line, the merides are scaled so that they have the values of the usual system, for which, for example, Ut of the first tetrachord has 0 merides and the middle La 32 merides. This is obtained by adding 39 and subtracting 43 (the octave) if necessary. The fifth line contains the names of the new system, which correspond to the previous merides, in which the names without clef mark the sounds of the middle octave, and the names with the clefs sem and sub mark those of the first octave and the first sub-octave. Finally, the sixth line contains the old names.

Table 5.17 Tetrachords genera; from [73], p. 328

	Division de chaque Tetrachorde			
Système Diatonique	0	4	$11^{\prime\prime}$	$18_{\prime}$
Système Chromatique	0	4	$7_{\prime\prime\prime}$	$18_{\prime}$
Système Enharmonique	0	2	4	$18_{\prime}$
Système Composé	0	2 4	$7_{\prime\prime\prime}$, $11^{\prime\prime}$	$18_{\prime}$

Each tetrachord is divided into three main genera, which form the diatonic, chromatic and enharmonic systems, differing in the arrangement of the movable notes. Each of these genera has been divided differently by different authors. Sauveur's division is shown in Table 5.17. The diatonic genus presented by Sauveur is the just diatonic, with the sequence STt; the chromatic genus has the sequence Ssα, where s is the minor semitone or sharp ($3_{\prime\prime}$) and α is the minor third ($11^{\prime\prime}$); the enharmonic genus has the sequence 2c2c β, where c is the comma (1 meride) and β is a major third ($14_{\prime}$).

Sauveur has drawn the internal division of the three genera from Mersenne, which is quoted below.

> The diatonic system divides its fourths into a semitone and two tones; the chromatic system into two semitones and a three-emitone or minor third, and the enharmonic system into two diesis and a ditone or major third [56].[93]

Figure 5.25 shows the Greek genera according to Mersenne, coinciding with those of Fig. 5.17. In the figure, the symbol 𝕏 is a sharp to raise the note by a minor semitone, and $\times$ means that the notes should be raised by 2 merides.

Sauveur tested the flexibility of his method for non-Western music also. For example, by considering, albeit very briefly, what he called the system of the Orientals, with reference to the Turks and Persians [61]. He obtained information from the Arabic text *Edouar*,[94] translated into French by Fraconçois Petis de la Croix (1653–1713), who had been his colleague since 1692, as professor of Arabic at the Royal college.

The system of the Orientals, given in Table 5.18 consists of 17 notes, chosen from a Pythagorean tuning of 19 notes (obtained by 18 cycles of just fifths); from this sequence, the first and last notes are discarded, leaving a sequence of 17 notes. The resulting scale contains major tones called *Tanini* (9/8) of 51 heptamerides, the limma (246/243) of 23 merides called *baya*, and the Pythagorean comma of 5 heptamerides

[93] Part I, Troisieme livre des genres de la musique, p. 172.

[94] *Ad douar*, which in Arabic means an aggregate represented by a circle.

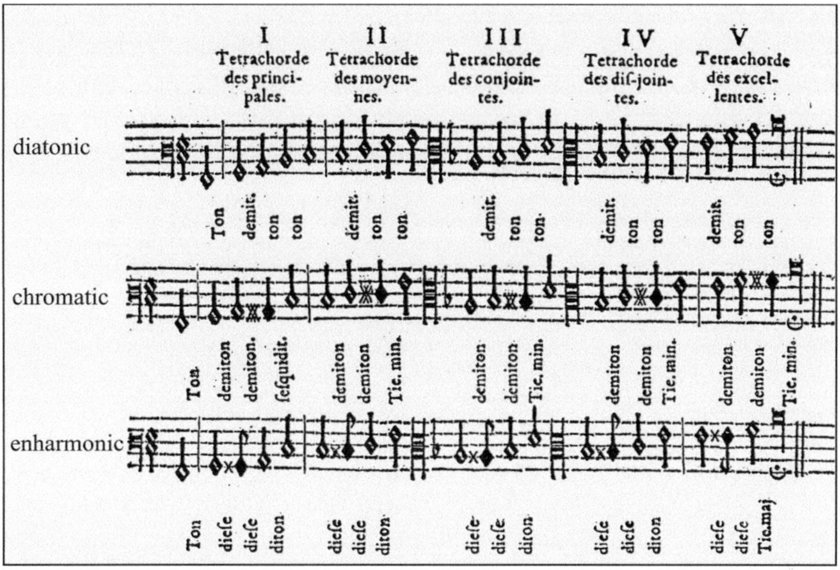

Fig. 5.25 Various systems of tetrachords. Redrawn from [56], p. 174

[60],[95] called *comma*. The scale begins with the fundamental S, contains the fifth T and the octave R.

In order to have the tempered system of the Orientals, Sauveur decided to divide the octave into 17 equal parts, as many as there are intervals. Each part is given in heptamerides measuring $17^{12}/_{17}$. The last column of Table 5.18 reports the tempered intervals from which it appears that the tanini is of 53 merides. Note that since each of the 17 parts of the octave is represented by a non-integer number, the baqya varies from 17 to 18 heptamerides.

Once these intervals have been found in heptamerides, their names and their notes according to the diatonic system are obtained from the general table. The following quotation briefly describes the characteristics of the system of the Orientals as seen by Sauveur:

> Although the Orientals have the comma and the baqya as elements of their system, they recognize in their intonation only the baqya, the minor and the major tone, which they call *tanini*, and they mark these 3 intervals with b, c, d, with which they divide the two fourths and the tone, which form the octave; for in the manner of the Greeks they relate their intervals to the fourth, and divide the fifth into the fourth and the major tone. According to the combinations of b, c, d in the fourth, they form several classes, of which there are 7 consonant fourths, and 9 fifths, also consonant; while the other classes of fourths and fifths are dissonant. The combination of the consonant classes of fourths with the consonant classes of fifths forms 37 major simple modes, each of which has its own mihast, or major chord. Finally, the Orientals have their own way of beating time on two nacarats, or small timpani,

95 pp. 33–34.

Table 5.18 The systems of orientals. Adapted from [73], p. 329

Arabs	Arabs	Intervals	Eptam.	Elemens	Eptam.
Zylcoul	yb	R	301	b	301
octave	yg	t	278	c	283
	yf	τ	273	b	266
	ye	r	250	b	248
	yd	t	227	b	230
	yc	s	204	c	212
	yb	r	199	b	195
Zylcoms	ya	T	176	b	176
fifth	y	s	153	c	159
	t	r	148	b	142
Zylarba	h	τ	125	b	125
fourth	g	s	102	c	106
	f	r	97	b	89
	e	τ	74	b	71
Tanini	d	s	51	comma	53
min. tone	c	r	46	b	35
Baqya	b	τ	23	baqya	18
	A	S	0		0

or on the dayré, which is similar to the Basque drum, with 3 beats of different measures, variously combined; but their music requires a special treatise [73].[96]

The quotation is not very clear, because in Table 5.18 there are only two kinds of elementary intervals, the baqya (b) and the comma (c), while in the quoted passage 3 kinds of intervals are named (b, c, d). For reasons of space we will not examine the reasons for this inconsistency, and even less will we go into the system of the Orientals; for this we refer to [61].

Application of the General Echometer and System to Voices and Musical Instruments The section begins by stating that in order to regulate the sounds of voices and all musical instruments, they should be compared to the sounds of an organ or a harpsichord, which are regulated in such a way that the fundamental sound Ut, which corresponds to the clef C Sol Ut, is found to be in the middle of all possible voices, which is nevertheless subject to change, because composers of music regulate it on the basis of the range of voices they intend to use for the performance of their pieces. When these voices change, they prefer to raise or lower the pitches of the instruments rather than change the notes of these pieces by transposing them.

In the first line of Fig. 5.26, the octaves are marked in whose compass the sounds of all the musical instruments are contained, that is, from the third sub-octave to the

[96] p. 329.

third octave. In the second line, the keys of the harpsichord are represented, the black ones marking the notes of the diatonic system and the white ones their sharps or flats; all together they form the chromatic system of the musicians.

In the third line we have placed under each black key the ordinary names which musicians give to them; thus the key of the fundamental sound, which is the first of the middle octave, is called C Sol Ut. Makers, mark the keys or strings by the letters C, D, E, etc., using B or B square for the last key of each octave, and *b* or *b* mol for the preceding white key.

To distinguish the octaves, the letters of the notes of the first sub-octave are capitalized, those of the second sub-octave are double-capitalized. The middle octave is lower case, the second octave is double lower case, and so on. The first names Sol, La, Si, etc. of each octave serve when *b* mol is used, and the second names Ut, Re, Mi, etc. serve when B square is used. Thus the first key of each octave, C Sol Ut, is called C by the makers, Sol by the musicians when they sing or play in *b* mol, and Ut when they sing or play in B square.

In the fourth line we have added a staff with all the notes that mark the black and white keys of the harpsichord, or the chromatic system of the musician; there are also added notes for part of the third sub-octave and the second octave. In the fifth row are the three common clefs of music: F Ut Fa, C Sol Ut and G Re Sol.

The sixth line contains in detail the ancient names of the chromatic system of the musicians, i.e. the names of the notes in the B-square with their flats or sharps.

The seventh line contains the clefs of the new system, that is, the syllables that must be added to the new names of the tones in order to distinguish their octaves. The middle octave has no clef, so PA means the PA of the middle octave, or the fundamental. Sem BO, is the BO of the first octave, or the octave of BO. Sub-ter-de, is the de of the third sub-octave, or the third sub-octave of de.

The eighth line contains the new names of our system. The ninth line contains the symbols for the new clefs. In the tenth line are the new notes, whose octaves are distinguished by the clefs above. These notes are taken from Table 5.10. In the eleventh row are the clefs of the intervals in merides. In the twelfth line are the merides, which mark the intervals of the tones of each octave, measured from the first sound of the same octave.

To obtain the intervals in merides for all the sounds from to the fundamental, multiply the clef by 43; thus the interval $2 + 18$ from the fundamental is $86 + 18$ or 104 merides, the subinterval $3 - 36$ is $129 - 36$ or 93.

The parts of all these lines correspond to each other to such an extent that we can have the expression of all the sounds of the chromatic system of the musicians within the compass of five or six octaves, as we wish; and thus we can apply to the expression of our system all the sounds of the harpsichord, and consequently all the sounds related to it.

Figure 5.27 marks the names and ranges of the voices and instruments. These voices and instruments are divided into two classes. The first includes voices and wind instruments, and the second includes stringed instruments.

Voices are generally divided into 6 parts.

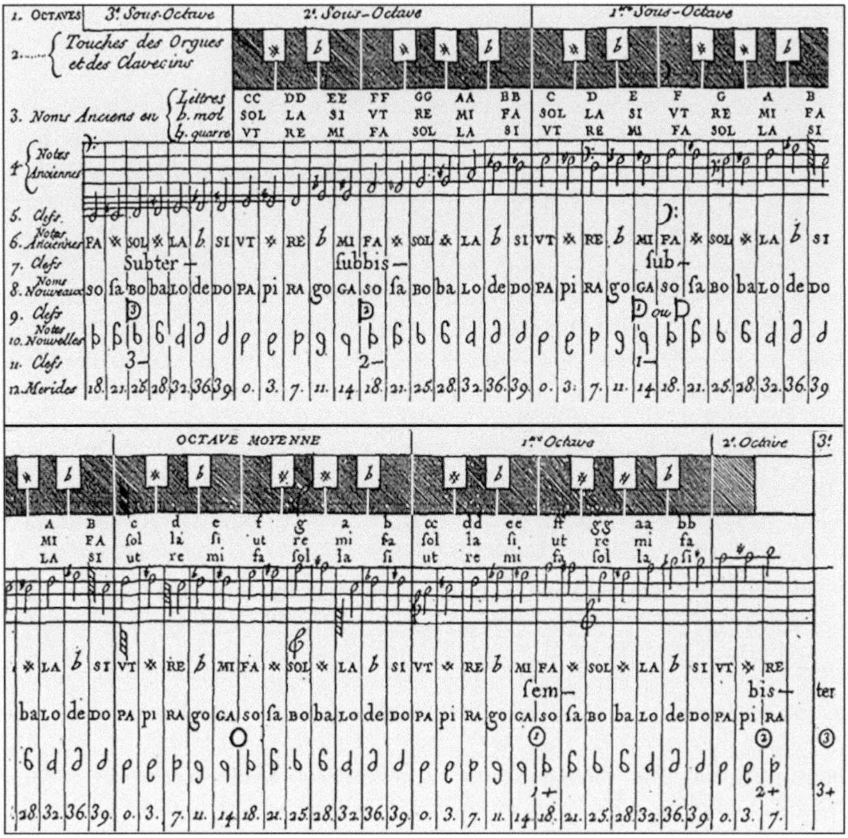

Fig. 5.26 Application of the general system to a keyboard. Drawn from [73], Planche II

1. The first treble, or high treble, marked *A*. A. a. *a*.
2. The second treble or low treble, marked *D*. D. d .*d*.
3. The countertenor *H*. H. h. *h*.
4. The tenor or high tenor *T*. T. t. *t*.
5. The baritone, under which one includes the concordant *C*. C. c. *c*.
6. The bass *B*. B. b. *b*.

In addition to these letters, dots are added to indicate the center of each part.

To these parts, Sauveur assigned five singers, whose names are given in the upper right corner of the figure: Mad.lle de Lalande, one of the two daughters of Michel-Richard de Lalande, either Marie-Anne (1686–1711) or Jeanne (1687–1711), both well known singers; Jean Jonquet (d. 1728; from 1682 singer of the Musique de la chambre of the French court); Antoine Dufour (from 1696–1732 basse-taille singer of the Musique de la chambre of the French court); Louis Bouteiller (1648–c. 1724, from about 1663 until his death maitre de musique at the cathedral of Le Mans);

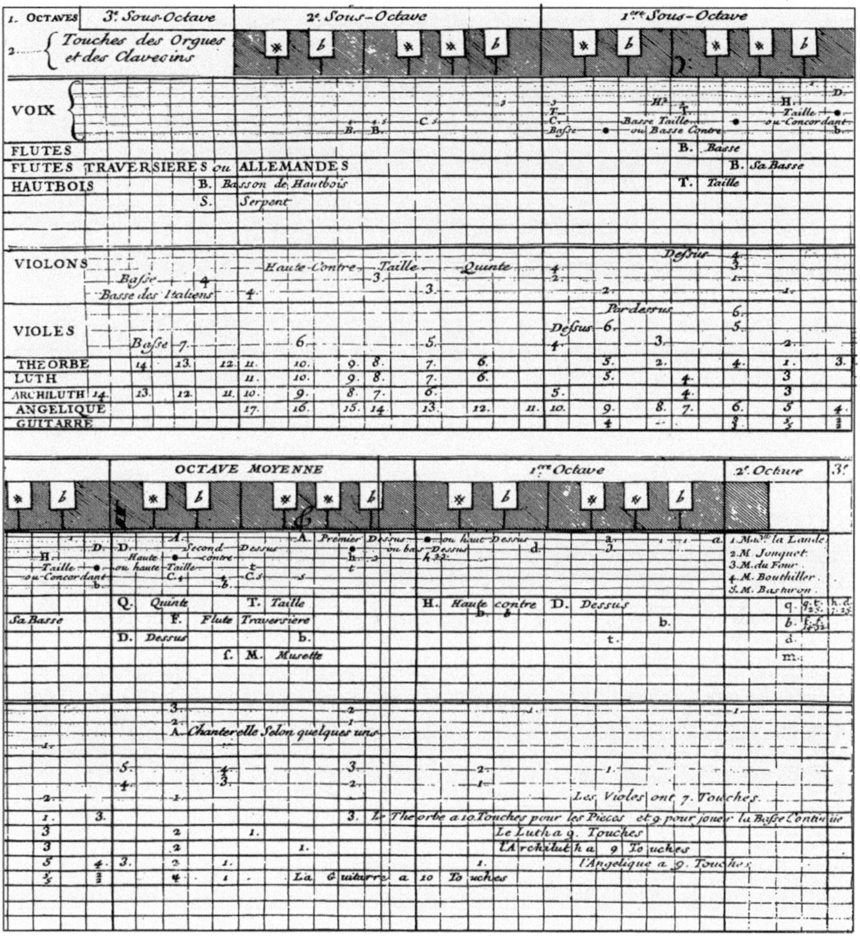

Fig. 5.27 Application of the general system to voices and musical instruments. Redrawn from [73], Planche II

and Jacques Bastaron (d. 1726, from 1697 basse-contre singer of the Musique de la chapelle at the French court) [12]. Figure 5.28 shows the ranges of the five singers.

The wind instruments are better mastered than the human voice; therefore they have more precise terms and extensions, which are generally divided into the following five parts.

1. The *dessus* marked D. d. *d.*
2. The *haute-contre* H. h. *h.*
3. The *taille* T. t. *t.*
4. The *quinte* Q. q. *q.*
5. The *basse* B. b. *b.*

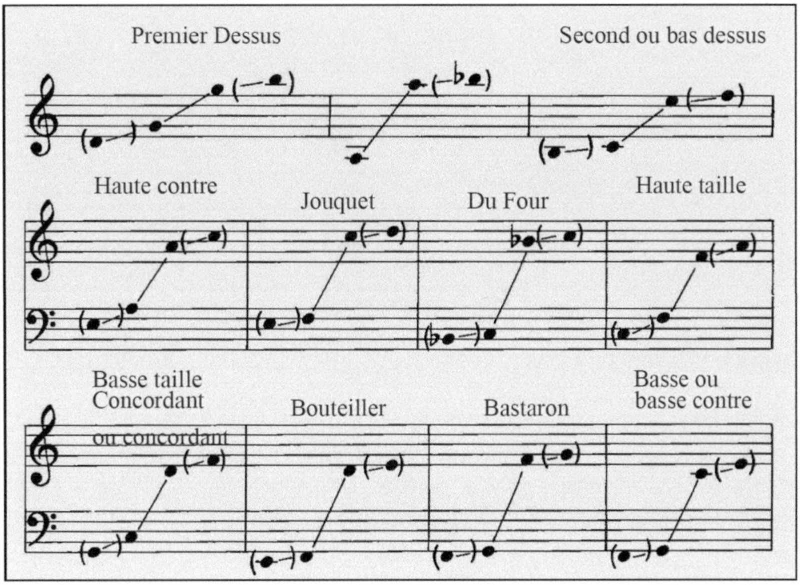

Fig. 5.28 The compass of the voices of five singers. Adapted from [60], p. 43

Les Touches des Instrumens à Cordes s'accordent de cette maniere	Touches	à Vuide	1.re	2.e	3.e	4.e	5.e	6.e	7.e	8.e	9.e	10.e
	Caracteres des Touches Selon les Tablatures	a.	b.	c.	d.	e.	f.	g.	h.	i.	k.	l.
	Interualles Diatoniques	I.	2.	II.	3.	III.	4.	IV.5.	V.	6.	VI.	7.
	Interualles par Merides	0.	4.	7.	11.	14.	18.	21½.	25.	29.	32.	36

Fig. 5.29 The frets of a stringed instrument. Adapted from [73], Planche III

The first capital letter denotes the lowest note of the instrument; the second lowercase letter denotes the highest note at which we can still play; and the third letter, written in italics, denotes where the instrument can rise in passing.

Sauveur claimed to have regulated the limits and compasses of most wind instruments according to the practice of Jean-Jaques Ripert (c. 1668–1724), a wind instrument maker in Paris, and Jean Hotteterre (1677–1729), composer, musician and musical instrument maker, who were among the most skilled makers in Paris. However, these limits may change according to the will of the player.

On stringed instruments with a neck, the numbers 1, 2, 3, 4, etc. mark the open strings, beginning with the highest. Repeated numbers or letters mark the ordinary interval of the highest string. Some instruments, such as the guitar, have their double and even triple strings, usually in unison but sometimes in octaves, marked with the same number repeated twice. Finally, the necks of all stringed instruments have 7, 8, 9 or 10 frets, except the violin, which has none. The sounds of the open strings of musical instruments, correspond mostly to the practice of Hurel,[97] considered one of the most skilled makers of stringed and necked instruments.

97 Many members of the Hurel family were violin makers in Paris at the turn of the 18th century.

Fig. 5.30 Plainchant. Adapted from [73], Planche III

Figure 5.29 shows:

1. The order of the frets according to the numbers 1, 2, 3, 4, etc.
2. The letters a, b, c, d, etc. by which they are designated in tablatures. The letter a indicates the open string, b the first fret, etc.
3. The diatonic intervals of the sound produced by the string at each fret, in relation to the sound of the whole open string.
4. The same intervals in merides.

Application of the General System to Plainchant

Plainchant is the kind of singing that uses only the sounds of the diatonic system, without any alteration of the notes other than being long or short. Sauveur did not introduce anything new in this section with respect to what has been said before, but he did present a more precise and somewhat interesting exposition.

In Fig. 5.30, the first line shows a staff with the notes and clefs of ordinary plainchant with the new clefs ○ and D, representing C Sol Ut and its sub-octave. The second line shows the old names of these notes; these two lines extend only to two octaves. The third line, gives the diatonic intervals in the compass of three octaves. The fourth line, gives the same intervals in merides (they are taken in the temperate system without heptamerides). The fifth line gives the names and the sixth the symbols of the notes in the new system, of which the white ones indicate the long notes and the black ones the short ones [73].[98]

I. The names of the ordinary intervals of the diatonic system are retained, second, third, fourth, etc., because they are very much in use among both ancient and modern authors: but to understand the exact relations of these intervals Sauveur had recourse to merides, from which we know that the major semitone or minor second contains 4 merides, the tone or major second 7, etc. Table 5.19 gives the tempered intervals as they are used in the practice of plainchant:

[98] p. 336. Both the new and old names of the notes in Fig. 5.30 are in uppercase for the lower octave, lowercase for the first octave up, and italic for the second octave up.

Table 5.19 The diatonic system of plainchant; from [73], p. 337

I.	2.	II.	3.	III.	4.	IV.	5.	V.	6.	VI.	7.	VII.	VIII.
0.	4.	7.	11.	14.	18.	21.	22.	25.	29.	32.	36.	39.	43.

To obtain the reciprocal intervals of the notes of the plainchant, for example, from DO to so, remove 39 corresponding to DO from 61 corresponding to so, leaving 22, corresponding to a false fifth; see Table 5.10.

At this point Sauveur began to enumerate the advantages of the new system:

1. The common names Ut, Re, Mi, etc., are rejected because they were chosen at random, without taking into account the properties of the intervals, which can only be known by these names through prolonged use.
2. The new names PA, RA, GA, SO, BO, LO, DO, indicate in a simple way all the properties of the tempered diatonic system, and are therefore very suitable for learning the intonation of the notes of plain chant. Because:

   a. The consonants P, R, G, S, B, L, D are used to distinguish the diatonic intervals of sounds or notes; since the interval from one consonance to the next always marks a second, such as P-R, or G-S, or D-P.
   b. The vowels A, O are used to indicate whether the preceding intervals are major or minor, according to the following two rules:
      i. The first is that if the two notes do not change vowels, the small intervals, that is, the second, third, and fourth, are major, because they contain only tones without any semitones; the corresponding large intervals, that is, the fifth, sixth, and seventh, are minor, because they contain two semitones: PA-RA, therefore, is a major second or tone; PA-GA, a major third. SO-DO is a major fourth or tritone; conversely, DO-SO is a minor fifth or false fifth; DO-BO is a minor sixth. Finally, DO-lo is a minor seventh. Their replicas follow the same rule.
      ii. The second is that when two notes change vowels, the opposite happens, that is, small intervals are minor because they contain a semitone, and large intervals are major because they contain only one semitone. So GA-SO or DO-PA are minor second or major semitones. GA-BO is a minor third. PA-SO is a minor fourth, or a just fourth. PA-BO is a major fifth or simply a fifth. PA-LO is a major sixth, and PA-DO is a major seventh. Their replicas follow the same rules [73].[99]

One can use the usual plainchant notes; but instead of the two clefs called F Ut Fa and C Sol Ut, it is easier to use our D and ○ keys to mark pa and *pa* (ut and *ut*; convention of Fig. 5.30).

Sauveur then emphasized the reasons why the new notes were preferable to the old:

99 p. 338.

1. The new notation immediately indicates the name and the sound that goes with it, whereas in ordinary notation it takes a long time to recognize the names and the sounds of the notes from the clefs, the lines, and the spaces.
2. From the new symbols we recognize with ease the major and minor intervals, as well as from the new names; for the names that have the vowel A correspond to the notes that have the stem at the bottom, and the names in O have in their notes the stem at the top. With ordinary tablature, it takes more time and practice to distinguish and tune the major and minor intervals than with ours.
3. The new symbols of the notes have some connection with the consonants of the names, especially with the four letters p, q, b, d. If we wanted to write a plainchant, instead of symbols we could use the consonants P, G, R, S, B, L, D, *p, r, g, s, b, l, d*, p, r, g, s, b, l, d. If we want to mark short notes, we put commas after them; for long notes, we put a dot or two after them. The old notes have no connection with their names, to make normal music convenient.
4. We have a small advantage with the heads of new notes, in that they allow us to distinguish the odd intervals III, V, VII, from the even intervals II, IV, VI. For in the same octave two notes which both have heads, or which both have none, form an odd interval; and two notes, one of which has heads and the other none, form an even interval. For two notes in two different octaves, on the other hand, the opposite occurs [73].[100]

Sauveur then listed the objections commonly raised against the new notes and names:

1. There is a general use for the old notes and old names, everyone is used to them, and it takes new work to learn a new way.
2. The names Ut, Re, Mi, etc. are softer and more energetic to express the sounds of the diatonic system than the names PA, RA, GA, etc.
3. The names are indifferent in themselves, because a skilled musician plays the notes without pronouncing or paying attention to the names.
4. Finally, the notes placed on 4 or 5 parallel lines help the imagination; for when the notes rise or fall, they warn us that the voice should rise or fall.

He then answered the objections in a way that seems convincing, although we know that he did not succeed in getting his notation accepted:

> 1. The history of music teaches that use has defined notes and their names, refining them little by little. For the names, we first used the letters A, B, C, D, E, F, G, as is still done in Germany, Turkey and Persia. Then Guido of Arezzo changed them to Ut, (or Do for the Italians) Re, Mi, Fa, Sol, La, and the usage, after encountering many difficulties with the change, added Si by only 30 years.[101] The symbols of notes were at first unknown, as they still are among peoples outside Europe, and letters were used; then we used dots, after which we invented plainchant notes, and finally those of music. It is not surprising, therefore, that after passing through all these stages, chant and music have reached a more perfect state than the previous ones.

---

[100] pp. 338–339.

[101] Sauveur dated the introduction of the syllable Si in the 1670s; probably referring to [3].

Fig. 5.31 Notes on the music. Adapted from [73], Planche III

2. We will find our notes PA, RA, etc., softer and more energetic than ordinary Ut, Re, etc., if we do enough to get used to them, because the vowels warn us whether we are going up or down a tone or a semitone.
3. I admit that names are useless for good musicians, but beginners need the simpler and more meaningful names, which even good musicians find easier to write.
4. I am convinced that the use of this system will make our notes far more visible than ordinary notes placed on parallel lines, because if we use a single line, the octaves will be as easily distinguished from each other as notes are distinguished in ordinary parts. Then, in each octave, the first 3 notes, which have their stems at the bottom, can be very well distinguished from the last 4, which have their stems at the top; moreover, the notes which have their stems before the body of the sign can be easily distinguished from similar notes which have their stems after; finally, even the heads will be well distinguishable, if care is taken to mark them well. Let us add that our sharps and flats are easier to recognize than ordinary ones. Our notes are known at once for themselves, and we avoid by them the difficulties which arise from clefs in transpositions of ordinary notes; finally, the language is uniform for voices and instruments [73].[102]

The lower part of Fig. 5.30 shows an example of plainchant. The key D means that notes without horizontal lines are from the first grave octave; white notes are long and black notes are short. The lines between notes mark word separations.

Application of the General System to Music
After framing the plain chant, Sauveur moved on to what he called music: "By music we mean that kind of singing which uses the sounds of the diatonic system with all the modifications of which they are capable, that is, with their sharps and flats, their different measures, their values, their durations, their pauses, and their use" [73].[103] That is, for him any kind of music, whether instrumental or vocal, is a song. It is called plain chant if only diatonic notes are played; it is called music if there are alterations of notes, even if there are only voices.

The salient features of the music are shown in Fig. 5.31. Specifically:

1. An octave of natural notes, that is, notes of the tempered diatonic system, expressed by merides, new names and symbols, and old names.

[102] pp. 339-341.
[103] p. 341.

2. All the notes with sharps are obtained by adding 3 to the number of merides of the natural notes; for the names, by changing A to i and O to a; for the symbols, by bending the stem to the right.
3. It can happen that composers, after having placed sharps immediately after the clef, place a new one on the note itself; in this case it is doubly sharpened. In the illustration, for example, we consider the note SO, which has 18 merides, its sharp 21; and its double sharp 24. The sharp in SO is sa, and its double is bi, which is obtained by taking the consonant b after s and the vowel i before o. With respect to the symbols, the sharp curves the stem to the right, and the double sharp has a double curvature of the stem.
4. All notes with a flat are obtained by removing 3 from the number of merides of the natural notes; from the names, by changing A to o and O to e; and from the symbols, by bending the stem to the left.
5. Sometimes double flats also occur, as for double sharps [73].[104]

According to Sauveur, comparing the new musical system with the old one, the following advantages are found in the new one, similar to those found in plain chant.

1. The merides mark exactly the intervals of sounds necessary for the music, as in plain chant.
2. The names of the natural sounds have the properties noted for plain chant, their sharps and flats do not change consonants but only vowels, taking the next 3rd meride for sharp and the previous 3rd for flat.

The new names are also used to distinguish the different types of diatonic intervals; to know the superfluous [augmented] and maximum intervals, which are greater than the major intervals of 3 and 6 merides; the diminished and minimum intervals, which are smaller than the minor intervals of 3 and 6 merides.

To distinguish the interval types with the new names, three things must be assumed:

1. Two sounds approach each other by one degree when the lowest is a sharp or the highest is a flat; and that they approach each other by two degrees when these two cases occur simultaneously. Conversely, two notes move away from each other by one degree when the lowest is a flat or the highest is a sharp; and that they move away from each other by two degrees when these two cases occur simultaneously.
2. Sounds do not move closer or farther together if they are both sharps or flats.
3. Neighboring vowels are equivalent to each other; thus e is equivalent to A, and i to O.

Assuming this, the following two rules apply:

1. At small intervals, when two names have the same vowels, or when one has a vowel and the other has its equivalent:

    a. Their intervals are major if the sounds are natural (i.e. without alterations) or if they are withdrawn by only one degree.[105]

[104] pp. 341–342.

[105] For the meaning of degree here, see point 1 of the previous list.

   b. Their intervals are maximum when they are withdrawn by two degrees.
   c. The intervals are diminished when they are drawn together one or two degrees.

2. At small intervals if the two names have different vowels or their equivalents:

   a. Their intervals are minor if the sounds are natural or if they are drawn together by one degree .
   b. They are minimum if they are drawn together by two degrees.
   c. They are superfluous if they are withdrawn by one or two degrees [73].[106]

3. For large intervals the opposite is true.

   In Table 5.20 there are examples of all kinds of fourths with:

1. The names of the fourth in all its species.
2. The merides.
3. Two examples of the names of the sounds that make up all the species of fourths, with the natural names in capitals, the sharp in lower case, and the flat in italics.
4. The rules to which these types of intervals refer.
5. The altered intervals to which the types of fourths refer, and which keyboard instruments use in their place.
6. The merides that these intervals should have in order to be just [73].[107]

   Figure 5.32a is relative to the value (duration) of notes and rests. It reports:

1. The ratio of the values of notes following a double inverse geometric progression.
2. The new notes for all these values.
3. The similar old notes.
4. The new rests denoted by O, i.e. the notes without stem or head.
5. The old rests in all the same values.

From this figure we can see that, assuming that a black O is worth one beat, a black O with a dash on it is worth $^1/_2$, with two dashes $^1/_4$, with a curve $^1/_8$. Conversely, a white O is worth two beats; one with a dash on it is worth four beats, and so on.[108] A dot placed after a note increases its value by half, as in ordinary music.

Figure 5.32b shows the new keys in which:

1. ○ marks the middle octave,[109] and 1, 2, 3, 4, etc. within ○ marks the first, second, third, and fourth octaves higher. D alone or with 1 marks the first sub-octave, and with 2, 3, etc. the second, third sub-octave.
2. The clef ○ placed on a line indicates the acute parts for voices and instruments, and means that the notes in this line are in the middle octave, those below are

[106] p. 343.

[107] p. 344.

[108] Notice the change of convention with respect to the *Traité* where there were not curves but only dashes.

[109] There is a slight change of symbol here; Sauveur had previously used ⊙ to indicate the middle octave.

Table 5.20 Different kinds of fourths; from [73], p. 344

1	2	3		4	5	6
Quartes	Merid.	Exemples		Regles	Altérés	Justes
					Intervalles	
Maximes	27	{se, da	......}	1ere	V. Super flue	28
Superflues	24	se, DO {SO, da sa, da	...... po, sa PA, sa}	2^{e}	V.	25
Majeures or Tritons	21	{SO, DO *se, de*	*po*, SO pi, sa}	1^{e}		
Mineures ou simplement Quartes	18	{sa, DO SO, *de*	pi, sa PA, SO *po, se*}	2^{e}		
Diminuées	15	{sa, *de*	*pi*, so PA, *se*}	1^{e}	III.	14
Minimes	12	{......	pi, *se*}	2^{e}	3.	11

Fig. 5.32 **a** Note values. **b** New clefs. Adapted from [73], Planche III

in the first grave octave, and the notes above are in the first acute octave. To go higher, a second line is added.

3. The clef D placed on a line is for the grave parts, and it means that the notes that are on the line are of the first grave octave. To go lower, a second line is added.
4. Notes without clefs mark the different octaves with dashes below the body, inside the body and above the body; those without a dash belong to the middle octave. Four octaves are sufficient for ordinary music. To go higher or lower, double the dashes. A clef at the beginning of the notes means that the notes without dashes belong to the octave of that clef [73].[110]

Figure 5.33 shows an example of two-part music developed in three different ways, taken from *Atys* by Jean-Baptiste Lully, the original of which is shown in Fig.

[110] pp. 345–346.

Fig. 5.33 Music example. Adapted from [73], Planche III

5.34. The first way, in which the notes are on a line with the key, where ○ marks the acute and D the grave. The second way, without clef, by placing the bodies of the notes on a line (actually not drawn in the figure, which is misleading). The third way, still without a clef, by placing all the notes at the same height. Experience will help you choose the best way.

Several observations can be made about this musical example:

1. At the beginning of each staff are the clefs ○ for treble and D for bass.
2. On the first clef ○ there may be a second clef with a number of merides in it to indicate that the fundamental, i.e. the C Sol Ut of the organ or harpsichord on which we are singing or playing this piece, is far from the fixed sound of this number of merides.
3. In front of the clef there are two notes, one below the part to be sung and the other above it. These two notes are used:
   a. To make the person singing this part alone understand the range of pitches from the lowest to the highest note.
   b. To make it clear to which person a part belongs.
   c. If there are several parts in a staff, these two notes let each musician know at a glance which staff to sing.
4. In the body of the piece, bars, repetitions, and various ornaments of the piece are marked as usual.
5. In front of the clef, on the left side above the first line, the bar is marked by a pause and a number; the number indicates whether the bar is of 2, 3 or 4 beats. In the present case, it is O 2.
6. Before the clef, below the first line, there is a pause, a letter, and a number to indicate the duration of a bar. O S14. O has the same meaning as before, S14 means that Sauveur's chronometer is used, which is expressed in twelfths (1/12 of

Fig. 5.34 Allons, allons, accourez tous. From Atys by Jean-Baptiste Lully [54], act 1, scene 2, f. 27r

a second), and that a white O (see Fig. 5.11) lasts $^{14}/_{12}$ of a second. When using Loulié's chronometer to mark duration, the letter P is used. For example, O P42 means that a white O lasts as long as the oscillation of a pendulum 42 inches long. Using Sauveur's chronometer in thirds ($^{1}/_{60}$ of a second), O T70 means that a white O lasts as long as a vibration of $^{70}/_{60}$ of a second [73].[111]

5.3.2.2 On Harmonic Sounds

That a sonorous body emits other sounds in addition to its fundamental sound has been known since antiquity, at least for the first octave. Mersenne in his *Harmonie universelle* of 1636 had amply testified to the existence of sounds at least five times higher than the fundamental. John Wallis [89],[112] Robert Plot [59] and Francis Robartes [64], had demonstrated experimentally that a string can vibrate not only in its entirety but also by $^{1}/_{2}$, $^{1}/_{3}$, $^{1}/_{4}$, etc.

Sauveur did a thorough analysis of sounds other than the fundamental, but he too remained vague in his explanation. That is, he was a little ambiguous in deciding whether or not a body, a string for example, can vibrate at different frequencies at the same time. At the end of this section on harmonic motion, he spoke as if there were an interaction between different modes, i.e. they coexist. In a writing on the procès-verbaux, commented on below, he was more explicit in assuming coexistence, as he did in the *Histoire* of 1702.

---

[111] pp. 346–347.

[112] vol. 2, p. 466.

Although his work does not in fact present any particular novelty, at least on the experimental level, it is completely new in presentation in the sense that Sauveur took up a phenomenology that had only just been glimpsed, and understood its fundamental importance for his new science, acoustics. In addition to new studies and experiments, he proposed a terminology that is still in use today, with some modifications; to name a few: harmonic sounds, nodes, bellies, fundamental sound.

The first definition he gave is that of harmonic sounds and fundamental.

> A harmonic sound of the fundamental is a sound that makes several vibrations, while the fundamental makes only one. A sound at the 12th of the fundamental is so harmonic because it makes three vibrations while the fundamental makes only one [73].[113]

Note that today the fundamental sound is also qualified as first harmonic sound, and that Sauveur's first harmonic sound is our second; and so on. Sauveur's harmonic sounds are now classified under the category of partials, which includes not only harmonic sounds whose frequencies follow the progression of natural numbers, but also the sounds of vibrating beams and bells, which have frequencies in irrational proportions.

Sauveur reported the principal harmonic sounds of a string in Table 5.21, which is not derived from experiments but from the application of his definitions. Note that the table is analogous to Table 5.12 in the *Traité*, though more precise. Col. 1 gives the number of oscillations a harmonic sound makes while the fundamental makes one; it also gives the order of the harmonic sound. Col. 2 assumes that each octave of the fundamental makes 10,000 vibrations and gives the number of vibrations made relative to the first note of each octave. This number is obtained by adding 0000 to the number in Col. 1 and dividing it by the first number of the octave; for example, 13,750 is obtained by dividing 110,000 by 8. Col. 3 shows the octaves and merides with the usual conventions. Col. 4 shows the diatonic intervals. Note that many of these intervals are not just; on the left are the missing merides, and on the right are the superfluous ones. Col. 5 shows the same diatonic intervals measured against the fundamental. Col. 6 contains the new names of the harmonic sounds. Col. 7 contains the usual names, where *b* means flat and *d* sharp. You can see that these names are not just in many cases. The column clearly shows that La and Fa cannot be obtained as just notes. Sauveur assumed that the 27th harmonic gives a just La; but this is not exactly the case, since this harmonic, reduced to the octave, gives the ratio (27 : 16) close to (5 : 3), the ratio of the major sixth corresponding to just La, but not coinciding with it.

Sauveur then went on to illustrate some experiments, real or thought. Let us divide a string of a monochord into equal parts, five for example, as in Fig. 5.35, pluck the empty string. It will produce a sound that is the fundamental of that string. Then place a light obstacle C, similar to the tip of a pen if the string is thin, on one division D; the perturbation of the string will be transmitted to both sides of the obstacle. The string will produce the fifth harmonic, a 17th.

To understand the reasons for this effect, note that when AB is plucked, it makes its vibration along its entire length, but when an obstacle C is placed on the first division

---

[113] p. 349.

Table 5.21 Tables of harmonic sounds

1. Rapports des vibrations au Son fondamental.	2. Rapports des vibrations au premier Son de chaque Octave.	3. Intervalles en Octaves, Merides & Eptamerides.	4. Intervalles Diatoniques au premier Son de chaque Octave.	5. Intervalles Diatoniques au Son fondamental.	6. Noms nouveaux.	7. Noms anciens.
1 1	10000	0	I	I	PA PA	VT
1 2	10000	1+ 0	I	VIII	ſem - PA	VT
1 3	15000	1+25′	V	XII	ſem - BO*r*	SOL
1 4	10000	2+ 0	I	XV	bis - PA	VT
1 5	12500	2+14,	III	XVII	bis - GA*n*	MI
1 6	15000	2+25′	V	XIX	bis - BO*r*	SOL
1 7	17500	2+35,,	7 7	7 21	bis - la*l*	*la . d*
1 8	10000	3+ 0	I	XXII	ter - PA	VT
1 9	11250	3+ 7″	II	XXIII	ter - RA*s*	RE
1 10	12500	3+14,	III	XXIV	ter - GA*n*	MI
1 11	13750	3+20,,	10 IV	10 XXV	ter - *ſul*	*fa* ½ *d*
1 12	15000	3+25′	V	XXVI	ter - Bor	SOL
1 13	16250	3+30′	6 7	27 7	ter - *lor*	*la* ⅔ *b*
1 14	17500	3+35,,	7 7	7 28	ter - la*l*	*la . d*
1 15	18750	3+39	VII	XXVIII	ter - DO	SI
1 16	10000	4+ 0	I	XXIX	quat - PA	VT
1 17	10625	4+ 4,,	2 2	2 30	quat - ro*l*	*re . b*
1 18	11250	4+ 7″	II	XXX	quat - RA*s*	RE
1 19	11875	4+11,,	4 3	4 31	quat - go*l*	mi *b*
1 20	12500	4+14,	III	XXXI	quat - GA*n*	MI
1 21	13125	4+17,	7 4	7 32	quat - *Sin* ou *gin*	*fa* ⅓ *b*
1 22	13750	4+20,,	10 IV	10 XXXII	quat - *ſul*	*fa* ½ *d*
1 23	14375	4+23,,,	5 5	33 5	quat - *bec*	*ſol . b*
1 24	15000	4+25′	V	XXXIII	quat - BO*r*	SOL
1 25	15625	4+28,,	10 6	10 34	quat - ba*l*	ſol . *d*
1 26	16250	4+30′	6 7	34 7	quat - *ler*	*la* ⅔ *l*
1 27	16875	4+32,,,	VI 5	XXXIV	quat - LO*t*	LA
1 28	17500	4+35,,	7 7	7 35	quat - la*l*	*la . d*
1 29	18125	4+37,	7 7	35 7	quat - *den*	*ſi* ⅔ *b*
1 30	18750	4+39	VII	XXXV	quat - DO	SI
1 31	19375	4+41	VII 14	XXXV 14	quat - *pu* ou *du*	*ſi* ⅔ *d*
1 32	10000	5+ 0	VIII	XXXVI	quin - PA	VT

D of the string and AC is plucked, the entire undulation[114] AB is first divided into two parts, AD and DB, and since AD is one-fifth of AB or one-fourth of DB, it tends to make its undulations five times faster than the whole AB or four times faster than the other part DB. The part AD then interacts with the neighboring part DE then forces part EF to follow the same motion, and so on, one part after another, until the end. All the parts make a undulation by crossing each other at points D, E, F, G.

[114] Note by Sauveur: A string undulation is the shape of a spindle as a result of the vibrations of the string.

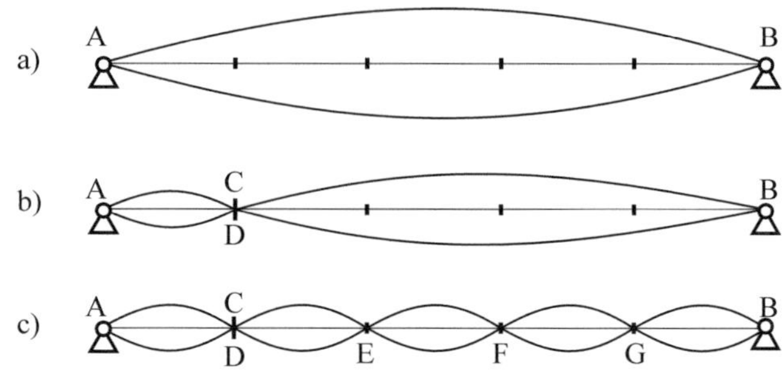

Fig. 5.35 Nodes and antinodes; Redrawn rom [73], p. 351

> I call the points A, D, E, F, G, B the nodes of this undulation, and the intermediate points of these undulations their bellies [73].[115]

This designation, nodes and bellies, is also current, although in the technical literature the latter are more commonly called *antinodes*; and we will use this term in the following.

If the obstacle C is placed on the second division E of the string as in Fig. 5.36, it will produce the same overtones as if C were in D, because:

1. The obstacle C first forces the string to make two undulations AE, EB, but the undulation AE, which is faster than the other, will push EG (Fig. 5.36c), its counterpart, to follow this movement.
2. The remaining part BG, which is half of GE and twice as fast, will force its counterpart GE to follow its movement, and this will involve the following FE, and so on, until the whole string is divided into a number of undulations equal to the greatest common factor of the parts AC and DC divided by the obstacle C.

One can be sure of the existence of these undulations:

1. By hearing, because those who have an acute sense of hearing distinguish a harmonic sound proportional to the parts that form the undulations, or even by using a monochord in unison with the harmonic in question.
2. By sight, because after having divided the string into equal parts, five for example, one can place a movable bridge in C or D and some small pieces of black paper in the nodes E, F and of white paper in between. After pinching part AC, you will see that the white pieces of paper on the antinodes of the undulations move and the black ones on the nodes remain stationary.

The results reported above were not new, at least as far as experiments are concerned. We have already commented on the writings of Wallis, Plot and Robartes on this subject. It is possible, however, that Sauveur did not know them, for they appeared in English journals, when he presented his relation on harmonic sounds at the Académie,

[115] p. 352.

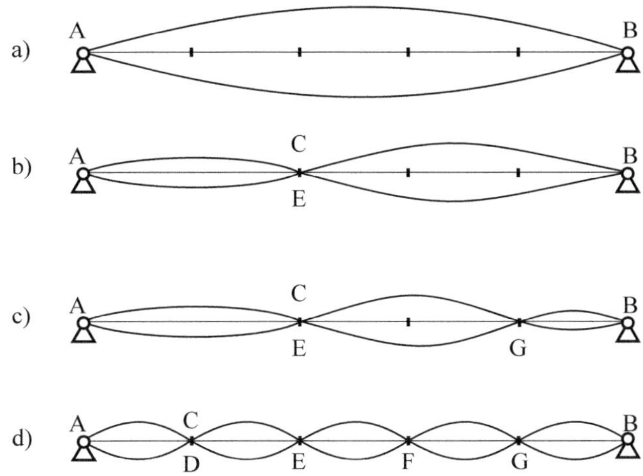

Fig. 5.36 Nodes and antinodes

and some of his colleagues pointed out that the same results had already been obtained by Wallis, he was surprised and "smoothly abandoned his claim to priority" [36].[116]

In the printed text, Sauveur explicitly acknowledged Wallis's priority (but did not refer to Robartes and Plot): "We were led to remark that this effect had already been observed in 1673 [...] See Johannis Wallis operum, in folio, tom. 2, p. 466, chapter CVII" [73].[117]

In any case, Sauveur approached the problem in a more systematic way and presented a plausible explanation for the formation of higher harmonic sounds. He drawn the following conclusions:

1. If a harmonic sound has been created by placing a thin obstacle, it will continue to exist if the obstacle is removed or another is placed on another node.
2. After generating a harmonic, for example the 5th, if you place a thin obstacle on the antinode of an undulation, dividing it into three parts, for example, it will form a third harmonic of the 5th harmonic, that is, the fifteenth harmonic of the fundamental.
3. Without placing a thin obstacle on a string, a harmonic sound is induced when another string is placed near it in unison with one of its harmonic sounds. Or, if the first string is not in unison with a harmonic of the other, it will be divided into undulations of length equal to the greatest common factor of the two fundamental undulation lengths; thus, if one is to the other, with a ratio and three to four, the shorter will form the third harmonic and the longer the fourth, which are in unison.

[116] pp. 131–132.

[117] p. 353.

4. An undulation generated by sympathy from a neighboring string or a thin obstacle is the more easily activated the greater the undulations. Thus, the third harmonic is better received than the fourth.
5. The undulations caused by sympathy soon become imperceptible, and those caused by a thin obstacle become indefinite. But as the smallest [the higher harmonic sound] is very difficult to distinguish, it will be assumed in the following that a three-foot string can be heard up to the 32nd harmonic, or up to the fifth octave [73].[118]

Sauveur idealized the fact that harmonic sounds with nodes less than 1/32 of the length of the entire string are not heard by making the stronger assertion that they do not exist at all. By assuming this, he was able to think geometrically about how different harmonic sounds differ.

Some of the following considerations can only be understood if the coexistence of multiple harmonic sounds is assumed; however, some doubts remain about Sauveur's thinking.

1. The nodes of the first harmonic sound,[119] that is, the ends of the string, are far enough away from the nodes of the next harmonic sound, since they are far at least 1/32 of the whole string. The nodes of the second harmonic sound are closer to the nodes of the next harmonic sound because their distance is half the 32th part of the whole string, and so on.
2. It is also the case that if the node of a small [higher] harmonic sound is in the neighborhood of two nodes of a large undulation, the smaller one will be deleted.[120]
3. Plucking an open string and moving the light obstacle along the string from left to right, one hears a chirping of harmonic sounds, the order of which will appear confused, but which may nevertheless be determined by the principles we have established.
4. From the same principles there are other consequences, so that, for example, lower harmonic sounds move as much air as large ones.[121]
5. Experience shows that long strings, when of good quality and harmonious, produce the first harmonic sounds, mainly those which are not at the octave; bells and other resonating bodies do the same.
6. Long wind instruments also divide their length into equal undulations. If a wave of air occupying the entire length from the mouthpiece through which the air enters to the first hole through which it exits is forced to go faster, it divides into two equal waves, then into 3, 4, and so on, according to the length of the instrument. By blowing slowly in a wind instrument the fundamental is obtained; by blowing

[118] pp. 353–354.

[119] Notice that Sauveur is inconsistent in his nomenclature here, as he equates the first harmonic sound with the fundamental.

[120] This statement, which is unclear in reality, seems to assume the simultaneous presence of several harmonic sounds.

[121] On this point see Fontenelle's accounts and the procès-verbaux, where an explanation of this fact is given. Probably Sauveur's written memoirs are very synthetic, because he had some doubts about his considerations.

harder, or, in the case of the trumpet, serpent, or oboe, by pressing the lips more, the sound changes into the 2nd, 3rd, 4th, etc. [73].[122]

Application of the general system to the trumpet marine
Sauveur had already dealt with the trumpet marine and the ordinary trumpet in his *Traité*, before he had fully developed his theory of harmonic sounds. Basing on it, Sauveur was rather brief in this section.

Interestingly enough, the following quotation, which expresses an ungenerous judgment of his predecessors.

> Since the principles of harmonic sounds were hitherto unknown, it is not surprising that explanations of the sounds of the trumpet marine and of instruments whose sounds proceed by leaps were imperfect. The discovery of harmonic sounds, will give rise to other discoveries to perfect acoustics and even invent new instruments corresponding to those of optics [73].[123]

A brief description of the trumpet marine follows, emphasizing the fact that the finger applied along the string plays the role of the thin obstacle considered in the previous section, and when it passes through the various positions of the board, it produces only the harmonic sounds up to the XVI, as indicated in Table 5.21.

On the Reciprocal Intervals

> A reciprocal interval is the interval of a sound or musical system or instrument with respect to another sound that follows it in the compass of an octave [73].[124]

According to this definition, any interval can be reciprocal. Sauveur narrows the meaning here by calling reciprocal the intervals of the notes of an octave that refer to any one of them.

The notion of reciprocal intervals is useful:

1. To know whether, taking a particular sound of a system or instrument as the fundamental, the intervals of the other sounds are consonant or dissonant, just or altered with respect to it.
2. To know which sound is best taken as the fundamental or the "final" (tonic) of a proposed musical piece.
3. Comparing the reciprocal sounds of one keyboard with another will reveal whether they are in tune and whether a sound can be made just by art if the difference is small.

Consider, for example, an ordinary harpsichord in which each octave has 12 keys; mark the merides of the tempered diatonic intervals of the keys of two octaves as in Table 5.22. Then Table 5.23 is constructed, in which the first line gives the old names of the notes, where the letters *d* and *b* mean sharp and flat, respectively. The second line gives the new names, and the third the intervals. The fourth line gives

[122] pp. 354–356.
[123] p. 356.
[124] p. 357.

Table 5.22 Diatonic and chromatic intervals; from [73], p. 358

1	2	II	3	III	4	IV	V	6	VI	7	VII
0	3	7	11	14	18	21	25	(28)	32	36	39
VIII	9	IX	10	X	11	XI	XII	13	XIII	14	XIV
43	46	50	54	57	61	64	68	(71)	75	79	82

Table 5.23 Reciprocal intervals; from [73], p. 358

		Ut	*d*	Re	*b*	Mi	Fa	*d*	Sol	*d*	La	*b*	Si	Ut
		PA	pi	RA	go	GA	SO	sa	BO	ba	LO	de	DO	PA
		1	2	II	3	III	4	IV (5)	V	6	VI	7	VII	VIII
Ut	PA	0	(3)	7	11	14	18	21	25	(28)	32	36	39	43
d	pi	0	4	(8)	11	(15)	18	(22)	25	29	(33)	36	(40)	43
Re	RA	0	4	7	11	14	18	21	25	29	32	36	39	43
b	go	0	(3)	7	(10)	14	(17)	21	25	−28	32	(35)	39	43
Mi	GA	0	4	7	11	14	18	(22)	25	29	32.	36	(40)	43
Fa	SO	0	(3)	7	(10)	14	18	21	25	(28)	32	36	39	43
d	sa	0	4	7	11	(15)	18	21	25	29	(33)	36	(40)	43
Sol	BO	0	(3)	7	11	14	18	21	25	29	32	36	39	43
d	ba	0	4	(8)	11	(15)	18	(22)	(26)	29	(33)	36	(40)	43
La	LO	0	4	7	11	14	18	−22	25	29	32	36	39	43
b	de	0	(3)	7	(10)	14	18	21	25	−88	32	−35	39	43
Si	DO	0	4	7	11	(15)	18	−22	25	29	32	36	(40)	43

the reciprocal intervals with respect to the tone PA, which are the same as those of the first octave of Table 5.22. The fifth line gives the reciprocal intervals with respect to the tone pi, which are obtained by subtracting 3 merides from the intervals in the previous line; the sixth line gives the reciprocal intervals with respect to RA, and so on.

Table 5.23 is misleading. For example, the fifth line contains the reciprocal intervals with respect to pi, but it shows the number 4 under pi of the second line, where it should be 0. To read the table correctly, you must imagine that when you go down one line, from r to $r+1$, the elements of the $r+1$ line move one column to the right, entering the upper octave. Thus, the 0 of the second line expresses the distance of pi from itself, while 43 is the distance of pi from pi one octave higher (the interval represented by the number 9 in Table 5.22). Diatonic intervals altered by one meride are in parentheses; the others concern just consonances and dissonances.

From the table according to Sauveur it is possible to see the following:

1. If we take RA or LO as the fundamental of a piece of music, all the diatonic intervals are just, or better, slightly tempered. On the other hand, if we take other tones, we will always have at least one altered interval.
2. One can see which sound of the harpsichord can be preferred to serve as the final for a piece that is in major or minor third; or which just or altered intervals can best emphasize the passion that is to be expressed.
3. A similar table can be made for an instrument whose sounds can be changed in different modes, to determine which sounds should be raised or lowered.
4. If two instruments accompany a piece of music, making the tables of their mutual intervals, one will recognize the sounds that are common and those that are different, and their difference, from which one will understand those to change or pass slightly and those that should be avoided [73].[125]

5.3.2.3 Way to Find the Fixed Sound

Sauveur presented two ways to find a fixed sound. The first way is that proposed by Mersenne, who started from a string 17 feet long (more than 5 m) which had a low frequency (8 Hz), to derive the frequency of strings of any length by using his law, for which the frequency of vibration is inversely proportional to its length, having the same tension, material and diameter.

> A string seventeen feet and a half long is sufficient to make the experience. Because it does not swing too fast, it allows to count its returns, as we can see with a string of a lute or viola of the size of those used in the rackets (of mutton intestines), which returns only twice in a second when it is stretched with half a pound, four times being taut by two pounds, *eight times being taut by eight pounds* [56].[126]

Sauveur argued:

> Assuming exactly this experience, it will be necessary to make this rule of three simple: since 100 vibrations which the fixed sound makes in one second stand for 8 vibrations which this string makes in the same time; the length of 17 feet or 204 inches of the string stands for the length of 16 inches 35/6 lines which the length of the string must have to render fixed sound of 100 Hz [77].[127]

A second method, which Sauveur considered more certain, uses the counting of the beats of two sounds, based on a simple, rather convincing, but fundamentally incorrect theory that works only in some special cases.

According to Sauveur, the superposition of two sounds with a ratio of frequencies in the radical form $f_2/f_1 = i = m/n$ with $m > n$ produces a beat when the first sound has completed m cycles and the second n, because at that moment the two sound pulses hit the ear simultaneously.

---

125 p. 359.

126 Part I. Livre troisieme du mouvement, p. 169.

127 p. 360.

So if the frequency of the more acute sound is f_2, there are f_2/m beats per second. For example, if $f_2 = 100$ Hz and $i = m/n = 25/24$ there are 4 beats per second, if $m/n = 3/2$ there are about 33 beats per second. According to Sauveur, in the first case the beats are both perceptible and measurable, in the second case they are not. The limit of perceptibility of the beats is set to 6 beats per second.

Sauveur did not pretend to give his own priority to this explanation. It is essentially the same as that given by Beeckman in 1628, who had learned of the phenomenon of the beats from an organist of his acquaintance, who described it as a succession of *wow, wow, wow* [11, 20].[128] Beeckman's explanation, based on his corpuscular conception of sound, is that there is a periodic superposition of strokes. However, there is no evidence linking Sauveur to Beeckman, not least because the latter's observations are written in Flemish in the pages of a journal that would not be published until the early 20th century. A theory similar to Sauveur's can be found in Loulié, probably inspired by him [80].

Sauveur's theory of beats is simple and is quite clearly reported in the minutes of the procès-verbaux of 1700:

> In searching for the causes of this beat, I found that it could not be attributed to anything other than the different number of vibrations which each of these sounds produces at the same time, and which are arranged in such a way that one, for example, produces 25 while the other only 24, and that they meet and unite only once in the space of these 25 vibrations, and that from the first to the 25th the vibrations are always separated. Now since it is certain that two vibrations reunited together strike the ear with more sounds than when they are separated, it follows that the sound of these two pipes will sound more at each 25th vibration than before the 24 others, and it is in this variety, sensitive to the ear, of loud and soft sounds that the beating of two sounds consists, and consequently whenever the ear makes us perceive that these two sounds are united, i.e., that they sound louder than usual, we can certainly conclude that the highest of these two pipes has made 25 vibrations and the lowest 24 [71].[129]

The phenomenon of beats was seen by Sauveur as a way to discriminate consonance from dissonances, anticipating the theory elaborated by Helmholtz more than a century after:

> From the foregoing, we can conclude that sounds beat in all possible intervals. But in the high octaves they beat so quickly that they are not noticed in the consonances, and this is what makes them more pleasant and what makes the real difference between consonances and dissonances. It seems that the thirds that beat in the lower octaves were not included among the consonances by the ancients, and that they are not included there now by the Orientals, and that it is only since we have invented the concerts that, being obliged to temper the consonances, the ear is accustomed to the beats and to indulge in those of the thirds that are very sensitive in the lower octaves and that habit has made people find them pleasant. It will also be by the nature of the beats that we will explain [71].[130]

The dissonance of two simultaneous sounds is explained by Sauveur with the same mechanism that is usually used to explain consonance: the theory of the coincidence of the pulses. Dissonance occurs when the ratio between the frequencies of two

---

[128] vol. 3, p. 51; pp. 144–145.

[129] f. 113v.

[130] f. 182v.

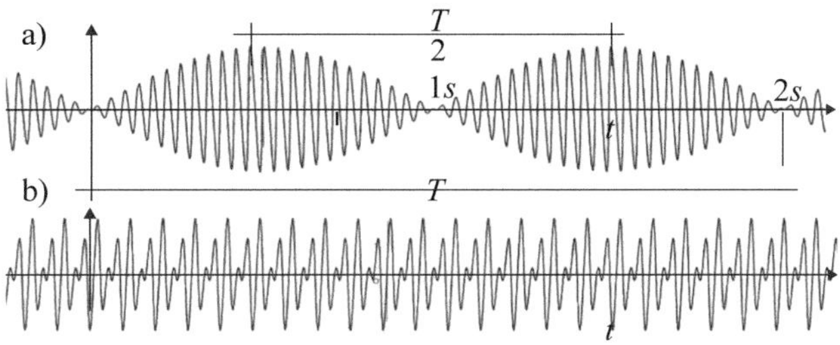

Fig. 5.37 Beats: **a** $f_1 = 24$ Hz, $f_2 = 25$ Hz, **b** $f_1 = 20$ Hz, $f_2 = 30$ Hz

simultaneous sounds, in the form $(n + 1)/n$, is sufficiently close to unity (that is, when n is sufficiently large), and the beats can be counted because they occur infrequently. The perception of beats causes an unpleasant sensation, i.e. discordance. Only when n is sufficiently small and the beats become very frequent, uncountable, and thus unperceivable, does consonance occur. The theory of coincidence comes to the same conclusion, but with a specular explanation. Consonance occurs when there is frequent coincidence of pulses (thus for n small) and discordance when the coincidence is rare. Sauveur must have been aware of this, because he did not use the coincidence theory to explain consonance.

Beats received a simple modern explanation framed into the interference phenomena in which sinusoidal sound waves are considered. If two waves of different frequencies are both directed toward the ear they interfere with each other forming a single complex sound.

By superimposing two sinusoidal sounds of close frequencies and comparable amplitude the resulting sound shows the pattern of Fig. 5.37a related to the case of two notes whose frequencies are 25 and 24 Hz. The superposition of the two sounds is perceived as a sound of approximately equal frequency to that of its two components and with an amplitude modulated with a frequency equal to the difference in the frequencies of the two sounds (in such a case 1 Hz).[131]

When the percentage difference between the frequencies is not small there is no beating, as illustrated in Fig. 5.37b where we consider the superposition of two sounds of frequencies 30 and 20 Hz; in this case we hear the two sounds as distinct (for some more comments on the beat phenomenon see Sect. 6.3.4.)

If we compare modern theory with Sauveur's theory, we have:

1. For two sounds with frequencies of 25 and 24 Hz, both theories give beats with a frequency of $25 - 24 = 1$ Hz, or one beat per second. Multiplying the frequencies

[131] For the two sounds of frequencies f_2 and f_1, the resulting sound has a frequency equal to the average of the frequencies $f_m = (f_1 + f_2)/2 = 24.5$ Hz, while the frequency of the modulation is $f = (f_2 - f_1)/2 = 0.5$ Hz. From a practical point of view instead of the period it is convenient to consider the half-period, which marks the distance between the instants for which the maximum amplitude occurs regardless of the sign. In this way the frequency of the modulation is simply equal to the difference of the frequencies, $f = f_2 - f_1 = 1$ Hz.

by 4 gives the two frequencies 100 and 96 Hz, and the beats have a frequency of four beats per second.
2. In the case of the frequency ratio $3/2$, according to modern theory, there are no beats at all. According to Sauveur's theory, there are beats, but they are not perceptible. For example, at $f_2 = 100$ Hz, there are about 33 beats per second, which are not perceptible because they are too numerous. So from an empirical point of view the two theories give the same result.
3. In the modern theory, the beats do not consist of pulses that follow each other periodically in a discrete manner, but have a character of continuity.
4. Sauveur's theory considers only the case $m - n = 1$. Modern theory considers arbitrary values of $m - n$.

Sauveur had already discussed his method for determining the fixed sound at the Académie, as reflected in Fontenelle's account in the *Histoire* of 1700 and the procès-verbaux of the same year. The procedure was based on counting the beats produced by the sounds of two organ pipes tuned to a minor semitone (25/24) and measuring the time interval between each beat; in this way he was able to play the fixed sound, which for him had to be 100 vibrations per second (i.e., 100 Hz). In this 1701 memoir, he was a bit more sophisticated; however, it is unclear whether his was simply a more complete exposition or a refinement of the previous procedure. Here is what he wrote:

> Use several organ pipes, which, being open, must be at least two feet long,[132] and tune the sounds of these pipes to the following diatonic intervals, which are so precise that the ear does not perceive the slightest beat in these sounds. 1^0. Tune the octave PA-pa (Ut-ut), 2^0 the fifth PA-BOr (Ut-Sol), or rather PA-BO*ra*; for in order to express the intervals of these sounds with due precision, it is necessary to make use of the decamerides. 3^0. The major third PA-GA*nā* (Ut-Mi). 4^0. The minor third PA-go*se* (Ut-Mi♭). After carefully checking these intervals by comparing these sounds with each other, the just interval go*se*-GA*nā* of the minor semitone can be obtained accurately [the difference between major and minor thirds], the ratio of whose vibrations is 24 and 25. If you play these two pipes, you will hear a beat in their sound at every twenty-fifth vibration of the higher GA*nā* (Mi) [...]. To measure the duration of these beats, a simple pendulum is used, each vibration of which is isochronous, that is, of a time equal [...] to four beats. Measure the length of the pendulum (from the point of suspension to the center of the sphere) and go to the fixed-sound scale of the general echometer marked "pendulum for fixed sound". If the pendulum falls on 0, the sound GA*nā* will be the fixed sound [of 100 Hz], since the pendulum marks one second and therefore there are four beats per second; otherwise, if the pendulum falls toward the acute sounds or toward the grave sounds of the scale, you have in merides and heptamerides the interval of the sound GA*nā* from the fixed sound. Then take the unison of GA*nā* on a monochord and move away from this sound toward the acute or toward the grave by the amount marked by the pendulum, then you have the fixed sound, which will finally have to be marked assigned to an organ pipe or to a whistle [73][133]

The first part of the procedure shows how to get two sounds that are a minor semitone apart (25/24). Start with five different organ pipes A, B, C, D, E, that are to be tuned

[132] A two-foot organ pipe has a frequency of about 250 Hz.

[133] pp. 360–361.

to each other; at the end of the tuning process (by ear?) they would produce the following sounds:

Pipe A: Note Ut
Pipe B: Note Ut, one octave higher
Pipe C: note Sol
Pipe D: note Mi
Pipe E: Note Mi♭

In principle it would have been sufficient to take only three pipes, A, D and E, because A-D form a major third (5 : 4) and A-E a minor third (6 : 5), so D-E form a minor semitone (25 : 24), which is what we wanted to find. The other pipes have only an instrumental function, and are useful for achieving correct tuning by checking the correctness of their relative intervals; for example, pipes C and D should produce a minor third.

In the memoir of 1701, Sauveur also considered the possibility of using intervals smaller than the minor semitone in order to have beats that are further apart in time and thus easier to measure. For example, with an intermediate pipe F between D (Mi) and E (Mi♭). Since E-D makes the interval $^{25}/_{24}$, or equivalently $^{50}/_{48}$, we can assume D at 50 Hz, E at 48 Hz, which implies F = 49 Hz.

If we measure the beats between D and F, whose frequencies are $^{50}/_{49}$, we will find a half number of them per unit of time with respect to the ratio $^{25}/_{24}$, and thus easier to count. Sauveur also suggested the possibility of another pipe G, intermediate between D and F, with frequencies as $^{100}/_{99}$, and beats for unit of time $^{1}/_{4}$ of the case of pipes D-E [73].[134] However, considerations of this kind are probably only speculation, and it is most likely that Sauveur considered the assumption of pipes D and E as the most convenient option.

The second part of the procedure concerns the actual determination of the fixed sound: "After having found the interval between a proposed sound and the fixed sound, it remains to find the fixed sound itself" [71].[135]

If the pipes D and E are tuned so that they are a semitone minor apart, we can say that when we hear one beat, it means that the sharpest pipe has made 25 vibrations—and the less sharp pipe 24 – and when we hear 4 vibrations, the sharp pipe has made 100 vibrations. But we do not know how long these vibrations last; in particular, we cannot say that they are made in one second. This indeterminacy of time has been pointed out by Semmens [82], the editor of the *Traité*, who finds Sauveur's explanation a bit convoluted. We agree only in part, and believe that a careful reading of Sauveur's writings allows to say that his approach is all in all quite clear. To determine how long it takes the more acute pipe to make 100 vibrations, it is enough to measure how long it takes to have four beats. This will give the frequency of the acute pipe. It will generally be different from 100 Hz, because there is no method yet to construct a pipe that vibrates 100 times per second, and this is precisely the problem that Sauveur set out to solve.

---

[134] pp. 360–361.

[135] f. 114.

The time between beats is measured with a pendulum by adjusting its length so that four beats are heard at each oscillation. If the duration of the oscillation is one second, it means that the acute pipe makes a sound of 100 Hz; if the duration is less than one second, the sound is higher, and if it is greater, the sound is lower. To find the actual frequency of the acute pipe, simply divide 100 by the duration in seconds required for four beats. Sauveur, as usual, avoided this calculation, which might be too complicated for a non-mathematician, and resorted to the use of the fourth scale of his echometer (Fig. 5.26).

However, Sauveur was not satisfied with having found the absolute frequency of an organ pipe; he wanted to be able to emit exactly the fixed sound of 100 Hz. To do this, he used a series of vibrating strings tuned in octaves. Among these strings, he chose the one that contained the sound of the acute pipe in its octave. Sauveur did not say so, but if he wanted to use the monochord division given by his echometer, he would have to use strings of the same material, about 60 inches in length, placed one octave apart by varying the tension and/or the diameter.

At this point Sauveur introduced a movable bridge on the chosen string and placed it so that the string made exactly the sound of the pipe. To obtain the 100 Hz sound, it is enough to move the bridge by the number of merides obtained with the fourth scale of the echometer.

The procedure can only work if the acute pipe emits a sound with a frequency not too far from 100 Hz.[136] It also requires sensitive and trained ears and a very quiet environment, as Sauveur pointed out:

1. Have a fine ear that can accurately judge the just diatonic intervals.
2. These intervals must be checked in every possible way, and the pipes must be played with the same wind.
3. When comparing the beats with pendulum vibrations, to be sure they are isochronous, we must compare these beats with more than 100 pendulum vibrations.
4. If we perform this experiment in warm countries where the pendulums of the seconds are slower than in Paris, on each line of difference we will have to add half a heptameride to the sounds that are more acute than the fixed sound, and to subtract it from those that are more grave [73].[137]

In the procès-verbaux of 1700, Sauveur claimed that his theory was supported by experiments: "With the presence of many musicians among the most able in Paris, and of some people skilled in theoretical music, it was found in all the experiments that were made, that the result corresponded to my reasoning with all the precision that could be hoped for" [71].[138]

---

[136] We do not know how Sauveur arrived at pipe D as a candidate for emitting the fixed sound. He probably used the procedure proposed by Mersenne and then made several tests with his method of beats, adjusting the length of the pipe until the procedure gave a number of beats per second close to four.

[137] p. 361.

[138] f. 114v.

However, the results of these experiments are nowhere documented in detail. What is documented instead is their failure in at least one situation; "when M. Sauveur [...] proceeded to make the experiments of beats [...] himself reported and avowed that for this time it had not been very successful" [35].[139]

And that experimental confirmation was not so certain even in 1701, is documented by the conditional form in which the beat experiments are referred to [36].[140] In Sauveur's last memoir of 1713, there is mention of four successful experiments carried out in 1699, 1700, and 1704 [77],[141] of which there is no trace anyway.

This uncertain result is echoed in [31], which refers to the experience of the musician Giuseppe Sarti (1729–1802), who repeated Sauveur's attempts with interesting modifications. The first announcement of Sarti's experiment dates from 1796 and is reported in the *New Acta* of the Petersburg Académie des sciences [6]. Sarti's apparatus consisted of two 5-foot organ pipes mounted on their somiere, with two bellows and two keys to play the pipes separately or together. Finally, there is a monochord and a pendulum that beats the second.

The two pipes are first brought into unison with proper devices, then the pitch of one of them is slightly changed until the beats that occur have a frequency of one per second, using the pendulum as a measuring instrument. At this point, the monochord is used to measure the interval between the sounds of the two pipes. Sarti stated that in his experience this interval is 100/99, which means that the sound emitted by the higher pitched pipe is 100 Hz.

Since this fixed sound is the same in all times and places, Sauveur commented:

1. Having found the interval from a sound to the fixed sound, we know how many vibrations this sound makes in a second; and if we call PA the fixed sound, we shall know the absolute sound of all sonorous bodies, and the name that represents it.
2. Having once found the absolute sound of a sonorous body, and the sound it has in another time, we can recognize from the change in the sound of this body, that of its substance; which is useful for physics.
3. Mechanics and acoustics also have their advantages; for it will be known how many vibrations the lowest and highest sounds that the ear can perceive make per second. The path that the string of an instrument takes with its vibrations in a second, when this sound is very weak or very strong, will be known by its interval to the fixed sound, and the magnitude of its vibration; and finally, the amount of air that this string moves per second.

5.3.2.4 Addition to Section VI

This section deals with the classification of voices. There are three things to consider in a voice: compass, degree of acute or grave, and part.

---

[139] p. 139.

[140] p. 129.

[141] p. 336.

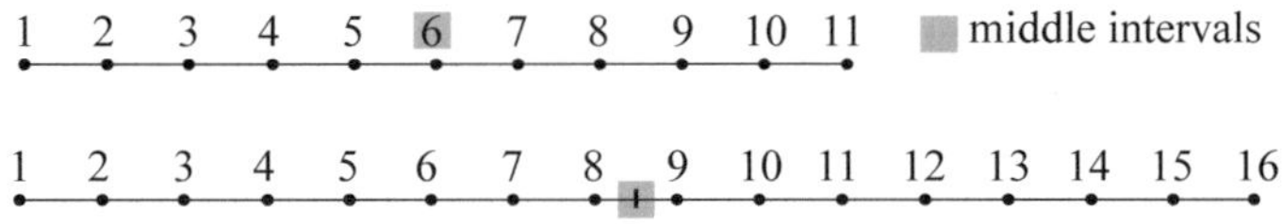

Fig. 5.38 Middle intervals

The compass of a voice is the series of notes in the diatonic system that it can go through while singing, thus one voice is said to have a compass of XI and another a compass of XVI. We denote the compass by twice the half of the interval covered. To find this half, for an odd interval add 1 and divide by 2; for an even interval add 2, divide by 2 and add the word half to the interval thus obtained. Thus the compass of the interval of an XI is a VI; that of an XVI is a half-IX, which means that an XVI is composed of an VIII and an IX. Figure 5.38 shows the rational of the previous statements.

The degree of a voice is the middle sound of that voice, indicated by the name of its note. Thus, a voice is sub BO (Sol) if the average sound of that voice is sub BO, and if it rises as much above this value as it falls below it.

The part of a voice means two things together: the degree and the half of its compass. So we can say that Miss de Lalande's part is LO (La) octave, which means that she can go one octave above or below LO. Similarly, Mr. Du Four's part is sub-DO half IX, which means that his voice can go up one IX from sub-DO (Si) and down one VIII.

It follows that you can have as many parts as there are notes in the diatonic system. To find these notes, we need to consider:

1. The highest human voice goes up to sem DO (Si), but we can admit that it goes a little further up to bis-PA (Ut); the lowest voice goes down to sub-bis GA (Mi), but we will admit that it goes down to sub-bis PA. Thus the middle of all possible voices is the fundamental PA (C Sol Ut).
2. The usual compass for plain chant is one IX or two Vs, and for music an XI or two VIs. Composers limit these compasses in order to prevent one part from being confused with its neighbor, or to use the voice in the most pleasing part, and finally to find voices that can perform both neighboring parts.
   In the new system, it is agreed that the normal compass of a voice is one XI or two VIs. Thus, we say that the part of a voice is BO (G Re Sol) to indicate the middle note BO and the compass of two VIs; we say that it is PA octave if its middle note is PA (C Sol Ut) and its compass one octave up and down; finally, a voice is sub LO (La) half VII if its compass is one VII up and one VI down.
3. Assuming that the highest voice rises to bis-PA and its compass is an XI or two VIs, subtracting one VI from bis-PA, the degree of the highest voice is sem GA. Similarly, assuming that the lowest voice is sub-bis PA, adding a sixth makes the degree of the lowest voice sub-bis LO. Thus the middle of all the voices is between sub-bis LO and sem GA; this covers a range of one XIX, or two Xs, or

finally 19 notes, and so we can conclude that we can have 19 parts, the extremes of which are very unusual [73].[142]

From what has been discussed above, the order and names of the parts according to the new system are:
1. sem-GA (Mi).
2. sem-RA (Re) very high treble.
3. sem-PA (octave of C Sol Ut) high treble, or first treble.
4. DO (Si).
5. LO (La) low treble or second treble.
6. BO (G Re Sol).
7. SO (octave of F Fa Ut).
8. GA (Mi) very high counter-tenor.
9. RA (Re) counter-tenor.
10. PA (C Sol Ut) very high tenor.
11. sub-DO (Si) high tenor.
12. sub-LO (La) tenor.
13. sub-BO (sub-octave of G Re Sol)
14. sub-SO (F Ut Fa) low tenor.
15. sub-GA (Mi).
16. sub-RA (Re) lower tenor.
17. sub-PA (sub-octave of C Sol Ut).
18. sub-bis DO (Si) bass.
19. sub-bis-LO (La) [73].[143]

Note that, as we have set each part in the compass of an XI, or two VI, if a voice has a greater compass, it can take as many parts as it has notes at its middle that are at least a VI away from the most acute and the most grave. The sub-Do half-IX voice can make all the parts from 8 to 13, because it can go up a sixth above the highest 8 and down a sixth below the lowest 13.[144] More specifically, it can make the following parts: the very high countertenor, the countertenor, the very high tenor, the high tenor, and the tenor.

When indicating the part of a voice, it is assumed that it can only reach the extreme notes in passing and that it can only dwell on the middle notes. Note, however, that the parts of a voice indicated above concern their absolute arrangement, but composers usually consider them in terms of the way in which they accompany each other, and so for treble and bass they may take counter-tenor and tenor, for instance.

---

[142] pp. 363–364.

[143] p. 364.

[144] Notice that the parts are ordered from more acute to more grave, so a lower number means a more acute sound.

5.3.3 Overtones and Harmonics

A modern reader who knows a little about music theory will be surprised to see that Sauveur distinguished between the two concepts of overtones—the presence, audible to the ear, of higher-order notes such as the twelfth and nineteenth, along with the fundamental of musical strings—and harmonic sounds—that is, the fact that a string deforms in a complex way with the presence of nodes and antinodes, and that in this situation the sounds produced by the third and fifth harmonic sounds correspond to the twelfth and nineteenth of the fundamental, respectively.

As we have already seen, in Sauveur's time there was strong resistance to accepting that a string could vibrate at different frequencies simultaneously, and overtones, known since antiquity, were not associated with the harmonic sounds that had recently been discovered.

But in Sauveur's presentation to the Académie in 1701, he explicitly admitted the derivation of overtones from harmonic sounds. This is clear in Fontenelle's account of the 1702 memoir (see below) on organ pipes, in which the concept of harmonic sound is fundamental. But it is even clearer in the report of the procès-verbaux of 1701.

Fontenelle in the 1702 report, which is almost entirely devoted to harmonic sounds, made the explicit statement that it was possible for several harmonic sounds to coexist, probably reporting what Sauveur said at a session of the Académie.

Fontenelle reiterated what was widely accepted at the time, namely that a harpsichord string, in addition to the sound it produces, depending on its length, cross section, specific weight and tension, also produces other sounds that can be heard at the same time by a fine and trained ear.

> These other sounds are more acute than those of the whole string, and are produced by a part of it which is somehow detached from the motion of the whole to produce particular vibrations. This mixture[145] of vibrations can be understood by referring to a string attached at both ends, like that of the dancers. While the dancer produces one large undulation, he can use his hands to give each half of the string its own undulation, and these in turn can give other undulations to the adjacent parts. That is to say, each half, each third, each quarter of a string of a musical instrument has its own vibrations over and above the total vibration of the whole string. It is the same with bells, if they are well made and harmonious [37].[146]

All these particular sounds produced by individual parts of the strings are harmonic in relation to the fundamental sound. The least acute of the perceived sounds is an octave of the fundamental sound; the next least acute is a 12th, then a double octave, then a 17th, until the sounds become too acute for the ear to perceive.

No sound is perceived that is not accompanied by either a fifth or a third, or by some chord that is included in the sequence of harmonic sounds with the whole sound.

In the procès-verbaux of 1701, Sauveur introduced the concept of *confuse undulations*, that is, the composition of different harmonic sounds. According to Sauveur,

[145] Fontenelle, as Sauveur uses the word complication.

[146] p. 91.

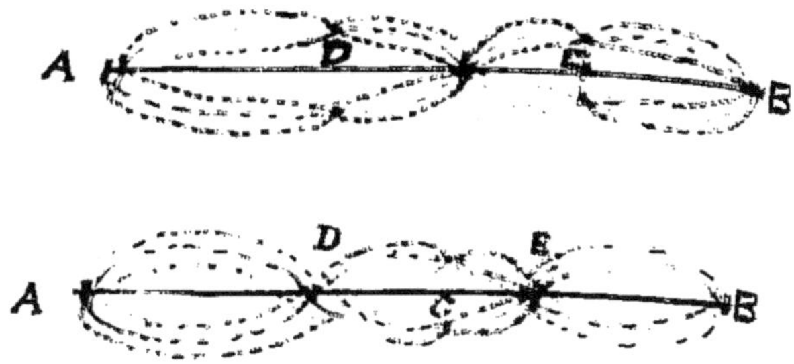

Fig. 5.39 Complex harmonic sounds. Redrawn from [72], f. 123v

if a rope AB (see Fig. 5.39) is divided into two equal parts at point C, each of which makes its own undulations, and if it is also divided into three equal parts at points D, E, each of which also makes its own undulations, then we say that the undulations of the second sound are confused with those of the third sound and more. It is obvious that in these confused undulations the nodes have less movement than the neighboring parts.

The ear confirms this in the long strings of the harpsichord, because we hear several overtones at the same time. But reason teaches us that if, among the confused waves, those of one harmonic have a greater force than those of the other, we hear the first and not the second. If one sound is an octave above the other, we hear only the most grave, which explains the oddity that occurs in the harmonic sounds of a sonorous body (see Sect. 5.8.3).

5.3.4 1701 Fontenelle's Account on the System of Music

In the *Histoires* of the Académie of 1701, along with Sauveur's memoir, there is a Fontenelle's account, probably based on the latter's presentation to the Académie, which treats arguments not reported in the memoir. Because of its relevance, a detailed summary is given below, as was done for the 1700 account.

Fontenelle opened his writing by emphasizing the innovative aspect of Sauveur's memoirs:

> After having given a glimpse of his acoustics in the *Histoire* of 1701, Mr. Sauveur here takes a more exhaustive approach and proposes an entirely new system of music, with new symbols and new rules that completely change the ordinary practice of musicians. *It is an amazing thing to see all music reduced to tables of numbers and logarithms, as one does for the sines, tangents, and secants of the circle* [emphasis added]. Perhaps the relationship between the sounds that strike the ear is only the sad fruit of a long and thorny search for mathematics? [36].[147]

Fontenelle then reported considerations, aesthetic or metaphysical, which reiterated the theses presented by Sauveur in his *Traité* of 1696. The soul, by its very nature,

[147] pp. 121–122.

loves both simple perceptions, because they do not tire it, and varied perceptions, because they protect it from the boredom of uniformity. If a perception is both simple and varied, it does not tire the soul, and by offering its variety it satisfies the soul's inclinations toward the two opposites. Any variety that pleases the soul, therefore, is limited by certain boundaries; it is necessary that the variety does not reach the point where it becomes difficult to perceive, confused, too complicated, and so on.

As far as sound is concerned, Fontenelle pointed out that unison and octave are the simplest chords and meet the criteria of unity, because in the first case there is a coincidence between the beginning and the end of the sound, and in the second case there is a coincidence every two cycles of the highest sound. As the intervals become more complex, when we move to the ratios (2 : 3), (3 : 4), (4 : 5), (5 : 6), the perception of the soul becomes more difficult because the coincidences of the vibrations of these chords become rarer and rarer. This is why chords associated with small ratios produce consonances, others dissonances [36].[148]

In addition to these explanations for why only ratios between small numbers are pleasing, Fontenelle also considered another circumstance that Sauveur mentioned in his memoirs without being quite explicit. According to Sauveur's theory of beats, when the ratios are between small numbers, such as (2 : 3), the beats are not perceptible because they are too frequent. Beats, on the other hand, are perceptible for larger numbers, such as (24 : 25). Since beats are unpleasant to the ear, it follows that ratios of large numbers produce unpleasant or dissonant intervals, and those of small numbers produce pleasant or consonant intervals.

However, Fontenelle continued, the causes of the pleasantness of chords are not precisely determined by nature, so that they could be rigorously proved. Perhaps there are no unpleasant chords at all to untrained ears, and the delicacy that leads to severe judgments may be a mixture of something natural and long habit, due to great attention and arbitrary prejudice.

In this mixture, which is not natural at all, reality can often be ignored, and hence the extreme difference in the musical tastes of nations. However, by consulting nature as far as possible, one can find fixed points; for example, the (5 : 6) chord is the last to be pleasant. That of (6 : 7) is no longer pleasant, nor that of (7 : 8). Moving on, we have (8 : 9) and (9 : 10), which are a bit more pleasant; this is because they can be obtained from consonant chords, e.g. (8 : 9) is the difference between a fifth and a fourth. In conclusion, the first natural numbers 1, 2, 3, 4, 5, 6 give pleasant chords, the most pleasant in fact, then there are gaps here and there interrupted by pleasant sounds, but only of the second order [36].[149]

Fontenelle explained the main purpose of Sauveur's general system. It was proposed in order to compare the various systems with each other and to judge of their merits and defects, to recognize the errors of tempered systems and to have an exact measure of them, to unify the whole at a glance, and to bring back to precise and fixed rules the infinite variety to which the relations of music are susceptible, it would be desirable to have a general system into which all other conceivable systems could

---

[148] p. 123.
[149] p. 125.

fit. Such a system would divide the octave into such small parts that it would be impossible or unnecessary to subdivide it further, so that the ear would have nothing to wish for as far as perfect chords are concerned.

In his system, Sauveur considered a meantone divided into 7 equal parts, of which he assigned 4 to the major semitone and 3 to the minor, so that the octave, containing 5 mean tones and 2 major semitones, has 43 parts, called merides. Each meride is divided into 7 parts, called heptamerides; consequently the octave is composed of 301 heptamerides, each of them defined by the ratio (434 : 435),[150] thus normally insensitive even to musicians.

Fontenelle devoted a great deal of space to the description of harmonic sounds, considering their discovery a fundamental contribution to music theory. The discovery is said to have resulted from the search for an explanation of the strange behavior of some musical instruments.

> As it often happens that one research, or even one discovery, is only the consequence of another, Sauveur, examining the theory of certain instruments that go from one sound to another in leaps, was forced to explain this fact by resorting to experiments that produced a phenomenon that at first surprised him. For what philosopher could have believed that a body set in motion in such a way that all its parts move would nevertheless retain some immobile parts? [36].[151]

Sauveur presented his discovery to the Académie and gained much attention by reporting experiments on vibrating strings.

> When Sauveur presented to the Académie the experiment whereby there were two equal pitches for a string divided into two different parts, it was received with the consideration that a new discovery deserved. But someone in the company remembered that this experience had been reported in a work by Wallis, although in truth it had not been emphasized as it deserved. Sauveur unceremoniously renounced the priority of the discovery [36].[152]

For the nodes and antinodes, then only a hypothesis, a way was found to verify their actual existence in the assembly. On the points of a vibrating string, which according to the hypothesis were the nodes or antinodes of the waves, small pieces of paper folded in half were placed, which would fall off at the slightest movement by pinching the string. It was immediately seen that the pieces of paper placed at the antinodes fell off immediately, while those placed at the nodes remained stationary. Later, to make them easier to distinguish, some were colored red, while the others were left white, and they were arranged alternately. The colored piece of paper could not be seen to fall in any way. To see these equidistant pieces remain fixed while all the other points moved in a body that should have moved in its entirety made a great impression on physicists who had not expected it [36].[153]

Fontenelle devoted more space than Sauveur to explaining the operation of the trumpet marine. He also avoided commenting on Robartes' results, which basically

---

[150] About equal to $\sqrt[301]{2}$.

[151] p. 129.

[152] pp. 131–132.

[153] p. 132.

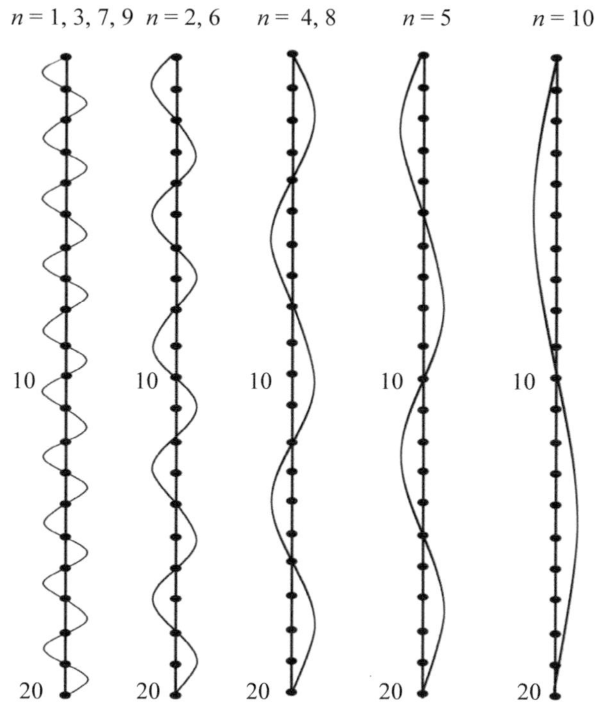

Fig. 5.40 Different intervals in the trumpet marine

considered only the case where the finger divided the chord into two parts, the larger of which was a multiple of the smaller, for example a $1/2$, $1/3$, $1/4$, etc., where the undulations being $1/2$, $1/3$, $1/4$, the frequencies are double, triple, quadruple. When the finger divides the string into two parts, such as $2/5$ or $3/5$, the string produces a fuzzy sound according to Robartes, but not according to Sauveur.

Fontenelle described the sounds that can be obtained with a trumpet marine equipped with frets that divide its string of length l into m equal parts. Placing the finger on the fret n produces waves whose nodes are l/m apart if m and n are prime to each other, or nodes that are m/p apart if p is the greatest common factor between m and n. For example, for $m = 20$, $n = 10$, the greatest common factor is 10 and the nodes are $1/2l$ apart. Fontenelle reported the sequence of musical intervals for $m = 20$ and $n = 1, 2, 3, 4, \ldots, 20$, in Fig. 5.40 and in Table 5.24.

> The other divisions, beyond 10 up to 20, give the same intervals, because it is the same to place the obstacle on 11 or on 9, on 12 or on 8, etc. It is clear that moving the small obstacle gradually from one side to the other produces an irregular sequence of sounds, sometimes the same sound, sometimes different. A musical instrument, where something similar happens, produces leaps by going from one pitch to another or returning to the same one without any obvious criteria, without gradualness and against every known rule [36].[154]

[154] pp. 134–135.

Table 5.24 Parts of strings and intervals of a trumpet marine. Translated into English from [36], pp. 234–235

Parts of strings	Intervals
1	Major third of the fourth acute octave
Because $\frac{1}{20}$ is lowered by four octaves, to get $\frac{16}{20} = \frac{4}{5}$, which is the major third	
2	Major third of the third acute octave
Because $\frac{2}{20} = \frac{1}{10}$ is lowered by three octaves, to get $\frac{8}{10} = \frac{4}{5}$, which is the major third	
3	Same interval as 1
4	Major third of the second acute octave
Because $\frac{4}{20} = \frac{1}{5}$ is lowered by two octaves, to get $\frac{4}{5}$, which is is the major third	
5	Second acute octave
Because $\frac{5}{20} = \frac{1}{4}$	
6	Same interval as in 2
Because $\frac{6}{20} = \frac{3}{10}$ and $\frac{3}{10}$ is the same as $\frac{1}{10}$	
7	Same interval as 1
8	Same interval as 4
9	Same interval as 1
10	Acute octave

The last part of the *Histoire* of 1701 brings back a very brief comment on Sauveur's plan for renaming the notes; this harsh comment is perhaps due to Fontenelle's skepticism about the outcome of this operation, although he later used words of great admiration for Sauveur. Fontenelle claimed that if languages had been invented by philosophers, they would undoubtedly have been much easier to learn. What Sauveur has done for music is a kind of philosophical language, composed of monosyllables, in which the meaning of the words depends only on the number, nature and combination of the letters, which give a meaning that is not ambiguous, as happens in ancient language, where, for example, thirds and fourths do not always have the same value. Fontenelle concluded:

> The signs in which it is used to write music have been artfully chosen and arranged, it is not left to musicians to decide on a wonderful practice, easier and not more difficult. However, the general acceptance of Si, which has occurred only 20 years ago, greatly facilitating the formation of scales, gives some hope [36].[155]

[155] p. 137.

5.4 1702. *Application des Sons Harmoniques à la Composition des Jeux d'Orgues*

Sauveur's memoir *Application des sons harmoniques à la composition des jeux d'orgues* of 1702 aims to capture the organization of organ sounds better than in the past by using the concept of harmonic sounds.

In this memoir, Sauveur assumed that the frequency of vibration of the sounds of organ pipes is inversely proportional to their length, as if this were a generally accepted result. He did not mention how he had arrived at this conclusion, whether on a theoretical basis or with reference to construction practices. The acoustic theory of organ pipes at that time did not provide definitive results. Vincenzo Galilei in 1589 [41] had proposed an inverse dependence of frequencies on the cube root of the volumes of air contained in the pipes. Mersenne, a possible source for Sauveur, explicitly denied the inverse frequency-length relation.

> As I have already shown in the second book of movements, it can be seen that cylinders of equal width and different lengths do not at all produce pitches that are in the same proportion to each other as their lengths, e.g. that what is twice as long does not at all make an octave, as many believe [56].[156]

Note that in denying the inverse dependence, Mersenne also said that many believed in it. More explicit statement from mathematicians supporting this belief came only with Millet De Chales, who in 1674 introduced the concept of the *aerial string* to explain the operation of wind trumpets: "All air in the trumpet vibrates as one string" [23],[157] [7]. Suggestions may also have come from Robartes, who assimilated the behavior of the ordinary trumpet, in which the vibrations depend on the air in it, and the trumpet marine, in which the sounds produced by one string are harmonic.

Ignace Gaston Pardies (1636–1673), through Pierre Ango (1640–1694), assumed that the elasticity of air due to atmospheric pressure played a decisive role in the acoustics of pipes: "the force of the column of air acting on the opening of the pipes" [7, 8],[158] [2].[159] Sauveur, almost certainly did not know Robartes' work, but he may have known those of De Chales and Pardies.

In fact, for a mathematical expression showing that the fundamental frequency of an organ pipe is proportional to the inverse of its length, one has to wait Leonhard Euler, 25 years after Sauveur's paper [9, 32]. Euler presented his theory in his *Dissertatio physica de sono* of 1727. Here he relied on the analogy between the string and the column of air contained in the pipe; the length of the string is for the length of the pipe, the tension of the string is for the atmospheric pressure, the specific weight of the string is for the specific weight of the air [32].[160]

---

[156] Part II, Livre sixiesme des orgues, p. 333.

[157] pp. 23–24.

[158] p, 175.

[159] p. 43.

[160] p. 9.

In the face of the uncertainties of the theorists, there was a construction practice that has been established for centuries which did not follow the reasonings of the theorists, but took into account the musical performance, and we have practical treatises on pipe sizing, in which the pitches of the pipes vary more or less in inverse proportion to the length of the pipes. Among these treatises, a 12th century manuscript *De commensuralitate fistularum et monocordi* by Gerbert d'Aurillac (c. 950–1003, Pope Sylvester II) is interesting [47, 83].

Obviously, the clues that came from the practice of making pipes were not easy to grasp, since pitches, as will become clear in the 18th century, do not depend exactly on the actual length of the pipes, but on what is called the equivalent length, which is always greater than the actual length and which also takes into account the width of the pipes; hence the various attempts to correlate pitches with volumes. Sauveur ignored this fact, perhaps out of a need for classification, on the assumption that it was merely a disturbing effect.

The memoir begins with the introduction of the object, the organ. It is a piece of furniture containing pipes grouped in homogeneous blocks, i.e. blocks with pipes of different length but of the same type and therefore of similar color, called *ranks*. There are usually as many pipes in each rank as there are keys on the organ keyboard. Ranks can be grouped together to form *stops*; by extension, stops also denote buttons or levers that activate the ranks of the stops. If there is only one rank, the stop is called simple, otherwise compound. Each rank is defined by the length of its longest pipe. This length cannot vary arbitrarily, but must be related to the length of the longest pipe of the organ according to their octaves or fifths, so there are organ pipes of 32, 16, 8, 4, 2 feet [21].

Going back to his work of 1701, Sauveur compared the sounds emitted by the different pipes of the organ with the sound of the longest pipe, which he called the first harmonic sound, modifying somewhat the designation used in the memoir of 1701 and adopting the modern convention. Thus, he could say that a 16-foot pipe produces a tone with a pitch equal to the second harmonic of the 32-foot pipe, rather than saying that it is one octave higher. The fifth of the second octave is the third harmonic, the second octave is the fourth harmonic, and so on. This is mainly a change in terminology that makes it easier to describe organ sounds, but not only that.

The name of an organ is derived from the length of the longest pipe; for example, if the longest pipe is 32 feet, it is called a 32-foot organ. A church organ has two parts; the large organ body, also called the *grand orgue*, and the *positif*, which is a sideboard usually placed in front of the main organ.[161] The organ may have only one keyboard if it is of one body, it has at least two if there is a positif. Large organs may have up to four or five keyboards or manuals (i.e. keyboard played by hands). The pedals also have their own keyboards (known also as the pedals).

---

[161] In current organ usage, positif and grand orgue also mean respectively: (a) the manual of a larger organ that resembles (and perhaps historically originated in such a smaller organ [87], Entry: *Positive*. Williams P, Thistlethwaite) (b) the main manual of the organ [87], Entry: *Great organ*. Williams P.

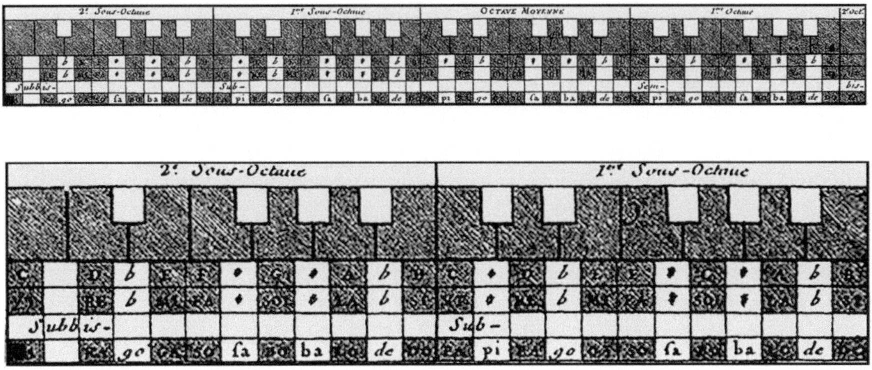

Fig. 5.41 Keyboard of an organ; detail from [74], p. 32, Planche XII

Figure 5.41 shows a keyboard divided, as usual, into four octaves, namely the second sub-octave, the first sub-octave, the middle octave, the first octave and a key of the second octave. For a better overview, the entire keyboard is shown in the upper part of the figure, and an enlarged image of the first two octaves is shown in the lower part.

> Each octave is divided into 12 keys of which the seven black ones mark the natural sounds and the five white ones mark the alterations, i.e. the sharps and flats. [74].[162]

In the second sub-octave the first alteration is missing, which is compensated by a third octave key, so that a normal keyboard has 48 keys. Some organists use one or two extra keys in addition to the second sub-octave. In addition, the pedals have two or three octaves at the organist's option, so the total number of keys is indeterminate.

Below the keyboard are the letters, sharps and flats, used by the makers to indicate the keys and the division of their ruler, which they call the diapason. The fourth row contains the names Ut, Re, Mi, etc., which are assigned to the keys when playing in B natural. The fifth row contains the keys of each octave according to the new symbology, and the sixth row contains the names of the new notes. Each key opens a valve that allows air to enter the holes of one of the pipes of the ranks housed in the wind-chest.[163]

The pipes that make up the organ ranks are of different types, as shown in Fig. 5.42. Some are labial, like a sweet flute, and some are reed. Labial pipes consist of the foot, AABB, which is shaped as a hollow cone (see Fig. 5.43). It is this cone that receives the air that makes the sound. Attached to the foot is the pipe body BBDD; between the foot and the body is a diaphragm EF, the *languid*, which lets the air out through a thin, wide slit, the *mouth*. Above and below the mouth there are upper and

---

[162] p. 309.

[163] The wind-chest is a structure (a box) traditionally parallel to the keyboard. It is the place where the air, generated under pressure by whatever winding system is used, goes. Pipes are located on the wind-chest and receive the air from it when a valve is open [16], p. 633.

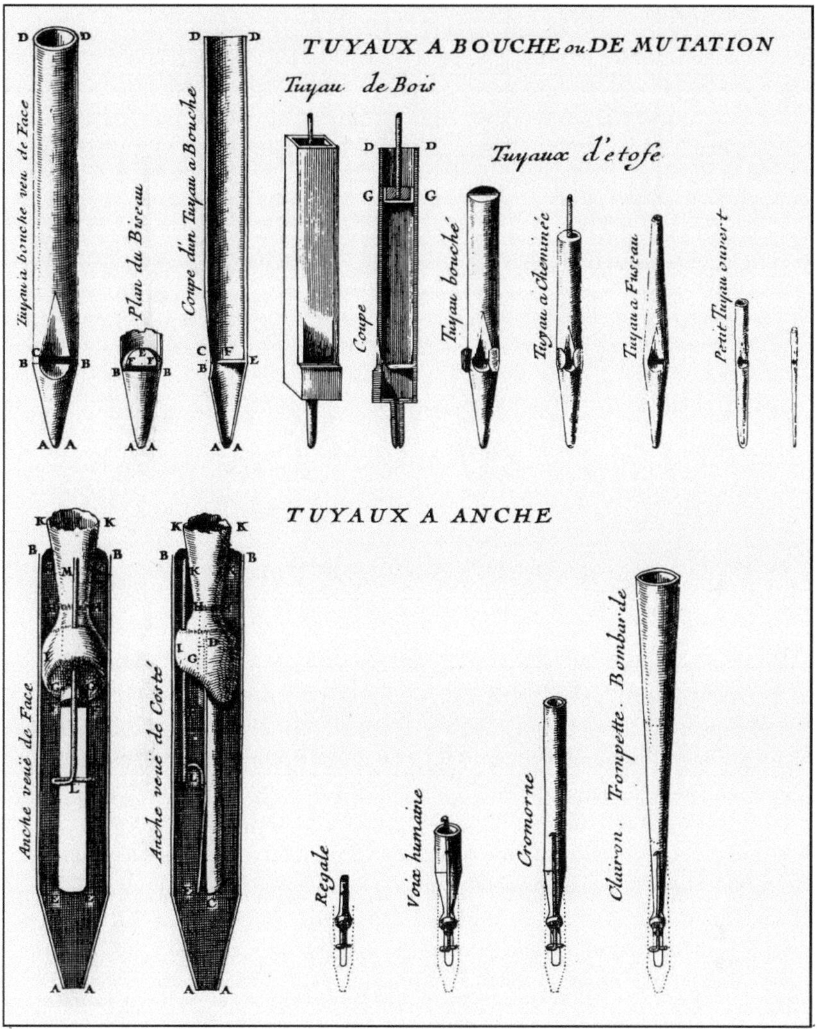

Fig. 5.42 Different kinds of pipes; from [74], p. 328, Planche XI

lower lips. The pipes can be made of pure tin, spotted metal (an alloy of lead with $^1/_{12}$ tin) or wood. Pure tin pipes are always open at the end DD, are very narrow and their sound is very bright and clean. Spotted metal pipes are wider than tin pipes. Wooden pipes are square and are closed by a leather-covered piston GG (see Fig. 5.42). The sound of wooden and spotted metal pipes is very sweet.

Long pipes produce the lowest notes, short pipes the highest. Closed pipes are only half as high as open pipes. The longest open labial pipe is usually 16 feet (a

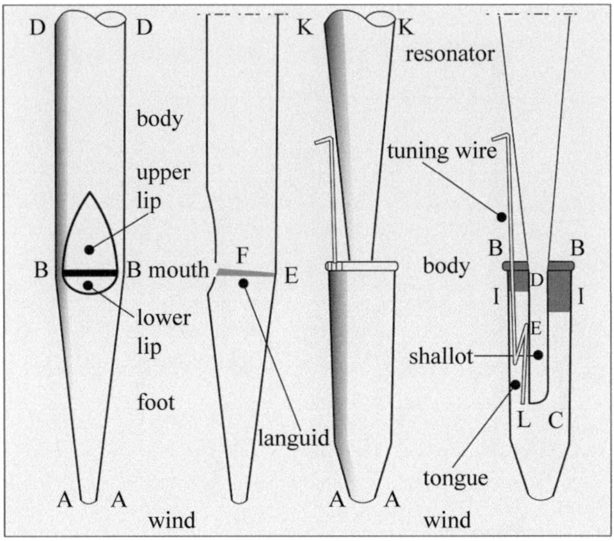

Fig. 5.43 Flue (left) and reed (right) pipes. Adapted from [14], p. 493

little less than 5 m), but can be up to 32 feet, the shortest 4.5 lines (about 1 cm). Pedal pipes are always open and are made of wood or spotted metal.

A reed pipe consists of a foot AABB, which directs the flow of air into the shallot, consisting of a hollow semi-cylinder CD, contained in a block II, which is open at the top to let the wind pass from the shallot. A metal foil fixed at the top of the shallot and free at the bottom is the tongue. Its length EL is adjusted by a tuning wire. From the top KK comes a tube, the resonator, which can have different shapes and lengths. The sound, high or low, of a reed pipe depends on the length of the tongue and the length of the resonator. The quality of the sound depends on the shape of the reed pipe, the material it is made of and its density, and the width of the resonator. Pedal reeds have stronger tubes, wider shallots, denser tongues, and also receive more wind.

5.4.1 *The Stops of an Orgue*

Table 5.25, divided into five parts separated by double lines, gives the names of all the simple stops of an organ, one for each row. The first part contains two columns; Col. 1 gives the octaves, thirds and fifths of the stop. Col. 2 gives the number of the harmonic sound associated with the longest pipe of the stop in large font, and the harmonic sound of the last and shortest pipes in small font.

The second part of the table gives the length of the longest pipes of the stops, in octaves only, since the makers refer to earlier octaves for thirds and fifths. Col. 1

Table 5.25 Different kinds of stops. Redrawn from [74], p. 328. Planche XI

Octaves	Sons Harmoniques du 1er et dernier Tuyau	JEUX A BOUCHE ou DE MUTATION. Longueurs des Tuyaux		Jeux dont les Tuyaux sont d'estain fin et Ouverts.	Jeux dont les Tuyaux sont de Bois ou d'étoffe				JEUX A ANCHE.				
		Ouverts	Bouchés		Par Octaves	Par Tierces	Par Quintes	Jeux de Pedales	Jeu dont le Tube est court et estroit	Jeu dont le Tube est court et large	Jeu dont le Tube est long et Cilindrique	Jeux dont les Tubes sont longs et Evasez	Jeux de Pedales
Son Fondamental	1. 16	32 pieds	16 pieds	le 32. Pied $\dot{1}$	le Bourdon de 16. Pieds 1								
1ere octave	2. 32	16 pieds	8 pieds	le 16. Pied $\dot{2}$	le Bourdon de 8. Pieds 2							la Bombarde 2	Pedale de Bombarde 2
Quinte	3. 48	leur quinte											
2e octave	4. 64	8 pieds	4 pieds	le 8. Pied $\dot{4}$	le Bourdon de 4. Pieds 4			Pedale de Flute de 8. Pieds 4	la Regale 4″	la Voix humaine 4′	le Cromorne 4	la Trompette 4	Pedale de voix humaine 4′ de Trompette 4
Tierce	5. 80	leur Tierce											
Quinte	6. 96	leur quinte					le Double Nazard 6						
3e octave	8. 128	4 pieds	2 pieds	le Prestant $\dot{8}$	la Flute 8			Pedale de Flute de 4 Pieds 8				le Clairon 8	Pedale de Clairon 8
Tierce	10. 160	leur Tierce				la Tierce 10							
Quinte	12. 192	leur quinte					le Nazard 12	Pedale de Nazard 12					
4e octave	16. 256	2 pieds	1 pied	la Doublete $\dot{16}$	la quarte de Nazard 16								
Tierce	20. 320	leur Tierce				La Petite Tierce ou Tiercette 20							
Quinte	24. 384	leur quinte					le Larigot 24						
5e octave	32. 512	1 pied	6 pouces	le Flageolet $\dot{32}$									

gives the length of the open pipes; Col. 2 the length of the closed pipes, which is half the length of the open pipes.

The third part of the table contains the names of the stops with tin pipes: the 32 feet, the 16 feet, the 8 feet, the *prestant*, the *doublette*, and the *flageolet*.[164] The harmonic sounds of tin stops are identified by a number with a dot on it: $\dot{1}$, $\dot{2}$, $\dot{4}$, $\dot{16}$, $\dot{32}$. The largest of these stops is called a montre.[165] Thus the montre may be a 32, a 16, a 8, and a 4 feet.

For a better understanding of the connection between the six tin stops, see Table 5.26, which shows the 4 octaves of each stop. Each stop spans 4 octaves, the 32 feet from 1 to 4, the flageolet from 5 to 9, for a total of 9 octaves.

The fourth part of Table 5.25, in Cols. 1-4 contains the names of flue stops with wooden or spotted metal pipes. Col. 1 contains the stops that go by octaves, i.e., the *bourdons*, the *flute* and the *quart the nazard*. Col. 2 contains the pipes that go by thirds of the previous octaves, i.e. the *tierce* and the *petite tierce* or *tiercette*. Col. 3 shows the pipes that go by fifths, namely the *double nazard*, the *nazard* and the *larigot*. These wooden or spotted metal pipes are marked 1, 2, 4, 8, 16; 10, 20; 6, 12, 24.

---

[164] In the following the names of the stops are not translated into English, because some of them are typical of French organs; on these points see [18].

[165] Sauveur's definition is quite different from the standard one, in French organ *montre* is in fact generally the same as *principal*, which is not always the stop with the longest pipe.

Table 5.26 Tin stop connections; from [74], p. 314

octaves	0	1	2	3	4	5	6	7	8	9
diatonic intervals	I	VIII	XV	XXII	XXIX	XXXVI	XLIII	L	LVII	LXIV
harmonics	1	2	4	8	16	32	64	128	256	512
	feet						inches		lines	
lenght	32	16	8	4	2	1	6	3	18	9
stops	the 32 feet (0–4)				the doublette (4–8)					
		the 16 feet (1–5)				the flageolet (5–9)				
			the 8 feet (2–5)							
					the prestant					

Col. 4 contains the stop pedals, i.e. the *pédales de flûte de 8*, the *4 pieds* and the *nazards*. The pipes of the pedals are open and larger than the others. They are indicated by numbers with a dash underneath, such as $\underline{4}$, $\underline{8}$, $\underline{12}$.

The fifth part of Table 5.25 concerns reed stops, the variety of which depends mainly on the shape of the pipes. Col. 1 contains the *regale*, whose pipes are short and straight. It is called $4''$. Col. 2 is for the *voix humaine*; it has short and wide pipes. It is called $4'$. Col. 3 is for the *cromorne*, which has long, cylindrical pipes. It is denoted by $\grave{4}$. Col. 4 is for the *bombard*, the *trompette* and the *clairon*, which have conical pipes. They are denoted by $\hat{2}$. $\hat{4}$. $\hat{8}$. In the fifth column are the pedal stops that are larger and have ticker reeds, the *pédales de bombard* (which is rare), the *trompette*, the clairon, and the *voix humane*.

Regale are usually found only in small organs. In large organs, a second trompette stop, controlled by a special keyboard, is sometimes added; it is called the *petite trompette*. To vary the sound of the organ stops, a *tremblant* is also added, which is a valve that allows the flow of air to come out in sections as it opens and closes. It can be powerful or sluggish.

Tables 5.27 and 5.28, which should be seen as a single table, list the harmonic sounds of the first pipe of each stop. Each table has three parts. The first part consists of three columns. Col. 1 shows the diatonic intervals (of the first pipe) of each stop, relative to the fundamental. Col. 2 contains the intervals in octaves, showing thirds and fifths. Col. 3 contains the lengths of the open pipes, from 32 feet to $4\frac{1}{2}$ lines.[166]

The second part contains the harmonic sounds of all the pipes of each stop. In Col. 1, corresponding to sub-bis-PA, there are the harmonic sounds 1, 2, 3, 4, 5, 6, 8, 10, 12, 16, 20, 24, 32 of the first pipes of the stops. For the compound stops, the harmonic sounds numbered 48, 64, 96, 128, 192, 256, 384, 512, 768, 1024 are used.

Combining this column with the previous three gives the link between the harmonic sound of each stop and the lengths of the pipes, octaves, and diatonic intervals. For example, the number 24, representing the 24th harmonic, corresponds to a pipe

[166] The ratio between the minimum and maximum lengths of the pipes is $32 \times 12 \times 12 : 4.5 = 1024$.

Table 5.27 Sub-bis and sub-octaves; from [74], p. 32, Planche XII

			2e Sous-Octave												1re Sous-Octave											
			C		D	♭	E	F	♯	G	♯	A	♭	B	C	♯	D	♭	E	F	♯	G	♯	A	♭	B
			UT		RE	♭	MI	FA	♯	SOL	♯	LA	♭	SI	UT	♯	RE	♭	MI	FA	♯	SOL	♯	LA	♭	SI
			PA		RA	go	GA	SO	la	BO	ba	LO	de	DO	PA	pi	RA	go	GA	SO	la	BO	ba	LO	de	DO
Son dame	Fon. neal	32 pieds	1												2							3				
VIIIe	1ere Octave	16 pieds	2							3					4				5			6				
XIIe	Quinte	10⅔ pieds	3					4							6					8		9		10		
XVe	2e Octave	8 pieds	4				5			6					8		9		10			12				15
XVIIe	Tierce	6⅖ pieds	5			6					8		9		10			12				15	16		18	
XIXe	Quinte	5⅓ pieds	6					8		9		10			12				15	16		18		20		
XXIIe	3e Octave	4 pieds	8		9		10			12				15	16	16⅔	18	19⅕	20	21⅓	22½	24	25⅗	26⅔	28⅘	30
XXIV	Tierce	3⅕ pieds	10			12				15	16		18		20			24	25			30	32		36	
XXVI	Quinte	2⅔ pieds	12				15	16		18		20			24	25	27	28⅘	30	32	33¾	36	38⅖	40	43⅕	45
XXIX	4e Octave	2 pieds	16		18	19⅕	20	21⅓	22½	24	25⅗	26⅔	28⅘	30	32	33⅓	36	38⅖	40	42⅔	45	48	51⅕	53⅓	57⅗	60
XXXI	Tierce	1⅗ pieds	20			24	25			30	32		36		40		45	48	50			60	64		72	75
XXXIII	Quinte	1⅓ pieds	24		27	28⅘	30	32	33¾	36	38⅖	40	43⅕	45	48	50	54	57⅗	60	64	67½	72	76⅘	80	86⅖	90
XXXVI	5e Octave	1 pied	32		36	38⅖	40	42⅔	45	48	51⅕	53⅓	57⅗	60	64	66⅔	72	76⅘	80	85⅓	90	96	102⅖	106⅔	115⅕	120
XLe	Quinte	8 pouces	48		54	57⅗	60	64	67½	72	76⅘	80	86⅖	90	96	100	108	115⅕	120	128	135	144	153⅗	160	172⅘	180
XLIII	6e Octave	6 pouces	64		72	76⅘	80	85⅓	90	96	102⅖	106⅔	115⅕	120	128	133⅓	144	153⅗	160	170⅔	180	192	204⅘	213⅓	230⅖	240
XLVII	Quinte	4 pouces	96		108	115⅕	120	128	135	144	153⅗	160	172⅘	180	192	200	216	230⅖	240	256	270	288	307⅕	320	345⅗	360
Le	7e Octave	3 pouces	128		144	153⅗	160	170⅔	180	192	204⅘	213⅓	230⅖	240	256	266⅔	288	307⅕	320	341⅓	360	384	409⅗	426⅔	460⅘	480
LIVe	Quinte	2 pouces	192		216	230⅖	240	256	270	288	307⅕	320	345⅗	360	384	400	432	460⅘	480	512	540	576	614⅖	640	691⅕	720
LVIIe	8e Octave	18 lignes	256		288	307⅕	320	341⅓	360	384	409⅗	426⅔	460⅘	480	512	533⅓	576	614⅖	640	682⅔	720	768	819⅕	853⅓	921⅗	960
LXIe	Quinte	12 lignes	384		432	460⅘	480	512	540	576	614⅖	640	691⅕	720	768	800	864	921⅗	960	1024						
LXIV	9e Octave	9 lignes	512		576	614⅖	640	682⅔	720	768	819⅕	853⅓	921⅗	960	1024											
LXVIII	Quinte	6 lignes	768		864	921⅗	960	1024																		
LXXI	10e Octave	4½ lignes	1024																							

length of 1 and 1/3 feet; it is the fifth of the fourth octave, or XXXIII with respect to the fundamental.

To obtain the harmonic sounds corresponding to the different keys, we move horizontally. For example, taking the fourth octave of nazard 16 (the 2-foot pipe marked with a dark gray box in Table 5.27), the harmonic associated with sub-bis-RA is 18, that associated with sub-bis-GA is 20. For sub-bis-go, the situation is more complicated. It is a sharp above sub-bis-RA and there is no harmonic that can give its value. Sauveur wrote the number 19 1/5, which is a sharp over 18, and claimed it could be approximated with the harmonic 19.

In fact not all notes can be represented by an appropriately chosen harmonic sound, as noted in the previous section when commenting on Table 5.21. However, Sauveur seems to ignore this fact, and when there is no harmonic sound for a note, he used a fractional number, with the idea that the integer part of this number gives a harmonic sound that approximates the note. For stops with low harmonic sounds no integer number is indicated as it should be meaningless.

Note:

1. In the organ, the pipes commanded by the key PA (Col. 1) produce, in all octaves, the exact harmonic sound with respect to the fundamental.

Table 5.28 Middle and first octave; from [74], p. 32, Planche XII

			Octave Moyenne															1re Octave												2e Oct.
			A LA LO	b b de	B SI DO	c ut PA	# pi	d re RA	b go	e mi GA	f fa SO	# ſa	g ſol BO	# ba	a la LO	b de	b ſi DO	c ut PA	# pi	d re RA	b go	e mi GA	f fa SO	# ſa	g ſol BO	# ba	a la LO	b de	h ſi DO	c ut PA
Son fondamental		32 pieds				4				5			6					8		9		10			12				15	16
VIII.e	1re Octave	16 pieds				8		9		10			12				15	16	16 2/3	18	19 1/5	20	21 1/3	22 1/2	24	25 3/5	26 2/3	28 4/5	30	32
XII.e	Quinte	10 2/3 pieds	10			12				15	16		18		20			24	25	27	28 4/5	30	32	33 3/4	36	38 2/5	40	43 1/5	45	48
XV.e	2e Octave	8 pieds			15	16	16 2/3	18	19 1/5	20	21 1/3	22 1/2	24	25 3/5	26 2/3	28 4/5	30	32	33 1/3	36	38 2/5	40	42 2/3	45	48	51 1/5	53 1/3	57 3/5	60	64
XVII.e	Tierce	6 2/5 pieds		18		20			24	25			30	32		36		40		45	48	50			60	64		72	75	80
XIX.e	Quinte	5 1/3 pieds	20			24	25	27	28 4/5	30	32	33 3/4	36	38 2/5	40	43 1/5	45	48	50	54	57 3/5	60	64	67 1/2	72	76 4/5	80	86 2/5	90	96
XXII.e	3e Octave	4 pieds	26 2/3	28 4/5	30	32	33 1/3	36	38 2/5	40	42 2/3	45	48	51 1/5	53 1/3	57 3/5	60	64	66 2/3	72	76 4/5	80	85 1/3	90	96	102 2/5	106 2/3	115 1/5	120	128
XXIV.	Tierce	3 1/5 pieds		36		40		45	48	50			60	64		72	75	80		90	96	100			120	128		144	150	160
XXVI.	Quinte	2 2/3 pieds	40	43 1/5	45	48	50	54	57 3/5	60	64	67 1/2	72	76 4/5	80	86 2/5	90	96	100	108	115 1/5	120	128	135	144	153 3/5	160	172 4/5	180	192
XXIX.	4e Octave	2 pieds	53 1/3	57 3/5	60	64	66 2/3	72	76 4/5	80	85 1/3	90	96	102 2/5	106 2/3	115 1/5	120	128	133 1/3	144	153 3/5	160	170 2/3	180	192	204 4/5	213 1/3	230 2/5	240	256
XXXI.	Tierce	1 3/5 pieds		72	75	80		90	96	100			120	128		144	150	160		180	192	200			240	256		288	300	320
XXXIII	Quinte	1 1/3 pieds	80	86 2/5	90	96	100	108	115 1/5	120	128	135	144	153 3/5	160	172 4/5	180	192	200	216	230 2/5	240	256	270	288	307 1/5	320	345 3/5	360	384
XXXVI	5e Octave	1 pied	106 2/3	115 1/5	120	128	133 1/3	144	153 3/5	160	170 2/3	180	192	204 4/5	213 1/3	230 2/5	240	256	266 2/3	288	307 1/5	320	341 1/3	360	384	409 3/5	426 2/3	460 4/5	480	512
XL.e	Quinte	8 pouces	160	172 4/5	180	192	200	216	230 2/5	240	256	270	288	307 1/5	320	345 3/5	360	384	400	432	460 4/5	480	512	540	576	614 2/5	640	691 1/5	720	768
XLIII.	6e Octave	6 pouces	213 1/3	230 2/5	240	256	266 2/3	288	307 1/5	320	341 1/3	360	384	409 3/5	426 2/3	460 4/5	480	512	533 1/3	576	614 2/5	640	682 2/3	720	768	819 1/5	853 1/3	921 3/5	960	1024
XLVII.	Quinte	4 pouces	320	345 3/5	360	384	400	432	460 4/5	480	512	540	576	614 2/5	640	691 1/5	720	768	800	864	921 3/5	960	1024							
L.e	7e Octave	3 pouces	426 2/3	460 4/5	480	512	533 1/3	576	614 2/5	640	682 2/3	720	768	819 1/5	853 1/3	921 3/5	960	1024												
LIV.e	Quinte	2 pouces	640	69 1/5	720	768	800	864	921 3/5	960	1024																			
LVII.e	8e Octave	18 lignes	853 1/3	921 3/5	960	1024																								
LXI.e	Quinte	12 lignes																												
LXIV.	9e Octave	9 lignes																												
LXVIII	Quinte	6 lignes																												
LXXI.	10e Octave	4 1/2 lignes																												

2. The pipes commanded by any key produce harmonic sounds that are all in the same proportions as those commanded by the PA key.
3. In each stop, the sounds of the key PA (or keys with the same name) of different octaves are harmonic because they are in double ratio.
4. The keys of a stop that have different names are not harmonic in the organ—no more than in a harpsichord—because they go through tempered intervals. Nevertheless, they are denoted by (fractional) harmonic sounds:

    a. Because harmonic sounds expressed in decimals are not far removed from tempered sounds.
    b. The lengths of pipes measured by the makers do not give accurate sounds.
    c. The use of these harmonic sounds is convenient in the case of compound stops.

5. To find the length of any pipe of a stop, divide 32 by the number expressing the harmonic of that pipe. For example, to obtain the length of the sub-BO pipe of doublet 16, divide 32 feet by 48, which is the number of the harmonic associated with sub-BO, and you have 8 inches [74].[167]

To measure the lengths of the pipes of each register, the makers use a divided rule called the *diapason*. To obtain the divisions of this diapason, the monochord of Fig. 5.21 can be used.

[167] pp. 317–318.

The flue pipes of an organ, as measured by the diapason, are generally out of tune; it is up to the ear of the maker to correct the defect. He begins by tuning the prestants, which, with pipes ranging from 4 feet to 3 inches, produce the sounds best perceived by the human ear.

To tune the prestant, the maker begins with the first sub-octave and proceeds only by fifths or common consonances (thirds, fifths, octaves). To tune by fifths, one must consider the order of the following 12 notes, which form a sequence of fifths [74][168]:

go. *de*. SO. PA; BO; RA; LO; GA; DO, sa, pi, ba.

Starting with any note, proceed by fifths in order forward to ba, then return to the selected note back to *go*. Since the fifths must be defective, some experience is required for tuning. Harpsichord players can provide this with the monochord.

To tune in thirds, fifths and octaves, the previous 12 notes are used again, still starting with any note, with the caveats:

1. Its octave must be just.
2. Its fifth is the one before or after the selected note in the list of 12 notes.
3. The third is major if there is a dot between two of the twelve notes, minor if there is a comma, and first major and then minor if there is a semicolon.

Once the partition is tuned, that is, after the first sub-octave has been tuned, the notes of the other octaves are tuned by octave. To tune the other stops, partitions are made by tuning one or more octaves of this stop with the octaves of the prestant, in unison with them. Table 5.26 may be helpful in finding the octaves of the stops that can be tuned [74].[169]

In general:

1. To tune closed wooden reeds, push the plunger down to raise the pitch and pull it up to lower it.
2. For flue or spindle pipes, widen or narrow the tip with the tuning cone, which is a concave or convex copper cone.
3. The flanged pipes by opening or closing the flanges.
4. The reed pipes by acting on the tuning wire.

So far we have only considered the simple stops and their overtones. In the organ there are also three compound stops, the *cornet*, the *fourniture* and the *cimbale*.[170]

The cornet is a stop composed of 5 half-ranks, beginning with the PA key of the middle of the harpsichord and contains only the middle octave, the first octave and a key of the second octave. The harmonic sounds of the 5 pipes of the PA key are 16, 32, 48, 64, 80. The harmonic sounds of the bis-PA key (two octaves above) are 64, 128, 192, 256, 320, proportional to 1, 2, 3, 4, 5. The pipe lengths corresponding to these sounds can be found in the column for the sub-bis-PA key.

---

[168] p. 319.

[169] pp. 319–320.

[170] Sauveur did not mention other types of compound stops.

There are three types of cornets: the grand cornet, the cornet separé, also called the recit, and the cornet d'écho, which is hidden at the foot of the organ. The three cornets differ only in the size of the pipes and the strength of the wind.

The fourniture is made up of many ranks. Starting with the first pipe, the height is decreased to a certain point, then resuming with a longer pipe, decreasing again, and so on for 6, 7 times. The first pipes of each rank, corresponding to the sub-bis-PA key, are defined by 16, 24, 32, 48, 64, 96, 128, 192, 256, 384 harmonic sounds, ordered by octaves and fifths of the fundamental; there are no thirds.

In smaller organs, the 16th harmonic and some of the sharper pipes are omitted. In all organs, the last pipe activated by the bis-PA key must be the 11th or 12th of the first pipe, or must produce a harmonic sound that is at most triple that of the first sound.

In general, there are four ways to make a compound stop in the fourniture, called repetitions.

1. Equal above, that is, by making equal the shortest (and therefore higher-pitched) pipes of each repetition, except the last; for example, 24 ⋯ 40; 27 ⋯ 40; 28 ⋯ 40; 32 ⋯ 40; 28 ⋯ 40; 32 ⋯ 40; 28 ⋯ 64.
2. Equal below, i.e. make the longest (and therefore lowest pitched) pipes of each repetition take equal except the first; e.g. 24 ⋯ 45; 32 ⋯ 40; 32 ⋯ 45; 32 ⋯ 40; 32 ⋯ 45; 32 ⋯ 64.
3. Alternating equal above and below, like 24 ⋯ 30; 32 ⋯ 45; 36 ⋯ 50; 36 ⋯ 51; 40 ⋯ 57; 40 ⋯ 64 (with 51 considered the same as 50).
4. Highest and lowest both ascending, like 24 ⋯ 40; 28 ⋯ 42; 33 ⋯ 50; 36 ⋯ 53; 43 ⋯ 60; 42 ⋯ 64.

Once the repetitions of the first rank have been selected, we proceed to the repetitions of the second rank, which have the same configuration as those of the first rank in proportional proportion.

To show how the makers indicate the ranks of each repetition, let us consider, for example, stop 24, with its equal above repetitions, referring to Fig. 5.44.

1. The first column gives the numbers of the harmonic sounds of stop 24 up to two octaves.
2. Use these numbers to go to the row of Table 5.27, marked by the darker gray box, corresponding to the fourth octave.
3. Search for the values of the letters corresponding to these numbers (approximate to integer if needed). This gives the letters of the diapason of the makers.
4. Indicate the six repetitions by writing 1 next to each repetition.
5. Similarly, 0 is written for the second rank.

The cimbale is similar to the fourniture, except that it begins one fifth or octave higher. So if the fourniture starts at 24, the cimbale starts at 32 or 48. Thus the cimbale has fewer ranks than the fourniture. The repetitions are similar.

So far we have given all the simple and compound ranks that can be placed in an organ. But the number of simple stops and the ranks of the compound stops are adjusted by the makers according to the size of the sideboard in which the organ is to

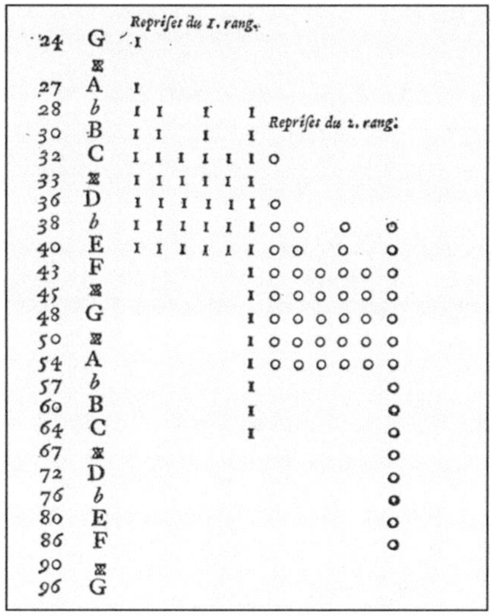

Fig. 5.44 Representation of the stop fourniture, from above repetition; from [74], p. 323

be placed, the nature of the places, the taste of the organist, and the cost. Therefore, nothing more can be done than to describe each individual organ.

5.5 1707. *Méthode Général pour Former les Systêmes Temperés de Musique*

In the *Méthode général pour former les systêmes temperés de musique* of 1707, Sauveur revisited the problem of possible divisions of the octave, which had already been addressed in the *Traité* and taken for granted in the memoir of 1701. He did so, however, with greater maturity and clarity; many concepts are explained again. For the sake of completeness, and especially because there are new nuances in the following, we will quote them in large part.

Sauveur began by stating that the diatonic system is the one followed in Europe because it is considered the most natural. It divides the octave into major semitones and minor and major tones, with notes called Ut, Re, Mi, Fa, Sol, Si, Ut, which, according to him, should be renamed PA, RA, GA, BO, TO, DA, PA. Consider that Sauveur did not insist on this point, and throughout the memoir he referred only to the 'traditional', or old notes, as he called them [203]. It is as if he had accepted that his systems had not received the attention he had hoped for, and he refrained from insisting on it.

The chords that give perfect consonances are the octave, the fifth, and the fourth; those that give imperfect consonances are the major and minor thirds and sixths. The dissonances are the major and minor second and seventh, the tritone and the false fifth. The smaller intervals that form the elements of the preceding intervals are the

Table 5.29 Diatonic intervals; from [75], p. 204

24	27	30	32	36	40	45	48	54	60	64	71	80	90	96
Ut	Re	Mi	Fa	Sol	La	Si	ut	re	mi	fa	sol	la	si	ut
T	t	S	T	t	T	S	T	t	S	T	t	T	S	

major semitone, the minor, and the major tones. As a possible representation of two octaves of the diatonic system, Sauveur considered the numerical values in Table 5.29.

These are, at least for the first octave, the same values reported in the 1701 memoir. Sauveur pointed out the arbitrariness of these numbers, which could be replaced by their multiples. From Table 5.29 we see that:

1. All the octaves are equal to each other: Ut-ut, Re-re, etc.
2. There are three types of seconds; one minor (S) as Mi-Fa and two major (T, t) as Ut-Re and Sol-La.
3. The minor thirds defined by the interval TS are just, such as Mi-Sol, La-Ut, Si-Re. Those defined by t S are lower of a comma, such as Re-Fa. Major thirds, defined by the interval Tt, are all equal and just.
4. The (minor) fourth (TtS) is just between Ut-Fa, Re-Sol, Mi-La, Sol-Ut, Si-mi. But it is larger of a comma between La-Re (TTS). The major fourth or tritone (TtT) is between Fa-Si.

The large intervals as the fifths, sixths, and sevenths are the complements of the small intervals and therefore are minor or major in inverse order to those. Hence we have that:

1. To obtain a large interval, the elements of the small interval must be subtracted from the octave (3T 2t 2S). Thus the fifth as the complement of the fourth (TtS) is (2TtS).
2. The same variety is encountered in large intervals as in the small intervals. Thus the minor second and the major third being just intervals give rise to a major seventh and a minor sixth, both just. Small intervals altered by a comma give rise to large intervals still altered by a comma but in the opposite direction so the fourth La–Re being too large by a comma. gives rise to a fifth Re–La smaller by a comma [75].[171]

Musical instruments are divided into three classes. In the first class there is the human voice, the violin and instruments whose accuracy of notes depends only on the musician's ear. In the second class formed by trumpets, flutes, theorbos, guitars, etc., the sound is regulated by projections, holes, frets but the musician has room for adaptation. In the third class formed by the organ, harpsichord and instruments where the sound depends only on the keys of a keyboard, the musician cannot make any adjustments.

[171] p. 205.

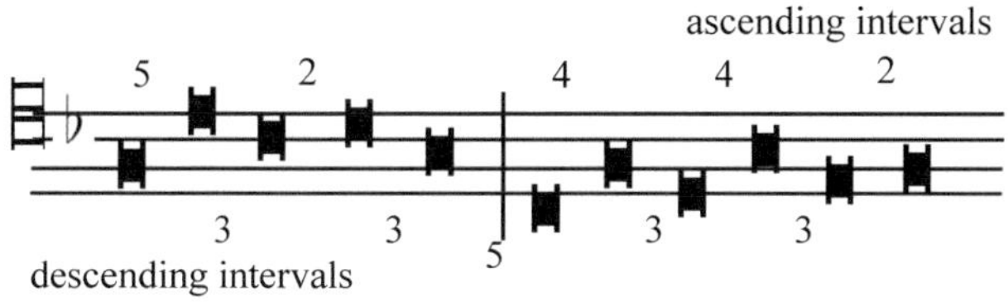

Fig. 5.45 Migration of pitches. Redrawn from [75], p. 207

The just diatonic system cannot be applied to all the three classes because:

1. If the instruments of the third class are tuned to the intervals defined by Table 5.29, when it is required to make an altered interval of a comma in order to have perfect consonance, such as when after La you want to go up a just fourth, it cannot be done, because the La–Re chord—which is a fourth—turns out TST, larger by a comma than the just one (TSt).
2. For instruments of the first and second classes one must reason in the same way when dealing with beginner or poor musicians, because for the first class they are used to thinking about the sounds Ut, Re, Mi Fa, etc.; while for the second class they cannot make the necessary adjustments.
3. But even the most experienced musicians cannot follow the just diatonic system without problems. In fact they usually begin and end a piece of music going up and down by various intervals, and it may happen that the last note is a little higher or a little lower than the one from which they started (what has ben called, in Chap. 3, the migration of pitches [75]).[172]

Consider, for example, an elevation of a litany beginning with Ut as shown in Fig. 5.45,[173] which, said Sauveur, a religious sent to R. P. Bussier.[174] Sauveur admitted that singers performed intervals that are always the same, although a strict reading of the staff seems to indicate otherwise. Thus, thirds are always minor, T S, fourths are just, T t S, and seconds are minor, S.

Since the fifth is repeated in ascents and descents it is not considered in the counts of intervals. This being the case, the intervals that go up are 2, 4, 4, 2, that is two minor seconds and two fourths, whose elements are 2T 2t 4S; the intervals that go down are 3, 3, 3, 3, that is four minor thirds, whose elements are 4T 4S; subtracting from these intervals the common ones gives 2t and 2T, or 2t and 2t 2c, respectively. So the intervals going up are two syntonic commas (2c) shorter than those going down, and thus the last note is two commas lower than the initial Ut.[175] If one were to repeat the litany starting from the last ut 55 times, for example, the last note would be 110 commas higher than the starting note, or two octaves.

---

[172] p. 206.

[173] The solmization chosen by Sauveur is that corresponding to B mol, see Fig. 5.6.

[174] We have not been able to obtain information about this musician.

[175] There is a typo in Sauveur's text, as he said that the pitches goes up and not down.

This fact seemed absurd to the clergyman who, taking it for granted that singers intone notes correctly, became convinced that musical intervals are different from what is currently believed, for example, that the fifth does not consist at all in the ratio of (3 : 2) but in some other value that would avoid the inconvenience of rising peaks. But, suggested Sauveur, it can be answered to him that the justness of singers' voices is not certain, and it is not certain that they cannot deviate from the accuracy of some part of a comma without noticing it. It would be better, therefore continued Sauveur, to say, with Huygens in his *Cosmotheoros*, see Sect. 3.2.3.3, that a singer preserving in memory the sound of the first Ut he naturally returns to it with an imperceptible alteration of the intervals. Here the need for a tempered system [207].

The Way of Forming Tempered Systems.
Sauveur recognized the difficulties of the pure diatonic system in the presence of two different tones and set out to introduce a system in which all tones were equal.

> The diatonic system is the basis of the chromatic and harmonic systems; it has 3T 2t 2S intervals as its elements. To make this system tempered, one must take a meantone T between T and t, so that the octave is composed of five mean tones and two major semitones [75].[176]

In the present memoir Sauveur, assumed, as in the third and fourth methods of temperament discussed in the *Traité*, that the octave has the form $5T + 2\mathrm{S}$, where T and S are respectively a tempered meantone and a tempered major semitone.

To understand Sauveur's reasoning, we must remember that he thought of intervals as logarithms. So to say that the interval of the octave is made up of five tones and two semitones, instead of the product $T^5\ \mathrm{S}^2$, he referred to the sum of the logarithms $5T + 2\mathrm{S}$. And if the octave is formed by 12 minor semitones s and 7 commas c, instead of the product $\mathrm{s}^{12}\ \mathrm{c}^7$ he referred to the sum $12\mathrm{s} + 7\mathrm{c}$. Furthermore, if Δ is the logarithm of the interval obtained by dividing the octave into n parts, that is, $n\Delta = 301,030$, the octave can be written as $12p\Delta + 7q\Delta = (12p + 7q)\Delta$, where p and q are integers. Sauveur simplifies everything by ignoring Δ and denoting p by s and q by c. He can say that the octave has the form $12\mathrm{s} + 7\mathrm{c}$, where s and c are integers.

In fact, Sauveur did not pose the problem very clearly; he began by defining the minor semitone s as the difference between the meantone and the major semitone; an ambiguous concept, since the meantone and the major semitone have not yet been calculated; only two boundary values can be assumed: the major and the minor. He then redefined the comma c as the difference between the major and minor semitones, which is generally different from both the syntonic and Pythagorean comma. With these definitions we have $\mathrm{S} = \mathrm{s} + \mathrm{c}$, while the meantone is $T = 2\mathrm{s} + \mathrm{c}$; so the octave will be $12\ \mathrm{s} + 7\mathrm{c}$.

A more traditional but equivalent approach, also suggested by Fontenelle (see below), to obtain the tempered systems is to imagine the tempered octave as consisting of 12 semitones, 7 major (diatonic semitones) and 5 minor (chromatic semitones): $7\ \mathrm{S} + 5\ \mathrm{s}$. This approach reduces to Sauveur's by introducing the comma $\mathrm{c} = \mathrm{S} - \mathrm{s}$.

[176] p. 208.

To obtain an octave division, one must choose values for c and s, but since they are two undefined quantities, one can only look for the interval in which their ratio can fall. For the time being, it is better to interpret s and c not as integers, but as the logarithm of the smallest semitone and the comma, respectively. We start with the relations $S = s + c = 28.0287$, $t = 2s + c = 45.7575$ or $T = 2s + c = 51.125$, depending on whether we take the meantone to be its minimum value t or its maximum value T. From these relations, Sauveur obtained two values, $c = 10.2999$ and $c = 4.9049$, which correspond to $s = 17.7288$ and $s = 23.1238$, respectively, giving the maximum of the ratio $s/c = 23.1238/4.9049 = 4^5/_7$ and the minimum $s/c = 17.7288/10.2999 = 1^5/_7$ [209]. For simplicity, the maximum and minimum values for the ratio s/c, $4^2/_3$ and $1^2/_3$, respectively are chosen.

The number of divisions of the octave depends on the values of s and c, or rather their ratio. For example, if c = 1, s = 2, 3, 4, the octave is divided into 31, 43, 55 parts, an arithmetic progression of reason 12. For c = 2 and s = 4, 5, 6, 7, 8 or 9, the octave is divided into 62, 74, 86, 98, 110, still an arithmetic progression of reason 12. For c = 3 there is an even finer division of the octave.

Table of tempered systems compared with the just diatonic system.
To divide the octave, we are going to use the following values: c = 1, s = 2, 3, 4, resulting in the values 31, 43, 55. No sequences are considered for higher values of c, since this would result in too large a number of parts. Instead, the case c = 0, which leads to the division of the octave into 12 equal parts is considered, "because it is very simple, it has its partisans" [75].[177]

To obtain the optimal system, the values of each interval of the just diatonic system must be compared with the corresponding tempered values; this comparison is made in Table 5.30.

1. Col. I of the (upper part of the) table contains the names of the diatonic intervals, consonances in upper case, dissonances in lower case; Arabic numerals on the left denote minor intervals, Roman numerals denote major intervals. The tritone is considered a major fourth and the false fifth is considered a minor fifth because of the similarity of these intervals to major and minor intervals.
2. Col. II contains the elements of the intervals of the just diatonic system.
3. Col. III contains the logarithms of the just intervals.
4. Col. IV contains the elements of the tempered system, which can easily be obtained from the second column by replacing S by s + c and t or T by 2s + c.
5. Col. V contains the tempered system of 12 equal semitones (that is, the equal temperament according to the modern nomenclature). This column, like the following three in the lower part of the table (Col. VI, VII, VIII), is divided into three sub columns, the first of which gives the number of semitones, the second the logarithms of the corresponding intervals, and the third the differences between the logarithms of the tempered and just systems, with a negative sign if the former is

[177] p. 210.

Table 5.30 Comparing tempered and just intervals; from [75], p. 212

Group	I	Intervalles diatoniques	II Syst. juste	III Logarith.	IV Elem. temp.	V Systèmes tempéres des 12 semitones moy. sem	c = 0, s = 1	diff.
	VIII	OCTAVE	3T 2t 2S	301.0300	12s 7c	12	301.0300	0.0000
			comma	5.3950	c	0		
petit intervalles	2	Seconde min.	S	28.0287	s c	1	25.0858	-2.9425
petit intervalles	II	Seconde maj.	t	45.7575	2s c	2	50.1717	4.4141
petit intervalles	II	Seconde maj	T	51.1525	2s c	2	50.1717	-0.9808
petit intervalles	3	TIERCE MIN.	T S	79.1812	3s 2c	3	75.2575	-3.9237
petit intervalles	III	TIERCE MAJ.	T t	96.9100	4s 2c	4	100.3433	3.4333
petit intervalles	4	QUARTE	T t S	124.9387	5s 3c	5	125.4292	0.4905
petit intervalles	IV	Triton	2T t	148.0625	6s 3c	6	150.2550	2.4525
grand intervalles	5	Fausse quinte	T t 2S	152.9675	6s 4c	6	150.2550	-2.4525
grand intervalles	V	QUINTE	2T t S	176.0913	7s 4c	7	175.6008	-0.4905
grand intervalles	6	SIXTE MIN.	2T t 2S	204.1200	8s 5c	8	200.6867	3.4333
grand intervalles	VI	SIXTE MAJ.	2T 2t S	221.8488	9s 5c	9	225.7725	3.9237
grand intervalles	7	Septiem. min.	2T 2t 2S	249.8775	10s 6c	10	250.8583	0.9808
grand intervalles	7	Septiem. minim	3T 2t S	255.2725	10s 6c	10	250.8583	-4.4142
grand intervalles	VII	Septieme maj.	3T 2t S	273.0013	11s 6c	11	275.9442	2.9429

	VI Systèmes tempéres des 31 parties: part	c = 1, s = 2	diff.	VII Systèmes tempéres des 43 merides: merid.	c = 1, s = 3	diff.	VIII Systèmes tempéres des 55 comma: com.	c = 1, s = 4	diff.
	31	301.0300	0.0000	43	301.0000	0.0300	55	301.0300	0.0000
	1	9.7106	4.3156	1	7.0000	1.6050	1	5.4733	0.0783
2	3	29.1319	1.1032	4	28.0000	-0.0287	5	27.3664	0.6623
II	5	48.5532	2.7957	7	49.0000	3.2425	9	49.2595	3.5020
II	5	48.5532	-2.5993	7	49.0000	2.1525	9	49.2595	-1.8931
3	8	77.6852	-1.4960	11	77.0000	2.1812	14	76.6250	2.5540
III	10	97.1064	0.1964	14	98.0000	1.0900	18	98.5188	1.6088
4	13	126.2382	1.2997	18	126.0000	1.0613	23	125.8853	0.9466
IV	15	145.6597	-2.4028	21	147.0000	-1.0625	27	147.7784	-0.2841
5	16	155.3703	2.4028	22	154.0000	1.0325	28	153.2516	0.2841
V	18	174.7916	-1.2997	25	175.0000	-1.0913	32	175.1447	0.9466
6	21	203.9236	-0.1964	29	203.0000	-1.1200	37	202.5112	-1.6088
VI	23	223.3448	1.4960	32	224.0000	2.1512	41	224.4042	2.5554
7	26	252.4768	2.5993	36	252.0000	2.1225	46	251.7705	1.8930
7	26	252.4768	-2.7957	36	252.0000	-3.2725	46	251.7705	-3.5020
VII	28	171.8981	-1.1032	39	273.0000	-0.0013	50	273.6336	0.6625

less than the latter.

Sauveur was very critical of this system, the equal temperament, which was not yet well established. Here is his judgment:

> This system has its use among unskilled musicians because of its simplicity and the ease of transposing notes on any key without any change between intervals, but as the difference of the intervals of this system with those of the just diatomic system is too great, it is rejected by the more skilled musicians [75].[178]

[178] pp. 201–215.

6. Col. VI is about the tempered system of 31 parts, which results from c = 1, s = 2. Sauveur referred to Huygens, who in his *Cycle harmonique* showed its excellence. Huygens had pointed out the closeness of his temperament to that of the 1/4 comma, in which the major third and minor sixth are just.
7. Col. VII concerns the system of 43 merides. As already commented, a meride is almost the same as 1/43 of the octave.
8. Col. VIII contains the tempered system of 55 parts obtained for c = 1, s = 4, which Sauveur called the system used by musicians.[179]

Sauveur also commented on the system of 19 and 67 parts, corresponding to s = 1 and s = 5, respectively; they should be discarded because the ratio s/c of the former is less than 1 2/3, and that of the latter is greater than 4 2/3. Sauveur said he did not consider either Zarlino's or the 1/4 of a comma systems, because in them the octave is not divided into equal parts and thus belongs to a different kind of temperament. What he did not realize was that he had taken them into account, albeit without being aware of it. This misunderstanding is due to the fact that he did not have a clear idea of the correspondence between the two ways of obtaining temperament: that by dividing the octave into n equal parts, and that by cycles of fifths narrowed by a fraction p. In particular, he did not show that he knew that Zarlino's temperament of 2/7 of a comma corresponded to dividing the octave into 19 parts and that of 1/4 of a comma corresponded to dividing it into 31 parts, values which he had discussed.

To judge the quality of temperaments, Sauveur referred to the differences found with the just system, highlighted in Table 5.30. For simplicity and better understanding, the errors should be compared with something known to musicians, for example the heptamerides, which according to Sauveur are now quite well known to musicians (one heptameride is worth about 5 syntonic comma). Or they should be compared directly with the syntonic comma or with the octave, even if it is a very large interval, or with the number of vibrations of the sounds. Only a small part of the differences is due to rounding, and this error does not exceed half a unit of the last significant digit considered.[180]

If we compare the larger intervals with the smaller ones, we find that

1. The sum of the values of the intervals equally distant from the extremes is an octave, 301.0300, except for Col. VII, where there is a small difference, since a meride is not exactly 1/43 of an octave.
2. The intervals and their complements have the same difference but opposite signs (although there are small differences for the seventh column).

[179] This attribution, according to Barbour, depends on the fact that the 1/6 of a comma temperament was widespread in France at the time, of which the division of the octave into 55 parts is a good approximation [10], p. 126. Along with the 1/6 comma temperament, especially in Italy and Germany, its variants were diffused as good temperaments, such as Vallotti's temperament, see Sect. 3.2.2.4.

[180] If we use all seven digits of the logarithms of Vlacq's small tables, the rounding error due to the logarithms is half the unit of the last digit on the 4 decimal places (the places reported in Table 5.30), which is less than 1/20000 of heptamerides or nearly 1/108000 of a comma (Sauveur used to multiply logarithms by 1000; thus, using tables with seven digits there are four decimals).

3. The major second has two values for difference, and the sum of the absolute values of these differences is worth a comma; the same happens for the minor seventh, which is its complement.

Obviously, to choose the best system, it is sufficient to examine the minor intervals, since the major have the same differences.

If we compare the same intervals of all tempered systems, leaving out the system of 12 semitones, we can make the following observations:

1. The difference which is the smallest in one of the extreme systems is the largest in the other, while in the system of 43 it is intermediate; so, for example, the difference in the fourth is the smallest in the system of 55 and the largest in the system of 31.
2. The differences are negative for the tritone and the minor third, and positive for the fourth and the major third. As for the minor second, it is in defect for the systems of 43 and 55 parts and in excess for the system of 31 parts. Of the two major seconds, one is in excess and the other is in defect.
3. The smallest differences are in the fourths and fifths for the system of 55, in the thirds and sixths for the system of 31; in the minor second and major seventh for the system of 43 [75].[181]

A comparison of the just and tempered systems for different values of s and c can be made by considering the limit as s increases. For simplicity, we will consider only the fifth and third major, whose tempered values are given in Table 5.30 as 7s + 4c and 4s + 2c, respectively. Dividing by the octave, 12s + 7c, gives:

$$\frac{7s + 4c}{12s + 7c} \qquad \frac{4s + 2c}{12s + 7c}$$

It can clearly be seen that the limits of these ratios for $s \rightarrow \infty$ do not depend on the value of c and tend to be 7/12 and 4/12, i.e. 7 and 4 parts of an octave, respectively; the same proportions as in the tempered system of 12 equal intervals, as shown in Table 5.30. This is because for $s \rightarrow \infty$ and c fixed everything is the same as for $c = 0$, because gradually the chromatic and diatonic semitones tend to be the same. In particular, we see that the difference of the tempered fifth tends to decrease, but not below that of the 12-division system (this is characteristic of the negative tempered systems, where the syntonic semitone is greater than the chromatic semitone), while that of the tempered major third tends to increase, but not above that of the 12-division system. The 12 equal division system thus appears as the limit of all tempered systems à la Sauveur. This fact shows the limitations of this type of temperament, which Sauveur was unable or unwilling to point out.

[181] p. 218.

5.5.1 The Choice of the Tempered System

1. We have seen that the extremes of the ratio s/c are 4²/₃ and s/c = 1²/₃. Thus, we can discard the systems that do not satisfy these constraints. The system of 12 equal semitones should also be discarded, in part because the differences between thirds and sixths are about ²/₃ of a comma.
2. A tempered system should be sufficiently simple, with a limited number of octave divisions, while maintaining good accuracy. For this reason we must consider systems for c = 1, s = 2, 3, 4 for which we have 31, 43, 55 divisions, and reject those for c greater than 1.
3. Usage shows that similarly tempered consonances give less discomfort than when some consonances are little altered while others are much altered. And that is why the system of 43, which lies between those of 31 and 55, is preferable. In fact, in the system of 43, the major third, the fourth, the fifth, and the minor sixth are quite accurate, differing from the just values by a little more than 1 heptameride. The minor third and major sixth, on the other hand, have a little more than 2 heptamerides, but experience shows that in the case of consonances where the interval is expressed by larger numbers, such as the minor third (6 : 5) and the major sixth (8 : 5), a large difference is more tolerable than in the case where the interval is expressed by small numbers, such as the fifth (3 : 2) [75].[182]
4. In justifying the choice of the system of 43 merides, the following reasons may be considered:

   a. Subtracting 0.0300 from the 301.0300 gives our system, and neglecting 0.0300 is of no consequence, since it is only 1/33 of a heptameride or 1/180 of a comma, which is negligible.
   b. The heptamerides defining the intervals are divisible by seven, which leads to expressing the intervals in simple numbers of merides.
   c. Heptamerides are the logarithms of intervals and make it easy to find the ratio of the vibrations of the highest note in an interval to the lowest note. For example, consider the tempered fifth, which is defined by 175 heptamerides, and assume that the lowest note has 10,000 vibrations. The logarithm of the vibrations of the fifth is 4.1750000 (the heptamerides are the logarithms, and adding 4 is equivalent to multiplying by 10,000). The antilogarithm of 4.1750000 is 14,963, so if the lowest note in a tempered fifth has 10,000 vibrations, the highest note has 14,963. According to Sauveur, for the other systems the calculation is more complicated; that is, it is really true only if one does not have the tables with the logarithms of the intervals.
   d. The system of 43 merides is such that adding one or two heptamerides to the tempered intervals gives the just ones with an error of no more than ¹/₆ of a heptameride ¹/₃₀ of a comma. Huygens could not provide his system of 31 parts with this advantage, which is not even found in that of 55 commas. So by

[182] p. 219.

adding a heptameride to the tempered fifth, which is 25 merides, it becomes just.

e. The temperament of our system is more natural than the other two, because the meantone is formed by the arithmetic mean of the logarithms of the major and minor tones [75].[183]

Sauveur therefore concluded that the system of 43 merides is the most perfect and the only one that should be considered in order to have all the possible advantages of tempered systems in music and acoustics.

The 1707 memoir ends with a defense of his system against attacks by an editor, perhaps the same editor-in-chief Louis Cousin, of the 1707 *Journal de sçavants*, in the course of a review of a work on musical acoustics by Louis Carré (1663–1711):

> He [Carré] tells us that his monochord is taken from the system of Mr. Sauveur, & that he found it cleaner than any other for a general system of music; he gives the details. We may say that this system is strictly an extension of that of Mr. Huygens, which he entitled *Harmonic cycle*. In truth, Mr. Huygens only divides the ottave into 31 equal intervals, which he does by looking for 30 proportional means between a whole string and its half; that is, between 2 and 1. Mr. Sauveur divides it into 40 [sic] parts, which he calls merides; but this difference is infinitely slight [22].[184]

Meanwhile, Sauveur denounced the inaccuracy of the quotation in dividing the octave into 43 and not 40 parts. Then he wonders why the system of 31 parts is called Huygens' system, when in fact Huygens himself attributes it to others, and in any case the system of 43 can be seen not only as an extension of that of 31, but also as a limitation of that of 55. Rather, it must be said that these three systems are the result of the formula $12s + 7c$, which Huygens did not think of [75].[185]

According to Sauveur, the editor wants to imply that the origin of the 43-part system is merely the result of a whim or personal preference, and that one can use as many systems as he wants:

> I leave that judgment to the reader. I will add, however, that those who speak of the works of others should exercise strict restraint in their judgment, lest they deprive the public of the benefits to be derived from the discoveries of science [75].[186]

5.5.2 *Fontenelle's Account*

Fontenelle's account in the *Histoire* of Sauveur's memoir, presented in the previous section, begins by stating that the division of the octave into 43 parts does not have the character of necessity, and is not even unique as regards the effect of obtaining the softening of dissonances with a small alteration of consonances. If the meantone

[183] pp. 220–221.

[184] p. 124.

[185] p. 221.

[186] p. 222.

is divided into five parts, and 3 are given to the major semitone and 2 to the minor, the octave of 5 tones and 2 major semitones will be divided into 31 equal parts. It will be divided into 55 parts if the middle tone is divided into 9 parts, of which 5 belong to the major semitone and 4 to the minor. However, the division of the octave is not at all arbitrary.

It seems strange that, knowing that it is composed of five mean tones and two major semitones, the octave can be divided indifferently into 31, 43 or 55 parts by varying the ratio of the minor to the major semitone. But, asks Fontenelle, shouldn't this ratio be determined by the diatonic system itself and therefore unique? This is not the case, however, because in the diatonic system there are two slightly different tones, a major and a minor, and consequently the major semitone can be considered as the larger half of one or the other of the two unequal tones. So the relationship between the semitones varies, not arbitrarily, but within limits that Sauveur easily determined.

If we want the difference between the major semitones to always be equal to one, the minor will be 2, 3, 4, the major will be 3, 4, 5, and the division of the octave will be 31, 43, 55; if we assume a greater difference, the division of the octave will occur in a greater number of parts; but this is inconvenient for the calculation.

Particularly interesting is Fontenelle's criticism of the equal tempered systems, which we report in full:

> However, if the octave were to be divided into 12 equal semitones, an alternative method of division would be possible. The octave, with only 12 equal parts, would have an insufficient number of parts to provide chords with a sufficiently just temperament. As a result, the thirds would be so altered that the ear could not tolerate them. This system is therefore rejected by musicians. Furthermore, the situation would be even more problematic if the octave were divided into fewer than 12 equal parts [34].[187]

5.6 1711. *Table Générale des Sistemes Temperés de Musique*

The memoir *Table générale des sistemes temperés de musique* of 1711 ties in with that of 1707 and is a deepening of it. At the outset, Sauveur reiterated the need to introduce regularity into music theory, along the lines of what had been done in mathematics and physics.

> The unanimous approval given by geometers and astronomers to the division of the circle into 360 degrees and of each degree into 60 minutes, as well as to the division of the whole sine into 10,000,000 parts, has shown the necessity in acoustics of agreeing on a division of the intervals of sounds accepted by all mathematicians, so that we can speak a uniform language, which is absolutely necessary here for the progress of this science.
> In the memoir of the Académie of 1701, I established the division of the octave into 43 merides, that of the merides into 7 heptamerides, and that of the heptamerides into 10 decamerides: I even gave names to all these intervals; then I applied them to all systems and all musical instruments. Since then, in the memoir of 1707, I have shown that the system of 43 should be preferred to those of 55, 31, and 12, which are the best known [76].[188]

---

[187] pp. 119.

[188] p. 307.

The memoir takes its cue from a work on music by Conrad Henfling (1648–1716) published in the Berlin academy memoirs of 1710 [4, 45]. Sauveur claimed that he had carefully read the memoir and found in it principles of theory based on "algebraic reasoning, a great deal of erudition, and a great deal of practice with musical instruments," and the statement that a rational division of the octave is that into 50 equal parts. From his reading Sauveur evidenced the principles of Henfling:

1. By means of sums and subtractions of octaves, fifths, and major thirds, Henfling found all the intervals that could be greater or smaller than a comma of the just diatonic intervals. And he decided that a correct musical system must have the tempered intervals between the just ones and those he had found.
2. Henfling established his system of 50 equal parts, starting with the minor tone t, which contains 458 decamerides. From the minor tone he subtracted the major semitone S, which he called *diatonum* and which contains 270 decamerides, the remainder t - S = 178 is a minor semitone which he called *chroma*. From the major semitone he subtracted the minor semitone, or 2S - t = 102, which he called *harmonie*. Subtracting harmonie from chroma leaves 2t - 3S = 76, called *hyperoche*; subtracting harmonie from hyperoche leaves the sound called *eschatum* 5S - 3t = 26. Finally, subtracting the eschatum from the hyperoche gives 5t - 8S = 50.
   At this point Henfling proposed the following ratio between the minor and major semitones t : S = 8 : 5, so that if t is 8 parts S is 5 parts, the minor and major semitones, whose sum is one tone, are in the ratio (3 : 5) and the octave 5t + 2S is of 50 parts, from which his system takes its name.[189]
   If we use the previous subtractions to define the ratio of the minor to the major semitone, for the *eschatum* 5S - 3t, we take t = 5, S = 3, the octave is made up of 31 parts; if we consider the *hyperoche* 2t - 3S, we have t = 3, S = 2, a system of 19 equal parts is obtained, and finally, if we take the *harmonie* 2 S - t , the system is made up of 12 parts. Since the system of 43 is not derived from this operation, it is ignored.

Figure 5.46 shows the possible divisions of the octave proposed by Henfling, as discussed above.

Sauveur was highly critical of Henfling work, stating that:

1. There are two errors in the first set of assumptions.
   a. Too great an interval, that of a comma, has been assumed between a tempered interval and a just interval; indeed, a tempered interval would be very defective even if it differed from the just interval by only half a comma.
   b. There is a certain arbitrariness in the principle. For example, having taken the just minor second and one exceeding it by a comma, the tempered value is assumed to be between the just and the exceeding. Why, Sauveur wondered,

[189] Note that the minor tone is considered here as a meantone; which is not specified.

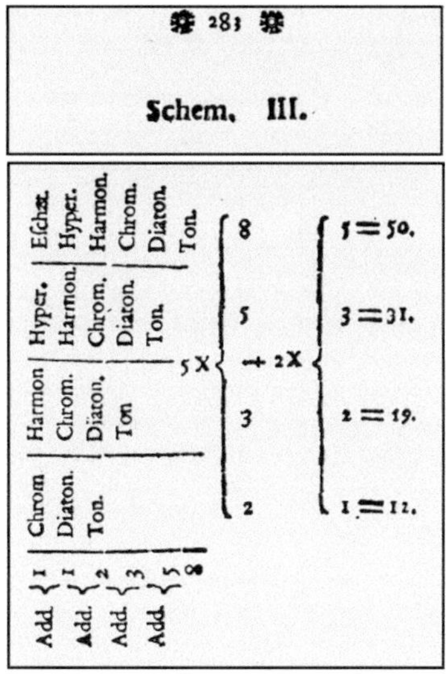

Fig. 5.46 The division of the octave; from [45], p. 283

could one not take it below the just interval? Since a tempered interval can approach the just interval both from above and from below.

2. There are two criticisms of the second set of assumptions ("I also have two things to say in response"):

   a. It would have made more sense to subtract the diatonum S from the major tone T of 511, since there are three major and only two minor tones in the octave. In this way we would have had the systems 12, 19, 31, 43, 55, and 67; but we would not have had the system of 50.
   b. Neither method should be followed, since the main purpose of a temperament is to obtain a tempered tone intermediate between major and minor tones [76].[190]

To avoid spending more time answering Henfling or anyone else who wanted to propose new systems, Sauveur provided Table 5.31 in 5 columns, which lists "all the tempered systems of music",[191] with their differences from the just intervals so that they can be compared with each other.

The intervals are labeled with a title and literary expressions to make evident the operations that must be performed to obtain the diatonic intervals of each system.

Col. V is for the major semitone and shows the letter S; Col. VI is for the mean tone and shows the letter t; Col. IV is for the minor semitone with the letters t - S;

[190] p. 308.

[191] Obviously, these are not all the tempered systems that can be devised, but only the most relevant of those generated by dividing the octave into equal parts, the only ones considered by Sauveur.

Table 5.31 Tempered systems of music compared with the diatonic just system

I	II	III	IV	V dem.	VI ton.	VII oct.	VIII parties	IX Demit. min.			X Seconde min.			XI Seconde maj.		
2S-t	t-S	2S-t	t-S		t	5t+2S	decam.	t-S	210	diff.	S	280	diff.		490	diff.
60	60	1	1	2	3	19	158.487	1	158	-52	2	317	37	3	475	-15
60	90	2	3	5	8	50	60.206	3	181	-29	5	301	21	8	482	-8
60	105	4	7	11	18	112	26.876	7	188	-22	11	296	16	18	484	-6
60	120	1	2	3	5	31	97.106	2	194	-16	3	291	11	5	486	-4
60	135	4	9	13	22	136	22.135	9	199	-11	13	288	8	22	487	-3
60	140	3	7	10	17	195	28.669	7	201	-9	10	287	7	17	487	-3
60	150	2	5	7	12	74	40.680	5	203	-7	7	285	5	12	488	-2
60	160	3	8	11	19	117	23.730	8	206	-4	11	283	3	19	489	-1
60	165	4	11	15	26	160	18.814	11	207	-3	15	282	2	26	489	-1
60	180	1	3	4	7	43	70.007	3	210	0	4	280	0	7	499	0
60	195	4	13	17	30	164	16.360	13	213	3	17	278	2	30	491	1
60	200	3	10	13	23	141	21.350	10	213	3	13	278	-2	23	491	1
60	210	2	7	9	16	98	30.717	7	215	5	9	276	-4	16	491	1
60	220	3	11	14	25	153	19.675	11	216	6	14	275	-5	25	492	2
60	225	4	15	19	34	208	14.473	15	217	7	19	275	-5	34	492	2
60	240	1	4	5	9	55	54.733	4	219	9	5	274	-6	9	493	3
60	255	4	17	21	38	232	12.975	17	221	11	21	272	8	38	493	3
60	260	3	13	16	29	177	17.007	13	221	11	16	272	8	29	493	3
60	270	2	9	11	20	122	24.675	9	222	1.2	1	271	-9	20	493	3
60	280	3	14	17	31	189	15.928	19	223	13	17	271	9	31	494	4
60	285	4	19	23	42	256	11.759	19	223	13	23	270	-10	42	494	4
60	300	1	5	6	11	67	44.930	5	225	15	6	270	-10	11	494	4
60	inf.	0	1	1	2	12	250.858	1	251	41	1	251	-29	.2	502	12
-60	291	-1	5	4	9	53	56.798	5	284	74	4	227	-53	9	511	21
-60	120	-1	2	1	3	17	177.076	2	354	144	1	177	-103	3	531	41

V dem.	VI ton.	VII oct.	VIII parties	XII Thierce min.			XIII Thierce maj.			XIV Quarte			XV Triton		
	t	5t+2S	decam.	t+S	792	diff.	2t	969	diff.	2t+S	1249	diff.	3t	1481	diff.
2	3	19	158.437	5	792	0	6	951	-18	8	1267	18	9	1426	-55
5	8	50	60.206	13	784	-8	16	963	-6	21	1264	15	24	1445	-36
11	18	112	26.876	29	780	-12	36	968	-1	47	1263	14	14	1451	-30
3	5	31	97.106	8	777	-15	10	971	2	13	1262	13	15	1457	-24
13	22	136	22.135	35	775	-17	44	974	5	57	1262	13	66	1461	-30
10	17	195	28.669	27	774	-18	34	975	6	44	1261	12	51	1462	-19
7	12	74	40.680	19	773	-19	24	976	7	31	1261	12	36	1464	-17
11	19	117	23.730	30	772	-20	38	978	9	49	1261	12	57	1467	-14
15	26	160	18.814	41	771	-21	52	978	9	67	1261	12	78	1468	-13
4	7	43	70.007	11	770	-22	14	980	11	18	1260	11	21	1472	-11
17	30	164	16.360	47	769	-23	60	982	13	77	1260	11	90	1472	-9
13	23	141	21.350	36	769	-23	46	982	13	59	1260	11	69	1473	-8
9	16	98	30.717	25	768	-24	32	983	14	41	1259	10	48	1474	-7
14	25	153	19.675	39	767	-25	50	983	15	64	1259	10	75	1476	-5
19	34	208	14.473	53	767	-25	68	984	15	87	1259	10	102	1476	-5
5	9	55	54.733	14	766	-26	18	985	16	23	1259	10	27	1478	-3
21	38	232	12.975	59	766	-26	76	986	17	97	1259	10	114	1479	-2
16	29	177	17.007	45	765	-27	58	986	17	74	1259	10	87	1480	-1
11	20	122	24.675	31	765	-27	40	987	18	51	1258	9	60	1480	-1
17	31	189	15.928	48	765	-27	63	988	19	79	1258	9	93	1481	0
23	42	256	11.759	65	764	-28	84	988	19	107	1258	9	126	1482	1
6	11	67	44.930	17	764	-28	22	988	19	28	1258	9	33	1483	2
1	2	12	250.858	3	753	-39	4	1003	34	5	1254	5	6	1505	24
4	9	53	56.798	13	738	-54	18	1022	53	22	1250	1	27	1534	53
1	3	17	177.076	4	708	-84	6	1062	93	7	1249	0	9	1594	113

Col. III shows the difference 2S - t . Col. I and II have the same letters as Col. III and IV.

Col. VII is for the octave 5t + 2S, Col. VIII for the value in decamerides of each of the parts into which the octave is divided. The letters of Cols. IX, X, XI are the same as those of columns II, V, VI. The letters of Col. XII, XIII, XIV, XV are taken from the second column of the Table 5.30, from the 1707 memoir, with t substituted for T. The just intervals, expressed in decamerides, between these letters are also taken from Col. III of Table 5.30.

In Col. I, 2S - t, which is the comma c of the 1707 memoir, is assumed to be of 60 parts (the choice of 60 for the first column is arbitrary, and serves to simplify passages), and in Col. II, t - S, which is the minor semitone s, is assumed to be multiple of 60, 30, 15, ranging between 105 and 285. They have the ratios between $1^{31}/_{43}$ and $4^{5}/_{7}$ with 60, which are approximately within the limits of the ratio s/c found in the 1707 memoir. Sauveur added to the second column at the top of the table the values 60 and 90, giving 1 and 1.5 for the s/c ratio—outside the permissible limits—to account for the systems of 19 and 50 parts, and added at the bottom the values 300, ∞, 291, 120 needed for the systems 67, 12, 53, and 17 parts, also with s/c ratios outside the limits. Note that s/c $= \infty$, corresponds to the equal temperament system.

The second line gives the system of Henfling; the last two lines concern octave divisions in which the chromatic semitone is greater than the syntonic semitone. They lead to systems called positive after Bosanquet (see Sect. 3.2.2.2), a distinction that was not meaningful to Sauveur. Among them is Mercator's system of 53 parts (it is unclear why Sauveur gave the pair -60 – 291 instead of -60 – 300, which would give the exact values -1 and 5 for the comma c and the minor semitone s) and that of 17 parts, typical of the Orientals.

The numerical values in Cols. III and IV are the same as those in Cols. I and II divided by 60 or by 30 or by 15, and these numbers are the values of the parts of 2S - t and t - S; values we will use for each temperament system. Adding the numbers of Cols. III and IV gives S of Col. IV, while adding those of Cols. IV and V gives t of Col. VI. Taking the values of Cols. V and VI , we obtain the values 5t + 2S in Col. VII, which indicate the number of parts by which the octave is divided in each system, and with these numbers we obtain the tempered system. For example, 31 in the fourth row will indicate the system adopted by Huygens. The line corresponding to 31 contains all the intervals and all the properties of this system.

In Col. VIII we find the values obtained by dividing the octave 3010.30 by the numbers in Col. VII, which correspond to a part of each system expressed in decamerides up to the third digit (thousandths of decamerides).[192] Thus, dividing 3010.30 by 31 gives $97^{106}/_{1000}$ decamerides for each of the 31 parts of Huygens' system.

The next columns, from IX to XV, which give the small intervals of the diatonic system, are divided into three sub-columns; the first sub-column gives the number of parts of the intervals obtained from Cols. V and VI. For example, in Col. IX, the system of 31 parts gives $5 - 3 = 2$ parts for the minimum semitone $t - S$; in the second sub-column the numbers of parts have been multiplied by the value of each part of the system, and so in the system of 31 parts it is $97^{106}/_{1000} \times 2 \approx 194$. Finally, subtracting 210 (which is under the heading of column IX) gives the differences between the tempered and just systems of the third sub-column. Thus, in the system

[192] In fact, numerical data is not always that accurate.

of 31 we have $194 - 210 = -16$, which tells us that the minor semitone of this system is 16 decamerides smaller than that of the just system. What was done in column IX for the system of 31 can be repeated for all systems; and what was done for this column can be repeated for all columns.

The large intervals, which are the false fifths, fifths, sixths, and sevenths, are not shown in the table, because they are the complements to the octave of the small intervals. Finally, neither diminished or minor nor superfluous or maximal intervals have been considered, since their use is rare.

The diatonic intervals are expressed in decamerides, i.e. ordinary logarithms [base 10], the last three digits neglected (we are talking about seven-digit logarithms), because half a decameride or (about) 1/108 of a comma is absolutely negligible for the ear. Thus, according to Sauveur, the 10 digits of the logarithms used by Huygens and Henfling are unnecessary; seven digits are sufficient for calculations, and four for expressing diatonic intervals.

If we look at the differences between tempered intervals and just intervals, we see that:

1. The differences are zero in the minor semitone, in the minor second and in the major second of the system of 43, in the minor third of the system of 19, in the major third of the system of 12, in the fourth of the system of 17, and finally in the tritone of the system of 189. Thus it can be said that there is no tempered system whose intervals are all equal to those of the purely diatonic system.
2. It is also noticeable that above or below each row where zero is found, the differences increase as one moves away from the row, and those that are positive on one side are negative on the other [76].[193]

It may be added that the best tempered system is that which, having divided the octave into a few parts, gives small and evenly distributed differences. Thus, in choosing the best systems, we must exclude those of 19 and 50, on the one hand, and those of 12, 17, 53, on the other, because, in addition to exceeding the limits prescribed in the memoir of 1701, the differences of their intervals are too great, and some are even greater than half a comma [27 decamerides]. The systems formed by the numbers in Col. IV greater than 4 must be excluded because their octaves, shown in Col. VII, have numbers that are too large. That leaves the systems 31, 43, 55.

System 55, the system of ordinary musicians, should be excluded because it is represented by too large a number and because the differences in thirds are too great. It is true, then, that system 31 is expressed by the smallest number, but system 43 is to be preferred because:

1. It preserves the minor second S of the just diatonic system.
2. It also preserves the major second and the minor semitone.
3. The differences of the other intervals are equal, except for the minor third, which is doubled. This equalization of the differences between tempered and just tones gives less discomfort to the ear.

---

[193] p. 313.

4. If we divide a single part of the system 43 into 7 heptamerides, we find differences from 1 or 2 heptamerides to 1 or 2 decamerides. This is fundamental to a system; such uniformity is not found in the system of 31.
5. In the system of 43, the fourth (Sauveur said: "the most perfect consonance") has a smaller difference than in the system of 31, and for the third, in which ratios –5/4 and 6/5—are expressed by larger numbers than in the fifth and fourth—3/2 and 4/3—at the same value, the differences are less sensitive to the ear.
6. The system of 43 is an average system among the legitimate extremes. It is similar to the tempered system, which has as its average tone the arithmetic mean of the two minor tones and the three major tones.
7. The sum of the differences in the intervals of system 43, when all values are considered positive (i.e., with their absolute values), is less than the sum of the differences in all other systems.
8. The heptamerides and decamerides of system 43 are the common logarithms (base 10) of the ratios of the sounds of each interval, which is not found in the other systems.
9. The harpsichord makers of the king and of Paris, who based the octave of their harpsichords on a monochord divided into 43 parts, found this temperaments very good. By dividing the octave of a monochord into 31, 43, 50, and 55 parts and putting it in unison with the C Sol Ut of a just tuned harpsichord, and then putting a bridge on the monochord and tuning it with each key of the harpsichord, the bridge is always found in the divisions of 43 or goes near it and away from the other divisions. There are, however, makers who mount their harpsichords with the more piquant tunings, and as a result the bridge may come nearer to other divisions [76].[194]

These observations suggest the adoption of the system of 43 merides, each part being divided into 7 hepatmerides or 70 decamerides. The extent to which an interval must be raised or lowered in this system to align with those of the makers, which does not always adhere to a strict tempered system, can be determined.

Sauveur could thus conclude that the above reasons seem sufficient to induce mathematicians to choose the system of 43.

5.6.1 Fontenelle's Account

Fontenelle's account is very brief, in keeping with the length of Sauveur's memoir [39]. He began by noting that Sauveur had not considered the system of 50 proposed by Henfling in his 1707 work because it was outside the permissible limits of the ratio s/c (here Fontenelle is careful to consider the limits as what "he believed were permissible for a system"). Henfling's system, in addition to being outside Sauveur's limits, has intervals that are also altered. Furthermore, it can be shown that the

[194] pp. 314–315.

algebraic way in which Henfling arrived at his results is not unique, and others can be found which do not lead to the system of 50. Fontenelle concluded by claiming that Sauveur had done again what he had done in the 1707 work comparing the system of 43 with those of 31 and 55; this time with the claim of considering all tempered systems, taking as possible those systems that divide the octave into not too large numbers.

5.7 1713. *Rapport de Sons des Cordes d'Instruments de Musique, aux Fleches des Cordes*

The *Rapport de sons des cordes d'instruments de musique, aux fleches des cordes, et nouvelle détermination des sons fixes* of 1713, published in 1716, the last of Sauveur's memoirs on music, is the most complex to interpret for several reasons. First, because of its extensive use of Calculus, which in the early 18th century relied more on intuition than rigor, and so its application is difficult to decipher for a modern reader. Second because the objectives are not clearly stated, and finally, numerous typos are present (Sauveur could not revise this memoir because of his death in 1716).

The memoir can be divided into four parts; in the first part, the problem is introduced by questioning the validity of some of the approximations that appear in the memoir. In the second part, using the laws of statics, Sauveur related the sag f of a horizontally arranged string to tension, weight for unit length, and length using Calculus, assuming that the sag-length ratio is small. In the third part, the frequency of the first harmonic sound of a string is calculated from its equivalence to that of the oscillations of the same string about its own axis. It all boils down to finding the ratio of the length q of an equivalent pendulum to the sag f; a geometric problem to be solved using Calculus. The ratio found was $q = 4/5f$. Sauveur thus succeeded for the first time in finding a formula that gives the absolute frequency of a vibrating string as a function of the parameters that determine it. This formula is used in the fourth part of the memoir to determine the fixed sound, as an alternative to the method proposed in the memoir of 1701.

In order to facilitate the reading of the memoir, it was decided to keep the numbering of all the paragraphs according to what Sauveur had done.

5.7.1 General Remarks on the Sonorous Strings

This section, along with the following three, is rather difficult to follow and presents a challenging and potentially tedious reading experience due to extensive use of mathematics; however, we have opted not to simplify or summarize very much Sauveur's writing in order to gain an appreciation for the effort required to achieve a result.

Let us begin with a detailed exposition of Sauveur's memoir.

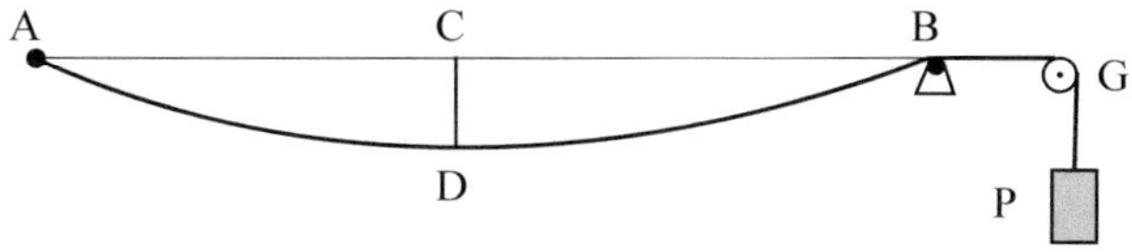

Fig. 5.47 The sag of a string

1. To understand the properties of sonorous bodies, consider a string AB stretched horizontally by a weight P passing through a pulley G, as shown in Fig. 5.47; this string will be deformed by its weight along a curve ADB. If one considers a straight line ACB, the vertical segment CD, drawn from the midpoint C of AB, is referred to as the *sag* of the curve. In the case where ADB represents the arc of a circle, CB would be the sine of the arc DB, while CD would be the versed sine.[195]

2. Experience shows that the string ADB, when stretched by the same weight P between two bridges A and B at the same distance, produces the same sound whether the instrument on which it is stretched is horizontal or inclined to the horizon or even vertical.

3. The difference in the sound when the string is horizontal and when it is vertical comes from the friction of the pulley to be used in the former situation. But this difference is small.

4. To measure the length of the strings, the astronomical inch is used, 36 of which give the length of the pendulum beating 1 second,[196] and this length becomes 36 inches 8½ lines of the reformed Paris foot, so that the ratio of the two measures is 52 to 51, and 40 astronomical inches are 40 inches 8½ lines.[197] Compared to the old Paris foot, the pendulum length is 36 inches 6½, so their ratio is 68 to 67.

5. In the following, the sag CD will not be represented with an accuracy greater than $1 : 10^6$ of the astronomical inch; the calculation will allow this accuracy, the ear will hear it, while it is 10,000 times beyond the reach of sight, as we are only able to appreciate it with 1 : 100 of the astronomical inch.

6. It is not necessary to know the unit of measurement of the weight p of the string, assumed to be 40 inches long, if we know the ratio of its weight to the weight that tends it. We will use as our unit the pound of Paris, which contains 9216 or $2^{10} \times 3^2$ grains,[198] and in any case we will denote in grains both the weight of the string and the weight that tends it.

---

195 The versed sine of the angle ϕ is vers $\phi = 1 - \cos\phi$.

196 The length l of the simple pendulum beating T seconds is given by the formula $l = gT^2/(4\pi^2)$, using standard symbols. Assuming $g = 9.81$ m/s^2, $T = 2$—since in Sauveur's time the period of a pendulum was considered for a single beat, and thus was half that of the value assumed today—we have $l = g/\pi^2 = 0.9934$ m, which divided by 36 gives 2.76 cm for the astronomical inch and 33.12 cm for the foot.

197 The Paris foot is thus about 32.5 cm.

198 Until the adoption of the metric system in 1795, the pound of Paris, or la livre du poids de marc, was worth 489.5 g. It was divided into 16 ounces of 8 *gros*; each gros into 72 grains, making a total of $16 \times 8 \times 72 = 9216$ grains [68], p. 115.

7. It is preferable to know the weight of the string of a given length to its diameter, because the diameter of the largest string is at most one 1/3 of a line, so that it is very difficult, if not impossible, to distinguish the strings by their diameter; whereas it is easy to measure the weight of a string of 40 inches, or, if one wishes, of 8 times this length, because the largest strings of the latter length weigh more than 1000 grains [about 53 g], and the smallest 40 grains. One can then determine the ratio of the diameters of strings of the same material from their weights, since the diameters are related to each other as the square roots of the weights.

8. In stretching a string, always of metal, the weight must not be too great, lest it should break; experience shows that a steel wire breaks at a weight 12,000 times greater than the 40-inch weight of this wire; one of yellow copper at a weight 9,000 times greater, and one of red copper at a weight 5,000 times greater. Thus, denoting by c the weight in grains of a 40-inch string, the weight required to break it must not exceed $4/3\,c$ pounds of Paris for steel, c pounds for yellow copper, and $6/11\,c$ for red copper.

9. In experiments it is useless to have more accuracy than the senses allow. Thus one can be content to give intervals of sounds in heptamerides, because the error of a half-heptameride is not perceived. Such an error is a difference of 1 in 870 for string lengths or 1 in 435 for weights.

10. In calculations, one can be a little more precise to avoid rounding errors, so one can round up to decamerides, since the error for a decameride is 1/54 of a comma and is absolutely negligible; in fact, the error of half a decameride is 1 : 8700 for lengths and 1 : 4350 for weights.

11. It is true that in calculations a small error may become large; but as the compass of perceptible sounds does not exceed the interval of 12 octaves, and experiments are made without the lowest sound being below the first four octaves, it is thus that the highest sound does not exceed the lowest by more than 8 octaves. For vibrations that can be multiplied by $2^8 = 256$, the error, for example in the weights, up to 1 in $4350 \times 256 = 1,113,600$, is not appreciable.[199] Using round numbers, we can say that an error of $1 : 2,000,000$ in string lengths and an error of $1 : 1,000,000$ in weights are not appreciable.

12. To find the error that may be due to friction on the pulley, remember that it takes a force equal to about one-fourth of the weight (i.e., the coefficient of friction is 1/4, a value derived from experience) to move a body on a horizontal plane when both the plane and the body are made of clean steel or copper. Assuming that the diameter of the spindle is 1 and that of the pulley is d, the resisting force opposed by the friction r of the spindle on the pulley is $f = r/d$.[200] The pulley of Fig. 5.48, is drawn horizontally along the directions GR and vertically along the direction GT by a force p equal to the weight P. If we draw the line GS from the center C of the pulley and the other lines ST, SR parallel to the two directions GR and GT, we see that the

[199] An error of 1 : 1,113,600, after multiplication by 256, can give at most an error of 1 : 4350, which is the maximum allowed in calculations made with decamerides (see Art. 10).

[200] There is a lever effect that reduces the force by the ratio of the arms, in this case the diameters of the spindle and the pulley.

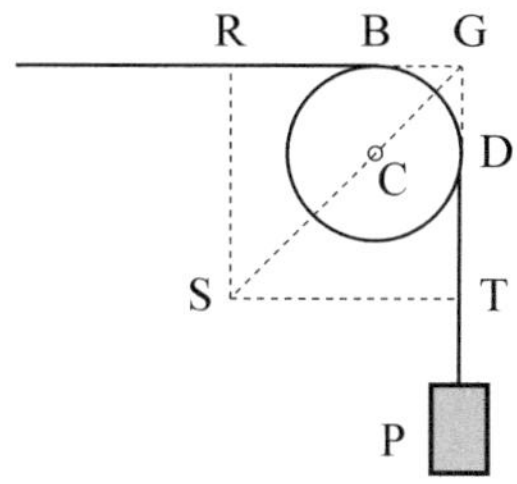

Fig. 5.48 Friction on a pulley

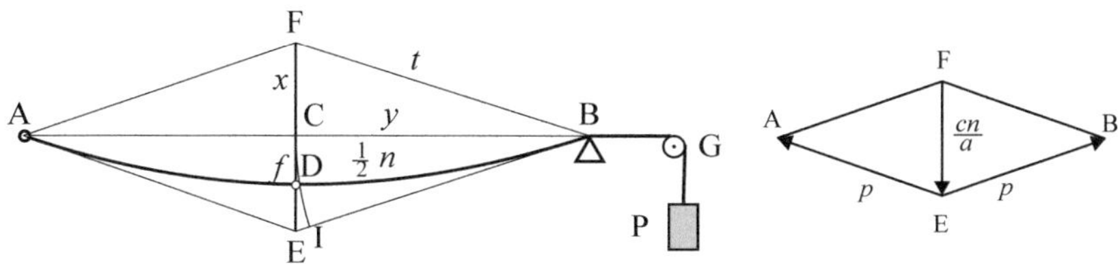

Fig. 5.49 Equilibrium of a string with the law of parallelogram

compression of the string against the pulley is $\sqrt{2}p \approx 7/5p$, by the parallelogram rule, and the force $f = r/d$ opposed by the friction $r = 7/20p$ is to p as 7 is to $20d$ and the error on sound is a half, i.e. as 7 is to $40d$. Since the diameter of the pulley used is 57 times that of the spindle, and thus $d = 57$, the error in tension due to friction will be 7 in 2280, or 1 in 326, while the error in sound will be 1 in 652. This error is less than 2/3 of a heptameride.[201]

5.7.2 *The Sounds of the Strings of the Instruments are in Inverse Ratio to their Sags*

In the first part of this section, Sauveur, using the laws of statics and Calculus, found the expression of the sag of an inextensible string object to a tensile stress and gravity. In essence, he was able to determine the tangent in B of the deformed string, as a function of the weight per unit length and the tension applied, and prove that the sag CD is half the sub-tangent CE.

13. Consider the situation in Fig. 5.49 and draw the tangents AE, EB (t) to the arc ADB (n) and the parallels AF, FB to the two tangents and extend the sag CD (f) to the tangents and their parallels. We will have the subtangent CE (x), the diagonal EF ($2x$), the brace CB (y). Let c be the weight of the string of length a, the weight of the string ADB of length n will be cn/a.

[201] There should be a mistake in Sauveur's calculations, as the error in tension is as 7 to $20d$, and thus as 7 to 1140 and not as 7 to 2280.

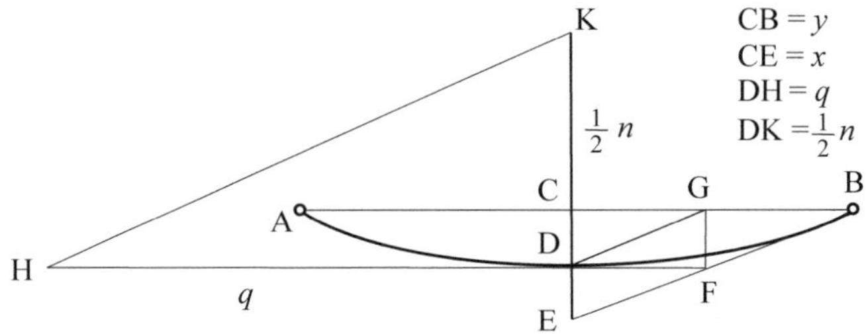

Fig. 5.50 Various parameters in the equilibrium of a string

14. For the laws of mechanics,[202] the weight cn/a of the string ADB is to the weight p tending it as the diagonal EF is to the tangent EB, so we have

$$cnt = 2apx$$

15. If we want to know the shape of the curve BD, with reference to Fig. 5.50, draw the horizontal HDF which meets the tangent BE at F, the vertical GF and the line DG parallel to EB. From mechanics we know that the center of gravity of the part DB of the string lies on the vertical FG (in fact, the equilibrium of the arc DB is determined by three forces; two are due to the tension of the string, a horizontal force at D and a force directed along EF acting at B, and a third force due to the weight of the string n. According to the parallelogram rule, all three forces must meet at one point for equilibrium to be possible. The point is determined by the intersection of the horizontal line for D and the inclined line for B, which is F).

Assume, as in Fig. 5.50, that the vertical DK is equal to the arc DB ($\frac{1}{2}n$) and KH is parallel to EB, and let DH = q; we have:

$$\text{DF} : \text{FG} = \text{BC} : \text{CE} = \text{HD} : \text{DK}$$

that can be rewritten as $y : x = q : \frac{1}{2}n$ or $qx = \frac{1}{2}ny$. It follows that if the arc $\frac{1}{2}n$ is infinitely small, it is equal to the brace y, and then we have $qx = \frac{1}{2}yy$.[203] Sauveur concluded by asserting that the curve CD degenerates into a circle $2qx - xx = yy$,[204] since xx is a fourth-order infinitesimal.[205]

[202] Here Sauveur, as in Art. 11, took for granted the law of the parallelogram, which had recently become the paradigm of statics after the publication of Varignon's *Project de la nouvelle statique* in 1687 [86]. Figure 5.49 shows on the right side the parallelogram in object.

[203] There is a typo here, the correct expression should be $qx = yy$.

[204] Actually $qx - xx = yy$.

[205] The expressions used by Sauveur are misleading and represent only an analogy. In fact, y is not a variable abscissa, but simply the abscissa of the point at which the sag f is evaluated, which is implicitly assumed to be half of AB.

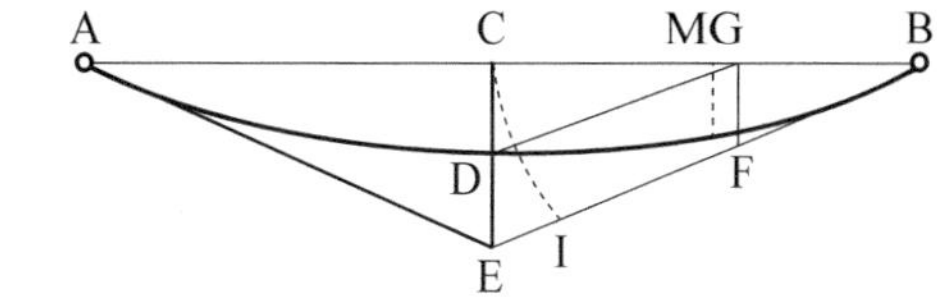

Fig. 5.51 Center of gravity of a string

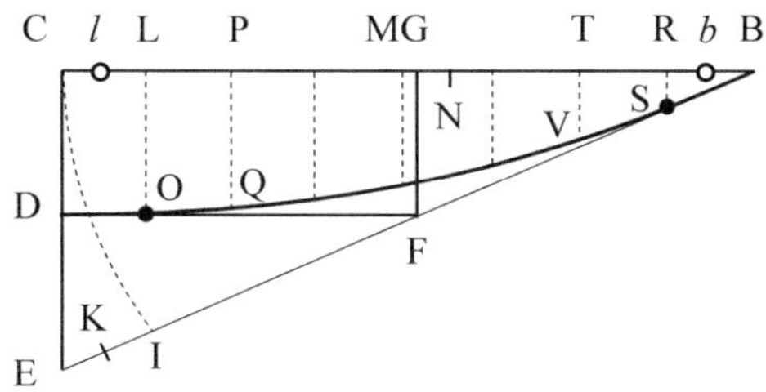

Fig. 5.52 Center of gravity of small portions of a string

16. From the above it follows that if DK ($\frac{1}{2}n$) denotes the weight of the string, then DH (q) denotes [the horizontal component] of the tension, which will be constant, and HK denotes the tension, i.e. the weight p tending the string.

17. Recall that in all experiments carried out by Sauveur the string ADB plays the sub-bis PA, that is the C Sol Ut, of the harpsichord bass or the sound of an 8-foot open organ pipe, and that the ratio of CE to CB is less than the ratio of 1/160 to 25, that is 1 to 4000, and if we draw the arc CI with center in B and radius BI (see Fig. 5.51), the part EI will be less than 1 to 8000 of CE (simple trigonometry can prove this), which is assumed to be of 4000 parts. Consequently, the ratio of the difference EI ($t - y$), is less than $1 : 32 \times 10^6$ of CB, which is absolutely negligible for Art. 11, and thus the difference of the arc DB and CB, that is $n - y$, is absolutely negligible.

18. It follows that we can take the three lines CB (y), DB($\frac{1}{2}n$) and EB (t) for each other and reduce the equality of Art. 14, $cnt = 2apx$, to $\frac{1}{2}cnn = 2apx$, or $cnn = 4apx$.

19. Remember that the center of gravity of the semi-string DB is on the vertical FG.

20. If we take CM to be equal to 1/2 CB, then MG will be less than 1/4 EI. To prove this, let CM and MB be divided into equal but infinitely small parts, as in Fig. 5.52. Drawing the verticals from these divisions, they divide the arc DB into unequal parts at points O, Q, ..., V, S, which parts will be the longer the closer they are to B. The first part DO will be equal to CL; the last BS will be part of the tangent BE; as is evident. Then we divide EI into two equal parts in K.

21. Sauveur demonstrated, in our opinion in an unnecessarily complex and unconvincing way, that the center of gravity N of the two extreme segments DO and SB of Fig. 5.52 is very little away from the center M of CB. He assumed the relation:

$$(\mathrm{DO} + \mathrm{BS}) : \mathrm{DO} = \mathrm{BC} : \mathrm{BN}$$

which he called an *analogy*, probably because it is only approximate, without giving any reason. It can be derived by imposing the equivalence around B between the resultant of the weights of DO and BS imagined at their common center of gravity N and the same weights imagined at l and b, respectively, with bB negligible and Bl equal to CB because of the infinitesimal value of DO and BS.

In the previous expression (taking into account that we are not dealing with infinitesimals in a very rigorous way), we can assume that DO = CL $\propto$ CB and BS $\propto$ BE, and write:

$$(\mathrm{CB} + \mathrm{BE}) : \mathrm{CB} = \mathrm{CB} : \mathrm{BN}$$

If we take half of the antecedents,[206] we have BK : BI = BM : BN, which divided becomes:[207]

$$\mathrm{BK} : \mathrm{IK} = \mathrm{BM} : \mathrm{MN}$$

But BM is less than half of BK, so MN is less than 1/2 IK or 1/4 EI. So MN is less than $1 : 64 \times 10^6$ of BK (EI less than $1 : 32 \times 10^6$ of BK, Art. 17), and even less of BC.

22. Since the arc OQ is greater than DO and NS smaller than SB, their centers of gravity are still closer to M. Taking two by two arcs equally distant from the vertical for M, we see that their centroids are getting closer to M.

23. From this it follows that the common center of gravity G of all the small parts, or of the whole arc DB, is between N and M, and consequently MG is much smaller than MN and can therefore be neglected, and thus G can be confused with M.

The reasoning developed in the previous articles to arrive at this conclusion seems too convoluted to a modern reader. For him, Sauveur could have simply said that since the sag is very small (infinitesimal), the arc of the curve DB can be confused with the brace and therefore its center of gravity coincides with M, the middle point of CB. However probably the confidence with Calculus was not so great at the time.

24. By similarity we then have BG : BC = BF : BE = GF : CE. But BG is about half of BC, and so CD is also half of CE, the difference being much less than $1 : 64 \times 10^6$ of CE; from which it follows that:

$$2\mathrm{CD}(2f) = \mathrm{CE}(x)$$

i.e., the sag f is half of the sub-tangent x, and the equality of Art. 18 $cnn = 4px$ becomes $cnn = 8apf$ and thus

$$f = \frac{cnn}{8ap}; \qquad \sqrt{\frac{1}{f}} = \frac{\sqrt{8ap}}{n\sqrt{c}}$$

From this we draw these consequences:

---

[206] If $a : b = c : d$, then $a/2 : b = c/2 : d$; BK = (CB+BE)/2; CB = BI; BM = CB/2.

[207] If $a : b = c : d$, then $a : (b - a) = c : (d - c)$; IK= BK-BI; MN= BM - BN.

25. Experience shows that the sounds (ν)[208] of a string stretched by the same weight are in reciprocal ratio of their lengths (n). Assuming in the previous relation $\frac{\sqrt{8ap}}{\sqrt{c}} = 1$, we have $\sqrt{\frac{1}{f}} \propto \frac{1}{n} \propto \nu$.

26. That the sounds ν of strings of the same length but stretched by different weights are proportional to the square roots of those weights, that is, $\nu \propto \sqrt{p}$; assuming $\frac{s}{n\sqrt{c}} = 1$, we have $\sqrt{\frac{1}{f}} \propto \sqrt{p} \propto \nu$.

27. That the sounds (ν) of strings of the same length but different diameters are proportional to the inverse ratio of those diameters and the square roots of their weights, that is, $\nu \propto \frac{1}{\sqrt{c}}$.[209]

28. From experiments made when strings of different matter are not significantly stretched by different weights, the sounds (ν) are still in inverse ratio of weight per unit length, so placed $\frac{\sqrt{8ap}}{n} = 1$, we have $\sqrt{\frac{1}{f}} \propto \frac{1}{\sqrt{c}} \propto \nu$.

29. Since the sounds of the strings are in direct proportion to the roots of the weights that stretch them, the inverse of the lengths of the strings, and the inverse of the square root of the weight per unit length, we have $\nu \propto \frac{\sqrt{p}}{n\sqrt{c}}$. Assuming $\sqrt{8a} = 1$, we have:

$$\sqrt{\frac{1}{f}} = \frac{\sqrt{p}}{n\sqrt{c}} \propto \nu$$

30. It follows that in all the cases above, the sounds of all the strings of musical instruments are inversely proportional to the square root of the sag.

31. And therefore, when the sags of the strings are equal, the frequencies are equal.

32. And when the sags are different, the frequencies are in inverse proportion to the square roots of these sags. This is confirmed by experience for all strings which are not stretched at all or stretched very little.

---

[208] For what we call frequency, a word not very common at the time, Sauveur used a locution, vibrations per unit second, and a term *sound*. The symbol he used for sound is an elongated capital S, which, when one considers the writing of the lowercase s and f, suggests an uncertainty between the two symbols s and f. Since it is difficult to reproduce Sauveur's symbol, we will adopt the letter ν, which is often used in modern literature to denote frequencies. This approach introduces a minimum of anachronism.

[209] This article seems to be useless.

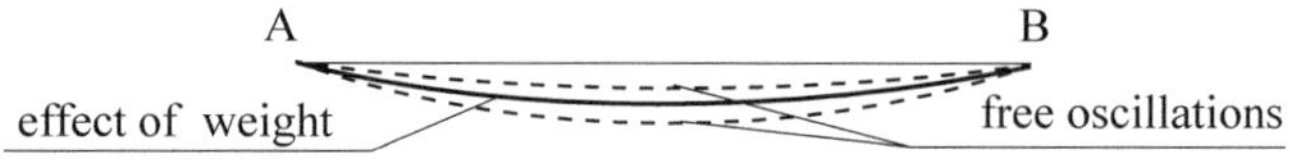

Fig. 5.53 Vibrations around static deflection

5.7.3 Simple Isochronous Pendulums with the Vibrations of a Sonorous String Have a Length of 4/5 the Sag of the Same String

This section is of paramount importance to the entire memoir, as it is here that Sauveur gleaned the formula for the absolute frequency of the vibrating string. He started from the idea that the frequency of the transversal oscillations of the string in the vertical plane, giving rise to curves of type ADB (Fig. 5.47), is the same as that of the oscillation of the compound pendulum formed by the string—considered as a single rigid body—out of the plane with respect to the axis AB.

This idea is contained, though not in a completely explicit form, in some of Huygens' writings not published in Sauveur's time, see Sect. 4.2.1. It is likely that Sauveur was more or less directly aware of these writings, even if he did not mention them anywhere. In any case, there is an original contribution by Sauveur. Indeed, in order to operationalize Huygens' ideas, one must consider a transversal deformation of the string to be taken as the reference configuration for the compound pendulum. Sauveur proposed for such a configuration that of the string deformed under its own weight, which we know today as a catenary, but which for Sauveur was an unspecified curve to be approximated by a parabola or a circle.[210]

A justification for this choice is offered in the procès-verbaux of 1713, which is summarized here:

> Though the vibrations of a string are very unequal, and those which it makes at the beginning of its movement are much greater than those which it makes toward the end of its movement, yet the sound which it makes is neither higher nor lower by this fact, though it is more or less easy to hear [...].
>
> However this string was first set in motion, its oscillations diminish to such an extent that they finally return the string to the shape which its weight naturally gives it, so that *the last oscillations only cause the string to move laterally* [emphasis added], as if the sag of this string had become a pendulum (see Sect. 5.8.5).

As can be seen, Sauveur's argument is not clear. He seems to state, as an experimental circumstance, that for small oscillations the string vibrates out of the vertical plane formed by the string and the brace, rather than moving in that plane, which is challenging to grasp. The subsequent results will fully validate his approach, however.

[210] Most likely Sauveur knew the solution to the catenary problem as it appeared in the *Acta eruditorum* of 1691, due to Leibniz, Huygens, and Johann Bernoulli [85], p. 66, but considered its equation too complex to be used in calculations.

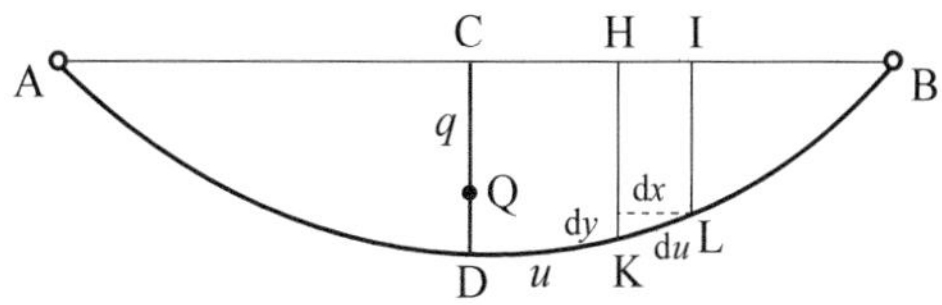

Fig. 5.54 Equivalent simple pendulum

To calculate the length of the simple pendulum equivalent to the compound pendulum formed by the deformation of the string, Sauveur used the rules laid down by Huygens in his *Horologium oscillatorium*.

Below Sauveur assumed that the deformation of the string was a arc of a circle, which makes most calculations easier (Fig. 5.54).

33. In Art. 24, it was shown that the sag CD (f) is less than half the subtangent CE (x) by an amount less than 1: 64×10^6 of CB. Assuming $f = 1/2$ CE $= 1/8000$ CB,[211] if $f = 1$, then CB is 8000 parts and the arc ADB is that of a circle whose radius is 32×10^6.[212] So we can take the arc DB of about 1/2 of minute degrees.[213]

34. It is a matter of finding the distance q[214] of the center of oscillation from the axis ACB of the small arc ADB, that is, the length of a simple pendulum isochronous with the oscillations of the arc ADB ($2a$) around the axis AB.

35. See Fig. 5.54, and along the abscissa CH (x)[215] draw HK (y) and IL infinitely close together. Let f be the sag CD and q the length of the equivalent pendulum, a the half of the arc ADB and u the part DK of this arc. Let the differentials of u, x, y be denoted by du, dx, dy.

36. At this point Sauveur called upon Christiaan Huygens' *Horologium oscillatorium* for the evaluation of the center of gravity and the center of oscillation of the string deformed under its own weight. For the fifth proposition of the *Horologium oscillatorium*, the center of gravity of the double of the arc DB (u) is given by:

$$\frac{\int y \mathrm{d}u}{\int \mathrm{d}u}$$

while the distance CP (p) of the center of oscillation of the same arc is given by [49][216]:

---

[211] For Art. 17, the string under consideration is such that CE/CB = 1/4000.

[212] With reference to Fig. 5.55, the relation $\mathrm{CB}^2 + \mathrm{NC}^2 = (\mathrm{NC} + f)^2$ holds, from which we get $\mathrm{NC} \approx R \approx \mathrm{CB}^2/2f$, which for CB = 8000 f gives $R \approx 32 \times 10^6 f$.

[213] Our calculations show a slightly larger angle.

[214] Sauveur in this part of his memoirs used different symbols to denote the same quantities. In particular, he used p to denote the length of the equivalent pendulum, whereas in the previous sections p was the tension of the string. To avoid ambiguity, in the following we will use q instead of p for the length of the equivalent simple pendulum, while for all other quantities we will follow Sauveur's nomenclature.

[215] The horizontal axis is denoted by x and the vertical axis by y, reversing the previous conventions.

[216] vol. 18, *Horologium oscillatorium*, Proposition V, p. 258.

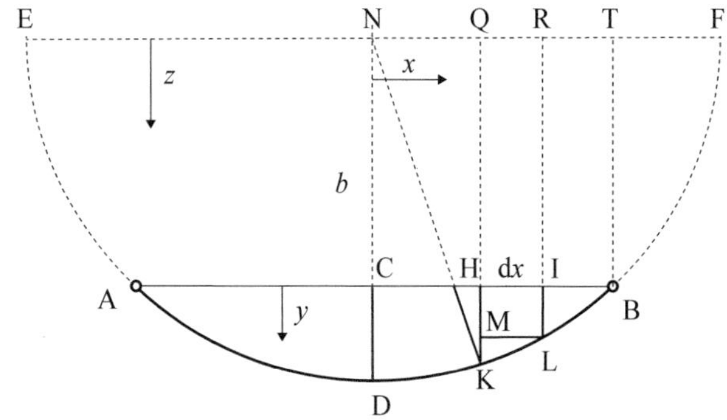

Fig. 5.55 The deformation of the string considered as an arc of circle

$$q = \frac{\int yydu}{\int ydu}$$

which agrees with what Johann Bernoulli wrote in his memoir of the Académie of 1703 [13]; note that the symbols used by Sauveur reported below are those of Bernoulli, since Huygens did not make use of the Calculus.

37. It follows that if AD and DB, instead of being two arcs of a circle, were two straight lines, then $du = kdy$ and:

$$q = \frac{\int yydu}{\int ydu} = \frac{\int yydy}{\int ydy} \left[= \frac{\int_0^f y^2 dy}{\int_0^f ydy} \right]$$

which, when solved, gives $q = \frac{2}{3}f = \frac{66,667}{100,000}$.

38. Referring to Fig. 5.55, assume ADB $(2a)$ as an arc of a circle with N at the center and EF the diameter parallel to ACB $(2c)$,[217] extend the sag CD and the straight lines KH, LI and draw BT. Let KN be the radius r, CN = HQ = IR = BT = b, KQ = z and KM = dz. We have $y = z - b$, dy = dz and $yy = zz - 2bz + bb$.

39. Substituting this value into the expression for the center of oscillation q, we have:

$$q = \frac{\int yydu}{\int ydu} = \frac{\int zzdu - 2b \int zdu + bb \int du}{\int zdu - b \int du}$$

[217] From here to Art. 46, c and a have a different meaning than in all other articles, where c is the weight of a string of length a.

40. Comparing the similar triangles NQK and KLM, we have that NK (r) : QK (z) = KL (du) : ML (dx) and thus $z\mathrm{d}u = r\mathrm{d}x$, which is substituted into the expression of q:

$$q = \frac{r\int z\mathrm{d}x - 2br\int \mathrm{d}x + bb\int \mathrm{d}u}{r\int \mathrm{d}x - b\int \mathrm{d}u}$$

41. Taking into account that $\int z\mathrm{d}x = \mathrm{NDKQ} = \mathrm{NDK} + \mathrm{QNK} = \frac{1}{2}ru + \frac{1}{2}zx$, we obtain:

$$q = \frac{\frac{1}{2}rru + \frac{1}{2}rxz - 2brx + bbu}{rx - bu} = \frac{rru + rxz - 4brx + 2bbu}{2rx - 2bu} =$$

Integrating up to $u = a$, $x = c$, $z = b$ and dividing by f we have:

$$\frac{q}{f} = \frac{arr + 2abb - 3bcr}{2cfr - 2abf}$$

42. Let us first assume DB is a quarter of circle, then $f = r$, $b = 0$, $a = \pi/2r$, $c = r$, which yields:

$$\frac{q}{f} = \frac{a}{2c} = \frac{1.5708}{2} = \frac{78,540}{100,0000}$$

a value not too far from 4/5.

If DB is an arc of a circle of 81 92/100, from the tables of *Arithmétique Britannique*,[218] we find $q/f = 79,563/100,000$. If DB is 40 96/100, we find $q/f = 79,921/100,000$. And finally, if DB is 20 48/100, we find $q/f = 79,982/100,000$, which is very close to $q/f = 4/5$.

43. To find the ratio for infinitely small sags, we can start from the equality:

$$\frac{q}{f} = \frac{arr + 2abb - 3bcr}{2cfr - 2abf}$$

where a, b, c, f, p are variables and r is constant. The differentials of the variable quantities are $\mathrm{d}a$, $\mathrm{d}b$, $\mathrm{d}c$, $\mathrm{d}f$. At this point a 'bold' assumption is made, namely that the ratio q/f varies little and can be considered as a constant: "Since the difference (the variation) of q/f is small". If the ratio of two generic functions g and h is constant, i.e. $g/h = k$, we still have $k = \mathrm{d}g/\mathrm{d}h$. So in the expression of q/f above, we can replace the numerator and denominator with their differentials, and we get the expression[219]:

---

[218] We have not found these tables, but they should be trigonometric tables.

[219] The second term of the denominator $2rc\mathrm{d}b$ should be $2rc\mathrm{d}f$, but $\mathrm{d}f = -\mathrm{d}b$.

$$\frac{q}{f} = \frac{rr\mathrm{d}a + 2bb\mathrm{d}a + 4ab\mathrm{d}b - 3cr\mathrm{d}b - 3br\mathrm{d}c}{2fr\mathrm{d}c - 2rc\mathrm{d}b + 2ab\mathrm{d}b - 2af\mathrm{d}b - 2bf\mathrm{d}a}$$

44. Since the triangles NBC, BLM (see Fig. 5.56) are similar, we have that BL (da) : BN (r) = BM (db): BC (c) = LM (dc) : NC (b); [i.e., $\mathrm{d}a/r = \mathrm{d}b/c = \mathrm{d}c/b = k$] so by introducing infinitesimals we can write $\mathrm{d}a = r/\infty,\ \mathrm{d}b = -c/\infty,\ \mathrm{d}c = b/\infty$.[220] Substituting into the previous expression and simplifying ∞, we have:

$$\frac{q}{f} = \frac{r^3 + 2bbr - 4abc + 3ccr - 3bbr}{2bfr + 2ccr - 2abc + 2acf - 2brf} = \frac{r^3 - bbr - 4abc + 3ccr}{2ccr - 2abc + 2acf}$$

Considering the relations $rr = bb + cc$ and $b = r - f$ we still have:

$$\frac{q}{f} = \frac{4ccr - 4abc}{2ccr - 2abc + 2acf} = \frac{4cr - 4ab}{2cr - 2ab + 2af} = \frac{2cr - 2ab}{cr - ar + 2af}$$

45. However, we do not know the value of a to replace in the expression we just obtained. To find it, we look for the distance g of the center of gravity G with respect to the axis ACB of all the infinitely small arcs dc = KL (Fig. 5.55), which is equal to the area of the circular segment ACBD divided by the length of the arc ADB. Since the area of the circular segment ADB is found as the difference of the circular sector NADBN and the triangle NACBN (see Fig. 5.56), we have:

$$g = \frac{NADBN - NACBN}{ADB} = \frac{ar - bc}{2c}$$

which, as $b = r - f$, gives:

$$\frac{g}{f} = \frac{ar - bc}{2cf} = \frac{ar - cr + cf}{2cf}$$

but one can also admit here that the ratio g/f varies very little; consequently, it is possible to employ the ratio of differences in the numerator and denominator.

Given that $rr - bb = cc$, taking into account that $cc = 2rf - ff$ and $bf = rf - ff$ and considering the expressions of da, db, dc as given in Art. 44, the following can be derived:

$$\frac{g}{f} = \frac{rda - bdc - cdb}{2cdf + 2fdc} = \frac{rr - bb + cc}{2cc + 2bf} = \frac{cc}{cc + bf} = \frac{2r - f}{3r - 2f}$$

Impose equality of the expressions of g/f obtained for finite and infinitesimal values:

$$\frac{g}{f} = \frac{2r - f}{3r - 2f} = \frac{ar - cr + cf}{2cf}$$

[220] A modern reader may find strange the use of the symbol ∞ as finite magnitude, here ∞ is for $1/k$, with k the common ratios $da/r = db/c = dc/b = k$ and $1/k$ is very very large.

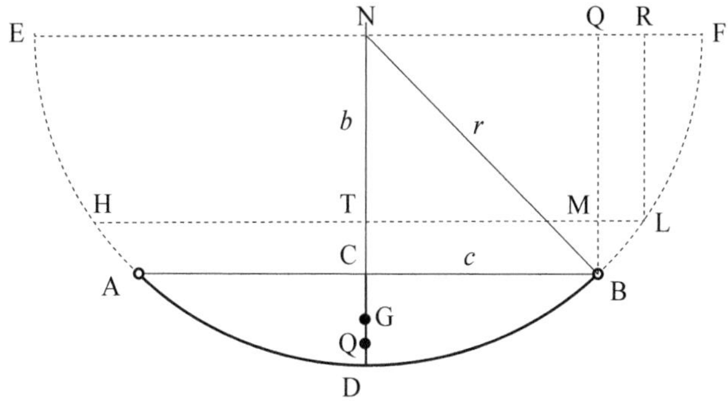

Fig. 5.56 The center of oscillation Q versus the center of gravity G

it is obtained:

$$a = \frac{3cr - cf}{3r - 2f}$$

46. Upon substituting the value of a, as previously determined, into the expression that provides the ratio q/f as outlined in Art. 44, the following expression is obtained:

$$\frac{q}{f} = \frac{4r - 2f}{5r - 2f}$$

which, since f is infinitely small, gives $\dfrac{q}{f} = \dfrac{4}{5}$.

5.7.4 *Way to Find the Number of Vibrations a Sonorous String Makes in One Second*

In this section its is reported the explicit expression of the absolute frequency of the vibrating strings as function of the parameters that influence it: length, tension and weight for unity length.

47. In Art. 24, the formula for the sag of a string was found to be:

$$f = \frac{cnn}{8ap}$$

so that the length q of the pendulum isochronous with the vibrations of the string is given by:

$$q = \frac{4}{5}f = \frac{cnn}{10ap}$$

Using Q[221] for the length of the pendulum beating the second, the ratio of the two lengths is given by:

$$\frac{Q}{q} = \frac{5}{4}\frac{Q}{f} = \frac{10apQ}{cnn}$$

This ratio gives the square of the sound ν of the string, because the frequency of a simple pendulum is proportional to the inverse of the square root of the length; so for two pendulums of length Q and q we have $\sqrt{Q/q} = \nu/\nu_0$, but $\nu_0 = 1$, so we have $\nu = \sqrt{Q/q}$. This will result in:

$$\nu = \frac{\sqrt{5Q}}{\sqrt{4f}} = \frac{\sqrt{10apQ}}{\sqrt{cnn}}$$

But since we always assume $a = 40$, Art. 6, we have:[222]

$$\nu = \frac{20\sqrt{pQ}}{n\sqrt{c}}$$

48. If we use the inches of the Paris foot, then $Q = 36^{17}/_{24}$, and so we have:

$$\nu = \frac{6744}{1000\sqrt{f}} = \frac{12,117\sqrt{p}}{100n\sqrt{c}}$$

49. Example. Consider a steel string of a harpsichord, 40 inches of which weigh $a = 20^1/_3$ grains, the distance n between the two bridges, that is, the length of the string, is 67 inches, and the weight p stretching the string is 10 pounds, or 92,160 grains. This string sounds sub-bis PA, or C Sol Ut, of the harpsichord bass, which corresponds to an open organ pipe 8 feet long. The value of ν, or the number of vibrations per unit of time, is asked. From the above equation we find $\nu = 121.75$.

50. Sauveur claimed to have found, by calculations derived from experiments, that the sag f of the strings playing the sub-bis PA is always 1/323 of a Paris inch.[223]

---

[221] Sauveur used the symbol q; here the symbol has been changed to avoid duplication.

[222] By using the modern symbols introduced in Chap. 3. i.e. $p = F, c/a = \gamma, n = l$, and considering that $Q = g/\pi^2$ (see footnote 196), the previous relations give:

$$\nu = 1.007\frac{1}{l}\sqrt{\frac{gF}{\gamma}}$$

which is numerically the same as the one proposed in Huygens' works (see Sect. 4.2.1), except for the coefficient 2, due to the different way of counting frequencies. This relation differs from the exact expression by less than 1%; but this is not a pity, because in real strings there is a bending stiffness not considered in the 'correct' theory, which increases the frequency of vibration in the real world.

[223] The relationship between calculation and experiment is not clear. An experimental approach would imply to measure the sag under static conditions. It is of the order of 1/300 inches, less than $^1/_{10}$ mm, not large enough to be measured accurately with the experimental apparatus of the

Assuming $f = 1/323$, then by Art. 48 we have

$$\nu = \frac{6744}{1000\sqrt{f}} = 121.76$$

51. This means that PA (C Sol Ut), two octaves higher, makes $121.76 \times 4 \approx 487$ vibrations in one second.

52. In the *Histoire* of the Académie of 1700, Fontenelle mentioned that the sub-bis LO (La of the harpsichord bass) makes 101 vibrations, so PA makes 242 2/5, exactly half the result given by the formula for vibrating strings. What found confirmation in several experiments made on four different days in the years 1699, 1704 and then 1701, in the house of Mr. Deslandes,[224] a very skilled organ maker, in the presence of Father Sebastien Truchet,[225] and many others. Using organ pipes between 4 and 2 feet long, it was found that sub GA (the Mi over F Ut Fa) made 152 vibrations, and consequently PA (a minor sixth above) made 243 1/5. These experiments, made very carefully and patiently, lead to the following observations.

53. Having found that one and the same sound makes twice as many vibrations in a string as in a pipe, it follows that in the case of the string both the forward and the backward vibrations are taken for its vibrations, while in the case of the organ only the forward vibrations are taken, because only these are heard.

54. Since sound travels at a speed of about 160 toises or 1080 feet per second, this means that the 243 1/5 vibrations of PA follow one after the other for a distance of 4 1/2 feet, almost twice the length of a pipe PA, which is about 2 feet (although Sauveur asserted that this phenomenon would be beneficial for investigating the characteristics of undulations in the air, no further inquiries were reported regarding this topic).

55. The agreement of the number of vibrations of sound obtained from two different principles leaves no room for doubt, and so we shall divide the values of Art. 47 and 48 by 2.

56. If we use astronomical inches, $Q = 36$, for Art. 47, we have the following expressions for the frequency divided by 2:

$$\nu = \frac{3\sqrt{5}}{2\sqrt{f}} = \frac{33,541}{10,000f} = \frac{60\sqrt{p}}{n\sqrt{c}}$$

which is written in logarithms as $\nu = 0.52558 - F$, or $\nu = 1.778015 - N + (P - C)/2$, with uppercase indicating logarithms.

57. If one were to utilize the foot of Paris, the result would be:

---

time. The reference to calculations seems to indicate a different approach; probably the use of the equation for sag under various practical situations for the parameters.

[224] Pierre-Francois Deslandes (1667–1709), who built the organ in the Chapelle de l'Hôpital de la Salpêtrière (1709). Here Sauveur spoke of experiments he had begun before publishing his first memoir on music in 1701,

[225] Jean Truchet (1657–1729), known as Father Sébastien, was a French Dominican priest.

$$\nu = \frac{338,695}{100,000f} = \frac{60,588\sqrt{p}}{10,000n\sqrt{c}}$$

which is written in logarithms as $\nu = 0.52981 - F$, or $\nu = 1.78238 - N + (P - C)/2$.

58. If we finally utilize the inches of the ancient Paris foot, we arrive at the following expression:

$$\nu = \frac{33,807}{10,000f} = \frac{60,475\sqrt{p}}{10,000n\sqrt{c}}$$

which is written in logarithms as $\nu = 0.52900 - F$, or $\nu = 1.78158 - N + (P - C)/2$.

The previous values of ν should be used with the following precautions.

59. It is preferable to avoid using strings whose material is not homogeneous and deforms differently when stretched by different weights. This includes gut strings, which should be avoided. Metal strings represent a superior alternative, even if they exhibit some degree of stretch. Of the available strings, those composed of steel and iron are particularly recommended.

60. The weight used to tension the strings should be limited.

61. The error in the calculations comes mainly from the error in the weight of the string of 40 inches; to reduce this error, one can weigh 2, 4, 8, 16, 32, or even 64 times the 40 inches.

62. To measure the length of a string, one must stretch it with a weight equal to about half the weight that will cause it to break.

63. In the event that the string in question exhibits a diameter that is larger at one end than at the other, it is necessary to utilize the central portion of the string in order to facilitate its use in a musical instrument.

5.7.5 New Determination of the Fixed Sound Table. Construction of an Echometer

In this section, Sauveur provided a renewed interpretation of the locution *fixed sound*. In the 1701 memoir, the fixed sound was a very precise pitch, a standard, 100 Hz, to be used for tuning musical instruments. And the value chosen was related to the particular way used to produce it: two organ pipes separated by a minimum semitone.

With the new procedure, Sauveur believed he could produce a sound of any frequency, not just 100 Hz. In this way, the term *fixed* was to be understood as *absolute* and not simply as standard. And all the sounds provided by the vibrating string equation he obtained are absolute or fixed sounds. The octaves chosen in the range of perceived sounds are also called fixed octaves.

64. Sauveur emphasized the fundamental of the middle octave, Ut = 256 Hz, with which he intended to replace the 100 Hz sound as the new standard against which to compare other sounds. As a power of two, it takes on a less arbitrary meaning.

Table 5.32 Double octave progression

1	2	4	8	16	32	64	128	
2^0	2^1	2^2	2^3	2^4	2^5	2^6	2^7	
256	512	1024	2048	4096	8192	16384	32768	65536
2^8	2^9	2^{10}	2^{11}	2^{12}	2^{13}	2^{14}	2^{15}	2^{16}

Today, the reference pitch is La = 440 Hz; by adopting Ut = 256, we would get La = 431 Hz in equal temperament, a little lower value. The pitch Ut = 256 is known as the *Sauveur pitch*, or philosophical pitch [44].[226]

65. The range of sounds between high and low is divided into octaves according to the double progression of Table 5.32.

66. By adjusting the fundamental of each fixed octave and the order of the octaves according to the double progression, we see that the sound that makes 256 vibrations per second is the fundamental of the eighth octave.

67. If we divide the numbers contained in the double progression, such as the tones between 256 and 512, into 43 proportional means, we divide each octave into merides, and if we divide each meride into 7 or 70 parts, we have a division of the octaves into heptamerides or decamerides.

68. To have the intervals of the fundamental sounds of each octave in logarithmic form, multiply 3010300 (the logarithm of the interval of an octave, 2) by the numbers of the double progression given above. Thus we find that the interval of the fundamental 256 of the octave is $0.3010300 \times 8 = 2.4082400$.

69. Subtracting the last three digits of this value gives 24,082 decamerides and 2,408 heptamerides. Dividing by 7 gives 344 merides, for the interval of this fixed octave.

70. By adding 1, 2, 3, 4, etc. to the merides of the fundamental of each octave, you get all the merides of that octave; if you add the products of these numbers by 7 or 70 to the heptamerides of the fundamental of each octave, you get the heptamerides and decamerides.

71. In Arts. 49 and 52, it was found that C Sol Ut has about 243 1/5 vibrations, and having then seen that the number 256 of the double progression is the one closest to it, Sauveur claimed that he would take 256 as the fundamental fixed sound of the middle octave as a new standard, so that the following numbers 512, 1024, 2048 are the fundamental tones of the first, second, third fixed octaves, and the previous 128, 64, 32, etc. are the fundamental tones of the first, second, and third sub-octaves.

Note that Sauveur did not explicitly say that Ut = 243 is no longer an Ut, although he probably thought so.

72. The fundamental of each fixed octave is called PA, and the other sounds are called by the names given in the 1701 memoir, distinguishing those of each octave by their clefs.

---

[226] p. 42.

73. The fixed sounds are indicated by Sauveur with the signs suggested in the said memoir.

74. As the organ makers identify the fundamental sound PA with that of an open pipe of 2 feet in length, which gives approximately this sound; the PA of the upper octaves are with pipes of 1 foot, 6 inches and so from half to half, and the sub-octaves with pipes of 4, 8, 16, 32 feet.

To facilitate comprehension and circumvent the burden of calculation, the following Table 5.33 presents a comprehensive listing of the fixed sounds. Due to space constraints, this table provides a selective representation of Sauveur's table, as documented in Planche 10 of his memoir, focusing on the results for the octaves 5, 6, 7, 8, and 9.

75. The table contains five columns of five-digit numbers; they range from the fifth to the ninth octaves, the more significant ones.

76. In each of these five columns there are 43 numbers in a continuous geometric progression, indicating the number of vibrations of the tones of the different octaves, from meride to meride.

77. Below these numbers are six lines:

The first is for the order of the fixed octaves.
The second indicates the decamerides of the intervals of the fundamental PA of each octave.
The third gives the same intervals in merides.
The fourth shows the sub-octaves and octaves with respect to the fundamental of the middle octave.
The fifth shows the names of the clefs that distinguish each octave.
The sixth, the lengths of the organ pipes that make up the PA sounds of each octave.

78. On the left side of the table are 5 columns; starting from the right:

The first column, next to the table, contains the merides of each octave.
The second contains the decamerides.
The third contains the new names of the sounds.
The fourth, the old names.
The fifth, the diatonic intervals.

79. Since the strings of musical instruments that give the sound [frequency] indicated in the table should be divided in reverse order, it is convenient to use an echometer that provides them. For the construction of this echometer, astronomical inches are used; the formula $\nu = \dfrac{60\sqrt{p}}{n\sqrt{c}}$ is assumed with $\dfrac{p}{c} = 4096$ and thus the formula:

$$\nu = \frac{60 \times 64}{n} = \frac{15 \times 256}{n}$$

is used.

80. On the wooden ruler of about 5$^1/_2$ feet shown in Fig. 5.57a, draw a line AB of 60 astronomical inches, or 61 inches and 2$^1/_6$ of the foot of Paris, and divide it into

Table 5.33 Tables of fixed sounds; from [77], Planche 10, p. 348

Intervalles diatoniques	*Noms Anciens*	*Noms Nouveaux*	*Decamerides*	*Merides*	TABLE GENERALE DES SONS FIXES				
	si✱	*py da*	2940	42	62. 977	125.953	251. 91	503. 81	1007.63
		pu dy	2870	41	61. 970	123.939	247. 88	495. 76	991.51
	ut b	*po du*	2800	40	60. 979	121.957	243. 92	487. 83	975.66
VII	SI	DO	2730	39	60. 004	120.007	240. 01	480. 03	960.06
			2660	38	59. 044	118.088	236. 18	472. 35	944.71
			2590	37	58. 100	116.200	232. 40	464. 80	929.60
7	ſi *b*	de	2520	36	57. 171	114.342	228. 68	457. 37	914.74
	la ✱	la	2450	35	56. 257	112.513	225. 03	450. 05	900.12
		ly	2380	34	55. 357	110.714	221. 43	442. 86	885.71
		lu	2310	33	54. 472	108.944	217. 89	435. 78	871.55
VI	LA	LO	2240	32	53. 601	107.202	214. 40	428. 81	857.61
		li	2170	31	52. 744	105.488	210. 98	421. 95	843.90
		lε	2100	30	51. 900	103.801	207. 60	415. 20	830.41
6.	la *b*	le	2030	29	51. 071.	102.141	204. 28	408. 56	817.13
	ſol ✱	ba	1960	28	50. 253	100.507	201. 01	402. 03	804.05
		by	1890	27	49. 450	98.901	197. 80	395. 60	791.20
		bu	1820	26	48. 660	97.319	194. 64	389. 28	778.55
V	SOL	BO	1750	25	47. 881	95.763	191. 53	383. 05	766.10
		bi	1680	24	47. 116	94.232	188. 46	376. 93	753.85
		bε	1610	23	46. 362	92.725	185. 45	370. 90	741.80
5	ſol *b*	be	1540	22	45. 621	91.242	182. 48	364. 97	729.93
IV	fa ✱	ſa	1470	21	44. 892	89.783	179. 57	359. 13	718.26
		ſy	1400	20	44. 174	88.347	176. 69	353. 39	706.78
		ſu	1330	19	43. 467	86.935	173. 87	347. 74	695.48
4	FA	SO	1260	18	42. 772	85.545	171. 09	342. 18	684.36
		si gi	1150	17	42. 088	84.177	168. 35	336. 71	673.41
		sε gε	1120	16	41. 415	82.831	165. 66	331. 32	662.65
		se ge	1050	15	40. 753	81.506	163. 01	326. 02	652.05
III	MI	GA	980	14	40. 101	80.203	160. 41	320. 81	641.62
		gy	910	13	39. 460	78.920	157. 84	315. 68	631.36
		gu	840	12	38. 767	77.533	155. 07	310. 13	621.27
3	mi *b*	go	770	11	38. 208	76.417	152. 83	305. 67	611.33
	re ✱	ri	700	10	37. 597	75.195	150. 39	300. 78	601.56
		rε	630	9	36. 996	73.992	147. 98	295. 97	591.94
		re	560	8	36. 405	72.809	145. 62	291. 24	582.47
II	RE	RA	490	7	35. 822	71.645	143. 29	286. 58	573.16
		ry	420	6	35. 250	70.499	141. 00	282. 00	563.99
		ru	350	5	34. 686	69.372	138. 74	277. 49	554.98
2	re *b*	ro	280	4	34. 131	68.263	136. 53	273. 05	546.10
	ut ✱	ri	210	3	33. 586	67.171	134. 34	268. 68	537.37
		rε	140	2	33. 048	66.097	132. 19	264. 39	528.78
		re	70	1	32. 520	65.040	130. 08	260. 16	520.32
Son fondamental	VT	PA	0	0	32. 000	64.000	128. 00	256. 00	512. 00
ORDRE DES OCTAVES FIXES					5	6	7	8	9
Decamerides des PA *fixes*					15051	18062	21072	24082	27093
Merides des PA *fixes*					215	258	301	344	387
Sous-Octaves *et* Octaves . . .					3.^e ſous-oct.	2.^e ſous-oct.	1.^e ſous-oct.	OCTAVE	1.^e Octave
Cleſs des Noms					*Subter–*	*Subbis–*	*Sub–*	MOYENNE	*Sem–*
Longueur des tuyaux d'orgues . .					*16. pieds*	*8. pieds*	*4 pieds*	*2. pieds*	*1. pied*

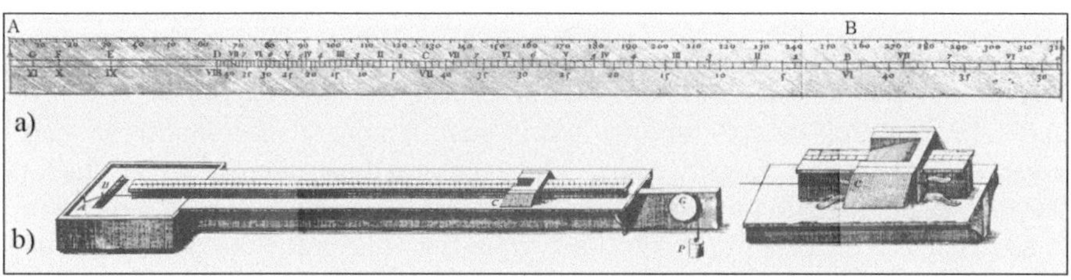

Fig. 5.57 The echometer for finding fixed sounds. Drawn from [77], p. 348

64, or rather 256, equal parts. Divide AB in half at C, then in half AD, AE, AF, AG, etc. Mark the octaves VI, VII, VIII, IX, X, XI, etc. on these divisions. Then divide each octave into 43 merides using Table 5.33. Use the numbers of the seventh octave, which are between 128 and 256, to divide BC, and the numbers of the sixth octave to divide DC, and so on for the other octaves. The divisions in merides can again be divided into 7 equal parts to mark the heptamerides. As clear from Fig. 5.57 the echometer can measure sounds from the VI (corresponding to a string length of 60 astronomical inches) up to the XI octave.

81. In each octave, the merides are marked with numbers 5 by 5; on the other side of the line, the diatonic intervals are marked. On this echometer you can add the names of the sounds and also the number of vibrations, although it is best to look for the latter values in Table 5.33.

5.7.6 Ways to Find Fixed Sounds

Below are several ways to find the fixed sound of a string, the first of which are simpler, but at the same time more complex for the calculations, and vice versa. A modern reader is puzzled by the absolute confidence that Sauveur seems to have in the values given by his formula, to the point of not doing any experimental verification. In contrast to the attitude of Taylor, see Sect. 4.2.1.

83. On a vertical board, fix a string to a nail A (Fig. 5.58a), which passes over bridges B and C; at the lower end of the string hang a weight P. Put a finger on bridge C and pinch the string. If you want to get a louder sound from this string, you must let the bridge B rest on the body of a musical instrument, as shown in Fig. 5.58b.

84. To find the fixed sound, we need to know:

1. The weight a in grains of the 40-inch string.
2. The weight p tending the string, reduced in grains.
3. The length n of the string BC in inches.

85. If one uses astronomical inches, you have the expressions of Art. 56.

86. If one uses the inches of the foot of Paris, one has the expressions of Art. 57.

87. In the case of the example in Art. 49 we have $\nu = 60.88$ vibrations per second (here Sauveur divided by two, the value of Art. 49 equals 121.76 vibrations per second) and in logarithms $\nu = 1.78449$.

88. To know its octave, we take the largest number of the double geometric sequence contained in 60.88; we have 32, so the sound is contained within the fixed fifth octave and its ratio to the fundamental of this fixed octave is 60.88 to 32.00.

89. To know the exact interval of this sound, it is necessary to perform a linear interpolation from the data of Table 5.33. Sauveur, as usual, provides the simple algorithm to perform this interpolation, without any explanation. Of course, if he had been writing for mathematicians, he would have simply said: interpolate!

Look in the table for the number closest to 60, 880; you will find 60, 004, which is the 39th meride of the fifth fixed octave or third sub-octave. Subtract 60, 004 from

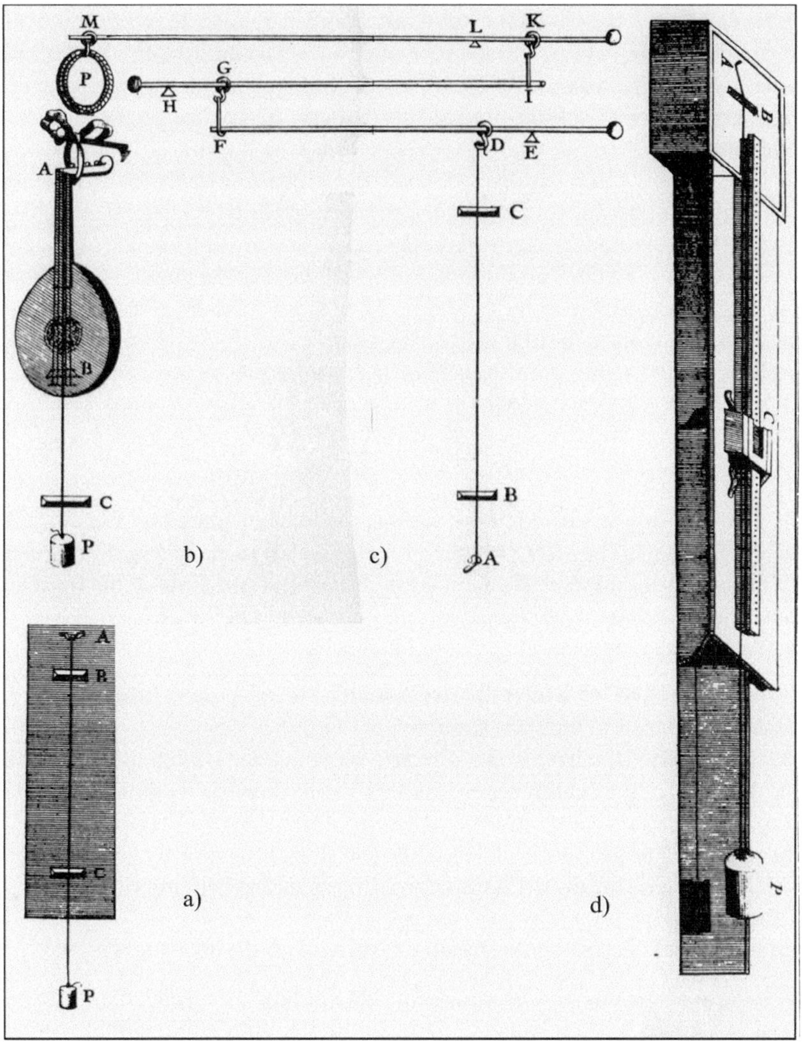

Fig. 5.58 Instruments to find fixed sounds. Adapted from [77], p. 348

60, 880; the remainder is 876. Then subtract 60, 004 from the previous number in the table (70 decamerides further), 60, 979 and you are left with 975. Using the rule of three, we have the ratio 975 : 876 = 70 : 63. We have 6 heptamerides and 3 decamerides, so the fixed sound of the string in question is 39 merides, 6 heptamerides and 3 decamerides.

90. Working directly with logarithms, the interval marking the sound is given by $\nu = 1.78449$; divide by 0.3010300 and you get 5 octaves, with a remainder of 2793 decamerides; divide by 70 and you get 39 merides, with a remainder of 63 decamerides, or 6 heptamerides and 3 decamerides, as above.

91. Having found the fixed interval of this sound in Table 5.33, we can now find its name. Since it is five octaves, 40 merides minus 1 heptamerides plus 3 decamerides, its name in merides is sub-ter-po or sub-ter-du, in heptamerides sub-ter-pon or sub-ter-dun, in decamerides sub-ter-poni or sub-ter-duni. In practice the name in merides is sufficient.

92. To reduce the computations, one could use *acoustic weights*. To do this, assume weights varying in double proportion, the smallest of which is 1/64 of 9 grains and the largest of 16 or 32 pounds. Denote the smallest by the number 0 and the others by 1, 2, 3, etc. The acoustic weights used for the string are those of the numbers 7, 6, or 4, that is, $2^k 9/64$ grains, with $k = 7, 6, 4$. Adding 12 to these numbers gives a weight of the numbers 19, 18, 16 used to stretch the string (so $p/c = 2^{12} = 4096$ times the weights of the string). Then, using astronomical inches, the frequency is given by:

$$\nu = \frac{60 \times 64}{n} = \frac{3840}{n}$$

93. To avoid this last calculation, we can use the echometer of Figs. 5.57b and 5.58a and by placing the string's bridge B (Fig. 5.58a) at the end A of the ruler; the length measured at bridge C will indicate on the ruler the interval of the fixed sound, in octaves and merides.

94. You can also use vertical or horizontal echometers consisting of:

1. A square box ABC to which the string is attached at A and a bridge at B.
2. A handle attached to the box, about 6 feet long.
3. Along and above the handle is a directrix on which a movable bridge can slide.
4. The divisions of the echometer are marked on this directrix, beginning with bridge B.
5. The movable bridge must be able to slide and stop at any time by means of springs; it will indicate on the directrix the octaves, merides and heptamerides of the sound of the string.
6. But for this it is necessary to stretch the string with the acoustic weight.

95. If the acoustic machine is horizontal, add a pulley G on which the string passes. This pulley should be well rounded, large and light, with a thin spindle of smooth steel turning in a copper bush to reduce friction as much as possible.

96. For use of the echometers:

1. Weigh the string as in Art. 84.
2. Attach the string at A and load the other end with the weight P determined as in Art. 84.
3. Place the movable bridge in the desired position on the graduation of the echometer.
4. Pull the string.
5. When you pluck the string, it will make a sound at the fixed interval indicated on the echometer, the vibrations of which can be obtained from Table 5.33.

97. Instead of weights, the acoustic balance shown in Fig. 5.58c may be used. Let there be three levers whose longest arms are EF, HI, LM, with fulcrums E, H, L. The shorter arms ED, HG, KL are one-eighth the length of the longer ones. Attach the string to be played to D.

99. In this arrangement, after attaching the string DA to the pin A, hang a wire eight times as long as the 40 astronomical inches of the same string at M, then turn the pin in A so that the arms of the levers are horizontal and thus in equilibrium. Then the interval BC between the two bridges is measured with the echometer; it will give the fixed sound.[227]

5.7.7 *Use of the Fixed Sound. Remarks*

100. In the *Histories* of the Académie of 1700 and in the memoir of 1701 many uses of the fixed sounds are indicated. Here the following can be added.

101. For Art. 71 the fixed fundamental, taken as standard for PA, makes 256 vibrations per second; while the PA (Clef G Sol Ut) of the two feet open pipe of the organ on which Sauveur used to adjust the strings, makes about 243¹/₅, Art. 52. The interval of this sound is about three merides lower, so the sounds of this organ are lower than their similar fixed sounds of three merides.

102. Therefore, it can be said that this whole organ is at the third sub-meridie, and the same can be said of the tones of the Chapel, the Opera, and other concerts after having experienced them.

103. This means that if you want to use Table 5.33 to find the number of vibrations to which a sound would correspond, after knowing the degree of the sound of an instrument, for example of BO (key of G Re Sol), one should look for BO in the middle octave of Table 5.33 and take three merides below.

104. Using the table of fixed sounds, one can find how many vibrations the epiglottis makes in one second for each voice pitch, and how many vibrations it makes at its lowest pitch; and also the lowest pitch of a low voice and the highest pitch of a high voice. The number of vibrations of the lips when whistling or playing the horn or trumpet will also be known, and finally the number of vibrations of the notes of any musical instrument.

105. This last observation leads us to conclude that since the lowest sound of the organ that can be heard is that of a 32-foot open pipe, which makes 16 vibrations per second, we can use this value to conclude that a sound that makes less than 16 vibrations per second cannot be heard. It can also be assumed that one cannot distinguish the sound above the 15th fixed octave, which makes 65, 536 vibrations per second, as a $1^1/_8$ line pipe would, because the smaller pipe of the organ is two octaves lower than this value.

106. By knowing the sounds of a sonorous body, that is, its degree of high and low, it is possible to know the name of this sound, which is identified by its merides. For example, the large bell of Notre Dame, called Emanuel, sounds sub-bis-LO (La).

[227] The tension p of the string, for the law of static, considering that each of the three levers amplifies by 8 the weight P, is $p = 8 \times 8 \times 8 \times 8c = 4096c$, i.e. the same value used for the ruler-echometer.

Subtract three merides and its name will be sub-bis-le. So we can say that the Emanuel is a sub-bis-le.

107. Using the memoir of the sounds of the bells given to Sauveur by Chastelain, honorary canon of Notre Dame in Paris,[228] preserving the old names of the notes, the names reported in Table 5.29 can be given, after the sounds have been corrected according to Art. 103.

108. See Table 5.34.

109. These names serve:

1. As terms of comparison to check other fixed sounds.
2. To understand their mutual agreements.
3. To know their relative weights and also their absolute weights.

For example, it is known that Emmanuel sub-bis-le weighs 30,000 pounds, the weight of the bells of St. Giovanni in Laterano sub-bis DO is asked.

Sauveur, as usual, gave an algorithm without any explication:

1. One looks in Table 5.33 of sounds for the sub-bis-le (19 merides) sub-bis-DO (29 merides) intervals, which is worth 10.
2. One triples this value, giving 30.
3. In Table 5.33 one should look for 30,000. However, as this value does not exist in our simplified table, we reduce it at the 9th octave by dividing 30,000 by $2^5 = 32$, which gives 937.5, corresponding to 37 merides.[229]
4. One subtracts 30 from 37, leaving 7.
5. One looks up 7 in the 9th octave and finds 573.16, which, multiplied by 32, gives 18,341, roughly the weight of the bell of St. Giovanni in Laterano.

To understand this algorithm, we must start with the assumption that the frequency of the bells varies in inverse proportion to the cube root of their weights.[230] That is, given p_1, p_2, ν_1, ν_2, the weights and frequencies of two bells, we admit that the relation holds:

$$\left(\frac{\nu_1}{\nu_2}\right)^3 = \frac{p_2}{p_1}$$

With logarithms, it is $\log_M p_2 - \log_M p_1 = 3(\log_M \nu_1 - \log_M \nu_2)$. The basis M can be chosen so that $\log_M 2 = 43$. In such a case, $\log_M \nu$ gives the merides corresponding to ν and Table 5.33 can be used as a normal table of logarithms. Thus, by indicating $\log_M x = Mx$, we can write: $Mp_2 - Mp_1 = 3(M\nu_1 - M\nu_2)$, and thus:

$$Mp_2 = Mp_1 + 3(M\nu_1 - M\nu_2)$$

In the present case $p_1 = 30,000$ and $(M\nu_1 - M\nu_2) = -10$; we need Mp_1. To find it we go to the table and look for the merides. But since 30,000 is too large for the

[228] Probably Claude Chastelain (1639–1712) French liturgist.

[229] Sauveur's procedure is adapted here to use the reduced version of his table of fixed sounds.

[230] This dependence can be inferred from Mersenne's *Harmonie universelle* [56], Part II, Livre septiesme des instruments de percussion, pp. 24–26, which Sauveur might know.

Table 5.34 Bells in Europe

Church	Bell	Nominal pitch	Sauveur pitch	Frequency Hz	Modern names
Notre Dame de	Emanuel	La	sub-bis-le	102.141	$A_2\flat$
Paris	Marie	Si	sub-bis-de	114.342	$B_2\flat$
	Gabrielle	Ut	sub-bis-DO	120.007	B_2
	Guillaume	Re	sub-ro	136.53	$D_3\flat$
	Pasquier	Mi	sub-go	152.83	$E_3\flat$
	Thibaut	Fa	sub-se	163.01	$F_3\flat$
	Jean	Sol	sub-be	182.48	$G_3\flat$
	Claude	La	sub-le	204.28	$A_3\flat$
	Nicolas	Si	sub-de	228.68	$B_3\flat$
Petit clocher:	Catherine	Re	ro	273.05	$D_4\flat$
	Magdaleine	Mi	go	305.67	$E_4\flat$
	Matiphas				
	Barbe	Fa	gu	310.13	E_4
	Anne	Sol	be	364.97	$G_4\flat$
St. Germain des	Germain	Ut	sub-bis-du	121.957	$C_3\flat$
Pres	Vincent	Re	sub-ro	136.53	$D_3\flat$
St. Paul		Re	sub-go	152.83	$E_3\flat$
		Mi	sub-SO	171.09	F_3
		Fa	sub-be	182.48	$G_3\flat$
		Sol	sub-de	228.68	$B_3\flat$
		La	PA	256	C_4
St. Jean en Grève		Fa	sub-se	163.01	$F_3\flat$
		Sol	sub-de	182.48	$G_3\flat$
		La	sub-le	204.28	$A_3\flat$
St. Victor		Re	sub-go	152.83	$E_3\flat$
		Mi	sub-SO	171.09	F_3
		Fa	sub-be	182.48	$G_3\flat$
		Sol	sub-le	204.28	$A_3\flat$
		La	PA	256.00	C_4
		Si	RA	286.58	D_4
		Ut	go	305.67	$E_4\flat$
Paris, St. Genevieve; Doüai, St. Amé		Re	sub-go	152.83	$E_3\flat$
		Mi	sub-SO	171.09	F_3
		Fa	sub-be	182.48	$G_3\flat$
Rome, San Giovanni Laterano; Gand		Ut	sub-bis-DO	120.007	B_2
Rome, San Pietro; Sens; Reims		Si	sub-bis-de	114.342	$B_2\flat$

table, it is reduced by 5 octaves, that is, divided by 32, giving 30,000/32 = 937.5 pounds[231]; we have about 37 merides, as shown in the 9th octave column. Thus $Mp_2 = 37 - 3(M\nu_1 - M\nu_2) = 7$, which corresponds to the Mp_2 of St. Giovanni in Laterano. Its anti-logarithm, p_2, is found by looking up 7 merides on Table 5.33 in the 9th octave column and find 573.16, multiplying by 32 to get $p_2 = 18,341$.

110. In the previous articles, the pitches of the notes are expressed in merides, but you can also have them in heptamerides or decamerides, which will give more precise values for the bells and ratios of their weights.

111. Certain bodies are known to tremble at loud sounds, such as the table of a musical instrument at the sound of certain strings; similarly plates, vessels, and even the bowels of a man. If one knows these sounds, or their fixed intervals or names, one also knows by what vibrations per unit of time these bodies can be disturbed by resonance, and this would help to know their nature.

112. If fixed sounds were established throughout the world, and if they were also known to the ancients, who were able to give appropriate names to the tones of the bodies in different situations, we might know from this the difference between men, animals, and the bodies of different places; their change in time, and the differences in the sounds of the ancients from ours.

113. The makers of musical instruments will find in the echometer a general diapason, which regulates organ pipes for organ makers, bell sizes for bell founders, monochords for harpsichord makers, and echometer machines for scholars who wish to know the number of vibrations of sonorous bodies, that is, the ratios of the intervals of the tones of all kinds of sonorous bodies.

114. We have seen in Art. 56 that this means that the sound ν does not vary with the diameter of the string, provided that there is always the same ratio between its weight c and the weight p that tends it. For Arts. 85 and 86 you can find out how much you can increase p before the string breaks.

115. Finally, the curious can produce notes to imitate the sounds of birds, animals, or savages, which go for small intervals that cannot be expressed by ordinary notes.

116. Thinking of other uses that may be found for the same subject, two observations should be added. The first is that established principles are of the same nature as the other principles of physics; they serve as a basis for experience, which are sometimes modified by new circumstances. For example, the parabola for shooting a projectile, the odd-numbered progressions for accelerating heavy bodies, and the cycloid for the pendulum.

117. The second observation is that Mr. Marius,[232] famous for the invention of the folding harpsichord to make it easier to transport, as well as for other inventions. After carrying out the experiments initiated by Sauveur to substantiate the preceding principles, Marius observed that the sounds produced by the thickest strings, separated by a bridge at distances of half to half, did not ascend in a precise manner from the first to the second or the third octave, but instead went higher than 1 or $1^1/_2$

---

[231] By dividing by a multiple n of 2, $\log_M$ differs by the amount $n43$.

[232] Jean Marius (d. 1720), harpsichord maker in Paris.

meride,[233] and that pipes of the same size, of which one is half or even a quarter of the other, with one half or even a quarter of the diameter of the other, produce sounds that do not reach the octave. These new circumstances would give rise to further researches.

5.7.8 Fontenelle's Account

In the first part of his account, Fontenelle described the sag-frequency relationship without any equation. He made the following observations: the number of vibrations of a string in a given time is all the greater, i.e., the sound a string makes is all the more acute, the greater the square root of the weight tending the string, the shorter the length of the string, and the smaller its diameter. A horizontal string, fixed at one end while the other passes over a pulley and is loaded by a weight, no matter how long it is, will bend at least a little by its own weight. We must therefore imagine the string between the two ends as an arc, the brace of which is the straight line drawn from the fixed ends; the line drawn from the center of the arc perpendicular to the brace is what Sauveur called *sag*. This being the case, the more a string bends, or which is the same, the greater the sag is, the less the weight tending the string, the heavier the string and the longer the string.

The first and second conditions are obvious to Fontenelle, who instead devotes a long and unconvincing argument to the third, which we will avoid reporting.

For a string, which is a cylinder, volume and weight are obtained from the product of its length by its circular base; since we do not consider other than ratios, and circles are like the squares of diameters, it can be said that the weight of a string is proportional to the product of its length by the square of its diameter.

So the sag is the greater:

1. As the weight it tends is the smaller.
2. As the square of the diameter and the length is the greater.
3. As the length is the greater.

Since the length of the string enters the sag expression twice, we can say that the sag is much greater because the weight tending the string is smaller, the diameter squared is greater, and the length squared is greater. Thus, the algebraic expression of the number of vibrations of a string in a given time, and the expression of the value of the sag are similar, differing only in two respects: They are in inverse proportionality and one is the square root of the other.

[233] This is an anticipation of what is now known as the octave stretching phenomenon, according to which listeners seem to prefer a slightly stretched octave, with the ratio between the two frequencies forming the octave being slightly greater than (2 : 1). The amount of stretching normally registered is 20–25 cents, corresponding to 0.7–0.9 merides, not very different from the values given by Marius [46].

> Although what has been said is sufficient to formulate this important proposition of acoustics; Sauveur proved it differently in a rigorous way based on geometry [40].[234]

From this theorem, Fontenelle said, Sauveur deduced a way to determine how many vibrations a sound makes in a given time. For the string that plays the C Sol Ut of the harpsichord bass, that is, the one with the lowest pitch and therefore the greatest sag.

Regarding the frequency of vibration of the string, Fontenelle was a little more explicit than Sauveur, but not too much. He admitted that the frequency of the transverse oscillations of the string coincides with the frequency of vibration of the string, considered as a rigid body, around the brace. To find this frequency, said Fontenelle, without mentioning Huygens, one must imagine each point of the arc as a pendulum hanging from the brace. There are an infinity of small pendulums parallel to each other, which together make their oscillations, the largest of which is that of the sag. The longer pendulums make their oscillations slower, the shorter ones faster, but since they are connected to each other, they influence each other and the whole arc takes on an average speed greater than that of the simple pendulum of the sag and greater than that of the smaller pendulums. This results in a center of oscillation, i.e. a simple pendulum of less length than the sag, which makes the oscillations at the same time as the whole string. Sauveur, by means of the theory of centers of oscillation, found that the length of this simple pendulum is $^4/_5$ that of the sag, following the same approach as found in a *Histoire* of 1711 presented by Fontenelle himself to find the center of gravity of an arc of circle [38].[235]

All that remains is to know the magnitude of the sag of a string playing a low C Sol Ut of the harpsichord bass. Sauveur found by 'experience' that the ratio of the sag with respect to the brace is 1/323. Knowing that a simple pendulum of 36 Paris inches and 8 $^1/_2$ lines, makes its oscillations in one second, and that the oscillations made at the same time by two pendulums are in the inverse ratio of the square roots of their lengths, we have that a pendulum of the length of $^4/_5$ of 1/323 makes about 122 oscillations per second. From this we can easily calculate the pitch of another string. For example, an upper octave of C Sol Ut, the harpsichord bass, makes 244 vibrations.

In the *Histoire* of 1700, Sauveur claimed to have found by experience that a five-foot open organ pipe made 100 vibrations per second. The same experience, repeated with greater care, led him to use 102 instead of 100. It is in unison with the La of the harpsichord bass, and consequently this La is thought to make 102 vibrations per second. It is known that the ratio of La to Ut is 6 : 5,[236] which should give (almost) 122 vibrations for the Ut of the harpsichord bass. If this were equal to the value found for the strings, it would be a confirmation of all the operations and reasoning of Sauveur, who would have arrived at the same result by following two different paths: determining the vibrations of organ pipes from their beats, and determining

---

[234] p. 71.

[235] Actually this memoir makes only general considerations on the applications of the Calculus to geometry and there are no hints about the calculation of the center of gravity of an arc of a circle.

[236] p. 73. Fontenelle erroneously wrote (5 : 3) which is the interval from Ut to La.

the vibrations of the strings of instruments from their sagging in comparison with pendulums.

But it is true that in the first case (organ pipes) 122 vibrations are found for Ut of the harpsichord bass, while in the second case (strings) 244 are found.[237] This difference may seem surprising at first, but it must be remembered that 122 is half of 244; and 244 are the oscillations of a pendulum, where one oscillation is counted for going and one for returning. In organ pipes, however, the vibrations are counted from the beats, that is, for the most sensitive effect on the ear. Consequently, only those were counted as vibrations that the ear perceived, and of the two counted by the pendulum, it perceived only the one for going out and not the one for returning in the opposite direction. So this difference, which was supposed to destroy Sauveur's system, actually confirms it and produces this new proposition: two *geometrical vibrations*, if one wants to call those of the pendulum worth only one *acoustic vibration*.

The Ut of the harpsichord in the middle octave makes 244 acoustic vibrations per second; this number would be convenient as a fixed sound from which to count the upper or lower octaves. But it lacks an essential convenience. The series of octaves follows the double progression 1, 2, 4, 8 or whatever is equal to the power of two. It would be good if the numbers expressing the fixed sound could be found at the beginning of an octave. The number 244 is not a power of two, but 256, not far from it, is. That is why Sauveur chose 256 for the fixed sound, abandoning the 100 vibrations per second he had chosen for the same purpose in 1700.

It is easy to find the ratios of the different tones to each other from the sound that makes 256 vibrations, chosen as the fixed and mean sound. Sauveur constructed a table from which we can say everything we want at a glance. Each sound has its own name, according to the language invented for music. If you know the sound of a sonorous body, i.e. the number of vibrations it actually makes in a given time, you can give it a name to distinguish it from all other sounds.

It is very common to experience that at certain loud sounds, such as those of trombones, bells, and violon basses, glasses and plates tremble, and sometimes the vibrations of the intestines are felt. The sounds of these bodies are in unison with those produced by the musical instrument, and as the number of their vibrations is known, so the vibrations of the intestines are also known. It will be seen that this new knowledge can be extended to the whole of physics, and it is no wonder, for music, especially when treated by philosophers, is nothing but the physics of sounds.

5.8 Minutes of the Procès-Verbaux for the Years 1696–1714

For each session of the Académie, minutes were drawn up to record the main topics discussed during the session, often including detailed summaries of the reports. The minutes are collected in volumes numbered annually or almost annually. Table 5.35

[237] In fact Fontenelle considered 61 and 122 Hz, but we have changed the values to be closer to what Sauveur discussed; there is no substantial difference, however.

Table 5.35 Reports in the procès-verbaux de l'Académie des sciences (modern spelling)

1696	T. 15b	Système général de musique	
1700	T. 19	Préface d'un Traité sur la découverte qu'il a faite du son fixe	f. 111v-114v, 179r-182v
1701	T. 20	Propositions d'acoustique	f. 44r-46v, 51r-60v and 77r-83v
1701	T. 20	Discours sur les noeuds des ondulations	f. 123r-124r
1703	T. 22	L'échomètre	
1704	T. 23	Système des orgues	
1706	T. 25.	Écrit sur une nouvelle manière d'écrire sa musique. Il l'appelle musique colorée	
1707	T. 26	Écrit sur les systèmes temperez de musique	
1707	T. 26b	Réflexions nouvelles sur son système de musique, compare a celuye Mr Huguens	
1711	T. 30	Résponse qu'il a faite à M. Henfling, auteur allemand qui a attaque son système de musique	f. 409r-409v
1712	T. 31	Remarques sur le nouveau système pour les claviers des instrumens de musique, inventé par Michel Bulyowski de Dulycz, conseiller de Bade, professeur en morale et en mathematique & organist. Inseré dans le Journal des Sçavans du 2 May 1712	f. 271r-274r
1713	T. 32	Propositions d'acoustique	
1713	T. 32	Nouvelle détermination du son fixe	
1714	T. 33	Écrit sur le centre d'oscillation des cordes sonores	
1714	T. 33	Rapport d'un pendule simple avec la fleche d'une corde d'instrument de musique, lorsque leur vibrations sont isochrones. Determination du nombre des vibrations de cette corde, et nouvelle maniere de regler le son fixe	f. 177v-184v
1714	T. 33	Remarques sur les demonstrations des flêches des cordes	f. 321r-325r

concerns the volumes that relate to Sauveur and music. The number of pages is given only for relevant contributions.

The minutes may differ from the published memoirs, either because they do not report some things or because they report others that have not been published. In the following, we will refer to the most important parts of the latter, because Sauveur expressed himself more freely here than in his official reports, and therefore they are in some respects more interesting. Given the relative brevity of the writings, we have decided to report their full translations, which do not seem to have been made by others until now.

The parts of the procès-verbaux we consider relevant are:

T 19 Determination of the fixed sound; 111v-114v
T 19 On the use of the fixed sound; 179v-182v
T 20 Complex harmonic sounds; 123r-124r

T 31 A comment on Michael Bulyowsky's design for a piano composed of five keyboards; 271r-274v
T 33 Some considerations on the fixed sound; 177v-178v
T 33 A comment on how to calculate the period of vibration of a string, plus some epistemological considerations; 321r-324v.

5.8.1 Determination of the Fixed Sound

Translation.
Of all the possible ways of dividing the octave into small equal intervals, the one I found in 1696 seems to me the simplest and most convenient. I divide the octave into 43 equal intervals, which I call merides, and each of these merides into 7 other equal intervals, which I call heptamerides, so that the octave is thus divided into 301 heptamerides.

But in defining the order of the octaves, we must first determine the sound at which the first octave must begin. We cannot take the lowest sound, because it is impossible to determine; it is the same as the highest; therefore, we must choose a middle sound, which we will call the fixed sound, and which will serve as the principle or fixed point from which we will begin to count the octaves, ascending for the high sounds and descending for the low sounds.

It is easy to choose this sound, because you are free to take it as you wish; but once you have chosen it, the difficulty is to fix it in such a way that you can always find it easily when you need it, and that you can never doubt that it is really the same one that you have chosen.

Musicians use a wooden or metal whistle of a certain length to determine the sound to which the voices and instruments must be tuned in a concert; this sound is the middle of the singing voices and instruments, but it cannot give the fixed sound we are looking for.

For in the first place we know that all pipes of four feet, though much more accurate than whistles, do not produce exactly the same sound, and that after having lost the pipes or whistles which we used to tune, we have no means left by which we can be sure of finding a similar one; secondly, experience shows that the material of which these instruments are made, being very liable to change, the use which we make of them, and a hundred other accidents or unforeseen events, change them in such a way that after a number of years the sound they make is noticeably different from that which they made at the beginning, and in the end, there is no one who does not know that the same pipe, without undergoing any changes or alterations, produces a lower or higher sound, depending on whether the wind given to it is more or less strong.

Let take a closer look at the nature of sound.

All musicians agree that if a sonorous body makes, say, 100 vibrations in one second, then that which makes 200 vibrations in the space of a second is one octave away from the first, and that which makes 150 vibrations is one fifth away from the first, and the same is true in proportion to all other musical intervals.

From this it is easy to conclude that in order to know exactly what these different intervals consist of, that is, consonances and dissonances, it is necessary to know the number of vibrations made simultaneously by the two sounds that make up these intervals, because it is in the different proportions of these that the intervals consist exactly.

Those who sometimes hear organs being tuned know that when two pipes approach unison and are, for example, a minor semitone apart, there is a beat between the two sounds which is very sensitive to the ear and which organ makers express by the same term, saying that the pipes beat. When I searched for the reasons of this beat, I found that it could not be attributed to anything other than the different number of vibrations that each of these sounds makes at the same time, and which are arranged in such a way that one, for example, makes 25 while the other makes only 24, and that they meet and unite only once in the space of these 25 vibrations, and that from the first to the 25th the vibrations are always separated. Now since it is certain that two vibrations reunited together strike the ear with more sounds than when they are separated, it follows that the sound of these two pipes will sound more at each 25th vibration than before the 24 others, and it is in this variety, sensitive to the ear, of loud and soft sounds, that the beating of two sounds consists, and consequently whenever the ear makes us perceive that these two sounds are united, i.e., that they sound louder than usual, we can certainly conclude that the highest of these two pipes has made 25 vibrations and the lowest 24. So every time these two sounds beat, the higher one made 25 vibrations.

This having been assumed, it seemed to me that there could be nothing more convenient for determining the fixed sound than to take two organ pipes which were a semitone minor apart and which had at least 3 feet open or 1$^1/_2$ pied closed. Since these pipes are grave enough for one to notice the duration of their beats by the vibrations of a pendulum, they afford us a means of measuring the time in which these pipes make a given number of vibrations, and on this basis I have found that there is nothing more suitable for determining a fixed sound, which should serve as a rule for all others, than to choose that which makes exactly 100 vibrations in one second.

I have chosen the minor semitone to define the fixed sound, and not any other interval, because it has been found to be the smallest, it is the difference between the major and the minor third, and therefore it consists in the ratio of 24 to 25, as all musicians agree, and which can be easily verified by the octave, the fifth and the third, which are the perfect chords that musicians distinguish best, and that finally 4 of these beats make 100 vibrations in the most acute of the two sounds that must regulate the fixed sound.

Having found the interval between a proposed sound and the fixed sound, it remains to find the fixed sound itself, and for this I had a monochord made on which

I placed several strings in octaves, one with the other.[238] On these strings I have drawn a line and divided it into merides and heptamerides, and in order to find the fixed sound, I place the movable bridges on one of these strings, I move it forward or backward until the sound of this string is in unison with the proposed sound, whose interval I know from the fixed sound; this bridge responds to a division of merides and heptamerides, I add or subtract the number found on the ruler with which the pendulum was measured to the number on which the bridge is, the sum or the remainder tells me where I must place the bridge to have the fixed sound.

We first used reasoning alone to determine the fixed sound, and then, having wished to verify it in the presence of some of the most skilled musicians in Paris, and of some people who are very knowledgeable in speculative and practical music, we found that, in all the experiments we made, they responded to reasoning with all the precision we could hope for.

P. Seb. Trucher, M. l'abbé Gallois, and de La Hire have been appointed to see the experiments by M. Sauveur on fixed sound.

5.8.2 *On the Use of the Fixed Sound*

Translation.

Start with Table 5.36. This table gives the number of vibrations in all the intervals of the acute octave beginning with the fixed sound, which we will call the fixed octave. So the number of vibrations of the 6th meride is 110 and $^{15}/_{100}$ in one second. If we divide the difference from one meride to another in 7, we get the number of vibrations from heptameride to heptameride and the just intervals.

To find the number of vibrations of all possible sounds within the 1st, 2nd, 3rd, and 4th octaves, multiply the numbers in this table by the terms of the double progression, i.e. 2, 4, 8, etc. For example, to obtain the number of vibrations of a sound that is more acute than the fixed sound of 2 octaves and 18 merides, multiply 133.66 by 4 and you will get 534.64, which means that this sound makes 534.64 or rather 535 vibrations per second. On the other hand, if the proposed sound is less acute than the fixed sound, divide the numbers in the table by the terms of the same progression, i.e. 2, 4, 8, 16, etc.

On this basis it is possible to determine how many vibrations per second a low sound must make in order to be heard; for example, knowing that an organ pipe of about 5 feet open makes the fixed sound, and that a pipe of 40 feet, in the low 3rd octave, is the last one that can be distinguished whose sound is octuple that of the fixed sound, I take the eighth part of 100 vibrations and conclude that the lowest sound that can be heard must make at least $12^{1}/_{2}$ vibrations in one second.

---

[238] There is a logical leap here; most likely the scribe did not report a part of Sauveur's account regarding the counting of beats per unit of time with the use of a pendulum, as exposed in the 1701 memoir.

Table 5.36 Table of the number of vibrations made in one second by the sounds of the fixed octave from meride to meride

[a]Merides	Vibrations	Merides	Vibrations
0	100.00	43	200.00
1	101.63*	42	196.78*
2	103.28	41	193.64
3	104.95	40	190.55
4	106.66	39	187.50
5	108.39	38	184.48*
6	110.15	37	181.55
7	111.94	36	178.65
8	113.76	35	175.79
9	115.61	34	172.98
10	117.49	33	170.22
11	119.40	32	167.49
12	121.34	31	164.82
13	123.31	30	162.18
14	125.31	29	159.59
15	127.35	28	157.04
16	129.42	27	154.53
17	131.52	26	152.05
18	133.66	25	149.62
19	135.83	24	147.23
20	138.04	23	144.88
21	140.28	22	142.56

[a] Values with an asterisk have a small error in the last digit

Suppose we want to know the path that the center of a metal string of about 3 feet in length makes, put it in unison with the fixed sound, i.e. with La below the tone of the Chappelle.[239] Measure the diameter of the string, that I suppose to be $^1/_6$, then pluck the string and look with a microscope at the path it makes, you will see that it will make a sound that is still sensitive but very faint when the path it makes is not more than a third of its diameter or $^1/_{18}$ of line and when this path is 4 lines or greater than 2. The path will be greater when the sound is stronger; so that the center of a 3-foot string, which will make the fixed sound, will be $^{100}/_{18}$ for a faint sound or almost 6 lines per second, and about 3 feet when the sound is loud.

Since low-pitched sounds have greater vibration but they make fewer vibrations per second, and high-pitched sounds have smaller vibration as they make more vibrations per second, we can assume that they always travel the same distance in a second,

[239] Here La is associated with the fixed sound. The pitch La of the Chappelle should be about 392 Hz. Here Sauveur considered a sound two octaves below [44], p. 116.

i.e., all strings 1/6 of a line in diameter that make a faint sound travel six lines per second, and three feet when they make a loud sound.

To find the amount of air that this string displaces in one second, note that the string makes two opposite excursions in each oscillation and contains two segments that can be thought of as those of a parabola.[240] Multiplying the length of the chord, which I will assume to be 3 feet or 36 inches or 432 lines, by the amplitude of the vibrations, which is 1/18 of a line in weak sounds, you get 24 square lines, of which two thirds, i.e. 16 lines will be the area of the two parabolic segments. Multiplying by the diameter of the string 1/6 of the line, we have 22/3 cubic lines of air that a faint sound moves in one vibration and 266 lines or 1/66 cubic inches in one second, and a loud sound will move 72 times more or 22/3 cubic lines.

Larger strings that sound in unison with smaller strings make their vibrations smaller, and it will be found that this sound corresponds to the ratio of their diameters. It can be established as a general principle that all the faint sounds of any string and close to the fixed sound move in one second 1/65 cubic inches or 266 cubic lines of air, and the loud sounds move 1/9 cubic inches or 19,200 cubic lines.

The experiments we have made to determine the fixed sound have shown that we can easily count the beats of a sound more acute than the fixed sound of a fifth, which makes 150 vibrations in one second, but since these beats are made of 25 to 25 vibrations, it follows that we can easily distinguish 6 beats in one second,[241] and we can even distinguish 8.

Since the fixed sound makes 100 vibrations in a second, of which the 6th part is about 16, it follows that we can distinguish a beat for any 16 vibrations, such as that of the 16/15 chord or the major semitone formed by two pipes which make the fixed sound.

It follows that in the first grave octave we distinguish beats any 8 vibrations and above, as in the major tone; in the second grave octave we distinguish those that go any 4 and above, as in the major third and fourth. In the third octave, those of the fifth and finally those of the octave.

To distinguish the beats in the high octaves it is necessary that in the first high octave the beats occur at least any 32 vibrations or that the interval between the two sounds is at most two merides. In the second high octave, the beats must be any 64 vibrations and above, like in the interval of a meride, and so on.

From the above we can conclude that sounds beat in all possible intervals. But in the high octaves they beat so fast that the beats are not noticeable in the consonances,

---

[240] Here Sauveur imagined the string deforming according to a parabola. The segment of the parabola has a height 1/18 of a line, equal to the sag f of the string, and a base of 3 feet, equal to the length l of the string (remember 144 lines, one foot) . The area of the segment of the parabola is obtained by the formula $2/3 fl$, which gives 16 square lines. To obtain the volume of displaced air, multiply the area of the segment by the diameter of the chord $d = 1/6$ of a line, which gives a volume of 22/3 cubic lines.

[241] Here there are inconsistencies in the text; indeed, for the interval of a fifth with frequencies of 150 and 100 Hz, there are 50 beats per second, because when the higher sound makes 3 cycles, the lower sound makes 2 cycles. One can only have 6 beats per second assuming an interval of a minor semitone (25 : 24), when the highest sound is of 150 Hz.

and this is what makes them more pleasing, and what makes the real difference between consonances and dissonances. It would seem that the thirds which beat in the lower octaves were not included in the consonances by the ancients, nor are they included at present by the Orientals, and that it is only since the invention of the concert that the ear, because of the temperament of the consonances, has become accustomed to the beats, and has been sobered by those of the thirds which are very sensitive in the lower octaves, and this habit has led to finding them agreeable. The nature of the beats will explain why the fourth is considered a dissonance in composition, why the sixths are harder than the thirds, and why the tritone and the false fifth are dissonances. This may justify the rules of composition.

5.8.3 Complex Harmonic Sounds

Translation.
If a rope AB is divided at point C into two equal parts, each of which makes its own undulation, and if it is also divided at points DE into three equal parts, each of which also makes its own undulations, then we say that the undulations of the second sound are confused with those of the third sound, and so on.

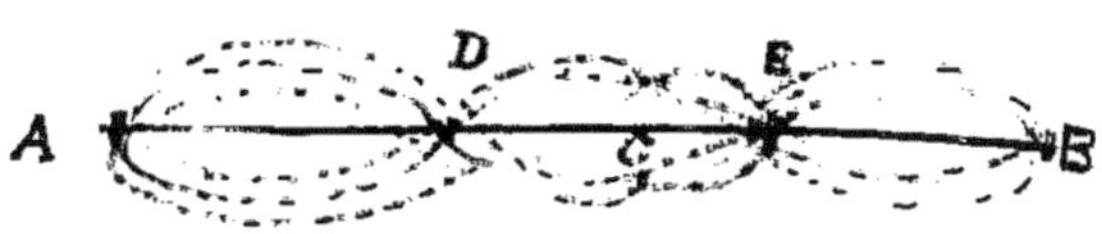

It is obvious that in these confused undulations the nodes have less movement than the neighboring parts. In the ordinary nodes of harmonic sounds, we feel that there are confused undulations because of their little tremors.

The ear confirms this in the long strings of the harpsichord, because we hear several harmonic sounds at the same time. But reason teaches us that if, among the confused undulations, those of one harmonic sound have a greater force than those of another, we hear the first and not the second; even if one is one octave above the other, we hear only the most grave, which serves to explain the oddity encountered in the harmonic sounds of a sonorous body.

5.8.4 A Comment on Michael Bulyowsky's Design for a Piano Composed of Five Keyboards

In this section, entitled *Remarques sur le nouveau systéme pour les claviers des instruments de musique, inventé par Michel Bulyowski de Dulycz, conseiller de Bade, professeur en morale et en mathematique. & organist*, Sauveur commented on the proposal of a piano with 31 keys per octave.

Michael Bulyowsky (de Dulycz) (c. 1650-c. 1711), a German organist and theorist, published several pamphlets in which proposed enharmonic keyboards to be tuned in multiple divisions of the octave. First he proposed a division into 19 parts; later he had a 31-tone harpsichord with five rows of keys, called the "fünfffaches, vollständiges-transponir Clavier", and described it in the Latin pamphlet *Tastastura quinquiformis panarmonico-methatetica* of 1711 [15]. Unfortunately, Bulyowsky's enharmonic keyboard has been lost [52].[242]

According to Sauveur, Bulyowsky had adopted Huygens' division of the octave into 31 parts, which essentially results in $1/4$ of a comma temperament. The division is achieved with five different keyboards per octave. The basic keyboard shows the tempered main notes, then above and below it, there are two keyboards with altered notes. For each octave, the sequence is as follows:

$$-2\ -1\ 0\ 1\ 2$$

where zero is the middle note and the other numbers are the notes shifted by 1 or 2 $1/31$ octaves. Sauveur proposed the notes of his division of the octave into 43 parts for the five keys.

It is possible, however, that Bulyowsky was inspired not by Huygens, but by Nicola Vicentino—an author unknown to Sauveur—and his archicembalo based on $1/4$ of a comma temperament [88, 90]; at least this is the opinion of Benedetto Bresciani (1568–1740), a music amateur who commented on Bulyowsky's keyboard after seeing it: "His is the division of Vicentino and the harpsichord is that of Nigetti, may be seen by him" [5].[243]

Translation.
Here is a new theme to mark the simplicity of my new music system. Mr. Bulgowsky's system may be new for the desire he had to correct the defects of keyboard instruments and for the layout of keyboards, but like Mr. Huygens, he had divided the octave into 31 parts which serve as the basis for his system. In the memoir of the Académie of 1701, in section XI of my system of intervals of sounds, I had noted the defects of the 12 keys of the clavichord and the organ, from which we can deduce the defects described by Mr. Bulgowsky. I can assure that, in the application that I made of my system to all musical instruments, I had not failed to think of a keyboard that would remedy the defects that M. Bulgowsky notes in the ordinary system, but because the

---

[242] p. 48.

[243] p. 185.

43 merides of which I compose my octave, multiplied with excess the strings of the keyboard, I had abandoned this view.

M. Bulgowsky makes the execution much simpler by subtracting 12 parts in each octave, that is 31 instead of 43, and what was less just in M. Huygens' system is recomposed by the defects he corrects in the ordinary keyboards.

Following this idea, I will try to explain the artifice of this keyboard by the names of my system, and to make clear its advantages as well as its inconveniences. Place in each octave the 7 main keys representing the natural sounds:

PA, RA, GA, SO, BO, LO, DO

and according to ordinary musicians:

Ut, Re, Mi, Fa, Sol, La, Si

even though the first names will give a much better idea of how to use the keyboard.

These 7 keys have above them the acute alterations in 2 rows, and as many grave alterations below them, which form the 5 rows of its keyboard in 3 kinds of levels, which are its acute alterations, the main keys and the grave alterations. And each note on the octave has 5 keys, so RA has 2 acute keys and 2 grave keys.

The 5 keys of a note have the same consonant as the main key, which is the middle one, but the vowels sound in the order AEIOU, OUAEI, so in the key RA which has the the vowel A, the first touch above will have E and the second I, the first below will have U and the second O; so the 5 keys of RA will sound:

ro (−3), ru (−1), RA, re (+1), ri (+3)

Similarly, the BO keys will be:

be(−3), bi (−1), BO, bu (+1), ba (+3)

In this arrangement, the keys between GA (Mi) and SO (Fa), which are only half a tone apart, are doubled, since ge and se are the same, as are gi and si. The same is true for DO (Si) and PA (Ut). Therefore, two of these keys must be suppressed as each must produce the same sound as the one to which it corresponds.

I leave to the makers to place the keys and to solve the problems of such a spinet, because it is not easy to make a harpsichord so that it is sonorous with 2 stops of strings, and even less with more stops. As for organs, we can distinguish between low and high sounds, but such an organ will have twice as many pipes as ordinary organs which have the same number of stops, since each octave of the new machine has 32 sounds, while ordinary ones have only 13.

As for the person who has to play a piece of music on such an instrument, the keys of which are so numerous, the following can be thought of:

1. The difficulty of stretching the fingers to make notes is not greater than on ordinary keyboards with cut keys, because the 8 main keys of an octave occupy the same width as on ordinary keyboards, although the alterations are arranged in 2 ways. The 3 levels formed by the 5 keys hardly take up any space, since the main keys

of the middle row can be shortened; moreover, the upper alterations, which form the sharps, are rarely used with the lower alterations, which form the flats.

2. The choice of the keys of such an instrument to play a transposed piece is as easy to observe as that of a theorbe or a bass viol, because there are only half tones between two notes of these instruments. The chords of these new keyboards are more uniform. If we write the vowels AEIOU on the keys in the order we have marked, and if we assign numbers to these letters in the following order:

A	E	I	O	U	A	E	I	O	U
1	2	3	4	5	6	7	8	9	10

if you have left a key that has, say, the vowel O, and you want to go up or down a tone, add or subtract 5 to the value of O, and you will still find O going up or down; and if you want to make only a major semitone, add or subtract 3, and you will have E going up and O going down. And if you want to have a minor semitone, add or subtract 2, and since these intervals or chords have a great uniformity, you will easily acquire the habit of forming them. It is true that when you go from one octave to the next, you have to subtract 1 from these numbers going up and add 1 going down.

Instead of marking the keys with letters, you can make the keys with 5 different colors, e.g. A white, I gray, O black, U yellow, which will have the same effect.

The way we have just explained for the major, minor tones, and semitones gives very great facilities for knowing all the major and minor intervals, distinguishing the small intervals, which are the second, third, and fourth, from the large intervals, which are the fifth, sixth, and seventh.

Smaller major intervals have the same vowel, and larger ones have a letter with the upper or preceding octaves. Larger major and minor intervals do the reverse. But when we go from one octave to another, instead of the second vowel, we must take the preceding vowel, when raising, the following vowel, when lowering, all of which is a continuation of Art. III of Section IV of the system of the intervals of sounds. The above can be summed up:

1. Mr. Bulyowsky's keyboard instrument is possible.
2. Its construction is much more difficult to make than ordinary ones.
3. We can only make spinets, the difficulty is too great for a harpsichord with 2 or 3 registers.
4. An organ can also be made, but it is necessary to subtract stops or to multiply the pipes in an extraordinary way.
5. On this keyboard you can transpose all kinds of songs to any key you want.
6. In all the transpositions it is possible to make all the intervals and the just chords (not counting here the small changes that are necessarily connected to the system of 31), avoiding thus the alterations of the ordinary keyboards.
7. One can get used to touching the chords as on ordinary keyboards or on the neck of theorbes.
8. The theory of chords is facilitated by our new system.

9. The patron one is looking for must be aware of the new system, because it is too long a way to go from the old music system to the theory and practice of this new keyboard.

5.8.5 Some Considerations on Fixed Sound

Translation.
In my previous researches, I said that a string stretched horizontally by its gravity forms an arc that could be assumed to be that of a parabola or a circle, since in ordinary strings these arcs do not exceed 4(?) minutes, that the curvature of these arcs was determined by the versed sine of its half, which I call the sag of the string, and that finally the ratio of the sounds of several strings was regulated by the square roots of their sags; so that two strings were in unison when their sags were equal, and when they were unequal, the sounds between them were like the square roots of these sags.

Though the vibrations of the strings are very unequal, and those which they make at the beginning of their movement are much greater than those which they make toward the end of their movement, yet the sound which they make is neither higher nor lower, though they are more or less easily heard.

I assume that these vibrations are not very forced, for if the string were pulled with too much force (this is not the place to determine what constitutes too much force), a sound would be produced at a higher pitch than the natural sound of the string.

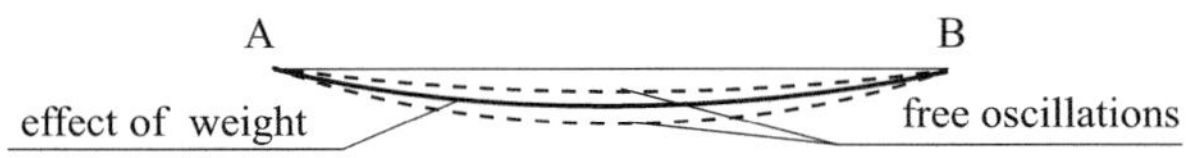

Anyway this string was first set in motion, its oscillations diminish to such an extent that they finally return the string to the shape that its weight naturally gives it, so that the last oscillations only cause the string *to move laterally* [emphasis added], as if the sag of this string had become a pendulum that would make the arc move in a plane (as Mr. Huygens says) around its brace.[244]

It is now a question of finding the center of oscillation of this arc, that is, of finding the length of a simple pendulum whose vibrations are isochronous with those of the string; this research will give us the opportunity to find out how many vibrations a sonorous string makes in one second, and to regulate the fixed sound in a new way.

[244] Here we have made a free translation because we could not follow exactly what Sauveur meant, although the general idea is quite clear.

5.8.6 A Comment on how to Calculate the Period of Vibration of a String, plus Some Epistemological Considerations

I learned that the demonstration that I gave on the relationship of the sags of the sonorous strings with the sounds that these strings make, was not clear enough to serve as a principle for the conclusions that I drew from it, although it is essential for the advancement of acoustics, which so far is a much more imperfect science than its companion optics.

It seemed to me that the difficulty of my demonstrations lay in the determination of the sag of a string, which makes the sound constant, when instead it appears that this sag varies and that the vibrations of the string vary, so that one does not see why one should regulate by the sag that the string forms by its gravity alone when it is stretched horizontally, since then it is at rest and does not produce any sound.

It also seemed that it was not clear that it was necessary to take the arc that forms a sonorous string for that of a circle or a parabola, as we could suspect which is a curve of a more compound degree.

Before entering into the matter, I am obliged to remark that I am in the physico-mathematical realm, and not in that of pure mathematics, in which the mathematician carries out his work, I mean that he proposes assumptions, either particular or general, or extremely general, and he derives from them truths which are necessary consequences of the assumptions he has chosen. These truths, which go far beyond what the ancients have left us in their treatises, have so aroused the curiosity of the scholars of a large part of Europe, that those who are distinguished in the matter have gained the right to decide on the correctness of all kinds of demonstrations.

It is not the same in the physical-mathematical realm. Nature has put the truths which she has deemed appropriate, and, in the most particular studies which we have made of various phenomena of nature, we have always found more assumptions than we had first imagined, so that the number which is indefinite belongs to the infinite essence of the author of nature.

The role of the physicist is to examine as many phenomena of nature as possible that appear to have the same cause. They seek to identify the relationship between this cause and formulas and general rules, which serve as a means of explaining these known phenomena and discovering new ones. Upon further examination of the formulas, the physicist often finds that nature does not adhere to the conclusions initially drawn. In order to rectify this discrepancy, the physicist must identify the underlying assumptions that correspond to the observed phenomena. This process often necessitates the introduction of corrections, or even an entirely new hypothesis, to account for the discrepancies.

It was the astronomers, the oldest examiners of natural phenomena, who first postulated that the Sun moved in a circle due to its annual movement. However, the inequality of this movement and the variety of its apparent diameter resulted in an eccentric circle. To achieve this, the ratio of the eccentricity to the radius of the circle was adjusted. However, this was not sufficient; it was necessary to make successive changes to obtain other curves. These other curves ensure a more precise regulation of the Sun's movement, but it is acknowledged that they always have some degree of error.

The observation of the acceleration of the motion of heavy bodies by mechanics has revealed that this phenomenon occurs in equal times in the proportion of odd numbers. The consequences for pendulums and the projection of bodies are noteworthy, yet the system had to be corrected due to the influence of air resistance. In optics, we are indebted to those who have observed the equality of the angles of incidence and reflection, as well as the constant ratio of the sines of incidence and refraction. We are continually impressed by the consequences that have resulted from these observations. However, it must be acknowledged that while there is consensus on the demonstration of equality of the angles of reflection, there is not yet unanimous agreement on the demonstration of the constant ratio of the sines of the angles of incidence and refraction. This is because the demonstration of the latter is based on principles that not all physicists accept or find satisfactory. Consequently, we are compelled to hypothesize the constant ratio of the sines of the angles of incidence and refraction, based on our initial observations, which were then validated by subsequent experiments, without any concern as to whether the principle was supported by empirical evidence alone.

Coming now to acoustics, we will find that the experiments make us agree with the following principles, which look at the relationships of sounds according to the acute and the grave:

1. We first agree that the ratio of two sounds according to the acute and the grave, consists in the ratio of numbers of vibrations that these sounds make at the same time.
2. Instrument makers and players agree that a stretched string produces a sound in the same degree of acute and grave whether it is made to make small vibrations, or large ones, on condition that these are not forced.
3. That a string of the same length between two bridges, stretched with the same weight, will produce the same sound (that is, in the same degree of treble and bass) whether we place the string vertically, horizontally, or inclined to the horizon, by passing the string over a very light pulley whose axis is very small and very polite, with no more difference than that produced by the friction of the pulley, which is of little weight in the circumstances I have just noted.
4. That two strings of the same size stretched by [the same] weights will form sounds in reciprocal proportion to their length; if one string is double the other, the former will be at the sub octave of the latter.
5. That two strings of the same length and stretched to the same weight, if they are of different thickness or weight, will produce sounds [frequencies] in the reciprocal proportion of their diameters and in the reciprocal proportion of the roots of their thickness or weight. Thus, a string that is four times heavier than another of the same length and stretched with the same weight will be at the sub-octave.
6. Two strings of the same length, thickness, and specific gravity, but stretched with unequal weights, produce sounds proportional to the roots of the weights that stretch them; thus, if the first string is stretched with a weight of one pound, and

the second with a weight of four pounds, the second string will produce a sound one octave higher than that of the first.

7. From these last principles it follows that if two strings are of unequal length, thickness, and tensioned by unequal weights, their sounds will be in direct proportion to the roots of the weights that tension them, and in inverse proportion to the lengths and roots of the sizes or their weights. Thus, if we call the sound or the number of vibrations ν, the length of the rope n, the weight of a given length of rope c, the weight that stretches it p, and α a constant quantity, then:

$$\nu = \frac{\alpha\sqrt{p}}{n\sqrt{c}}$$

8. A horizontal string ADBG, fixed at A, passing over a bridge B, over a pulley G and tensioned by a weight P, forms a curve ADB, the brace of which is ACB, the tangents AE, BE, their parallels AF, BF, will form a rhombus of which EF is the small diagonal and EC is the half of it.

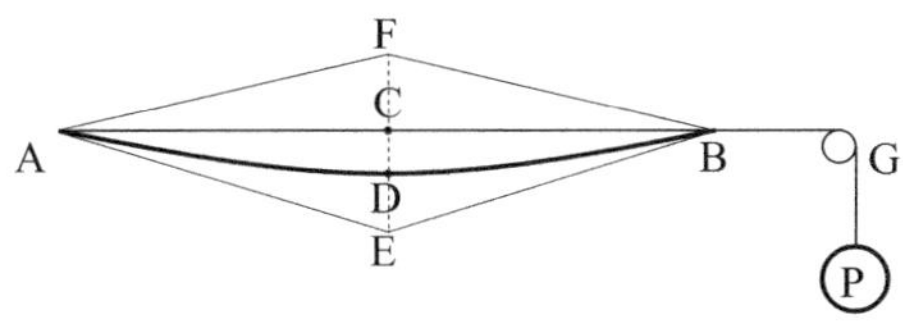

Note that for strings of musical instruments, where CE is only 1/14 of a line, we can assume that EB is approximately equal to BC or half of AB. We will see below that the difference is not more than 1 in 5,760,000, which is absolutely insensitive to sound.

Let c be the weight of a string of constant length ADBG, which I will call a, and the string of variable length ADB, which I will call n; its weight will be cn/a; we will assume that this weight is at its center of gravity D or at another point E of the vertical of this center of gravity, where the two tangents meet.

According to the rules of mechanics, the weight cn/a of the string ADB, is at the weight p, which stretches it, as the small diagonal EF, which I call $4f$, is at EB, which is essentially equal to CB or 1/2 AB. Thus:

$$\frac{cn}{a} : p = 4f : \frac{1}{2}n$$

So:

$$\frac{cnn}{2a} = 4pf$$

Or:

$$f = \frac{cnn}{8pa}; \qquad \sqrt{f} = \frac{n\sqrt{c}}{\sqrt{8pa}}; \qquad \frac{1}{\sqrt{f}} = \frac{\sqrt{8pa}}{n\sqrt{c}} \propto \nu$$

9. We are now concerned to prove that the sag CD is equal to f (that is a half of CE), at least very closely.

5.9 Sauveur's Legacy

The examination of Sauveur's writings discussed in the previous sections allows a modern reader to assess his contribution to music theory, but it does not allow him to know the views of his contemporaries or immediate successors. In the following, we will attempt to illustrate this aspect.

Sauveur lived in France during a transitional period in French science, after the death of Descartes and during the early years of the Académie des sciences of Paris. Not born rich, he had to work for a living, teaching mathematics to members of wealthy families. He was little concerned with the more abstract aspects of mathematics, but very adept at developing applications. And he was curious enough to keep up with the Calculus that was becoming popular at the turn of the 18th century by attending the so-called Malebranche circle.

From the time he joined the Académie in 1696, his interests were directed toward music, or more precisely, speculative music, as he made clear in his writings:

> Music may be classified into two principal categories: the speculative and the practical. The speculative variety of music investigates the properties of sound and the manner in which it may be pleasing to the ear. This treatise will focus on this aspect of music.

At a time when most scholars had a wide range of interests, Sauveur elected to concentrate his efforts on a single, specific field: music. The motivation behind this decision is unclear, particularly given the suggestion that Sauveur lacked a keen musical sensibility. Some music history texts perpetuate the false rumor that he was even deaf.

Additionally, Sauveur was not especially prolific in his field of interest. His writings on music are limited to a single surviving manuscript, dated 1697, and five memoirs published in the journal of the Académie between 1701 and 1716.

Nevertheless, his contribution represents a pivotal moment in the history of speculative music, or at the very least, became emblematic of that turning point. It is challenging to ascertain the extent to which the growing interest in the study of acoustics, both in general and musical acoustics in particular, that emerged during the 18th century can be attributed to Sauveur. He was a product of his era and reflected its spirit and outcomes. While he may have been a passive subject, he was undoubtedly more than that.

Modern historians acknowledge his contributions to the following issues:

1. Use of logarithms (base 10). Sauveur introduced a rational measure of musical intervals using logarithms. His unit of measurement was the heptameride, basically the logarithm of the interval multiplied by 1000. His methodology bears resemblance to the contemporary one, which employs cents.

2. Octave division. He was one of the first to study the properties of an important class of temperaments, those obtainable by dividing the octave into equal parts. He provided tables of intervals without distinction between consonances and dissonances, analogous to the trigonometric or logarithmic tables used in mathematics.
3. Standard frequency. Sauveur recognized the importance of measuring a standard frequency, referred to as the fixed sound, which enabled him to categorize the entire range of sounds into absolute octaves and, among other things, to identify the frequencies that delineate the domains of sound perception.
4. Beats. Developed a fairly satisfactory theory of this phenomenon, at least from a practical standpoint, and used it to determine the standard sound by means of the sound emitted by two organ pipes separated by a minor semitone.
5. Determination of an analytic expression of the frequency of vibration of a string.
6. The discovery of higher harmonic sounds in vibrating strings and the hypothesis that they exist in other cases.
7. Terminology still in use today. The term acoustics to refer to the science of sound as opposed to optics; the terms harmonic, fundamental, nodes, and bellies. He did not use the term frequency, but spoke of the number of vibrations per unit of time. He also used the generic term sound to indicate frequency, which also indicates pitch or note.
8. The use of special graphical scales, mainly to avoid tedious calculations, referred to generically as echometers. As a practical mathematician, Sauveur placed significant emphasis on these instruments, which were also quite challenging to construct, with dimensions reaching 2 m at considerable expense.

Missing from the above list is the contribution that Sauveur considered perhaps his most important, although it was not followed up. This is the introduction of a new system of musical notation, comprising names and symbols for notes corresponding to 301 partitions (heptamerides) of the octave, as well as a redefinition of the clefs. It is noteworthy that while the proposed renaming of individual notes ultimately proved unsuccessful due to the considerable labor involved, Sauveur's concept of reformulating the clefs by aligning them with octaves instead of specific voices did achieve partial success.

As is often, perhaps always, the case in the history of ideas in general, and of science in particular, every 'new' idea is matched by another similar one formulated earlier. As a result, it is often impossible to assign the name of an inventor or discoverer to a law or an instrument. Nevertheless, this practice persists, often with the false assumption that it is a factual assertion rather than a simplifying convention. This is evident in the case of Sauveur, who nevertheless left a significant legacy, warranting his association with the aforementioned innovations. However, the following clarifications are necessary.

1. The introduction of logarithms in music came shortly after their invention in 1614 or 1624 for common logarithms in base 10. Some use appeared at least in 1653 in the works of Nicolaus Mercator, followed a few years later by those of Huygens.

However, Sauveur was the first to make logarithms an essential tool for comparing musical systems.

2. The concept of temperament is perhaps as old as music itself. Sauveur studied what happens when the octave is divided into any number of equal parts, proposing a methodology that was revolutionary for its time.
3. The need to introduce a sufficiently precise absolute frequency to be used as a standard arose with the spread of increasingly complex musical groups, with the exchange of music and composers between different nations, and various attempts can be attributed to Mersenne, Huygens and Hooke. Sauveur is the first to succeed, albeit with some difficulty at first.
4. Beats. The phenomenon of beats was well known to early organ makers, who heard a variation in the amplitude of the sounds when playing consonant intervals that were not perfectly tuned. Beeckman had already provided an explanation based on the recognition of the periodic nature of sound, which was very similar to that of Sauveur. The latter stated the theory more clearly and provided an application.
5. Absolute frequency of the vibrating string. Apart from that of an experimental nature by Mersenne, the attempts of Huygens should be considered, who gave the expression of the frequency of vibration of the string in the case of lumped mass. Sauveur's was the first to publish a substantially correct expression. However, the priority was short-lived; in the same year, 1713, in which he provided his expression, came out the work of Brooke Taylor, who, using Newtonian mechanics, provided a correct formula not only of the value of the frequency but also of the shape of the deformed string: a sinusoid.
6. Higher harmonic sounds. The existence of overtones had been known since ancient times, at least for the octave, and Mersenne insisted that they also existed for the third and fifth. The possibility of deforming a string in complex ways with nodes and bellies was not documented until the 1600s. Wallis, with Plot, was among the first to make this behavior known in 1677, and Robartes, in 1692, provided a detailed discussion of trumpet operation based on harmonic sounds. From an experimental point of view, therefore, Sauveur did not introduce any innovations. He did, however, clarify phenomenology better, fully understood its importance for the whole of music theory, and clarified that it is responsible for overtones.
7. Most of the terms used by Sauveur already existed. However, he selected them from many alternatives and used them systematically until they became standard.
8. The use of graphical scales did not, of course, originate with Sauveur. Indeed, logarithmic scales had already been introduced shortly after the discovery of logarithms. However, Sauveur developed scales that were particularly useful in musical matters. Amongst the scales proposed by him, the simplest and arguably the most successful is the inverse quadratic scale, which is based on simple pendulum length and period of oscillation.

To better understand the innovations introduced by Sauveur, it is necessary to consider the authors who preceded him and to examine the influence on those who followed him. Among the former, there is Mersenne, not very far from him.

In terms of formal aspects, a very important difference lies in the expository style of the two authors. Sauveur wrote without the frills and repetitions, and published his work in the form of journal articles, so necessarily short and single-topic writings. In terms of content, it should be noted that Sauveur made no mention of the celestial or human music; furthermore, he devoted almost no commentary to compositional aspects.

Mersenne wrote an encyclopedic treatise; he studied the nature of sound and its propagation, he listed and tried to explain the functioning of all the musical instruments of the time, and he also addressed aspects of composition and harmony of spheres. A modern reader, tempted by labels, would consider the works of Mersenne and Sauveur as both treatises on acoustics; the first a substantially complete text, but written in a language that is imprecise and redundant, clearly a not original compilation, and also containing arguments of no interest to acoustics. The second is a concise text, good only as an introduction to a larger work. It is, however, a very well-written introduction which, if continued, promises to result in a text of considerably higher quality than Mersenne's, with a character of originality.

To make a comparison with authors following Sauveur, we can begin by noting that after his death many treatises on musical subjects inspired by him were sent to the Académie des sciences, especially in what concerns the measurement of intervals and the division of the octave, such as *La théorie des sons* [43] and *Arithmétique des musiciens* [42], both of 1754, by Jean-Edme Gallimard, and *Memoire théorique et pratique sur les systèmes temperes* of 1758, by Jean-Baptiste Romieu [65]. To signal an unpublished memoir by Giuseppe Sarti, presented at the Académie des sciences of Pétersbourg in 1796, in which Sauveur's technique of calculating the fixed sound by means of beats is revisited [6].

The study of temperaments obtained by equal divisions was carried out by many scholars, among them Leonhard Euler, Robert Smith (1689–1768), and Giordano Riccati (1709–1790). It should be noted that in this type of research, the use of logarithms was complemented by that of continued fractions, which were very useful for approximating real numbers. Euler in his *Tentamen novae theoriae musicae* considered the ratio between the logarithms of the octave and the fifth:

$$\frac{\log 2}{\log 1.5} \approx \frac{0.30103}{0.17609} \approx 1.7095$$

as equivalent to the ratio between the number of equal parts into which the octave and the fifth are divided. Representing this ratio as a continued fraction he proposed the following approximation:

$$1.7095 \approx 1 + \cfrac{1}{1 + \cfrac{1}{2 + \cfrac{1}{2 + \cdots}}}$$

which gives the sequence: (2 : 1), (3 : 2), (5 : 3), (7 : 4), ··· , (12 : 7), (29 : 17), ··· , (53 : 31) "the last of which is closest to the true" [33].[245] Note the presence in this sequence of ratios of that corresponding to the equal temperament (12 : 7) and Mercator's division of the octave (53 : 31).

Robert Smith was the first to relate the numerical value of the fraction by which fifths are reduced to the number of parts by which octaves are divided [84].[246] Giordano Riccati, the leading Italian acoustician of the 18th century, took up Sauveur's work and explored divisions beyond those of 43 and 55 parts by comparing the tempered values with the systems to be tempered using more sophisticated procedures than those employed by Sauveur, arriving at the proposal of the optimum value of 74 parts corresponding to the narrowing of fifths by 3/14 of a comma [62, 63].

It must be said, however, that these studies of the temperament mark a profound break with those of the previous two centuries; whereas those were motivated by musical needs, these are purely speculative. This is also the opinion of Giordano Riccardi, who wrote: "and in the meantime I conclude that general tempered systems will never move from theory to practice, remaining only as a figure of ingenious speculation not to be overlooked in a comprehensive treatise on counterpoint" [63].[247]

Sauveur had a great influence on the French Enlightenment, especially on Rousseau and Diderot. Of the former, the *Dictionnaire de music* of 1768 [67] listed entries strongly inspired by Sauveur, and the *Dissertation sur la musique moderne* of 1743 [66], proposed a system of musical notions inspired by Sauveur. The latter [Diderot] discussed Sauveur's ideas on standard frequency and harmonic sounds in the *Principes généraux d'acoustique* of 1748 [26].

It is more difficult to assess the direct influence of Sauveur on musicians. We have already noted that the nature of musical treatises changed after the Renaissance. Traditional speculative music has given way to the exposition of the rules of counterpoint, exposition which is called *theory*; but this is a very different theory from that represented by speculative music. In the new theory there are very few principles of a physical nature, and the use of mathematics is also very limited. However, speculative music has not completely disappeared. As a rule, all treatises contain a very brief preface referring to the fundamentals of music. With the gradual establishment of musical acoustics, these sections have undergone significant changes; in particular, intervals are measured in terms of ratios of string frequencies, and the phenomena of beats and harmonic sounds are explained in terms of modern acoustic theories.

Meanwhile, acoustics, for Sauveur basically synonymous with speculative music, began to change its status. It increasingly became the science of sounds and vibra-

---

245 p. 75.

246 pp. 156–160.

247 p. 75.

tions in general, even those beyond the threshold of audibility. As the science of sound, its nature and transmission in all media became the main subjects, along with studies of sound insulation, reverberation elimination, and so on. Acoustics, as conceived by Sauveur, became a short chapter, musical acoustics, the study of which, in addition to knowledge of physics and mathematics, also requires some of music theory understood in the modern sense.

Currently, Sauveur is acknowledged and referenced in all texts pertaining to the history of acoustics, particularly in relation to his contributions to the determination of absolute frequencies and the discovery of harmonic sounds. His studies on them influenced treatises on violins that had not mentioned the phenomenon before 1700. Interesting and curious enough, according to Rush [60], is Scherchen's view that Sauveur is one of the three founders of modern music along with Bach and Beethoven [79].[248] In contrast, it is noteworthy to mention the perspective of Clifford Ambrose Truesdell (1919–2000), a renowned historian specializing in mechanics, who allocated a mere 15 lines to Sauveur's biography in the Grove Dictionary of Music and Musicians. This segment was characterized by a certain level of ambivalence, as evidenced by the following statement: "His papers, through not so original as he may have thought were fairly clear and descriptive; they were very widely read" [87].[249] Truesdell is however renowned for his sharp criticisms directed at mathematicians of lesser repute throughout history.

References

1. Adkins C (2001) Trumpet marine, https://doi.org/10.1093/gmo/9781561592630.article.28494, Grove music on line. Accessed 25 July 2022
2. Ango P (1682) L'optique divisée en trois livres. Michallet, Paris
3. Anonymous (1666) Méthode facile pour apprendre a chanter la musique. Ballard, Paris
4. Bailhache P (1992) Leibniz et la théorie de la musique. Kincksieck, Paris
5. Barbieri P (1983) I temperamenti ciclici da Vicentino (1555) a Buliowski (1699): Teoria e pratica 'archicembalistica'. L'Organo 21:129–208
6. Barbieri P (1986) Giuseppe Sarti. Fisico, acustico e teorico musicale. In: Baroni M, Gioia TM (eds) Giuseppe Sarti musicista faentino. Modena, Mucchi, pp 221–240
7. Barbieri P (2007) The Jesuit acousticians and the problem of wind instruments (c.1580-1680). Analectica Musicologica 38:155–204
8. Barbieri P (2013) Physics of wind instruments and organ pipes 1100–2010. Il Levante, Latina
9. Barbieri P (2017) Canne d'organo ad anima: La faticosa gestazione di una teoria matematica, 1727–1960. Rivista Italiana di Acustica 36(4):5–24
10. Barbour JM (1951) Tuning and temperament. Michigan State College Press, East Lansing, A historical survey
11. Beeckman I (1939–1953) Journal tenu par Isaac Beeckman de 1604 à 1634 (4 vols). Nijhoff, The Hague
12. Benoit M (1971) Musiques de cour: chapelle, chambre, écurie. Picard, Paris
13. Bernoulli Johann (1703) Extrait d'une lettre de M. Bernoulli contenant l'application de sa regle du centre de balancement à toutes sortes de figures. Mémoires de l'Académie Royal des Sciences de Paris, pp 68–75

[248] p. 15.

[249] Entry: *Sauveur, Joseph*. Truesdell, CA.

14. Bucur V (2019) Handbook materials for wind musical instruments. Springer, Dordrecht
15. Bulyowsky M (1712) Tastastura quinquiformis panarmonico-methatetica. Review. Le Journal des Sçavants, pp 247–251
16. Bush D, Richard K (eds) (2006) The organ. An enyclopedia, Routhledge, New York
17. Cajori F (1910) History of the logarithmic slide rule. Astragal Press, Blue Ridge Summit, PA
18. de Celles B (1766–1778) L'art du facteur d'orgues (3 vols). Saillant and Nyon and Deisant, Paris
19. Cohen A (1981) Music in the French Royal academy of sciences. A study in the evolution of musical thought. Princeton University Press, Princeton
20. Cohen H (1984) Quantifying music. Reidel, Dordrecht
21. Corno S (2019) L'arte organaria. The author, Unknown
22. Cousin L (1707) Sur la theorie generale du son, sur les differens accords de la musique, & sur le monochorde. Le Journal des Sçavants Supplement, pp 115–225
23. Dechales MCF (1674) Cursus seu mundus mathematicus, 3 vols. Ex Officina Anissoniana, Leiden
24. Descartes R (1650) Musicae compendium. Ackersdijck and Zijll, Utrecht
25. Devie D (1990) Le Tempérament musical: philosophie, histoire, théorie et pratique. Société de musicologie de Languedoc, Béziers
26. Diderot D (1748) Principes généraux d'acoustique. In: Diderot D (ed) Mémoires sur différens sujets de mathématiques, Durant & Pissot, Paris
27. Dodart D (1700) Mémoire sur les causes de la voix de l'homme, & de ses différens tons. Mémoires de l'Académie Royal des Sciences de Paris, pp 244–292
28. Dostrovsky S (1975) Early vibration theory: physics and music in the seventeenth century. Arch Hist Exact Sci 14(3):169–218
29. Du Hamel JB (1701) Regiae scientiarum academiae historia. Delespine, Paris
30. Duverney J (1683) Traité de l'organe de l'ouie, contenant la structure, les usages & les maladies de toutes les parties de l'oreille. Michallet, Paris
31. Ellis AJ (1881) On an improved bimodular method of computing natural and tabular logarithms and anti-logarithms to twelve or sixteen places, with very brief tables. Philos Trans R Soc Lond 31(206–211):381–398
32. Euler L (1727) Dissertatio physica de sono. Thurnisiorum, Basel
33. Euler L (1739) Tentamen novae theoriae musicae ex certissismis harmoniae principiis dilucide expositae. Academiae Scientiarum, Saint Petersburg
34. Fontenelle de B (1707) Sur les systemes temperés de musique. Histoire de l'Académie Royal des Sciences de Paris, pp 117–120
35. Fontenelle de BB (1700) Sur la determination d'un son fixe. Histoire de l'Académie Royal des Sciences de Paris, pp 134–143
36. Fontenelle de BB (1701) Sur un nouveau systeme de musique. Histoire de l'Académie Royal des Sciences de Paris, pp 121–137
37. Fontenelle de BB (1702) Sur l'application des sons harmoniques aux jeux d'orgues. Histoire de l'Académie Royal des Sciences de Paris, pp 90–92
38. Fontenelle de BB (1711) Sur la quadrature des courbes. Histoire de l'Académie Royal des Sciences de Paris, pp 62–67
39. Fontenelle de BB (1711) Sur les systemes temperés de musique. Histoire de l'Académie Royal des Sciences de Paris, pp 80–81
40. Fontenelle de BB (1713) Sur les cordes sonores et sur une nouvelle détermination des sons fixes. Histoire de l'Académie Royal des Sciences de Paris, pp 68–75
41. Galilei V (1589) Discorso particolare intorno alle forme del diapason, Biblioteca Nazionale Centrale di Firenze, Manoscritti Galileiani, Anteriori a Galileo, III, 1589 c., ff. 44r-54v, https://bibdig.museogalileo.it/tecanew/opera?bid=18011&seq=90, accessed 3 July 2023
42. Gallimard JE (1754) Arithmétique des musiciens. Unknown, Paris
43. Gallimard JE (1754) La théorie des sons applicables à la musique, où l'on démontre dans une exacte précision les rapports de tous les intervales diatoniques et chromatiques de la gamme. Ballard, Paris

44. Haynes B (2002) A history of performing pitch. A story of A, Scarecrow, Lanham, Maryland
45. Henfling C (1710) Epistola de novo suo systemate musico. Miscellanea Berolinensia ad incrementum scientiarum, pp 265–294
46. Hubbard TL (2022) The Pythagorean comma and preference for a stretched octave. Theoretical Critical Paper 50(2):670–683
47. Huglo M (2000) Gerbert, théoricien de la musique, vu de l'an 2000. Cahiers de Civilisation Médiévale 43:140–160
48. Huygens C (1690) Traité de la lumiere, où sont expliquées les causes de ce qui luy arrive dans la reflexion, & dans la refraction. Et particulierement dans l'etrange refraction du cristal d'Islande. Vander, Leiden
49. Huygens C (1888–1950) Oeuvres complètes de Christiaan Huygens (22 vols). Nijhoff, The Hague
50. Huygens-Fokker (Unknown) Logarithmic interval measures, https://www.huygens-fokker.org/docs/measures.html, Accessed 2 January 2025
51. Jedrzejewski F, Papadoulos A (2021) Joseph Sauveur. Écrits sur la musique er l'acoustique, Hermann, Paris
52. Kipnis I (2007) The harpsichord and clavichord. An encyclopedia. Taylor & Francis, New York
53. Loulié E (1696) Elements ou principe de musique. Ballard, Paris
54. Lully JB (1675–1676) Atys, https://s9.imslp.org/files/imglnks/usimg/c/c2/IMSLP535900-PMLP44600-Atys_Lully_Jean-Baptiste_btv1b10551132f.pdf, Manuscript, n.d.(ca.1670-99), Bibliothèque de l'Opéra. Accessed 25 July 2022
55. Maxham RE (1976) The contributions of Joseph Sauveur (1653-1716) to acoustics (2 vols). PhD thesis, University of Rochester
56. Mersenne M (1636–1637) Harmonie universelle (2 vols). Cramoisy, Paris
57. Pajot LL (1732) Description & usage d'un métrometre, ou machine pour battre les mesures & les temps de toutes sortes d'airs. Mémoires de l'Académie Royale des Sciences de Paris, pp 182–195
58. Palisca CV (ed) (1989) The Florentine camerata. Yale University Press, New Haven, Documentary studies and translations
59. Plot R (1677) The natural history of Oxford-shine. Millers, Oxford
60. Rasch R (ed) (1984) Joseph Sauveur. Collected writings on musical acoustics (Paris 1700-1713). The Diapason Press, Utrecht
61. Raymond JM (1826) Des principaux systèmes de notation musicale usités ou proposés chez divers peuples tant anciens que modernes. Memorie della Accademia delle Scienze di Torino Sezione di Scienze Morali Storiche e Filosofiche 30:1–153
62. Riccati G (1762) Saggio sopra le leggi del contrappunto. Trento, Castel Franco Veneto
63. Riccati G (1987) Le leggi del contrappunto dedotte dai fenomeni e confermate dal raziocinio: Saggio sopra le leggi del contrappunto (1751-1754). In: Barbieri P (ed) Acustica accordatura e temperamento nell'illuminismo veneto, Torre D'Orfeo, Rome, pp 1–133
64. Robartes F (1692) A discourse concerning the musical notes of the trumpet, and trumpet-marine, and of the defects of the same. Philosophical Transactions of the Royal Society of London 17(195):559–563
65. Romieu JB (1758) Mémoire théorique & pratique sur les systèmes tempérés de musique. Mémoires de l'Académie Royal des Sciences de Paris, pp 483–519
66. Rousseau JJ (1743) Dissertation sur la musique moderne. Quillau, Paris
67. Rousseau JJ (1768) Dictionnaire de musique. Duchesne, Paris
68. Saigey JF (1834) Traité de métrologie ancienne et moderne. Hachette, Paris
69. Sauveur J (1692) Explication des echelles pour les calculs de marine. Sevestre, Paris
70. Sauveur J (1697) Traité de la théorie de musique. Ark:/12148/ btv1b1 0090014n. Bibliothèque Nationale de France. Département des Manuscrits. NAF 4674
71. Sauveur J (1700) Procès-verbaux de l'Académie des sciences de Paris, https://gallica.bnf.fr/ark:/12148/bpt6k55695c.image#, tome 19. Accessed 25.01.2024
72. Sauveur J (1701) Procès-verbaux de l'Académie des sciences de Paris, https://gallica.bnf.fr/ark:/12148/bpt6k55696q.image, tome 20. Accessed 25.01.2024

73. Sauveur J (1701) Systême general des intervalles des sons, et son application à tous les systêmes & à tous les instrumens de musique. Mémoires de l'Académie Royal des Sciences de Paris, pp 299–366
74. Sauveur J (1702) Application des sons harmomques à la composition des jeux d'orgues. Mémoires de l'Académie Royal des Sciences de Paris, pp 121–137
75. Sauveur J (1707) Methode generale pour les systêmes tempérés de musique, et du choix de celui qu'on doit suivre. Mémoires de l'Académie Royale des Sciences de Paris, pp 203–222
76. Sauveur J (1711) Table generale des systemes temperés de musique. Mémoires de l'Académie Royal des Sciences de Paris, pp 307–315
77. Sauveur J (1713) Rapport des sons des cordes d'instruments de musique, aux flêches des cordes; et nouvelle détermination des sons fixes. Mémoires de l'Académie Royal des Sciences de Paris, pp 324–350
78. Sauveur J (1754) Geometrie élémentaire et pratique. Edited by Le Blond. Jombert, Paris
79. Scherchen H (1972) The nature of music. Scholarly Press, Saint Clair Shores
80. Semmens R (1981) An early eighteenth-century discussion of musical acoustics by Étienne Loulié. Can Univ Music Rev 2:177–208
81. Semmens R (1987) Joseph Sauveur "Treatise on the theory of music". A study, diplomatic transcription and annotated translation. Faculty of Music. University of Western Ontario, London Canada
82. Semmens R (1990–1991) Sauveur and the absolute frequency of pitch. Theoria 5:1–41
83. Sigismondi C (2012) Gerberto e la misura delle canne d'organo. https://www.researchgate.net/publication/232805214_Gerberto_e_la_misura_delle_canne_d'organo, accessed 12 June 2024
84. Smith R (1759) Harmonics or the philosophy of musical sounds. Merril, London
85. Truesdell CA (1960) The rational mechanics of flexible or elastic bodies. In: Leonhardi Euleri Opera omnia (1911-, in progress), 2, vol 1, part 2, Teubneri GB; [then] Fussli Turici; [then] Birkhäuser, Basel
86. Varignon P (1687) Projet d'une nouvelle mechanique avec un examen de l'opinion de M. Borelli, sur les proprietez des poids suspendus par des cordes. Martin & als, Paris
87. Various (2025) New Grove Dictionary of Music and Musicians. https://doi.org/10.1093/gmo/9781561592630.article.27643, accessed 1 January 2025
88. Vicentino N (1555) L'antica musica ridotta alla moderna pratica. Barre, Rome
89. Wallis J (1695–1699) Opera geometrica (3 vols). Theatro Sheldoniano, Oxford
90. Wild J (2014) Genus, species and mode in Vicentino's 31-tone compositional theory. Music Theory on Line 20(2):1–19
91. Zaragoza J (1675) Fabrica, y uso de varios instrumentos mathematicos. de Zafra, Madrid
92. Zarlino G (1558) Istitutioni harmoniche. The author, Venice

Chapter 6
Epilogue

Abstract After Sauveur's death, acoustics can be seen as a evolving discipline. Musicians and mathematicians were confronted with the new results, in particular the existence of higher harmonics. Musicians include Jean Philippe Rameau and Giuseppe Tartini, two of the most famous composers (and performers) of the 18th century, while mathematicians include Jean d'Alembert and Leonard Euler, two of the most famous mathematicians of the 18th century. However, key achievements occurred in the 19th century, particularly with the work of Hermann Helmholtz, considered the father of contemporary acoustics. In his masterpiece *Die Lehre von den Tonempfindungen als physiologische Grundlage für die Theorie der Musik* of 1863, he developed an important set of arguments. In this text, Helmholtz proposed a new theory of consonances based on beats, generalizing Sauveur's ideas, a detailed analysis of Tartini's results on the third sound and, finally, a sufficiently convincing explanation of the meaning of Ohm's law, according to which all the properties of sounds are explained by the presence of elementary tones described by sinusoidal waves of given frequency; the amplitudes of these tones can be determined by Fourier's analysis.

6.1 Music Trend in the 18th Century

Around the turn of the 18th century, changing musical tastes glimpsed an alternative to counterpoint: harmony theory. Harmony, like counterpoint, is concerned with simultaneous sounds; but unlike counterpoint, it is based on groups of sounds called chords.

In previous chapters, especially the one devoted to Sauveur, the term chord has been used as a synonym for interval, for two notes produced simultaneously. In this chapter, we will use the term chord as it is generally used in modern harmony, that is, as a set of three or more simultaneous notes; in particular, we will speak of triads [96],[1] referring to chords formed by three notes with the superposition of two

[1] Entry: *Triad*.

D. Capecchi and G. Capecchi, *Joseph Sauveur and His Time: Music and Mathematicians*,
https://doi.org/10.1007/978-3-032-04707-6_6

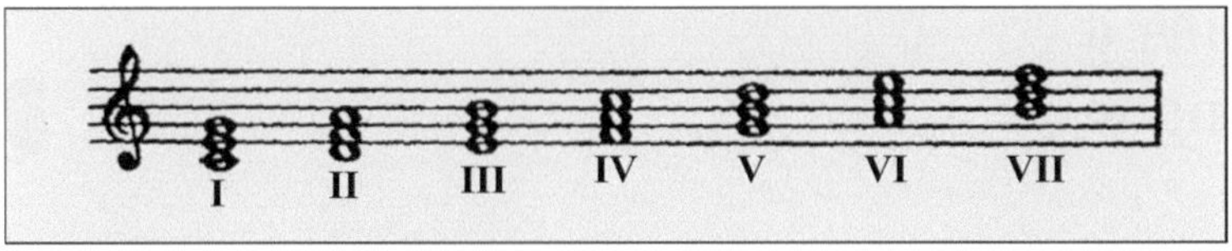

Fig. 6.1 Example of triads; from [80], p. 32

Fig. 6.2 Triad Inversion; from [80], p. 55

thirds. In order, major and minor for the major triad, minor and major for the minor triad [96].[2]

Figure 6.1 shows some examples of basic forms of triad built on different degrees in the major scale. Triads of the first type, major triads, are found on degrees I, IV and V; those of the second type, minor triads, are found on degrees II, III and VI; a third type, the diminished triad, is only found on degree VII.

Triads can also occur in forms that are all almost the same as the basic forms, through inversion. The term *inversion* means placing a low note of a chord one octave higher, [or] a high note one octave lower, while the other notes remain in their places. Figure 6.2 illustrates the situation for the major third chord in C_4. Basically, from the basic form C_4 E_4 G_4 one goes to E_4 G_4 C_5 (Fig. 6.2a) or to G_4 C_5 E_5 (Fig. 6.2b).

The concept of triads, obviously, is not new of the 18th century, we find it in Zarlino's *Istitutioni harmoniche*; the name of triad was taken up by Johannes Lippius (1585–1612), who in his *Synopsis musicae novae* (1612) coined the term *trias harmonica* [53].

With the introduction of harmony in the 18th century there was also a change, or rather a stabilization, in the way the octave was divided into semitones and tones. In earlier centuries the octave was organized in fairly free variations of the Renaissance modes. Gradually, only two of these modes were chosen, called the major mode—corresponding to the Ionian mode (CDEFGAB)—and the minor mode—corresponding to the Aeolian mode (ABCDEFG).[3] This structure of tones and semitones in the octave is defined by TTSTTTS or TSTTSTT respectively, where for simplicity we have referred to equal temperament, which does not distinguish between major and minor tones.

The tension of musical compositions, previously produced by the mode, is now defined by the tonic (initial note of the reference scale), the dominant (one fifth above)

---

[2] Entry: *Harmony*. Dahlhaus C.

[3] The names Ionian and Aeolian are derived from Heinrich Glarean (1488–1563), in his *Dodekachordon* of 1547 [34].

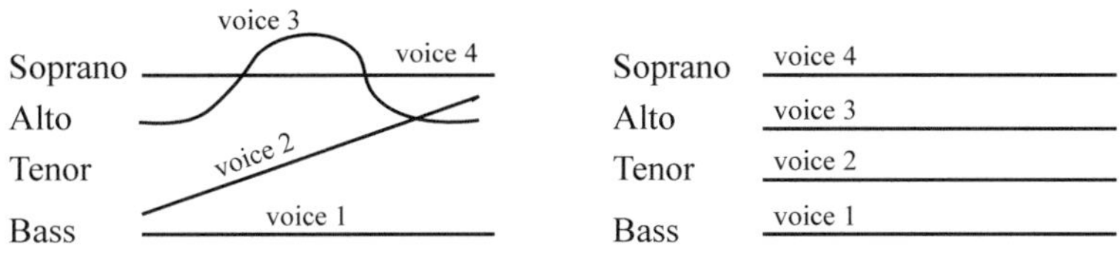

Fig. 6.3 Example of counterpoint on the left, example of harmony on the right

and the subdominant (one fifth below), and above which triads are built according to major or minor mode.

Rameau, as we shall see later in this chapter, saw triads and major and minor modes as natural products. According to him, the natural way of producing sound is by vibrating strings, in which, whatever the excitation, there are always harmonic sounds of a higher order in addition to the fundamental. The first three odd harmonic sounds are best heard; their frequencies are in the proportions (1 : 3 : 5), the same as in the major triad. The physical derivation for the minor triad is more difficult to justify and will be discussed in Sect. 6.1.1.1.

In harmony, the different parts of a staff are less melodically independent than in counterpoint. Renaissance vocal counterpoint involved crossing parts, 18th-century harmony relies in some ways on the practice of basso continuo, where the parts move as little as possible to emphasize the melody they accompany. Figure 6.3 shows intuitively what has just been explained. Later, in the 19th century, with Hugo Riemann, harmony became a concept inseparable from that of tonality.

6.1.1 Chords Between Musicians and Mathematicians

We have seen that starting from the 18th century, traditional speculative music has hardly existed. Much of it had been replaced by musical acoustics, which had become a complex discipline that required a good mathematical and physical education and could not be generally mastered by composers and performers.

Although acoustics and the remaining part of speculative music, hereafter referred to as music theory, had become separate disciplines, they were not completely independent of each other; scholars of music theory needed a minimum of notions of acoustics, and scholars of acoustics needed a minimum of notions of music theory. There were then scholars, specialized in one of the two disciplines, who had a strong interest and expertise in the other. Of particular interest are those concerning two of the greatest musicians of the time: the Frenchman Jean Philippe Rameau and the Italian Giuseppe Tartini, and two of the greatest mathematicians, Leonhard Euler and Jean d'Alembert.

Both Rameau and Tartini were engaged with the field of acoustics, and thus partially disregarded the issue of temperaments as a topic worthy of scientific inquiry, even though this was a subject actively pursued by the majority of musicians at the time. Rameau used Sauveur's findings on the nature of harmonic sounds, developed some new acoustical experiments of his own, and based his theory of harmony on them. Tartini was aware of Sauveur's and Rameau's findings and discovered—probably as the first—a new and very interesting acoustic phenomenon: the perception of an unexpected sound when playing any musical interval (the so-called third sound).

Both authors, although among the forerunners of the new theory of music, were still partially rooted in Greek speculative music. Rameau, like Zarlino, one of his references, saw the monochord as the fundamental source of chords in his early writings. Its invocation was somewhat unusual for the time; few theorists at the time used this ancient device. Firstly, the traditional function of the monochord was increasingly taken over by acoustics. Second, the variety of complex dissonant ratios, tuning systems, and temperaments of 18th-century theorists exceeded the monochord's practical capacity for easy presentation. Tartini, for his part, presented arguments that were essentially rhetorical in nature, based on suggestions derived from traditional speculative music. From geometry, when he considered the relationship of the circumference of the circle to its diameter as the fundamental principle of harmony. From arithmetic, when he tried to legitimize the empirical evidence of the third sound by proposing analogies between harmonic and arithmetic progressions.

Euler and d'Alembert moved in the tradition of mixed mathematics, which had not yet been fully exhausted their role in the 18th century music. Although neither seems to have had formal musical training, at least as far as performance was concerned, they played an important role in the new musical trend.

D'Alembert became involved in musical matters in 1749, when Rameau submitted a manuscript containing the elements of his musical theory to the Académie des sciences of Paris for approval.[4] He continued his interest in music and wrote about thirty musical articles for the *Éncyclopédie*; a small number compared to the approximately 1600 entries that bear his march, but important nonetheless, in the period 1750–1757. Moreover, in his role as editor, he made significant interventions with the articles presented by the 'musician' of the *Éncyclopédie*, Jean Jacques Rousseau (1712–1778) [8, 11].

D'Alembert was impressed by Rameau's writings on music and held him in such high esteem that in his *Discourse préliminaire de l'Éncyclopédie* he gave him a prominent place among the geniuses of science and art [20].[5] After a few years, however, the relation between the two men changed from friendly to hostile; for a discussion on this point see [4, 7].

Euler was a prolific writer. The publication of his writings, still incomplete, counts today about eighty volumes. His writings on music amount to only about 500 pages, not much compared to the rest. Nevertheless, their influence on music theory was great. The first relevant writing on music was quite early, the *Tentamen novae theoriae musicae ex certissimis harmoniae principiis dilucide expositae* (hereinafter *Tentamen novae theoriae musicae*), published in 1739 but written in 1731 when Euler was 24

---

[4] It was published in a revised form as *Démonstration du principe de l'harmonie* in 1750 [67].

[5] p. 122.

years old [25]. He remained interested in music for the rest of his life, perfecting his theory with some papers published in scientific journals [27, 28] and some articles of his well-known book *Lettres à une princesse d'Allemagne* [29].

There is no evidence of Euler's formal training in music. However, he was in Berlin, as a member of the local academy of sciences, in the years 1746–1766. During this time, famous musicians such as Carl Philipp Emanuel Bach (a son of Johann Sebastian) and Johann Joachim Quantz frequented the Berlin court. There was also a visit from Bach the Elder in 1747. It is hard to imagine that Euler did not interact with these great musicians. But also in St. Petersburg, where he moved in 1766, Euler had the opportunity to make very interesting musical encounters [30]. He corresponded with Tartini and Rameau. These acquaintances, if they do not serve to justify the writing of the 1731 work, do, however, justify Euler's interest in music.

6.1.1.1 Jean Philippe Rameau's *Corps Sonore*

Very little is known about the academic education of Jean Philippe Rameau (1683–1764). In essence, he learned to play first and then to write; attended the Jesuit college at Godrans, where he intended in vain to study law [33].

Rameau was (one of) the greatest French musician of the 18th century. His first major work in music theory is the *Traité de l'harmonie réduite à ses principes naturels* of 1722 [64] (hereinafter *Traité de l'harmonie*), translated into English as *Treatise on harmony* [72]. Here Rameau conceived of music as a purely rational discipline and lamented the lack of rigor in its theoretical developments, with explicit references to the rationalism of Descartes, whose ideas about science were still in vogue in the French academies in the early 1700s.

> Music is a science that must have certain rules; these rules must be derived from a self-evident principle, and this principle cannot be known to us without the use of mathematics. I must confess that, in spite of all the experience I have been able to acquire in music, after practicing it for so long, it is only with the use of mathematics that our ideas have taken shape, and that light has followed darkness which I did not perceive in the past [64].[6]

Rational music, then, is not for Rameau a traditional physical-mathematical discipline, as acoustics has become. This is clear from the brief discussion he presented in a few pages on the speculative aspect of music. Although he considered it irrelevant whether the musical intervals were associated with ratios of lengths or of frequencies, he chose to move in the traditional stream, preferring the reference to strings, mirroring the presentation of Descartes in his *Musicae compendium*. Although the theoretical basis of the *Traité de l'harmonie* was still of a metaphysical nature, based on Zarlino's senario, an empirical imprint was present. Music is the science of sound, and consonances are valid because the ear considers them as such.

Fundamental to Rameau's entire theory of harmony is the notion of major and minor triads discussed in the previous section. And a main principle is the *octave equivalence*, for which the addition or subtraction of one octave does not change the

[6] Preface. Translation into English in [97].

'substance' of a sound. A principle that is not new (it is implicit in the organization of notes in octaves since the 10th century CE), but which Rameau made more precise and stringent [14].[7] Rameau's move was extremely fruitful and allowed him to classify chords in a simple way, considering some of them as different forms of a fundamental type, the premise for the fruitful idea of the *fundamental bass* [83]. He used it to justify doubtful consonances, for example as the fourth [92], following the same reasoning developed by Descartes, for example, but using a simpler argumentation. Consider the perfect consonance $(f_2 : f_1) = (3 : 2)$, which is a fifth, and assume that it is associated with the frequencies $f_2 = 393$ Hz (G_4), $f_1 = 262$ Hz (C_4). If we replace the lowest frequency f_1 by a frequency one octave higher, $f_3 = 524$ Hz (C_5), we get the interval $(f_3 : f_2) = (4 : 3)$, i.e. the fourth G_4–C_5. This interval is considered consonant by Rameau because $(f_2 : f_1)$ was consonant and f_2 is equivalent to f_3 for the equivalence of octaves [64].[8]

We have shown in Sect. 4.1 that the paradigm shift in French science, with its emphasis on the empirical aspect, following England and Holland—thanks largely to the success of Newton's new natural philosophy—had a profound influence on music theory. Rameau, too, was deeply influenced, which led him to seek a physical explanation as an alternative to the widespread metaphysical suggestions about the pleasantness or unpleasantness of musical intervals. This explanation was first presented in the *Nouveau système de musique theorique* (hereinafter *Nouveau système*) of 1726, after Rameau had become acquainted with the recent results in acoustics of Mersenne and especially Sauveur.

In the *Nouveau système* Rameau introduced what he called the principle of the *corps sonore* [12]: an elastic body (a sonorous body), when set in vibration, invariably plays the major perfect chord, consisting of a major third (5 : 4) followed by a minor third (6 : 5) to give a fifth (3 : 2), such as C_3–E_3–G_3. This empirical fact seemed a revelation to Rameau, to the point that he considered it the true principle of harmony [65].[9] Note that this principle is not only of an empirical nature; it is supplemented by a metaphysical truth, according to which what corresponds to nature should be treated as a guide; and since nature says that by exciting a sonorous body a major chord is obtained, this chord must be considered perfect in the highest degree, and the musician's ear can only agree.

To be precise, it must be said that the experience cited by Rameau (see also below), as he himself admitted, did not directly yield the perfect major chord C_3–E_3–G_3, represented by the proportions (4 : 5 : 6).[10] Rather, it provides the ratios (1 : 3 : 5), which define a twelfth (1 : 3) and a seventeenth (1 : 5), that is, the chord C_3–G_4–E_5, a result already known at least since Mersenne [54].[11]

However, Rameau easily manages to go from (1 : 3 : 5) to (4 : 5 : 6) using the principle of octaves established in the *Traité de l'harmonie*. In fact, the chord C_3–G_4–

[7] p. 70.

[8] Table of terms, article: Consonance.

[9] Preface, first page.

[10] This triad consists of a major third (5 : 4) and a fifth (6 : 4) = (3 : 2).

[11] Part I, *Traité des consonances*, pp. 231–232.

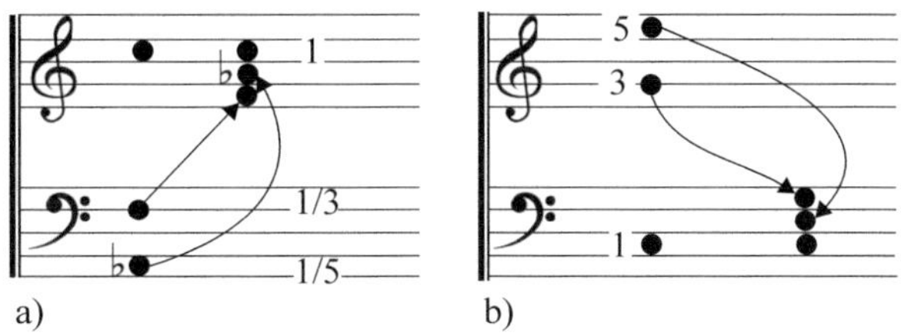

Fig. 6.4 Minor perfect chord **a**. Major perfect chord **b**

E_5 can easily be changed to C_3–E_3–G_3 by lowering G_4 by one octave and E_5 by two octaves (see Fig. 6.4b). But this passage is not just presented as a simple deduction. Rameau said that although it is known that the ratio of the frequencies produced by the sonorous body is (1 : 3 : 5), our ear is not trained to recognize the intervals of twelfths or seventeenths because they are too large. And therefore, instead of the twelfth, we will hear an interval one octave lower, that is, a fifth, and instead of the seventeenth, an interval two octaves lower, that is, a major third [67].[12]

Rameau developed the idea of the natural origin of harmony much more fully in his two works *Génération harmonique*, ou *Traité de musique théorique et pratique* (hereinafter *Génération harmonique*), 1737, and *Démonstration du principe de l'harmonie*, 1750. The first work is largely a treatise on acoustics in which music is presented as a mathematical-physical discipline, where sound is the physical object and the relationships between different sounds are the mathematical object [66],[13] emphasizing that the physics-mathematics relationship is easier to grasp for music than for other disciplines [66].[14]

The first chapter of the *Génération harmonique* presents 12 propositions and 7 experiments. Among the propositions, Proposition XII, which considers the ear as a sonorous body that transmits the sensation of sound and harmony to the soul [66],[15] is particularly interesting, since it anticipates Helmholtz, for example.

> *Proposition XII.* What has been said of the sonorous bodies must also be understood of the fibers which cover the bottom of the ear conch; these fibers are so many sonorous bodies to which the air transmits its vibrations, and from which the feeling of sound and harmony is carried to the soul [66].[16]

Among the experiments, some are based on auditory sensations and require a trained ear, while others are based on visual observation, as is common in physics. Probably the most interesting is Experience III, which is based on hearing. It reproduces in a clearer way the results of the sonorous body of the *Nouveau système*.

> *Experience III.* Scratch one of the largest strings of a viola or cello, and you will hear, with the fundamental, those of its octave, double octave, perhaps even triple octave, of the twelfth

[12] pp. 15–16.
[13] p. 30.
[14] Epistle, second page.
[15] p. 7.
[16] p. 7.

> and seventeenth major, which have the proportions 1, 1/2, 1/3, 1/4, 1/5, 1/8. You will hear them sometimes together, sometimes one after the other, sometimes only one, according to the attention you give to each other, and also according to the way you get the sound from the string, which should be done with gentleness rather than force, and sometimes also closer to the bridge than usual; you can also distinguish the sound of 1/7, not to say other [66].[17]

Note that Rameau used the ratios of the lengths of a string rather than the ratios of the frequencies to indicate harmonics. So he referred to the harmonic sounds with the sequence 1, 1/2, 1/3, 1/4, 1/5, 1/8 instead of 1, 2, 3, 4, 5, 8, as is more common today. We however, do not follow Rameau's usage.

Of the other experiments, Experience II is also interesting. It uses two strings of different lengths, l_0 and $l_0/3$.

> *Experience II.* Take a viola or a cello, two strings of which are tuned to the twelfth from each other; scrape the grave, you will see the acute vibrate, and perhaps even resonate, & you will undoubtedly recognize this if you touch it with your nail while it vibrates. Then scrape the acute, you will not only see the grave tremble in its entirety, you will also see it divide into three equal parts, forming three antinodes [ventres (bulges)] of vibration between two nodes or fixed points [66].[18]

The sense used in this experiment is sight, not hearing. The presence of overtones is revealed by the visible vibrations of the two strings, one moved by a bow, the other by resonance. Only the case of higher and lower harmonic sounds of orders 3 and 1/3 is reported. Experience II has a very high evidential value because it can be performed by a non-musician. In the first part of the experiment, when the longest string (l_0) of natural frequency f_0 is rubbed, the shorter string ($l_0/3$) begins to vibrate. This is an index of the presence of the higher harmonic of order 3 in the longer string, which causes the shorter string ($l_0/3$)—whose natural frequency is $3f_0$—to resonate (see Fig. 6.5a).

In the second part of the experiment, the shorter string is rubbed. The result of the experiment is not quite clear. It is indeed different to say that the string vibrates as a whole, as shown in Fig. 6.5b$_1$, or that it vibrates in three parts, as shown in Fig. 6.5b$_2$. In the first case it would mean that the string vibrates according to its fundamental frequency, i.e. $1/3 f_0$, in the second case that it vibrates with the same frequency f_0 of the rubbed string. The first situation is not possible for substantially elastic strings, and very difficult even for strings that exhibit nonlinear behavior. It is therefore very likely that Rameau actually saw what is shown in Fig. 6.5b$_2$ and imagined that it was also possible for the whole string to vibrate.

In essence, he wrongly assumed the length $3l_0$ of the string as a parameter and then used its natural frequency $1/3 f_0$, as the frequency of vibration, referring, incorrectly, to Sauveur's experiences.[19] In the *Démonstation du principe de l'harmonie* of 1750,

[17] p. 10.

[18] pp. 8–9.

[19] It is as Rameau would say: if a string with natural frequency $1/3 f_0$ is forced to vibrate by an excitation of frequency f_0, it will actually vibrate at the frequency $1/3 f_0$, i.e. according to a subharmonic. Analytical dynamics leaves no doubt about the existence of subharmonics for forced vibrations of strings in their normal use: they are not possible (they only can exist with

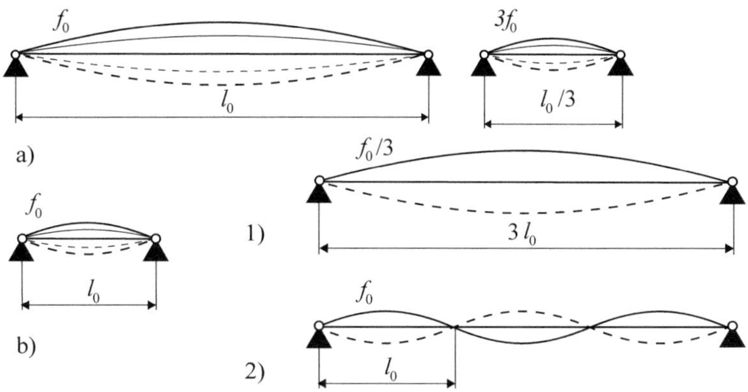

Fig. 6.5 Resonant higher harmonic of order 3 **a**. Resonant lower harmonic of order 1/3 **b**

he recognized that the vibration of the whole string as in Fig. 6.5b_1 was not possible [67],[20] without considering the consequences of this fact for his musical theory.

The result of Experience II, generalized to cover also the case of the 1/5 minor harmonic, allowed Rameau to consider as natural also the perfect minor chord defined by the triad (10 : 12 : 15), that is, a minor third and a fifth. The sonoursonorousus body would have given the proportions $(1 : 1/3 : 1/5)$, corresponding, for example, to the chord C_5–F_3–$A_2\flat$. But using the principle of the octave, F_3 can be changed to F_4, $A_2\flat$ to $A_4\flat$, so that the chord is C_5–$A_4\flat$–F_4, i.e. F_4–$A_4\flat$–C_5, which is a perfect minor chord, since the first interval is a minor third (see Fig. 6.4a).

Having shown that music has a physical foundation, Rameau goes on to reverse the situation: physics has a musical empirical foundation. Indeed, music has an empirical foundation (the sonorous body), music is at the basis of the theory of proportions, physics is based on proportions, therefore physics has an empirical foundation in music. Also, according to Rameau, the first 'scientific' investigations of nature came from music.

> Convinced of the necessity of this universal principle, the first philosophers sought it in music: Pythagoras, according to the Egyptians, applied the laws of harmony to the motion of the planets; Plato made it preside over the composition of the soul; Aristotle, his disciple, having seen that music is a heavenly thing & divine, added that we find there the reason for the system of the world [69].[21]

Rameau gave music an ontological and epistemological primacy, on the assumption that the principle of all the fine arts resides most surely and most meaningfully in

---

medium-strong nonlinearities) [57]. This does not mean, however, that subharmonic sounds cannot be perceived. They are instrumental to modern psychoacoustic accounts of harmony and consonance [9, 93]. So Rameau was correct in describing an existing phenomenon, even if his justification was completely wrong.

[20] pp. 20–21.

[21] p. 189.

music. “If one examines the matter more closely, one will see that it is not for nothing that nature has chosen the sense of hearing above all others. to judge her most general rules, *superbissimum auris judicium*” [68].[22]

Even temperament, which contradicts strict Pythagoreanism, is regarded by Rameau as a value, not a limit. Indeed, the need for temperament does not concern music alone: “How many other arts use it in compromises, borrowings, repayments, approximations?” [68].[23]

Rameau had a particular view of the harmony of the spheres. Instead of referring to a general harmony of music, he considered only the harmony determined by the sonorous body, namely the major and minor triads. Moreover, in his last writings, Rameau expressed his hostility to the strict Pythagoreans. For him, it was not a property of numbers that gave the proportions of music their ontological status. Rather, it is that these proportions arise as a fundamental sensory experience of a natural phenomenon; they can be described by numbers, but cannot be attributed to them [14].[24]

In one of his mature works on speculative music, *Origine des sciences* of 1761 [71] (followed by a commentary of 1762 [70]), Rameau presented what he considered to be the historical evidence for the priority of the sonorous body principle, relying on Montucla’s recently published *Histoire des mathématiques* (1758) [56].

Rameau suggested that a cabal of Egyptian priests must have possessed the knowledge of the sonorous body and been inspired by it to develop the sciences of geometry and algebra. In particular, the Egyptian priests were the first to discover the Pythagorean theorem, which states that the square of the hypotenuse of a right-angled triangle is equal to the sum of the squares of its two sides, based on the arithmetic series 3 : 4 : 5 of the sonorous body: “This triangle consists in the square of 3 added to that of 4, whose sum 25 is equal to the square of 5” [71].[25] Rameau also drew inspiration for his theory from reports on Chinese music written by the Jesuit missionary Joseph Marie Amiot in a widely circulated manuscript that arrived in Paris in 1754.

In his last work, *Vérités également ignorées et intéressantes tirées du sein de la nature* of 1764, the *sonorous body* assumed almost divine powers in a pantheistic view:

> It seems to me that I see in the sonorous body (I hope you will forgive me this prevention in favor of the most striking image of a Creator, as we shall see) a Sun suspended above our heads to illuminate us [79].[26]

Was Rameau fully convinced of his idea of the supremacy of music? Probably not, but deep down there was a personal ambition to be part of that scientific world in tumultuous evolution, and a need to declare that his field of research was an important

---

[22] p. 51.

[23] p. 58.

[24] p. 297.

[25] p. 3.

[26] p.12.

part of science. He had several contacts with the members of the Académie des sciences; the most interesting were those with Mairan, Castel, Euler and d'Alembert.

The criticisms of Rameau's theories by mathematicians, philosophers, and musicians were and are many; although many of his proposals have actually been accepted [15, 62]. One of the most serious criticisms is that Rameau's position on the natural origin of harmony is arbitrary. It is not true that nature only suggests harmonies like the higher harmonic sounds 3 and 5. It also suggests harmonies of a higher order, including the 7th harmonic. Rameau's position that this last harmonic sound is dissonant—which it is—is therefore unjustified. Furthermore, there are sonorous bodies other than the monochord in which the ratio between the higher harmonic sounds does not follow the series of natural numbers (such as vibrating beams), see for example [42].[27]

6.1.1.2 Giuseppe Tartini's Third Sound

Giuseppe Tartini (1692–1770), a citizen of the Republic of Venice, was a contemporary of Rameau and received an education of a similar caliber. Although he studied law, he soon devoted himself entirely to music. Like Rameau, Tartini owed his scientific training to himself and to sporadic interactions with mathematicians and natural philosophers, the most important of whom were Giordano Riccati (1709–1790), perhaps the greatest Italian scholar of acoustics, and Leonhard Euler, with whom he came into contact thanks to Francesco Algarotti (1712–1764); it must be said that, compared to Rameau, he lived in a less fertile cultural environment: Padua versus Paris.

Today, Tartini is best known as a *violin virtuoso*, the *maestro* of nations of this instrument. Less known is his role as a music theorist. This is not due to the lack of relevance of his contribution, but rather to sociological reasons. The same reasons that led to ignoring the contribution of many thinkers of the past, who were then rediscovered and recognized as great. In his time, he had a great admirer in Jean Jacques Rousseau, who wrote in his *Dictionaire de musique* of 1768: "The system of M. Tartini, although in our opinion superior, is not generally known and does not enjoy, at least in France, the same authority as that of M. Rameau" [78].[28] D'Alembert's position was more nuanced. On the one hand, in his 1762 edition of the *Élémens de musique*, he stated that Tartini wrote in an incomprehensible manner for which he could not give an explicit explanation; then, with a touch of nationalism, he denied Tartini the priority of the discovery of the third sound, attributing it to his compatriot Jean-Baptiste Romieu (1723–1766), who had announced it in 1753 [77]. Here is the positive opinion of a modern music historian: "Tartini one of the most gifted theorists and musicians, and who, intellectually at least, was Rameau's superior" [83].[29]

---

[27] p. 232.

[28] Preface p. VIII.

[29] p. 292.

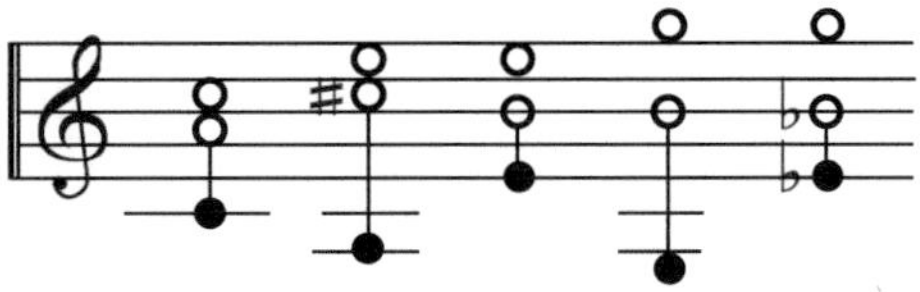

Fig. 6.6 Simple instances of third sound (notes in black indicate the third sound); from [87], p. 14

Tartini's main works on music theory are the *Trattato di musica secondo la vera scienza dell'armonia* of 1754 (hereinafter *Trattato*) [87], *De' principi dell'armonia musicale contenuta nel diatonico genere* (hereinafter *De' principi*) of 1767 [89] and *Risposta di Giuseppe Tartini alla critica del di lui trattato di musica di M. Le Serre di Ginevra*, still from 1767. The correspondence, especially that with Riccati and Euler, is also interesting.

The *Trattato* is a long text in six chapters plus an introduction.The main concerns of speculative music are summarized in a short premise:

> In the short premise, a small part of the ancient science of arithmetic, the common operations of addition, subtraction, multiplication, and division are taken as known [87].[30]

The first chapter of the treatise, largely repeated in *De' principi*, is of an empirical nature, in fact of acoustics. Here Tartini reported, quoting Rameau, on the presence of higher harmonic sounds in the vibrations of sonorous bodies, especially strings. Compared to Rameau, he gave greater importance to the multiples of the octave and criticized the fact that they were practically ignored by all musicians.

Tartini expanded the field of known acoustics with the introduction of a new phenomenon, designated as the *third sound*, or *third tone*. He claimed to have discovered this phenomenon in 1714 but did not publish it until 1754, when it was included in his *Trattato*. This was his first documented instance of this phenomenon being presented in print: when two different tones are played together on a bichord—but the results are of a general nature—a third sound is heard, which seems to appear from nowhere [87].[31]

> A new harmonic phenomenon has then been discovered which admirably proves the same and much more. Given two sounds of any musical instrument, which can prolong and amplify the sound as long as you want (trumpets, hunting horns, bowed instruments, oboe, etc.), there is a third sound produced by the impact of the two volumes of air moved by the two given sounds. For the time being, the physical explanation of how this third sound is produced is not important; the fact itself is enough, and it must be illustrated. Suppose a violinist plays the following perfectly tuned intervals simultaneously with a strongly sustained bow (Fig. 6.6).
>
> A completely distinguishable third sound will be heard, and it will be the marked [filled circle] one in the musical sheet [87].[32]

The above quotation claimed that the third sound was a matter of fact whose explanation could be postponed.

[30] p. 1.

[31] p. 13.

[32] pp. 13–14.

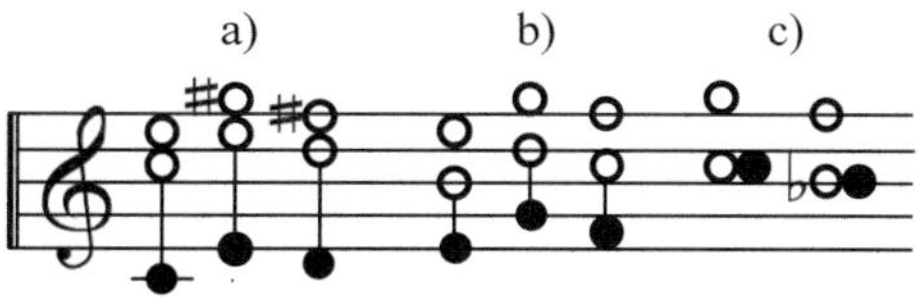

Fig. 6.7 Selected examples of third sounds generated by bichords: major third **a**, fourth **b**, and fifth **c**; from [87], p. 14

Already in the first chapter, Tartini presented various combinations of bichords played on a violin, and for each of them he gave the third sound; the only measuring instrument he used was the ear. In Fig. 6.7a the bichord is set to the intervals of a major third (5 : 4): C_5–E_5, E_5–$G\sharp_5$, D_5–$F\sharp_5$. The third sound is one octave below the lowest notes played, i.e. in C_4, E_4 and D_4. In Fig. 6.7b we are dealing with intervals of a fourth (4 : 3): B_4–E_5, D_5–G_5, C_5–F_5. The third sound is found one octave below the highest notes played, in E_4, G_4 and F_4. In Fig. 6.7c we are dealing with intervals of a fifth (3:2): C_5–G_5, $B\flat_4$– F_5. The third sound is given by a note equal to the lowest of those played, i.e. in C_5 and in $B\flat_4$.

The second chapter of the *Trattato* is of a mathematical nature and is motivated by the attempt to justify the rule obeyed by the third sound. This chapter is particularly complex and was subject to many criticisms [3], so that Tartini drastically reduced it in *De' principi*.

According to Tartini, for a musical (and scientific) theory to be valid, there must be physical evidence, mathematical evidence, and agreement with the expert's judgment at the same time [89].[33] Everything seems to go smoothly and in accordance with the physical and mathematical approach of the time. In reality this is not the case, in Tartini, unlike what happens in mathematical physics, the physical and mathematical parts are essentially independent of each other, there is neither a theoretical deduction from physical principles nor a hypothetical deductive approach. The mathematical part leads to the same results as the physical part, but does not confirm it. It is a free fruit of Tartini's imagination, which seeks the unity of harmony in mathematics, after having found it with physical tests. Moreover, in a mathematical-physical epistemology, how can an expert's opinion be taken into account? This is alien to science, even that of the 18th century.

A few clarifications are important at this point. With few exceptions, no violinist today is able to predict exactly the third sound produced by a bichord, and even after the performance the situation can be rather unclear: sometimes the same musicians, using different violins, declare to perceive different third sounds. Tartini himself, in a later writing, the *Rule of the third sound* [88], will make several additions to what he originally stated in his *Trattato*. For example, regarding the interval of fifth, he wrote that the third sound can be one octave lower than the lowest note.

In the *De' principi*, Tartini gave a general rule, a mathematical principle in fact, for finding the third sound:

> Of this third sound there is the formula, valid for two co-prime numbers, which defines the form [the interval] of two given tones. The form of the double, or octave, is (1 : 2).

[33] Preface.

> The multiplication of 1 by 2, the product 2, is the third sound that is in unison with the given sound 2 of the two sounds of the octave (1 : 2); and therefore, where experience is ambiguous, where *the hearing does not succeed in distinguishing the third sound 2 in unison with the given sound 2, the formula should be used* [emphasis added], by which the reality of the third sound 2 is known, although it does not fall under the sense. The form of the sesquialtera, or fifth, is (2 : 3). When 2 is multiplied by 3, the product 6 is the third sound, which is the subdouble of the given tone 3 [one octave below]. The form of the super 2 particular third, or major sixth, is (3 : 5). Multiplied 3 by 5, the product 15 is the third, which is the inverse of the triad of the given tone 5 [89].[34]

Tartini's principle can be stated in a more compact form as follows: given any interval of the form $(m : n)$, the third sound is associated with the ratio $(1 : nm)$, i.e. with the least common multiple of m and n (note that this is only true if the ratio $m : n$ is in its radical form, i.e. if m and n are co-prime) [2]. The principle applies to all intervals, i.e., according to Tartini, the third sound always exists, even though we cannot hear it.

The modern reader can reformulate Tartini's principle by using frequencies to measure intervals, and now it is $m > n$, in the form valid also for non-co-prime numbers:

> [*GCF principle*]. Given a musical interval defined by the two frequencies f_1 and f_2, the third sound is given by the frequency f that is the greatest common factor of f_1 and f_2.

To obey his principle, Tartini modified the positions of the third sound as given in the *Trattato*, moving it down one octave. Applying the GCF principle to the examples in Fig. 6.7, for the major third C_5–E_5, the third sound should be sought two octaves below C_5, in C_3, and not in C_4 as in the figure. And in the case of the fifth C_5–G_5, the third sound is in C_4 and not in C_5.

An interesting result found by Tartini, first in the *Trattato* and then in the *De' principi*, is that if we consider bichords tuned to the frequencies of the ratios: (2 : 1), (3 : 2), (4 : 3), (5 : 4), (6 : 5), as shown in Fig. 6.8, where two consecutive notes are imagined as a bichord. The third sound is always the same, corresponding to the fundamental frequency 1. In this way, Tartini somehow reversed Rameau's result. The latter stated that a body vibrates not only with the fundamental frequency, but also with its higher harmonic sounds (from unity to diversity), Tartini stated that the various higher harmonic sounds are naturally related to the fundamental frequency (from diversity to unity), which is heard even when it is not played.

Tartini is often accused of being a poor mathematician; he himself confessed to knowing only geometry and not algebra (it is not clear what he meant by algebra): "I do not know algebra" [87].[35] Reading his writings actually reveals an author who has some confidence with mathematics. What he lacks is the rigor of the professional mathematician who takes nothing for granted. Instead, he bends mathematics to his needs, mixing Platonic mysticism with Archimedean rigor. But he was not the only one; almost all the greats did so, including Galileo.

[34] p. 5. Tartini, as usual for him, measured ratios $(m : n)$ of musical intervals with $m < n$.
[35] p. 27.

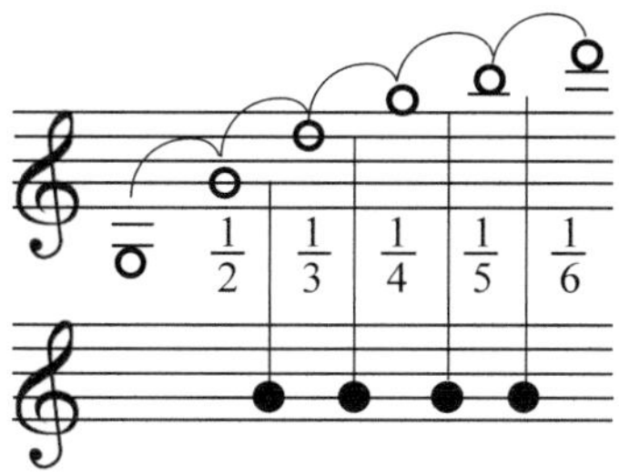

Fig. 6.8 Third tones of the first six harmonics; from [87], p. 18

The explanation of the phenomenon of the third sound is still partly controversial; a minority thinks that a neurological or psychological component is important, the majority points to an essentially mechanical explanation [5]. Section 6.3.3 will discuss the possible explanations in some detail.

The third sound can function as a harmonic bass and replace a physical bass in musical composition and performance.Tartini exploited this in some of his violin works [51].[36] A perhaps more interesting use is that made by Pietro Mascagni (1862–1945) in 1898 for his opera *Iris*. Here the third sound was used to allow the violin to play lower notes than it could naturally produce ($G_3 = 196$ Hz) [5].[37] Tartini's third sound is also used by organ builders to produce grave tones while avoiding too long pipes.

In a work of his maturity (1767), of a distinctly philosophical nature, the manuscript of which has recently been found, edited and published [90, 91], Tartini expressed his epistemological position on music, which is also evident in his musical treatises [36, 63]. In essence, he argued, in line with the many proponents of a *prisca sapientia*, that there is a form of knowledge superior to physical mathematics, equally grounded in experience, and that music is the discipline best suited to reveal it. This science was "once apparent to the ancient Egyptians, from whom Pythagoras and Plato learned it, but kept hidden with inviolable secrecy" [90].[38]

According to Tartini, the best way to achieve true science is to refer to music.

> Reason alone forms science: reason combined with music forms science and something more, and it is clear that Plato here does not mean material and ordinary music: but the intelligence of universal harmony.
>
> [...]
>
> There is a considerable difference between the science of harmony and any other demonstrative science. All the demonstrative sciences of quantity are purely for the material discovery of quantity and formal relations, or if there are reasons between the terms of the discovered material quantity: nothing more. The relations discovered by the other sciences are the material subject of the harmonic science.
>
> [...]

---

[36] p. 92.

[37] p. 35.

[38] p. 1.

> In retrospect, this principal science agrees with its subordinate ministries in all the materiality of the demonstrative genus; but this is its surface, in respect of which it may well happen (and does happen) that there may be possessors of the common demonstrative sciences who are incomparably better than a possessor of the harmonious science at the material discovery of what is in action [...]. A watchmaker who collects each integral part of a clock from several smiths, each of whom works his part excellently, and knows that it is part of a clock, but does not then know how to construct it with all its integral parts, is an adequate picture of the thing [90].[39]

6.1.1.3 Epistemology of Music in d'Alembert

When d'Alembert read Rameau's *Memoire* in 1749, he had little theoretical knowledge of music. In any case he was shortly able to understand and greatly appreciated it. However, even though d'Alembert praised Rameau's theory he did not like his prose, and for this reason he felt himself impelled to write a more readable treatise than that proposed by Rameau himself, which originated the *Élémens de musique* of 1752, written anonymously as a form of respect [16], which was followed in 1762 by a second edition, this time with the name d'Alembert, in the cover [19]. The new edition also incorporated some significant changes in the text, consisting mostly of critical commentary on Rameau's theories. In some cases, d'Alembert supplied alternatives to Rameau's explanations of musical phenomena.

The *Élémens de musique* was not simply a didactical work; it was rather an occasion for spreading d'Alembert's rational epistemology. "By reformulating Rameau's brilliant but ineptly articulated theory into a rigorous writing d'Alembert, was able to provide both a vindication as well as an advertisement for his own peculiar scientific epistemology" [13].[40] A treatise on music was a perfect means to illustrate the merits of his professed empirical-rational methodology of physical sciences because more simple of his *Traité de dynamique*, for instance [17].

D'Alembert in the *Élémens de musique* made recourse to more than one experience or empirical principle. Besides that referred to by Rameau (Experience I in the following), which remained the fundamental one, he considered two more experiences in the edition of 1752 and only one in the edition of 1762. In the following, for the sake of space, reference will be given to the 1752 edition only.

> Experience I. If one excites a sonorous body, beyond the fundamental sound and the octave, two others sounds are heard, one of which is the perfect twelfth, that is the octave of the fifth, the other the major seventeenth, that is the double octave of the major third [16].[41]

D'Alembert, added that this experience is particularly evident by making the larger strings of the cello vibrate. The fundamental sound is called generator, the others, the octave included, its harmonics.

---

[39] p. 4.

[40] p. 419.

[41] p. 12.

> Experience II. If two other bodies are tuned with the sonorous body, one of which is resonant at the twelfth and the other at the seventeenth, these two bodies will vibrate in their totality [16].[42]

This is an experiment of acoustics, for which two resonators, one tuned to a frequency three times that of the sonorous body and another tuned to a frequency five times are activated by the vibrations of the sonorous body.

> Experience III. Anybody notices the similarity between a sound and its octave. These two sounds are perceived by the ear as almost equal [16].[43]

In the 1762 edition of the *Élémens de musique*, d'Alembert eliminated the second experiment, perhaps because he considered it as contained in the first.

Basically what Experience I says is that together with the generating or fundamental frequency f_0, one hears also notes corresponding to frequencies $3f_0$ and $5f_0$. The result could not surprise d'Alembert, who knew dynamics of vibrations very well, only that for him the experience in question was not of dynamics; that is, it did not concern the knowledge of the frequencies that a body can potentially emit, but only those that the ear perceives in the ordinary conditions in which a musician operates.

D'Alembert, exposed the general ideas of his epistemology of music in the *Discourse préliminaire* to the second edition of the *Élémens de musique* and in the *Essai sur les elemens de la philosophie* [18], in a period in which his conceptions of philosophy of science were mature. To start with he declared music a physical mathematical science, like hydraulics and astronomy and unlike mechanics which for him was purely rational like mathematics. Thus for music too, as for the other physical mathematical sciences, one should not seek a striking evidence, which is the characteristic of the works of mathematics alone. According to d'Alembert, there will always be into the theory of musical phenomena "a kind of metaphysics, which these phenomena implicitly suppose and which carries its natural obscurity there" [19].[44] It is for this reason that one should not expect to find in this matter what is called demonstration; it is enough to have established on simple foundation the most well-known rules of music: "At the same time we doubt that it is possible to bring a greater light to these matters" [19].[45]

Although most of the phenomena of musical art seem to be deduced in a simple and easy way from the resonance of sonorous bodies, continued d'Alembert, one should perhaps not yet hasten to affirm that this resonance is demonstratively the unique principle of harmony. At the same time, it would not be less injurious to reject this principle, because a few phenomena do not seem to be deduced from it as easily as others. From this it is can only be concluded, either we will perhaps be able, by new research, to reduce these phenomena to a single known principle; or that there should be some other unknown principle, more general than that of the

[42] pp. 12–13.

[43] p. 16.

[44] p. XIII.

[45] p. XIV.

resonance of the sonorous body, of which the latter is only a branch. Or finally, that we should perhaps not try to reduce all musical science to one and the same principle.

There are indeed physical mathematical sciences which demand only one experience, or one principle; there are others which necessarily suppose several, the combination of which is necessary to form an exact and complete system and music belongs to this last category of sciences [19].[46] This is in contradiction with what claimed by Rameau, who asked for one principle only.

D'Alembert, warned philosophers not to waste their time looking for physical explanations of musical phenomena, explanations always vague and insufficient; still less must they consume themselves in efforts to rise in a region more distant from their gaze, and to lose themselves in a labyrinth of metaphysical speculations on the causes of the pleasure that harmony makes us experience. It is not a question here, of the physical principle of the resonance of sonorous bodies, which at the moment are unknown and which perhaps we will seek for a long time in vain; it is even less a question of the metaphysical principle of the feeling of harmony, a principle even less known, and which according to all appearances will always remain covered with clouds. It is only a question of showing how it is possible to deduce from a single experience the main laws of harmony, which musicians have found only by trial and error [19].[47]

D'Alembert, ended his *Discours préliminaire*, by warning against an abuse of mathematics in music, even though it is largely a mathematical science. "As a professional geometer, I believe I have some right to protest here (if I may be allowed to express myself so strongly) against this ridiculous abuse of geometry in music" [19].[48] And noticed that the perfect number used in the musical theory should be interpreted with prudence. The ratio of the octave as 1 to 2, that of the fifth as 2 to 3, that of the major third such as 4 to 5, etc. may not be the true relationships of nature; but only approximate ratios such as experience has made them known. Because experience gives them only approximatively. But luckily these ratios are sufficient, even though they are not exactly true, to give reason of phenomena that depend on the ratio of sounds [19].[49]

6.1.1.4 Euler's Suavitas

Although Euler's interest in music continued throughout his life, it is his treatise *Tentamen novae theoriae musicae* of 1731, that has received the greater attention from historians [25]. From the preface it is clear what logical status Euler ascribed to music; for him the judgment of its meaning and beauty does not lie in the arbitrariness of men or in their customs: "Already Pythagoras, who first laid the foundations of

---

[46] p. XVII.

[47] p. XXVII.

[48] p. XXXX.

[49] pp. XXX-XXXII.

music, recognized that the cause of the pleasure produced by consonances is inherent in the proportions which are the object of perception" [25].[50]

The *Tentamen novae theoriae musicae* consists of 14 chapters plus a preface and deals with many of the problems of contemporary music. It is based entirely on mathematics, with few references to the physics of sound, thus going against the trend of the mathematicians of the time who were developing acoustics (the year of its writing, 1731, is only shortly after Sauveur published his last memoir). To some extent, Euler's work can be said to still be in the footprint of traditional speculative music, not without numerological metaphysics, and is one of the latter of this type.

It is somewhat surprising how little attention Euler paid to the nature of sound, the wave aspect of which is not emphasized, reducing it to a sequence of pulses, as was typical in the 17th and part of the 18th century: "It has thus been made clear that the hearing of a simple sound is nothing more than the perception of several successive strokes for equal intervals of time, and that the difference between the low and the high lies in the frequency of these strokes" [25].[51] Euler will later return to the study of the nature of sound by emphasizing its wave-like nature in a landmark paper of 1748 [26].

Euler thought of music as the science of combining different sounds in such a way that they form a harmony pleasing to the ear. In light of this definition, it can be argued that the goal of music is to evoke positive sensations. This assertion is in line with the prevailing aesthetic principles of the time, although it differs from the position of Descartes, who undoubtedly influenced Euler. The latter, also considered the role of emotional involvement in the musical experience [21].[52]

For Euler, music theory required a dual foundation:

> One consists in the exact knowledge of sounds, which are the proper object of natural science; the second principle, on the other hand, must rather be demanded by metaphysics, since it is necessary to determine what things cause several sounds, whether simultaneous or successive, to please or displease the ear that perceives them [25].[53]

A musical composition is pleasing if we perceive an order in it, which must be contained in two aspects, the degree of acute and grave and the rhythm. Note that throughout his work Euler distinguished these two aspects of music, although he developed only the first.[54] Order is given by the existence of relations between sounds, that is, between their frequencies, which must be expressed between integers in a way that is somehow rational.

According to Euler, order in sounds is easy to perceive even when it is expressed by complex ratios:

> It is therefore much easier to distinguish the ratio of two tones when listening than when observing two lines with the same ratio. The ratio of tones to lines would be similar if we

[50] p. 3.

[51] pp. 9–10.

[52] p. 1.

[53] p. 11.

[54] The theory of rhythm is taken up again in [35].

> had only two recordings of each tone and tried to judge the ratio of their intervals. However, since there are many strokes [...] in a short time even in not very fast tones, it is much easier to know the ratio of the tones. For this reason, in music, complex proportions can still be used, which, if they were the same in lines, would be very difficult for the eye to recognize. [25].[55]

Music is divided into two parts, one theoretical and the other practical; the first is usually called harmonics and the other is properly called music; the two parts, however, should not be independent.

Chap. 4 introduces the concept of consonance, which has a different meaning from the traditional one:

> Several simple sounds that sound together form a composite sound, which we will call consonance [consonantia]. Others, however, interpret the term consonance in a narrower sense, as meaning only a compound sound that is pleasing to the ear and has much sweetness in it; and they distinguish consonance from dissonance, which for them is a compound sound that has little or no sweetness. And as it is in part difficult to define the limits of consonance and dissonance [...] we shall ascribe the name of consonance to all sounds composed of several simple sounds which sound together [24].[56]

Thus, Euler used the term consonance with a meaning similar to interval, implying that consonances are defined by ratios of integers, so that, according to his theory of sound, after a certain period of time the impulses of different sounds coincide with each other. Note that the definition is not limited to just two tones.

Consonances, however, are not all the same; they differ in their degree of sweetness, *suavitas*, [gradus suavitas]. Euler's way of calculating this suavitas is neither easy to follow nor convincing [1]. In what follows, we will avoid commenting on the explanation for the sake of space, in order to present the algorithm introduced for calculating suavitas.

First, considering the consonances of two sounds in the ratio $m : n$, reduced to the radical form in which m and n are prime to each other, we define the *consonance exponent* as the least common multiple (LCM) p between m and n, which, being prime to each other, is given by $p = mn$. The exponent p is considered in its k prime factors, $\{p\} = \{p_1, p_2, p_3, \ldots, p_k\}$, where for simplicity we consider p_i different even if they are equal, so for example 2^3 is treated as $\{222\}$. The suavitas g is defined by the relation

$$g = \sum_1^k p_i - k + 1$$

For example, in the consonance expressed by the ratio $(8 : 15)$, with the exponent $p = 120$ and $\{p\} = \{22235\}$, $k = 5$, we have $g = 14 - 5 + 1 = 10$. Note that different consonances can correspond to the same exponent, all with the same degree of suavitas; for example, the exponent 6 can correspond to the two ratios $3 : 2$ and $6 : 1$.

---

[55] Chapter 2, Sect. 19, p. 35.

[56] Chapter 4, Sect. 1, p. 56.

Table 6.1 Order of suavitas for different intervals; from [24], Chap. 4, Sect. 11, p. 63

Suavitas	Ratio	Interval	Exponent
II	2 : 1	Octave	2
IV	3 : 2	Fifth	6
V	4 : 3	Fourth	12
VII	5 : 3	Sixth M	15
VII	5 : 4	Third M	20
VIII	6 : 5	Third m	30
VIII	8 : 5	Sixth M	40
VIII	9 : 8	Tone M	72
X	10 : 9	Tone M	90
X	15 : 8	Seventh M	120

Using his formula, which gives an order similar to that proposed by Benedetti, based essentially on the exponent p, Euler prepared Table 6.1 of suavitas for the most common consonances.

Notice in the VIII degree the coexistence of the major tone, (8 : 9), which is usually classified as dissonant, and the minor third and sixth, which are classified as consonant.

The degree of suavitas in the case of more than two notes is obtained in a similar way. Given the sequence of ratios $a : b : c$, considered in radical form, i.e. with all coefficients divided by their greatest common factor, we determine the exponent by calculating the least common multiple p of a, b, c. Then we determine the degree of suavitas of p as above.

Also in Chap. 4, Euler introduced two interesting definitions, that of *complete and incomplete consonance* and that of *pure consonance*. A complete consonance is such if it is impossible to add another sound to the sounds forming it without changing their degree of suavitas. Otherwise the consonance is incomplete. For example, the consonance formed by the sounds 1 : 2 : 3 : 6 is complete with exponent 6. In fact, by adding a sound, say 4, the exponent changes to 12 and thus the degree of suavitas changes. In contrast, the consonance 1 : 2 : 3, with exponent still 6, is incomplete because 6 can be added without changing the exponent.

Euler believed that the difference between complete and incomplete consonances was important from the point of view of the perception of their sound:

> The difference that the ear recognizes between complete and incomplete consonances is that the former are understood much more clearly, while the latter are understood less clearly. In fact, when all the sounds are perceived at the same time, the relationships between the individual sounds are more clearly perceived than when the exponent is perceived by a few sounds. Thus, the consonance 1 : 2 : 3 : 6 allows us to perceive its exponent, which is 6, much more clearly than the sounds 1 : 6 alone [24].[57]

[57] Chapter 4, Sect. 30, p. 69.

It must be said, Euler specified, that although in truth a complete consonance offers itself much more clearly to the ear than an incomplete one, nevertheless, unless they are very simple, complete consonances should not be used. In the first place, unless musical instruments are very accurately tuned, such a large number of notes will dazzle the ear with a confused noise rather than a clear harmony. Moreover, some sounds, either because they are too grave or too acute, cannot even be perceived [24].[58]

A consonance is said to be pure if no sound can be removed without reaching a lower degree of suavitas. All consonants of two sounds are pure, simply because by removing one sound the consonance no longer exists. For example, the consonance 3 : 4 : 5 with exponent 60 is pure because removing any of the three numbers decreases the exponent.

Quite interesting is the measure proposed for intervals, i.e. the distances between two sounds; it is defined by the logarithm of the ratio of the numbers expressing the interval, so for example the interval defined by the ratio $a : b$ is given by $\log a - \log b$; thus the interval $a : a$ has measure zero.

In Chap. 5, Euler applied his theory of suavitas to the sequence of several consonances; and thus to an entire musical work. This is a step that no one had taken before. In fact, Euler considered sequences of multiple consonances as a single sequence, with certain caveats that we will not discuss here for reasons of space.

In Chap. 8, he introduced the concept of genus, with some assonance with the Greek meaning. An instrument can produce only a limited number of notes, all of which form a genus and correspond to a certain exponent. Otherwise, a genus is a series of notes that are periodically distributed over several octaves.

Each genus is characterized by some consonances, which in turn have their own exponents. If A is such an exponent, Euler generalized it as 2^m A to cover more than one octave, and explored the division of the octave for different kinds of A. To simplify, he left in A only odd numbers, since even numbers can be supplied by the coefficients 2^m. Euler assumed that A can only contain multiples of 3 and 5 by quoting Leibniz: "Thus Leibniz' statement is fully confirmed: in music it is not usual to count beyond five" [24].[59] Euler nevertheless developed a genus in which the factor 7 intervenes, but by way of pure scientific hypothesis.

As an example, we report the case of the exponent $2^m 3^3 5$, which is the exponent of the series of ratios: 128 : 135 : 144 : 160 : 180 : 192 : 216 : 240 : 256, which, according to Euler, agrees with the Ptolemaic diatonic system except for the number 135. In fact, if we do not consider 135, the previous series of numbers give ratios of the kind (9 : 8); (10 : 9); (16 : 15), which are those of the diatonic system. The number 135 introduces the ratio 135 : 128, known as the major chroma [6].[60]

An interesting application of Euler's measure of suavitas is found in *Conjecture sur la raison de quelques dissonances généralement reçues dans la musique* of 1766

[58] Chapter 4, Sect. 33, p. 72.

[59] Chapter 10, Sect. 19, p. 163.

[60] p. 94.

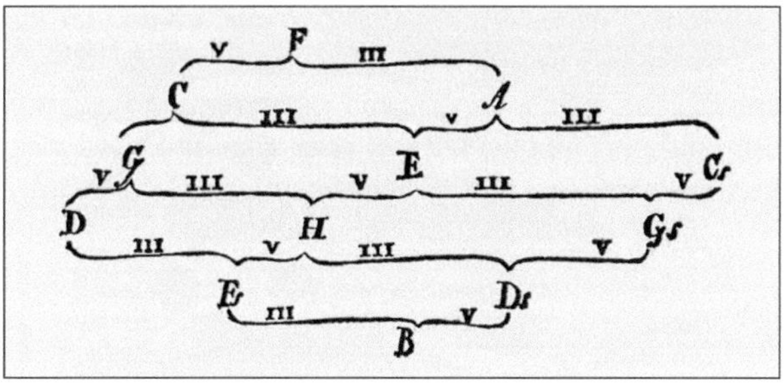

Fig. 6.9 First version of the tonnetz; from [25], Chap. 9, Sect. 13, p. 147

in the discussion of the dominant 7th chord, G B D F [27].[61] It corresponds to the numbers 36, 45, 54, 64, whose exponent (that is, its LCM) is $8640 = 2^6 3^3 5$. Without F, the chord is represented by the numbers 36, 45, 54, and considering the co-prime numbers, $4 : 5 : 6$, its exponent is 60. We can see that the addition of F spoils the beautiful harmony of this consonance. However, Euler added:

> Nevertheless, in the judgment of the ear, this dissonance is not very unpleasant, and one can use it in music with success; it even seems to give the musical composition a certain power, without which it would be too plain. Here, then, is a great paradox, where theory seems to contradict practice, about which I shall try to give an explanation [27].[62]

Euler, interestingly and mysteriously, recalled that the ear tolerates slight variations in the proportions of consonances: "I think so that when one hears the sound 36, 45, 54, 64, he imagines to hear those 36, 45, 54, 63, Consequently, nothing forbids to suppose that the ear *substitutes* the number 63 for the number 64, that is 4, 5, 6, 7, considering that the effect is absolutely the same" [27].[63] And the exponent 420 of the latter sequence is much smaller than 8640 of the former.

Of some interest for future development of speculative music is Fig. 6.9, known as the *tonnetz*, which shows in a simple way the relation of fifths, indicated by the Roman number V, and the major third, indicated by III. The tonal network is a conceptual grid diagram representing tonal space. Various visual representations can be used to show traditional harmonic relationships in European classical music [95].

6.2 Music Trend in the 19th Century

6.2.1 Progress in Acoustics

Modern acoustics is a complex field of study encompassing the investigation of the fundamental principles governing the nature and propagation of mechanical waves

---

[61] Euler wrote this chord as G H d f, where H is the German symbol for B [27], p. 166.

[62] p. 160. Translation into English in [1].

[63] p. 172

in a variety of material media. It is challenging to adequately summarize its recent advancements in a limited space. From the perspective of the layperson, the most apparent advancements are those pertaining to technology. These include the invention of the telephone, the recorder, and Wi-Fi systems, among others. However, the field of acoustical theory has also witnessed significant advancements, particularly with the advent of sophisticated recording devices and computational resources.

There are numerous sub-disciplines of acoustics, including:

1. Archaeoacoustics
2. Aeroacoustics
3. Signal processing
4. Bioacoustics
5. Electroacoustics
6. Noise reduction
7. Speech analysis
8. Musical acoustics

    8.1 Physical acoustics
    8.2 Physiological acoustics
    8.3 Psychological and neurological acoustics

As can be observed, acoustics as seen by Sauveur represents a branch of musical acoustics, which in turn constitutes a subfield of general acoustics. Musical acoustics can be conceptualized as comprising three distinct domains: physical acoustics, physiological and neurological acoustics, and psychological acoustics.

Physical acoustics is the oldest of the three branches and has its roots in the studies of the 18th and 19th centuries, especially those of Sauveur. The field of physiological acoustics emerged during the 19th century, while the disciplines of psychological and neurological acoustics are relatively recent developments, dating primarily from the 20th century. Their objective is to elucidate phenomena that remain inexplicable on the basis of purely material causes.

Subsequent to Sauveur's passing, acoustic studies continued without a unified objective. Extensive, though largely independent, studies were conducted on the nature and propagation of sound, the acoustics of musical instruments, the vibrations of bodies other than strings, and the interaction of different sounds on the functions of the ear. The chief proponents of this synthesis were John William Strutt (1848–1919), better known as Lord Rayleigh, and Hermann von Helmholtz (1821–1894), the former an Englishman and the latter a German.

The publication by the 'mathematician' Rayleigh of *The theory of sound* in 1877–78 [85], marked a significant turning point in the field of acoustics. The work is comprised of two distinct sections, which are presented in two separate volumes. The first section addresses the subject of mechanical vibrations in their various forms. As a scientist with a robust mathematical background, Rayleigh emphasized the formulation of overarching principles and devised effective techniques for addressing vibration-related issues. One notable approach he developed is the Rayleigh-Ritz method, which is utilized for identifying the natural frequencies and modal shapes of vibrating bodies. The second section of *The theory of sound* addresses the propagation of sound through diverse media, including the diffraction of sound waves around obstacles and the scattering of sound when traversing media with suspended particles. The second enlarged edition of Rayleigh's treatise, published in 1894 and 1896, was perceived as a comprehensive account of the field of acoustics as a branch of physics.

The modern reader views *The theory of sound* as the conclusion of classical acoustics and the advent of a new field of mathematical physics, namely, the dynamics of vibrations. Indeed, it would be reasonable for our reader to suggest that the title of *The theory of sound* should be changed to *Analytical dynamics of vibrations*, given that vibrations are increasingly being regarded as a purely mechanical phenomenon of interest in a number of fields, including the mechanics of structures, where the frequencies involved are often below or above the minimum audibility threshold of the ear. This progression from music to acoustics to dynamics serves to reinforce the significant impact of music on the advancement of contemporary science.

The publication by the 'musician' Helmholtz of the *Die Lehre von den Tonempfindungen als physiologische Grundlage für die Theorie der Musik*, first edition 1863, fourth and last edition 1877 [40], had an even greater impact; it was considered by both scientists and musicians as the culmination of musical acoustic. Since its first edition in 1863, however, in addition to the appreciation of most, Helmholtz had received important criticism from his German colleagues [81].[64] This may have had its roots in the changed philosophical taste of the time, with the crisis of positivism and the attempt by the new protagonists in philosophy to minimize the role of science in all human activities, including music. But it also stemmed from the realization that many aspects of music, including the appreciation of consonances, could not be fully explained mechanistically. Helmholtz's attempts to rely on memory to justify some phenomena, such as the consonance of non-simultaneous sounds, while somehow introducing psychology, are unsatisfactory. Stumpf, beginning with his seminal *Tonpsychologie* of 1883–1890 [86], placed psychic acts in the center of music theory in their direct evidence.

To date, a consensus remains elusive regarding the explanation of consonances. The research conducted by Helmholtz and Stumpf on the physiological and psychological aspects of consonance continues to be regarded as providing significant insights, though not as conclusive as desired [22, 23].

6.2.2 Progress in Theory

The interest in music by non-performing musicians was not limited to mathematicians, who, as previously discussed, transformed a significant portion of speculative music into acoustics. Additionally, philosophers, historians, and enthusiasts demonstrated a keen interest in music. They undertook a fresh examination of historical

[64] Chapters 1.4, 2.5.

texts, employing new analytical techniques, and developed innovative aesthetic theories. Toward the end of the 19th century, scholars from a variety of academic backgrounds converged on a new field of study: *musicology*, the academic discipline devoted to the study of music. The field is subdivided into three branches: music history, systematic musicology, and ethnomusicology. The field of music history encompasses the study of musical traditions, the origins of works, and the biographies of composers from different historical periods. Ethnomusicologists employ the methods and theoretical frameworks of anthropology to gain insight into the motivations and practices underlying musical expression across diverse cultural contexts. Systematic musicology encompasses a range of disciplines, including musical analysis, music theory, aesthetics, pedagogy, elements of musical acoustics, organology, and the musical implications of physiology, psychology, neurology, sociology, philosophy, and computer science.

From our perspective, the aspect of musicology that holds greater interest is systematic musicology, particularly music theory. It is a field of study that describes the fundamental elements of music and includes the development and application of methods for composing and analyzing music. Some music theorists endeavor to elucidate the techniques employed by composers by establishing rules and models. Others direct their attention to the experience of listening to and performing music. Despite their disparate interests, music theorists are unified in their conviction that the processes of composing, performing, and listening to music can be elucidated through a meticulous and logical approach, as opposed to a notion of musical expression as inherently inexpressible, except in the form of musical sounds.

The field of music theory frequently addresses abstract musical elements, including tuning and tonal systems, scales, consonance, and dissonance. Additionally, there is a body of theory that addresses more practical aspects of music, including the creation and performance of music, orchestration, ornamentation, improvisation, and electronic sound production. Pratictioners and composers may engage with musical theory to gain insight into the structural relationships inherent to musical compositions. Similarly, theoreticians may utilize musical performance to inform their creative processes and achieve desired structural outcomes.

Hugo Riemann (1849–1919) is unanimously considered the founder of musicology. He was first exposed to music by his father. Following studies in law, philosophy, and history at the university of Berlin, he commenced his musical studies at the Leipzig conservatory and soon became the most prominent musicologist in Germany. Notwithstanding national and international awards, he was appointed to the position of full professor only in 1901. This may have been a fortuitous circumstance, as it allowed him the opportunity to focus on his studies.

Riemann lived in a time and nation in which the natural sciences enjoyed considerable prestige, and he was undoubtedly fascinated by them. His education, however, did not qualify him to become a scientist in the strict sense, even though he could understand and utilize scientific methodology in a comprehensive manner. Just as the scientist seeks to identify the fundamental principles that govern the natural world, an entity that is independent of the observer, so Riemann sought to identify the underlying laws that govern musical phenomena, which are also not considered to be a mere human invention, subject to change according to the prevailing aesthetic preferences of a given era.The assumption is that music, like nature, has an unchanging essence, and that there is only one true description of it.

For Riemann, music was not merely a matter of sound; rather, it was a phenomenon that engaged both the ear and the brain. Consequently, psychology and psychoacoustics played a pivotal role in the development of his theory, which sought to identify the intrinsic laws of music, particularly harmony, through a rational approach. Riemann was the author of numerous papers and books on a vast array of musical subjects, encompassing theory, analysis, and history. Among his writings we will cite only a few of his writings. These include the *MuskLexikon* of 1882 [74], a comprehensive dictionary of music and musicians, which saw five notable German editions, the *Geschichte der Musiktheorie* (1898), a comprehensive and meticulously documented history of music in Europe, and the *Vereinfachte Harmonielehre* (1893), translation into English in 1896 as *Harmony simplified* [75].

Riemann's positions, although fundamental to modern music theory, will not be discussed here for reasons of space. Instead, in the following sections we will refer with some detail to Helmholtz's contribution.

6.3 Helmholtz's Physiological Acoustics

Over a period of 50 years, Hermann Helmholtz (1821–1894) made significant contributions to the advancement of various scientific disciplines. He served as a teacher, researcher, and science mediator in several German cities, including Königsberg, Bonn, Heidelberg, and Berlin. During this time, he made notable progress in the fields of medicine, physiological optics, physiological acoustics, thermodynamics, hydrodynamics, electrodynamics, and mathematics. His reflections on science and its philosophical dimensions represent a significant contribution to the epistemological discourse on the natural sciences [44].[65]

Before beginning his studies in acoustics, he had published the *Handbuch der physiologischen Optik* in three volumes in 1856, 1860, 1867 [43]. His first relevant work on sound, which complemented research in optics, was on combination tones, the *Ueber Combinationstöne* of 1756 [41]; his masterpiece is *Die Lehre von den Tonempfindungen als physiologische Grundlage für die Theorie der Musik* (hereinafter *Die Lehre von den Tonempfindungen*), first edition 1863, fourth and last edition 1877 [40].

Parallel to his scientific life, Helmholtz cultivated an interest in music from his early years spent in the family home. The interest in music continued in his later years in Bonn, Heidelberg, and Berlin. In Helmholtz's writings on musical acoustics, the piano and harmonium were used extensively as experimental tools in the study of sound sensations as a basis for music theory. Helmholtz explored the mechanisms of these instruments and the vibrational characteristics of their strings and vibrating

[65] p. 3.

reeds in terms of the production, quality, intensity, and decay of sound. His studies of consonance, dissonance, partials, and combination tones in relation to the purity of sound led to the construction of fixed-tone keyboards designed to approximate just intonation.

When Helmholtz wrote about keyboards as experimental research tools, he had been playing the piano for about twenty years. Letters to his parents and friends reveal that by the time he began university studies, he was well versed in traditional classical keyboards and other instrumental music, sacred music, and the music of contemporaries such as Johannes Brahms, Giacomo Meyerbeer, Gioachino Rossini, Felix Mendelssohn-Bartholdy, Robert Schumann, and Richard Wagner.

Helmholtz's masterpiece, *Die Lehre von den Tonempfindungen*, is divided into three parts. The first part (Chaps. 1–6) contains general considerations about *tones*[66] and their perception, the second part (Chaps. 7–12) deals with the interaction of different tones, especially the phenomena of beats, combination tones, and the third part (Chaps. 13–19) deals with topics of a more strictly musical nature in the modern sense: the construction of scales, and also the aesthetic dimensions of musical language.

In the following, we will refer to the English translation of the fourth edition, *On the sensation of tone as a physiological basis for the theory of music*, better known simply as *On the sensation of tones* of 1885, edited by the eminent mathematician and acoustics scholar Alexander John Ellis (1814–1890). This for two main reasons; it is the most diffused and appreciated edition of Helmholtz's treatise and is supplemented by very rich notes by the editor.

In Chap. 1, *On the sensation of sound in general*, Helmholtz reported on a system for naming notes that echoed the traditional one, but made it more precise by also specifying pitches. Considering the C major scale, the notes of the reference octave (the 4 foot octave on the organ) are labeled c, d, e, f, g, a, b. This corresponds to the notes labeled C_3, D_3, …B_3 in the international notation system we have adopted. Notes one octave higher are marked with one dash, such as c′, d′, e′, f′, g′, a′, b′; those two octaves higher with two dashes. Notes one octave lower than the reference are marked C, D, E, F, G, A, B. Those one octave lower are marked with a subscript dash, such as C_I, D_I, E_I, F_I, G_I, A_I, B_I; those two octaves lower with two dashes. Figure 6.10, summarizes the situation.

The following Table 6.2 lists the pitches of all notes in just intonation, as approved by the German Association of natural philosophers at a meeting in Stuttgart in 1834. Note the value a′ = A_4 = 440 Hz, known as the Stuttgart pitch, which has become the modern standard [39].[67] After Helmholtz wrote his paper, the Académie des sciences of Paris proposed the frequency of 435 Hz for the same note.

In Chap. 2, *On the composition of vibrations*, what Helmholtz calls Ohm's law is introduced:

---

[66] Note that Helmholtz used the word *tone* very often in his treatise, generally to indicate simple tones, i.e. sinusoidal waveforms.

[67] p. 349.

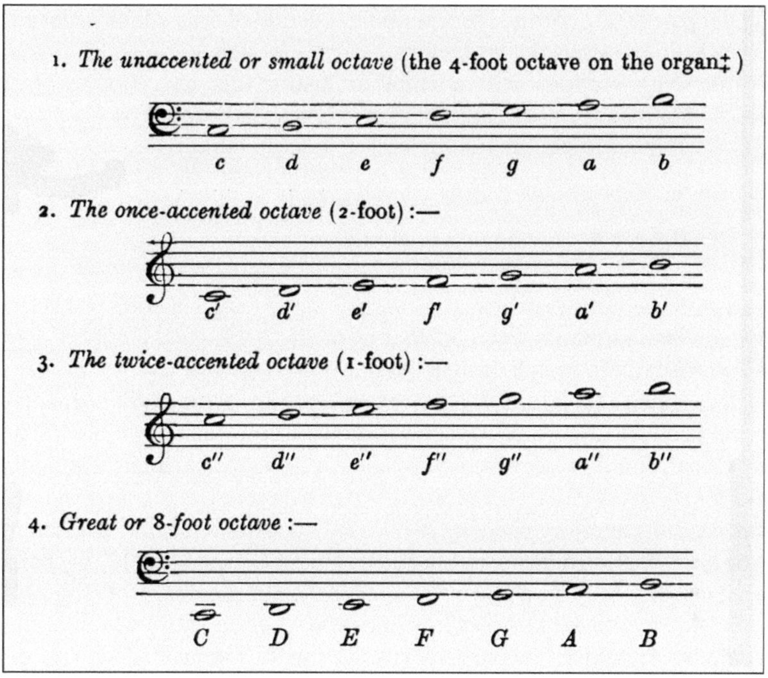

Fig. 6.10 Notation of the notes in different scales. Redrawn from [42], pp. 15–16

Table 6.2 Pitches of the different notes in just intonation. Redrawn from [42], p. 17

Notes	Contra Octave $C_{,}$ to $B_{,}$ 16 foot	Great Octave C to B 8 foot	Unaccented Octave c to b 4 foot	Once-accented Octave c' to b' 2 foot	Twice-accented Octave c'' to b'' 1 foot	Thrice-accented Octave c''' to b''' $\frac{1}{2}$ foot	Four-times accented Octave c'''' to b'''' $\frac{1}{4}$ foot
C	33	66	132	264	528	1056	2112
D	37·125	74·25	148·5	297	594	1188	2376
E	41·25	82·5	165	330	660	1320	2640
F	44	88	176	352	704	1408	2816
G	49·5	99	198	396	792	1584	3168
A	55	110	220	440	880	1760	3520
B	61·875	123·75	247·5	495	990	1980	3960

> Every motion of the air, then, which corresponds to a composite mass of musical tones, is, according to Ohm's law, capable of being analyzed into a sum of simple pendular vibrations, and to each such single simple vibration corresponds a simple tone, sensible to the ear, and having a pitch determined by the periodic time of the corresponding motion of the air [42].[68]

The simple pendular vibrations are what we call sinusoidal waveforms, or harmonic vibrations. The meaning of Ohm's law is that what we consider to be a simple tone,

[68] p. 33.

i.e. an upper partial, is described by a periodic function of time which is not arbitrary, but exactly a simple sinusoidal waveform.

The following sections will attempt to justify this law. Before examining Helmholtz's approach, we will present Ohm's law as introduced by his proponent.

6.3.1 Ohm's Law

In the early 19th century, there was considerable agreement that notes were composed of a fundamental harmonic and a number of higher harmonic sounds or overtones, or *upper partials* as Helmholtz called them .

This description was not easy to accept, however, because it did not explain how a note, the product of several superimposed harmonic sounds producing a single acoustic signal, could be decomposed by the ear into its elementary harmonic components.

In any case, the auditory phenomenon was not questioned. Until 1819, the year in which the siren was introduced by Charles Cagniard De La Tour (1777–1859),[69] the musical notes, on which the acoustic theories of the 18th century were founded, were those produced by stringed or wind instruments. After the introduction of the siren, notes could be reproduced more easily and accurately. Sirens made it easy to obtain sounds of desired frequency and intensity, replacing similar instruments developed by Hooke, Huygens, and Stancari, (see Sect. 4.14).

In 1843, Georg Simon Ohm (1789–1854) announced a fundamental law of auditory perception, published in the *Annalen der Physik* under the title *Ueber die Definition des Tones, nebst daran geknüpfter Theorie der Sirene und ähnlicher tonbildender Vorrichtungen*. In this paper he analyzed the operation of the siren and came to the conclusion:

> If we seek out the common elements in these investigations as well as of those of Savart on the tones of lowest and highest pitch that are audible, as well as those of Cagniard de Latour concerning the use of the siren, we find that the production of tones is brought about only by the regular repetition of any type of impulse affecting the hearing mechanism. In all cases the same dependence of the pitch of the tone on the number of impulses per second is found to prevail [58].[70]

Ohm tried to explain this perceptual phenomenon by assuming that the ear perceives as sound only vibrations of elementary (sinusoidal) type of assigned frequency. For this purpose he called for the use the Fourier series. "As a means of judging whether or not a given impulse contains the form $a \sin 2\pi(mt + p)$ as a real component, I use the

[69] The siren consists of a disk with evenly spaced holes through which air is blown as the disk rotates. It provides, by direct counting, the frequency (pitch) of the sound emitted. The instrument was perfected in stages, taking shape in Helmholtz's experiments in the 1850s and 1860s. It dominated the acoustic scene for a century. The polyphonic siren has several rings of holes arranged concentrically on the bowl, making it possible to produce more than one note whose pitch can be varied in a known ratio.

[70] pp. 513–514. Translation into English in [52].

theorem of Fourier, which has become famous for its many important applications" [58].[71]

It was not easy to develop the sound of the siren in Fourier series with the tools that Ohm had at his disposal (for example, he did not have a computer). He had to make some simplifying assumptions about the nature of the impulse emitted by the siren. Once he found the coefficients of the Fourier series—only a few are needed, since those corresponding to higher order harmonic sounds are negligible—he was able to verify the 'true' characterization of the sounds.

Ohm's paper was essentially ignored by the acoustics community, with the sole exception of August—Ludwig Friedrich Wilhelm—Seebeck (1805–1849), whose experiments on sirens formed the basis of Ohm's paper. This led to an interesting controversy which is well documented in [47, 94]. The controversy was only recognized as such in retrospect, and only because it was extensively reported by Helmholtz in his *Die Lehre von den Tonempfindungen*.

Seebeck admitted that the Fourier analysis of the siren could yield the series of tones that the ear actually heard, but he claimed that this analysis failed to correctly predict the intensities of such tones. In particular, Fourier analysis produced intensities of the upper partials that were too high compared to what the ear actually heard. Seebeck showed that Ohm had made a serious mathematical error in his work and performed a new Fourier analysis of the siren signal, free of any assumptions about the nature of the air impulses. Nevertheless, he concluded that, even under the most general assumptions, the Fourier analysis could not give the correct intensities of the various harmonic sounds as perceived by the ear, and that the fundamental harmonic was much higher than the Fourier analysis gave. This would falsify Ohm's law.

Ohm rejected Seebeck's objections on the assumption that the ear was the object of an auditory illusion:

> But before I resort to such assumptions, which incessantly murmur their ghostly I am and I am not painfully into my ear, I would rather set the egg on end, bold as Columbus, and claim that the contradiction which Seebeck perceives results from an auditory illusion (Gehorstauschung) to which our ear is susceptible, just as our eye is subject to optical illusions [...]. I assume specifically that our ear involuntarily regards the primary tone as stronger than it really is and its overtones (Beitone) as weaker than they really are [59].[72]

With this reply, Ohm withdrew from the controversy. Seebeck, however, did not and continued to publish on the subject until 1949, before his death. His research led him to suggest that the relative strengths of the upper partials determine timbre or color of sound, a result usually attributed to Helmholtz, who certainly proved and popularized the hypothesis. But the idea itself appears repeatedly in Seebeck's work, and there seems little doubt that Helmholtz got the idea from this source [94].[73]

---

[71] p. 519.

[72] p. 15. Translation into English in [94].

[73] p. 10.

6.3.1.1 Proof by Helmholtz

The starting point of Helmholtz's argumentation is represented by the introduction of the Fourier theorem [32]. Given Helmholtz's purpose to reduce analysis to the minimum, his statement of the theorem is simple:

> Any given *regular* periodic form of vibration can *always* be produced by the addition of simple vibrations, having pitch numbers which are once, twice, thrice, four times, as great as the pitch numbers of the given motion [42].[74]

To avoid complications and to make the exposition clearer, we assume that our reader is a bit more mathematically versed than Helmholtz's reader was presumably, and explain the meaning of the Fourier theorem with formulas. If $f(t)$ is the periodic time history of the vibration, T_0 is its period, and $f_0 = 1/T_0$ is its frequency, then the well-known relationship holds:

$$f(t) = \sum_{1}^{\infty} a_i \sin 2\pi \, (f_i \; t + \phi_i) \qquad\qquad f_i = i f_0$$

where $a_i \sin 2\pi(f_i + \phi_i)$ expresses simple oscillations (wave forms) with amplitude a_i, frequency f_i and phase ϕ_i.

In Chap. 3, *Analysis of musical tones by sympathetic resonance*, Helmholtz said, on the basis of a fairly well-established theory of differential calculus, that in order to determine whether a given periodic input would produce significant vibrations in a simple oscillator, we usually have to do complex calculations for the prediction of the result. Theoretical mechanics, however, would prescribe the following process to be pursued experimentally: consider the motion of a support (he supposed a hand) of a simple linear oscillator of natural frequency f_0; the oscillations of this simple oscillator are amplified, with respect to those of the support, if the Fourier analysis of the motion of the support reveals a simple (not negligible) waveform of frequency f_0. To conclude, "Hence the analysis of the motion of the hand into pendular swings has a real meaning in nature, producing determinate mechanical effects, and for the present purpose no other analysis of the motion of the hand into any other partial motions can be substituted for it"' [42].[75]

This circumstance shows that the Fourier series is not only an elegant mathematical tool, to be compared with other methods of series development, but has it is also a precise physical tool: the mathematical decomposition of a generic periodic function into Fourier series corresponds to a physical decomposition into its elementary waveforms.

Helmholtz imagined the use of a series of elementary oscillators attached to a rigid base that moves horizontally according to the law of the periodic sound signal to be analyzed. Each elementary oscillator begins to vibrate at its own frequency f_i and with an amplitude proportional to the coefficient a_i of the Fourier series. This

[74] p. 34.
[75] p. 38.

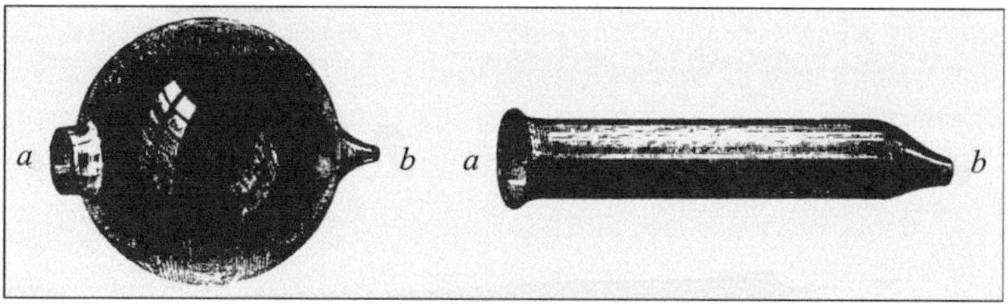

Fig. 6.11 Various shapes of resonators. Redrawn from [42], p. 43

makes the decomposition of a sonorous signal into its elementary tones visible. The simple oscillator system is in fact a mechanical spectroscope, performing a function analogous to that of the crystal prism in breaking down white light into its elementary components.

> We must therefore not hold it to be an illusion of the ear, or to be mere imagination, when in the musical tone of a single note emanating from a musical instrument, we distinguish many partial tones [...]. If we admitted this, we should have also to look upon the colours of the spectrum which are separated from white light, as a mere illusion of the eye [42].[76]

If it could be shown that there are structures in the human ear that can be compared to elementary oscillators, then Ohm's law would be proven without the need for any acoustic experiments. In Ohm's time, such a demonstration was not possible; what could be done was, on the one hand, to experimentally prove the plausibility of the assumption that the ear behaves like a mechanical spectroscope and, on the other hand, to identify which parts of the ear could possibly perform a Fourier decomposition of sound signals.

6.3.2 *Resonators and Overtones*

The first step to prove the plausibility that the ear behaves like a mechanical spectroscope requires the design and construction of acoustic devices capable of amplifying the intensities of the upper partials of notes, known as Helmholtz resonators. These are spherical bulbs or hollow cylinders of glass or bronze, as shown in Fig. 6.11, provided with two openings. The opening *b* is funnel-shaped and adapted for insertion into the ear. " I usually coat with melted sealing wax, and when the wax has cooled down enough not to hurt the finger on being touched, but is still soft, I press the opening into the entrance of my ear. The sealing wax thus moulds itself to the shape of the inner surface of this opening, and when I subsequently use the resonator, it fits easily and is air-tight" [42].[77]

[76] p. 48.
[77] p. 43.

The opening a is where the sound enters the resonator. The resonator has a well-defined natural frequency (with unimportant overtones), which essentially depends on the volume of air in it and the diameter of the opening a. Helmholtz, without any justification, gave the following formula for spherical glass resonators

$$n = aa\sqrt{\frac{2r}{8\pi^3 R^3}}$$

where aa is the speed of sound in air, r is the radius of the aperture a, R is that of the cavity, and n is the frequency in Hz if the time is in seconds. Resonators can be quite bulky; for example, for a resonance to the note g (198 Hz) the diameter of the sphere is about 154 mm; for g' (396 Hz) the diameter is reduced to 79 mm.

Resonators make it possible to detect the absence or presence of certain frequencies in the acoustic signal they receive, especially the presence of overtones. An untrained ear cannot hear the presence of overtones, even the most easily perceived ones such as the third or the fifth. Nor can the trained ear easily detect even harmonic sounds such as the first and second octaves.

For example, if the fundamental sound of a compound tone is c, with a resonator of frequency g′ you will hear the harmonic of order 3 (the fifth of the upper octave, or the twelfth); with a resonator of frequency e″ you will hear the harmonic of order 5 (the third of the second octave, or the 17th); with a resonator of frequency c′ you will hear the octave of c. The use of the resonator allows a first verification of Ohm's law. Although the information about the intensities of the overtones is imprecise, it is possible to verify the presence or absence of certain upper partials.

In Chap. 4, *On the analysis of musical tones by the ear*, Helmholtz subjected Ohm's law to direct verification by the unaided ear in order to better emphasize the mutual relationship of the amplitudes of the upper partials. For example, to hear the upper partials g' in the note c produced by a piano string, he suggested the following approach:

> First gently strike on a piano the note g', as marked above, and after letting the digital rise so as to damp the string, strike the note c, of which g' is the third partial, with great force, and keep your attention directed to the pitch of the g' which you had just heard, and you will hear it again in the compound tone of c. Similarly, first stroke the fifth partial e'' gently, and then c strongly [42].[78]

The upper partials are often clearer as the sound fades because they lose intensity more slowly than the fundamental.

Another way the unaided ear can hear overtones is through the use of resonators. When the resonator corresponding to any upper partial of a compound note c is applied to the ear, it, for example g′, becomes much more powerful when c is sounded. Now, hearing and discriminating g′ in this case does not at all prove that the ear alone and without this apparatus would hear g′ as part of the compound c. But the increase in loudness of g′ caused by the resonator can be used to direct the ear's attention to the tone it is required to discriminate. As the resonator is gradually removed from the ear, the intensity of g' will decrease. But the attention, once directed to it, remains

[78] p. 50.

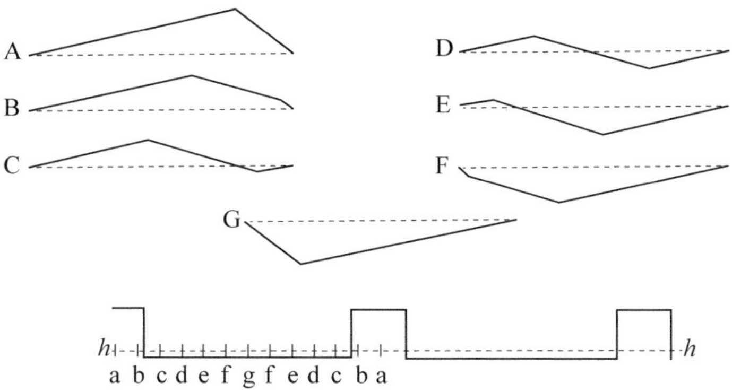

Fig. 6.12 Visual aspect of a vibrating string. Adapted from[42], p. 54

more readily fixed upon it, and the observer continues to hear this tone in the natural and unchanged compound tone of the given note, even with his unaided ear. The only function of the resonators in this case is to direct the attention of the ear to the desired tone [42].[79]

According to Helmholtz, his results led to an appreciation of the upper harmonic sounds that was quite different from that previously held by musicians and even physicists. The upper harmonic sounds were known, almost only in such composite tones as those of strings, where there was a favorable opportunity to observe them and they appear in earlier physical and musical works as an isolated accidental phenomenon of low phenomenon, a kind of curiosity that was certainly occasionally to lend some support to the notion that nature prefigured the construction of our major chord (see Rameau), but on the whole has been almost entirely ignored.

On the other hand, it can be asserted, as will be shown in Chap. 5, that upper partials are, with few exceptions, a general constituent of all musical notes, and that a certain stock of upper partials is an essential condition for a good musical quality of sound. Finally, these upper partials have been erroneously regarded as weak because they are difficult to observe, whereas, in fact, for the best musical quality of sound, the loudness of the upper partials is essential. The loudness of the first upper partials is not far inferior to that of the fundamental itself [42].[80]

At the end of the chapter, it is commented the difference between the perception of the eye, or rather the visual imagination, and that of the ear for the vibrations of a string. A string pushed aside by a sharp object assumes the shape of Fig. 6.12A before it is released. It then passes through the series of forms B, C, D, E, F, until it reaches G, which is the inversion of A, and then returns to A again. Thus it alternates between the forms A and G. The string itself transmits only a minimal amount of its own motion directly to the surrounding air. When both ends of a string are attached to immovable supports, the resulting lack of movement produces an almost

[79] p. 51.
[80] p. 58.

undetectable sound. In a stringed instrument the sound reaches the air through the end of the strings, which rest on a bridge, which in turn rests on an elastic sounding board. The sound due to the strings, therefore, depend essentially on the motion of this extremity, because of the force which it exerts on the sounding board. The magnitude of this force changes periodically with time, as shown in the lower part of Fig. 6.12, where the height of the vertical lines crossing the line *hh* corresponds to the force exerted on the bridge by the end of the string when it is at rest. The pressure of the string on the sounding board alternates, as the figure shows, between a higher and a lower value. The higher pressure remains unchanged for a while, then the lower pressure suddenly appears and also remains unchanged for a while. The letters a to g in the figure correspond to the moment when the string assumes the forms A to G. "We cannot but be astonished that a motion produced by means so simple and so easy to understand should be analyzed by the ear into such a complicated sum of simple tones. To the eye and mind the action of the string on the sonorous board can be figured with extreme simplicity. This is one of the most striking examples of the different ways in which the eye and the ear comprehend a periodic motion" [42].[81]

In Chap. 5, *On the Differences in the Quality of Musical Tones*, the subject of the timbre, or color, of a note is treated. The chapter begins with the statement that timbre must depend on the waveform of the sound signal; in fact, strength depends on amplitude, pitch on frequency, so that all that remains for timbre is form.

In fact, certain characteristic peculiarities in the sounds of various instruments depend on the mode in which they begin and end [67], but Helmholtz's analysis was limited to the peculiarities that continue uniformly. He considered tones produced by different instruments:

1. Musical tones without upper partials.
2. Musical tones with inharmonic upper partials.
3. Musical tones of strings excited by striking.
4. Musical tones of strings excited by bows.
5. Musical tones of flute pipes.
6. Musical tones of reed pipes.
7. Vowel tones.

Particularly interesting is the case of the bowed strings of a violin, for which a peculiar method of observation can be used, proposed by the French physicist Jules Antoine Lissajous (1822–1880), which allows to observe the whole movement of the string and to calculate the intensity of the upper partials. Using a sophisticated microscope, Helmholtz was able to observe the motion of some points on the string due to the action of the bow.

The result of the experimental observations is shown in Fig. 6.13, where the ordinate represents the displacement of the string and the abscissa represents the time in which this displacement occurs. As it moves from α to β, the string clings to the bow and is carried by it. Then, at β, it suddenly detaches and rebounds, whereupon it is grabbed by other points of the bow and carried forward again. The

[81] p. 54.

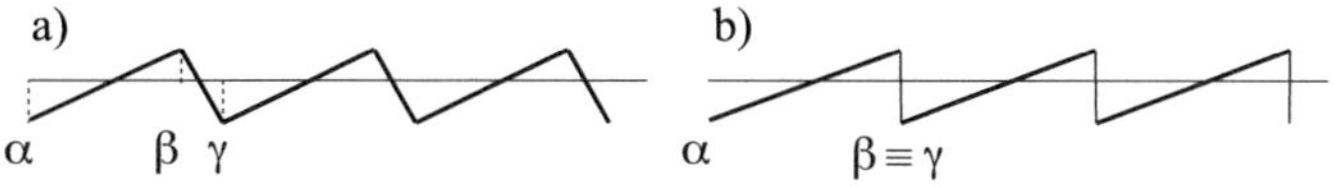

Fig. 6.13 Displacement of a point of a bowed string; from [42], p. 83

relative position of β and γ depends on the point on the string where the bow is used. If it is taken approximately at 1/4 of the length of the string from its extremity, then $\alpha\beta : \beta\gamma = 3 : 1$; see Fig. 6.13a. If the point is close to the end of the string, Fig. 6.13b comes into play.

Once the time history of the motion of a point on the string was known, it was not difficult for Helmholtz to develop it into Fourier series (he did so in Appendix 6). When the violin string responds well, upper partials of order $i = 1, 2, \cdots, n$ are present and their intensity decreases in the ratio $1 : i^2$. If the bow is placed too close to the fingerboard, the 5th and 6th partials, which are generally audible, are absent; the sound becomes dull.

The previous observations refer only to the vibrations of the strings of the instrument and the intensity of their high harmonics. Partial tones of different pitches, however, are not transmitted to the air in the same way and therefore do not reach the listener's ear with the same intensity as they do on the string itself. In fact, as we have seen, the vibrating strings of a violin do not directly transmit a significant part of their motion to the air. They primarily move the bridge on which they are stretched; the bridge has two feet and is located above the most mobile part of the body, between the two ∫ holes. One foot of the bridge rests on a relatively solid support, the *soundpost*, which is a solid rod inserted between the two plates, back and belly, of the instrument. Only the other leg moves the elastic wooden plates and, through them, the enclosed air mass. The air, bounded by the violin's elastic plates, contains some tones of its own, which can be evoked by blowing through the openings or ∫ holes. The violin treated in this way produces the c′ note, according to Savart,[82] who studied instruments made by Stradivari; Zamminer[83] found the same constant tone even in imperfect instruments.

The consequence of this particular resonance relationship is that the tones of the violin strings that are close to the actual natural frequencies of the enclosed air body must be amplified proportionally more. "Since the lowest tone on the violin is g, the only upper partials of its musical tones which can be somewhat reinforced by the resonance of the higher proper tone of its inclosed body of air, are the higher octaves of its three deepest notes" [42].[84]

In Chap. 6, *On the Apprehension of the Qualities of Tones*, Helmholtz reconsidered the case of color (timber). It depends on the waveform, which in turn depends on

[82] p. 86. Félix Savart (1791–1841), French physicist and mathematician, best known for the Biot-Savart law of electromagnetism. However, his main interest was in acoustics and the study of vibrating bodies. A particular interest in the violin led him to create an experimental trapezoidal model.

[83] Friedrich Georg Karl Zamminer (1817–1858), a German physicist and chemist.

[84] p. 87.

the amplitudes of the upper partials and their phases. Sophisticated experiments also show that "the quality of the musical portion of a compound tone depends solely on the number and relative strength of its simple partial tones, and in no way on their phase differences" [42].[85]

Before moving on to examine the anatomical structure of the ear, Helmholtz concluded the analysis of tones by again comparing the different ways in which the ear and eye see vibrations. When the vibratory motion is made visible, the eye is able to distinguish every possible different form of vibration from one another, even such as the ear cannot distinguish. But to break down each wave form into simpler elements according to a definite law, the ear is superior to the eye, which is quite incapable of doing so [42].[86]

A large part of Chap. 6 is devoted to the physiology of the ear [48]. In addition to summarizing known results, and many scientists are quoted, Helmholtz added his own observations. He postulated that there are fibers in the basilar membrane that can be compared to simple oscillators. However, this is a theory that cannot be proven with certainty: "Now we cannot precisely ascertain what parts of the ear actually vibrate sympathetically with individual tones. We can only conjecture what they are at present in the case of human beings and mammals" [42].[87]

Helmholtz's theory is a positional (or tonotopic) theory. According to this theory, each location along the basilar membrane has its own natural frequency, which is related to the morphological and mechanical properties of the organic tissue. Specifically, the thickness of the membrane increases from the oval window to the helicotrema, while its stiffness decreases.

In Helmholtz's original hypothesis, it was assumed that the individual transverse fibers of the membrane acted as independent oscillators, whereas today, thanks to the careful experiments of George von Békésy in the 1980s (who also took advantage of artificial systems built to resemble the basilar membrane), it is known that it is the membrane as a whole that oscillates, but with a maximum of deformation at different positions for each frequency [31].[88]

6.3.3 Combination Tones

Combination tone, the subject of Chap. 7, is the label Helmholtz gave to the phenomenon whereby, given two simple tones, the ear can hear one more. This is how Helmholtz presented the problem in his first paper on the subject, *Ueber Combinationstöne* of 1856 [41]:[89]

---

[85] p. 126.

[86] p. 128.

[87] p. 145.

[88] p. 113.

[89] The paper was published in two parts in the *Monatsberichte der Akademie* and in the *Annalen der Physik und Chemie* in 1856, and reprinted in *Wissenschaftliche Abhandlungen*.

Fig. 6.14 Differential tones for main consonances; from [42], p. 154

> Given m and n, integer numbers having no common factor, it long has been known that two tones with frequencies $m\lambda$ and $(m+1)\lambda$ produce the combination tone λ. On the other hand, for two tones with frequencies $m\lambda$ and $n\lambda$, Weber and Ohm put forward the idea in a general way that the combination tone, likewise, has a frequency λ, whereas Hällström designated $(m-n)\lambda$ as the first combination. At the same time he postulated a number of higher-order combination tones with frequencies $(2n-m)\lambda$, $(3m--2n)\lambda$, etc. The higher order combination tones were said to form lower-order combination tones with the original tones [41].[90]

Helmholtz assumed that combination tones are of two kinds: "The first class, discovered by Sorge and Tartini, I have termed differential tones, because their pitch number is the difference of the pitch numbers of the generating tones. The second class of summational tones, having their pitch number equal to the sum of the pitch numbers of the generating tones, were discovered by myself" [42].[91]

> Thereby I discovered an until now unknown class of combination tones that do not fit the existing theories [...]. I will designate these new tones with the name summation tone [Summationstöne] in contrast to the already mentioned and long known tones that we can call differential tones [Differenztöne] because their frequencies are equal to the difference in frequency of their primary or third lower-order combination tones [41].[92]

Helmholtz thus attributed the discovery of summation tones to himself, while differential tones are attributed to George Andreas Sorge (1703–1778) [84][93] and Tartini; he neglected to mention Jean-Baptiste Romieu here, but quotes him in another section [42].[94]

Differential tones for different consonance intervals are shown in Fig. 6.14, where the generating tones are written as minims and the differential tones as crotchets. As can be seen, the results of the major and minor sixth intervals, when the difference is 2 and 3, respectively, do not obey the Tartini rules (see Sect. 6.1.1.2). More precisely,

---

[90] p. 256. Translation into English in [44].

[91] p. 153.

[92] pp. 282–283. Translation into English in [44].

[93] pp. 40–41.

[94] p. 233.

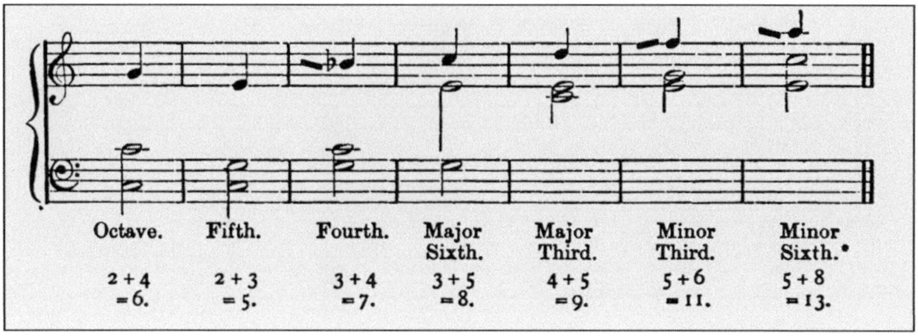

Fig. 6.15 Sum tones for main consonances; from [42], p. 156

the difference is one octave higher for the major sixth and a twelfth higher for the minor sixth.

Combination tones are in general easier to hear from the quite simple tones of tuning-forks and stopped organ pipes than from compound tones where a number of upper partials complicate the phenomenon. In the violin and harmonium, secondary differential tones are often heard because of the overtones.

Figure 6.15 shows summation tones, where the generating tones are still written as minims and the summation tones as crotchets. A sloping thick line downwards or upwards in front of a note indicates that it represents a tone that is slightly lower or slightly higher, respectively, than the note in the scale that precedes it. The summation tones are generally weaker than the differential tones, and can be heard only under peculiarly favorable circumstances.

Helmholtz briefly proposed explanations for the phenomenon of combination tones, in contrast to the more common explanations of the time (limited to differential tones), for which they were purely subjective, due to the beats that usually result from the interaction of generating tones. In these explanations, the frequency of the beats, equal to the differences of the generating tones, would correspond to the frequencies of the combination tones.

Essentially, the theory of beats was supported by Lagrange (1736–1813) in a memoir dated 1759, who explicitly assimilated the phenomenon of the third sound to that of the beats:

> We have just seen that the particle of air which finds itself at the concourse of two sounds receives a shock different from that produced by each sound separately; therefore, if the sounds are of such a nature that their vibrations always coincide after a certain time, the continuous and regular impression of these composite vibrations will be distinguishable from other particular impressions, and a fairly trained ear will hear a third sound, whose relation to the others will be found by comparing the number of particular vibrations which each of them reaches between two successive competitions. [...] It is therefore probable that Mr. Tartini's third sound is produced only by a series of strokes [49].[95]

[95] p. 42.

This perspective was likewise espoused by Thomas Young (1773–1829) in his theory of wave interference [55, 61]. Nevertheless, alternative theories have been put forth. Young himself posited that two tones sounding together could yield a tone higher than the third sound. In 1805, Gerhard Vieth (1763–1836) was the first to introduce the term *combination tone* [98].[96]

In the early years of the19th century, the phenomenon of combination tones began to capture the attention of physicists, including the renowned Wilhelm Weber (1804–1891) [99]. Weber supported the beat theory of the differential tone, as Young and Lagrange had done earlier, but suggested that more than one beat tone was audible. After reading Weber's paper, Gustaf Gabriel Hällström (1775–1844) proposed in 1832 as a challenge to Weber's beat theory that not only the fundamental frequencies of tones but also their harmonic partials should produce differential tones and spoke of primary, secondary, etc. differential tones [37].

> Example.Take two simple tones c′ and e′, ratio (4 : 5), difference 1, differential tone of the first order C. This with the generators gives the ratios (1 : 4) and (1 : 5), differences 3 and 4, differential tones of the second order g, and c′ once more. The new tone 3, gives with the generators the ratios (3 : 4) and (3 : 5), differences 1 and 2, giving the differential tones of the third order C and c, and the same tone 3 gives with the differential of the first order 1, the ratio (1 : 3), difference 2, and hence as a differential of the fourth order c once more, and so on. The differential tones of different orders which coincide when the interval is perfect, as it is supposed to be in this example, no longer exactly coincide when the generating interval is not pure; and consequently such beats are heard, as would result from the presence of these tones [42].[97]

But, according to Helmholtz, first, these assumptions, which are limited to the differential tones, do not explain the origin of the summation tones; second, it can be shown that, under certain conditions, the combination tones exist objectively, independently of the ear; and third, this assumption cannot be reconciled with Ohm's law, confirmed by many experiments, that the only tones heard by the ear correspond to simple vibrations of the ear [42].[98]

Thus, Helmholtz developed a new theory, also known as *distortion theory*, based on sound mathematical arguments framed in Newtonian dynamics, which says that when an oscillator behaves not linearly, as happens when it is subjected to strong impulses, it does not work as a mechanical spectroscope, showing the frequencies contained in the excitation, but gives rise also to oscillations of frequencies not contained in the excitation impulses, higher or lower. A detailed analysis, summarized in Appendix 12 (Theory of combination tones), is given in [41]. Here we consider a simple case in which the restoring force of the elementary oscillator f as a function of the displacement x is of the form $f = kx + bx^2$, i.e. there is a quadratic term in addition to the linear elastic term kx. If p and q are the frequencies of the generating tones, the combination tone $(p - q)$ will generally have the greatest intensity of all the tones. The $(p + q)$ tone will be much weaker. Other combination tones for the case

[96] p. 265.

[97] p. 155.

[98] p. 156.

Table 6.3 Combination tones for quadratic nonlinearity

First order	Second order	Third order
$\lvert p\rvert$ $\lvert q\rvert$	$\lvert p\rvert$ $\lvert q\rvert$ $\lvert p \pm q\rvert$	$\lvert p\rvert$ $\lvert q\rvert$ $\lvert p \pm q\rvert$
	$\lvert 2p\rvert$ $\lvert 2q\rvert$	$\lvert 2p\rvert$ $\lvert 2q\rvert$ $\lvert p \pm 2q\rvert$
		$\lvert 3p\rvert$ $\lvert 3q\rvert$ $\lvert p \pm 3q\rvert$
		$\lvert 4p\rvert$ $\lvert 4q\rvert$ $\lvert 2p \pm q\rvert$
		$\lvert 2p \pm 2q\rvert$
		$\lvert 3p \pm q\rvert$

in question are $2p + q, 2p - q, p + 2q, p - 2q$, etc. [42].[99] In [60], Helmholtz's calculations are repeated with a modern perturbation approach; the results are shown in Table 6.3.

In general, depending on the nature of the oscillator restoring force, combination tones of p and q are defined by a linear combination of the kind $|np \pm mq|$, where n and m are integers.

Appendix 12 is very important because it illustrates how an acoustic phenomenon discovered by professional musicians is at the origin of important physical theories. In this case, it is the physical-mathematical theory known as nonlinear dynamics, which studies interesting new phenomena such as (deterministic) chaos [57].

The combination tones are particularly evident in polyphonic sirens, where the mass of air is violently stirred by two tones simultaneously. The combination tones can be as powerful as the generating tones. On the other hand, when the two tones come from two singers, two separate wind instruments, or two violins, the amplification of the combination tones revealed by the resonators is small and doubtful. Here, then, as there is no clearly perceptible pendular vibration in the air corresponding to the combination tone, we must conclude that such tones, which are often powerfully audible, are produced in the ear itself [42].[100]

Different ideas about the origin of combination tones have been expressed from Helmholtz until today, with the scientific community still divided between supporters of the intra-aural origin [76] and the extra-aural origin [51]. Many experiments have been performed to understand the role of the physiology of the ear (see [10] for a review).

Psychoacoustics suggests the existence of virtual pitches created by our brain when there is no discernible fundamental sound. For example, if tones are produced by a musical instrument with frequencies of 780, 1040, 1300 Hz, a virtual pitch of 260 Hz is suggested, which is a natural fundamental of the three tones. Virtual pitches may or may not coincide with differential tones, as in the previous example [82].[101]

[99] p. 413.
[100] pp. 157–158.
[101] p. 83.

6.3.4 *Theory of Consonances*

Chapter 8, *On the beats of simple tones*, considers the theory of the beats of two simple simultaneous tones and lays the foundation for the theory of consonance, which for Helmholtz had both physical and physiological bases. He took up Sauveur's hypothesis, which associated consonance with the absence of beats. In fact, Sauveur's theory explains only the dissonance of small intervals where beats are clearly audible, in which case Helmholtz spoke of *slow beats*. In the case of intervals larger than, say, the minor third, Sauveur's hypothesis failed; he could not explain why, for example, a just fifth is more consonant than a seventh, since neither interval produces beats.

Sauveur's theory of consonances, although primitive because based on a very crude assumption about the propagation of sound, is actually not entirely wrong from an empirical point of view. Relatively recent experiments have been made with simultaneous simple tones [45, 46]. Consider two simple tones of equal intensity tuned to a certain frequency f, and vary the frequency of one of them up to $2f$ so that all intervals of the octave are formed. By recording the level of agreeableness for different ratios between the frequencies of the two elementary tones, a peak of dissonance is found when this difference is about 10%, or a little less than a major tone, as shown in Fig. 6.16. After the dissonance peak, the agreeableness of the union of the two tones increases until it reaches a maximum at the octave. without registering a local minimum or maximum at the consonances of the third, fourth, fifth, or sixth, as might be expected (Fig. 6.16).

The explanation for this somewhat surprising result and the difficulties of Sauveur's consonance theory must be sought in the nature of the tones produced by musical instruments; they do not consist of pure tones, but contain numerous upper partials and sometimes combination tones. Helmholtz believed that he could formulate a

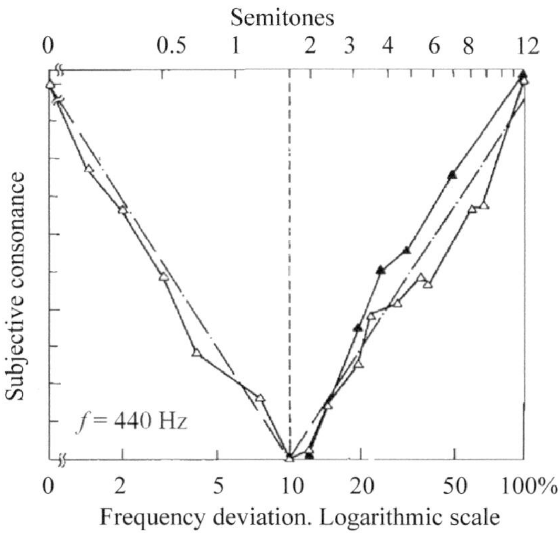

Fig. 6.16 Subjective perception of consonance for simple tones. Redrawn from [45], p. 1452

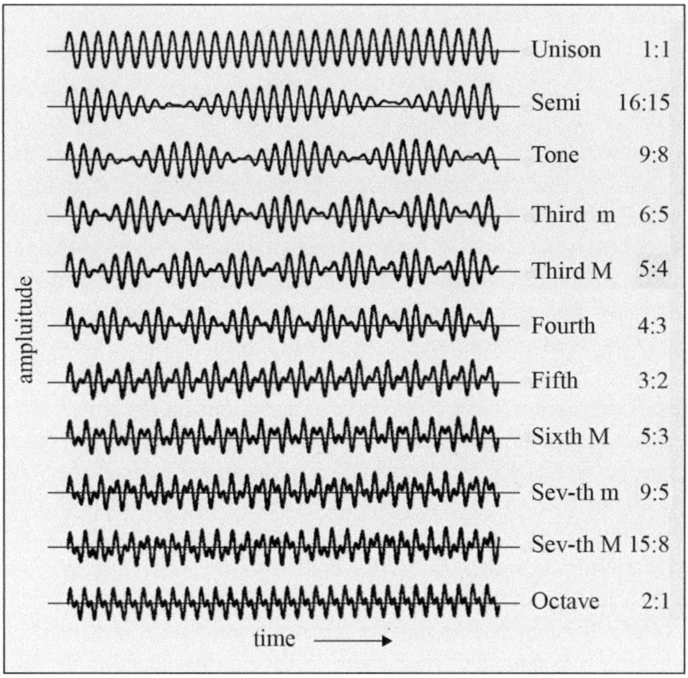

Fig. 6.17 Beats for different musical intervals. Adapted from [31], p. 175

fully satisfactory theory of consonance by admitting, as Sauveur did, that it is related to the absence of beats: "Beats produce intermittent excitement of certain auditory nerve fibers. The reason why such an intermittent excitement acts so much more unpleasantly than an equally strong or even a stronger continuous excitement, may be gathered from the analogous action of other human nerves" [42].[102]

Sauveur believed that when beats become too frequent, they are no longer heard because the ear cannot follow them. The proponents of the explanation of the combination tones with beats, on the other hand, believed that although they were not heard as an increase or decrease in amplitude (the classic wow-wow-wow), they left their own mark on the listener as a note with a frequency equal to the frequency of the beats.

Helmholtz opposed these two views; the criticism of the second view has already been made in the previous section on combination tones. As for the first thesis, Helmholtz believed that the perception of beats depended not so much on the absolute difference between the frequencies of the generating tones, but on their relative frequency, that is, their intervals.

Figure 6.17 shows the waveforms resulting from the interference of two pure tones. They depend only on the interval and not on the absolute value of the frequencies

[102] p. 169.

Fig. 6.18 Upper partials for some consonances; from [42], p. 180

of the tones; for example, if we consider the interval (16 : 15), a major semitone, if the major tone is 160 Hz, the minor tone is 150, and their interference results in 10 beats per second. On the other hand, if the major tone is 800 Hz and then the minor tone is 750 Hz, there are 50 beats per second, which Sauveur would no longer perceive. Of course, by increasing the absolute value of the frequencies, the phenomenon of beats is somewhat more difficult to perceive, but Helmholtz gave an example of 132 beats per second and argues that if one is attentive in listening, even in this case, although one cannot count the individual beats, one still has the impression of amplitude modulation.

Figure 6.17 shows that as the interval increases; although there is always amplitude modulation, it becomes increasingly difficult to perceive because the frequency of the modulation is of the same order as the frequency of the oscillations. Helmholtz assumed that the threshold at which two frequencies are still perceived as beats is less than the interval of a minor third; above this value, the two interfering sounds are perceived as distinct.

In Chap. 10, *Beats of upper partials*, after discussing the limits of the lowest tones that can be heard in Chap. 9, Helmholtz reported on his theory of consonances for compound tones. The chapter begins with the observation:

> The beats heard when the two generating tones are separated by more than a minor third in the scale are the result of upper partial and compound tones [...]. When two compound tones are played at the same time, it is easy to see from the foregoing that beats can occur whenever any two upper partial tones are sufficiently close together, or when the prime of one tone approaches an upper partial of the other. The number of beats is, of course, as before, the difference in the vibration numbers of the two partials to which the beats are due. When this difference is small, and the beats are therefore slow, they are relatively most distinct to hear and to count and to examine, just as for the beats of prime tones [42].[103]

Figure 6.18 shows the upper partials for some intervals. The minims denote the generating tones, the crotchets the corresponding upper partials. Example 1 shows the interval of an octave, C_2 C_3, the upper partial of the first tone C_2 is in unison with the second tone C_3. Example 2 shows the interval of the major seventh, C_2 B_3, the upper partial of C_2 makes a semitone with B_3 and thus beats occur. In the third example, the interval is C_2 $D_3\flat$, the upper partial of C_2 makes the semitone C_3 $D_3\flat$, and similarly beats occur. The fourth and fifth examples refer to equal tempered intervals. The upper partial A_5 is slightly mismatched with the second tone in example 4 (a twelfth) and with the second upper partial in example 5 due to the temperament; slow beats occur.

---

[103] p. 180.

Table 6.4 Beats for slightly mistuned consonances; from [42], p. 184

Interval	Beat partial tones	Beat partial tones	Beats
300 Unison	$1 \times 300 = 300$	$1 \times 301 = 301$	1
600 Octave	$1 \times 600 = 600$	$2 \times 301 = 602$	2
450 Fifth	$2 \times 450 = 900$	$3 \times 301 = 903$	3
400 Fourth	$3 \times 400 = 1200$	$4 \times 301 = 1204$	4
500 Major sixth	$3 \times 500 = 1500$	$5 \times 301 = 1505$	5
375 Major third	$4 \times 375 = 1500$	$5 \times 301 = 1505$	5
360 Minor third	$5 \times 360 = 1800$	$6 \times 301 = 1806$	6

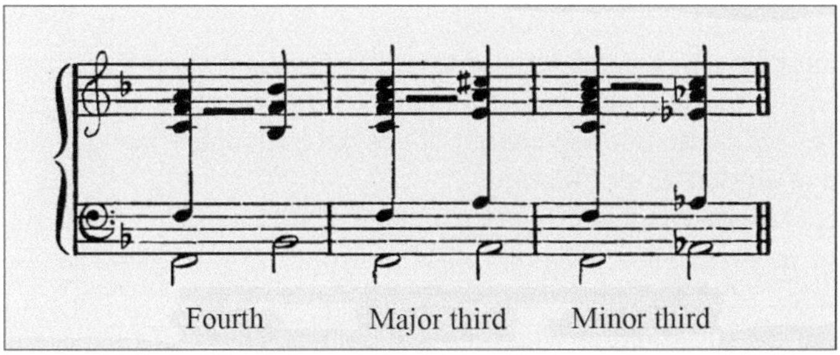

Fig. 6.19 Beats due to upper partials; from [42], p. 186

Helmholtz presented a detailed analysis of the consonance of various intervals; only a brief summary is given below.

Table 6.4 shows the beats that occur with a note of fixed pitch f, in this case 300 Hz, with other notes f_i forming consonant intervals with it, when f is slightly changed, in this case raised to 301 Hz, while the other notes remain unchanged. The first column shows the intervals that the notes f_i form with f. Since $f = 300$ Hz, f_i take the following values to respect the indicated intervals: 300, 450, 400, 500, 375, 360 Hz. The second column shows the number of orders and the frequencies of the partials of f_i that coincided with the partials of f before its modification (for example for the interval of the fifth, it is the second partial of f_i that coincides with the third partial of f); the third column shows the partials of f after the modification. The fourth column shows the beats that occur in one second in the different situations. As can be clearly seen, the beats become important for the sixths and thirds. Helmholtz could conclude: "For sixths and thirds, therefore, the pitch numbers of the tones must be much closer to the normal ratio, if we wish to avoid slow beats, than for octaves and unisons" [42].[104]

Figure 6.19 illustrates the interaction of higher harmonic sounds for some adjacent consonances, the fourth and the major and minor thirds; horizontal tick lines marks the coincidence of upper partials. In some cases, upper partials may be a semitone

[104] p. 185.

apart, which can create beats and disturb consonance. In the first example, for the interval of the fourth, there is a coincidence in F_4 between the 3rd of F_2 and the 2nd of B_2. In the second example, for the major third, the 2nd partial of the second note passes from F_4 to E_4, which is separated by a major semitone from the 3rd partial of F_2, just as it is the case for the 5th partial of the first note and the 4th of the second. In the third example, a minor third, there is a semitone between the 4th upper partial of F_2 and the 3rd upper partial of $A_2\flat$.

Helmholtz's analysis leads to the following conclusions, which are quite consistent with the current ones.

> When two musical tones are sounded at the same time, their united sound is generally disturbed by the beats of the upper partials, so that a greater or less part of the whole mass of sound is broken up into pulses of tone, and the joint effect is rough. this relation is called dissonance.
>
> 1. The most perfect consonances are those that have been here called absolute, in which the prime tone of one of the combination tones coincides with some partial tone of the other. To this group belong the octave, twelfth, and double octave.
> 2. Next follow the fifth and the fourth, which may be called perfect consonances, because they may be used in all parts of the scale without any important disturbance of harmoniousness. The fourth is the less perfect consonance and approaches those of the next group. It owes its superiority in musical practice simply to its being the defect of a fifth from an octave, a circumstance to which we shall return in a later chapter.
> 3. The next group consists of the major sixth and the major third, which may be called medial consonances. The old writers on harmony considered them as imperfect consonances. In lower parts of the scale the disturbance of the harmoniousness is very sensible, but in the higher positions it disappears, because the beats are too rapid to be sensible. But each, in good musical qualities of tone, is independently characterized, by the fact that any little defect in its intonation produces sensible beats of the upper partials and consequently each interval is sharply separated from all adjacent intervals.
> 4. The imperfect consonances, consisting of the minor third and minor sixth, are not in general independently characterized, because in good musical qualities of tone the partials on which their definition depends are often not found for the minor third, and are generally absent for the minor sixth, so that small imperfections in the intonation of these intervals do not necessarily produce beats [42].[105]
>
> The order of consonances proposed here is based upon a consideration of the harmoniousness of each individual interval independently of any connection with other intervals, and consequently without regard to key, scale, and modulation [42].[106]

In Chap. 11, *Beats due to combination tones*, Helmholtz considered beats due to combination tones, which are relevant in many situations, but for reasons of space we avoid commenting on them.

[105] p. 194.

[106] pp. 194–196.

6.3.5 *Major and Minor Chords. The Principle of Harmonic Dualism*

Chapter 12, *Chords*, is devoted to chords, that is, the simultaneous reproduction of more than two notes. Although Helmholtz developed an interesting and thorough analysis, we will limit our account to the case of triads and the condition under which they are considered as consonant.

The possible consonant triads in the octave, which we assume to begin with C, are summarized in the following table:

C E G	C E♭ G
C F A	C F A♭
C E♭ A♭	C E A

The two triads of the first row are known as major and minor triads, respectively; they span a fifth and are the union, in order, of a major and a minor third, and a minor and a major third. The triads of the second row are called, according to their composition, *chords of the sixth and the fourth*; the triads of the third row are called *chords of the sixth and the third*, or simply chords of the sixth. All of them can be obtained from the two chords of the first row by transposing and inverting [42].[107] This is Helmholtz's judgment of the degree of agreeableness of the major and minor triads:

> Judging merely from the intervals we should expect that the minor triad C E♭ G would sound as well as the major C E G, as each has a fifth, a major and a minor third. This is, however, far from being the case. The minor triad is very decidedly less harmonious than the major triad, in consequence of the combination tones, which must consequently be here taken into consideration [42].[108]

Figure 6.20 illustrates the situations. The fundamental tones are marked as minims, the combination tones resulting from them are represented as crotchets, those resulting from the fundamental and the first upper partials as quavers and semiquavers. A downward sloping thick line in front of a note indicates a tone slightly lower than the preceding note in the scale.

In the major triads, the first and even the second order combination tones (indicated by crotchets and quavers) are merely doubles of the triad tones in lower octaves. The higher combination tones of the second order (written as semiquavers) are extremely weak, because the intensity of the combination tones decreases as the interval between the generating tones increases. In the minor triads, on the other hand, the first-order combination tones, which are easily audible, begin to disturb the harmonious effect. They are not close enough to beat, but they are not part of the harmony.

[107] p. 213.

[108] p. 214.

Fig. 6.20 Combination tone for the major **a** and minor third **b**; from [42], p. 215

Helmholtz concluded:

> Modern harmonists are unwilling to acknowledge that the minor triad is less consonant than the major. They have probably made all their experiments with tempered instruments, on which, indeed, this distinction may perhaps be allowed to be a little doubtful. But on justly intoned instruments and with a moderately piercing quality of tone, the difference is very striking and cannot be denied. The old musicians, too, who composed exclusively for the voice, and were consequently not driven to enfeeble consonances by temperament, shew a most decided feeling for that difference. To this feeling I attribute the chief reason for their avoidance of a minor chord at the close [...]. Mozart is certainly the composer who had the surest instinct for the delicacies of his art. Among his vocal compositions the *Ave verum* corpus is particularly celebrated for its wonder fully pure and smooth harmonies. On examining this little piece as one of the most suitable examples for our purpose, we find in its first clause, which has an extremely soft and sweet effect, none but major chords, and chords of the dominant [42].[109]

Helmholtz's position was challenged by some musicians and physicists. Since the 18th century, musicians struggled to justify the minor triad. Event though most of them considered it to be inferior to the major triad, there were, however, some who considered minor and major triads are both equally important; an approach known as *harmonic dualism*, based on the fact that both major and minor chords contain a fifth, a major third, and a minor third.

In 1896, Arthur Joachim Oettingen (1836–1920) countered Helmholtz's acoustic criticism of the minor chord with acoustic arguments. Figure 6.21b shows the case of the minor chord C_4 $E_4\flat$ G_4. All tree notes have a common upper partial in G_6 (the sixth partial for C_4, the fifth for $E_4\flat$, the fourth for G_4), which is consonant with

[109] pp. 217–225.

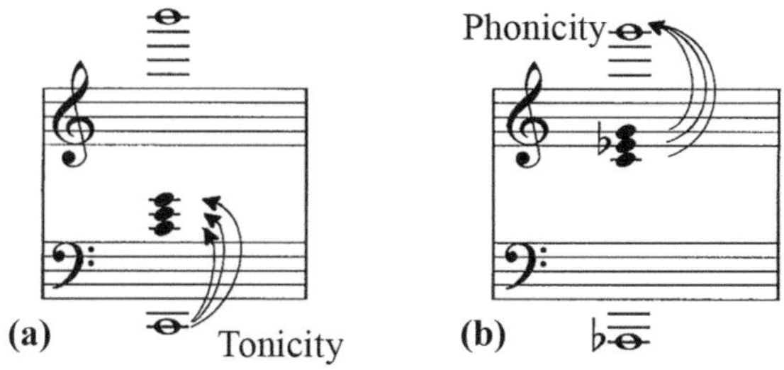

Fig. 6.21 Oettingen's tonicity and phonicity; from [73], p. 22

the chord. Figure 6.21a instead shows the major chord C_4 E_4 G_4; its notes still share a common upper partial coinciding with B_7, but it is not consonant with the chord. However, the chord has a potential fundamental in C_2 (the notes C_4, E_4, G_4 are the 4th, 5th, 6th upper partial of C_2, respectively). An equivalent fundamental also exists in the case of minor chord example—in this case A_1♭. However this is dissonant with the minor triad above [38].[110]

From this analysis, Oettingen might conclude that the minor chord is more consonant than the major. But this, by reversing the order of Helmholtz's chord preference, would not have solved the problem of duality, which requires that the two chords be on the same foot. Oettingen solved this difficulty somewhat by introducing two consonance criteria, the *phonic consonance*, in the case of the minor chord, for consonance with the upper partial G_6, and the *tonic consonance* for the presence of a common fundamental.

Helmholtz at first contrasted with Oettingen's conclusions, but from the third edition of *Die Lehre von den Tonempfindungen* on he appreciated them, though retaining his own ideas about the privileged role of the major mode:

> Von Oettingen [...] has carried out, in a most interesting manner, the complete analogy between the [...] minor and the major mode [...] I wish emphatically to recommend this book to the attention of musicians. On the other hand, it seems to me that it is necessary to shew by musical practice, that the new principle [...] really suffices for the construction of great musical pieces [42].[111]

Riemann took a more radical position. Like Oettingen, he felt it necessary to pair the major and minor chords on a physical, i.e., natural, basis, but he followed a different path, returning to the justification provided by Rameau in his *Génération harmonique*.

Riemann postulated the existence of undertones for a vibrating string, based on his own (mystical?) auditory experience.

> During a silent night in 1875, the young musicologist Hugo Riemann struck a key on his grand piano. He was listening for undertones, which he believed to exist in the sound wave. His nocturnal experiment seemed successful—his aural experience confirmed his experimental

[110] pp. 242–250.

[111] pp. 308–309, footnotes.

Fig. 6.22 Overtone and undertone series with corresponding triads; from [73], p. 16

> hypothesis. These undertones, he would explain later, relate to one sounded tone exactly in the manner of the harmonic or overtone series but extending in the opposite direction [73].[112]

That is, Riemann imagined the possibility shown in Fig. 6.22. But his experiment was not verified in any way by other musicians; simply because the sequence of undertones is impossible. As already noted, analytical dynamics shows that in the case of ordinary piano strings, i.e. with sufficiently linear elastic behavior and relatively small amplitudes, only the overtone series can exist. In the case of strong nonlinearities and large amplitudes, undertones can indeed occur, similar to what happens with combination modes, but they are hardly present as harmonic series.

Throughout his life, Riemann defended his position, going from the naive suggestion that one should use an apparatus similar to his own to answer his experiment, that is, the same brand of piano, to accepting that the overtones exist but cannot be heard because of acoustic interference, to finally accepting that an experimental proof is not necessary to credit dualism, still believing in the existence of the undertones.

Helmholtz had words of appreciation for Riemann's ideas, but was perplexed as to their justification:

> I have not been able to convince myself of the correctness of the fact adduced that the undertones of a tone strongly struck on the piano sound when the corresponding dampers are raised. Perhaps the author has been deceived by the circumstance that with very resonant instruments (especially older ones) any strong shake, and therefore probably a violent blow on the digitals, will cause some one or several of the deeper strings to sound its note [42].[113]

However, he did devote an appendix to an attempt to explain the formation of Riemann's undertones in the basic structure of the ear [42],[114] but concluded that if some

---

[112] p. 15.

[113] p. 365, footnote.

[114] Appendix XII. *Vibration of the membrana basilaris in the cochlea*, pp. 406–411.

mathematical model could furnish valid arguments for the presence of undertones, there is no known observable such a type of model in the ear.

6.3.6 *Music Beyond Acoustics*

Chapter 3 begins the third part of the *Die Lehre von den Tonempfindungen*, with the observation that up to then a satisfactory basis has been provided for the elementary rules of musical composition, while we are now entering new territory which is no longer subject to physical laws alone, although the acquired knowledge of the nature of hearing still has numerous applications. According to Helmholtz, we are passing through a problem that by its very nature belongs to the realm of aesthetics. When, in the theory of consonance, we spoke of pleasant and unpleasant, we referred exclusively to the immediate impression made on the senses when an isolated combination of sounds reaches the ear, and, as you know, we paid no attention to artistic contrasts and means of expression; we thought only of sensual pleasure, not of aesthetic beauty. The two aspects must be kept strictly separate, for Helmholtz, although the former is an important means of achieving the latter. At every step we encounter historical and national differences in taste. Whether one combination is rougher or softer than another depends solely on the anatomical structure of the ear and has nothing to do with psychological motives. But the degree of roughness that a listener is willing to endure as a means of musical expression depends on taste and habit; thus the boundary between consonance and dissonance has often been modified.

> The system of scales, modes, and harmonic tissues does not rest solely upon inalterable natural laws, but is also, at least partly, the result of esthetic principles, which have already changed, and will still further change, with the progressive development of humanity [42].[115]

Despite the idea of dealing with issues that are not strictly acoustic, acoustic theory intervenes heavily in Helmholtz's explanation of early musical scales, particularly those of the Greeks. The general idea is that the tones of a scale have some harmonic relationship to each other. A first difficulty arises, though not explicitly. Greek music was homophonic, and the pleasantness of the tones, i.e. consonance, cannot be explained by the phenomenon of beats, as had been done previously for simultaneous sounds.

A second definition of consonance between two sounds applies here, which is based on a coincidence of overtones. Indeed, according to Helmholtz, the memory is capable of establishing a straightforward association between the overtones of the initial sound and the common overtones of the second sound. This coincidence can be direct if it is between the overtones of the same sounds or indirect if it is between the overtones of a third sound.

> We shall consider musical tones to be related in the first degree which have two identical partial tones; and related in the second degree, when they are both related in the first degree

[115] p. 235.

> to some third musical tone. The louder the coincident in proportion to the non-coincident partials of compound tones related in the first degree, the closer is their relationship and the more easily will both singers and hearers feel the common character of both the tones [42].[116]

In contrast to the negative conception of consonance (*sensory consonance*) as the absence of beats, to which Helmholtz devotes the most space in the theoretical part of his treatise, this second conception (*harmony consonance*), which becomes very important in applications to musical pieces, defines consonance positively. The two conceptions of consonance, although contradictory, can be seen as complementary [50][117] [93].

Take the octave, for example, where the relationship to the fundamental is most remarkable. Let any melody be played on any instrument of good musical quality; the listener must have heard not only the fundamental tone but also its overtones, especially the octave. If then a higher voice plays the same melody an octave higher, that is, whose frequency is in the ratio (2:1) with the first, we hear again a part of what we heard before (i.e. the octave of the first sound), what we remember, and we say that there is a consonance, but if the ratio of the frequencies of the two tones is not exactly (2:1), the first overtone (the octave) of the second tone will be different from the second, and we can say there is no concordance, i.e. the two sounds are discordant. If, on the other hand, the two tones were heard at the simultaneous time, if their ratio is exactly (2:1), we have no beats and can say that there is consonance, but if the ratio of the frequencies of the two tones is not exactly (2:1), as the first overtone of the second tone will be different from the second, we have beats, and therefore discordance.

After a survey of the various principles of musical styles in different times and countries, Chaps 14–17 made a detailed study of the Western tonal system. The treatise ends with numerous appendices, in which Helmholtz gives mathematical justifications for some of the statements.

References

1. Bailache P (1997) La musique traduite en mathématiques: Leonhard Euler. Communication au colloque du Centre François Viète, “Problèmes de traduction au XVIIIe siècle”, Nantes
2. Barbieri P (1992) Tartinis Dritter Ton Eulers Harmonische Exponenten. Mit einem unveröffentlichten Manuskript Tartinis Musiktheorie 7(3):219–234
3. Barbieri P (1994) Il sistema armonico di Tartini nelle censure di due celebri fisico matematici: Eulero e Riccati. In: Bombi A, Massaro N (eds) Tartini, il tempo e le opere. Il Mulino, Bologna, pp 321–343
4. Bernard JW (1980) The Principle and the Elements Rameau’s “controversy with d’Alembert”. Journal of Music Theory 24(1):37–62
5. Bini C, Lami G (2019) I suoni di combinazione ed il terzo suono di Tartini: fisica, storia e musica. Il Campano, Pisa

[116] p. 256.

[117] pp. 99–100.

6. Blackwood E (1985) The structure of recognizable diatonic tunings. Princeton University Press, Princeton
7. Capecchi D (2020) Epistemologia della musica negli Élémens de musique di d'Alembert. SISFA 40th Annual conference. Pisa University Press, Pisa, pp 65–70
8. Capecchi D (2021) Epistemology and natural philosophy in the 18th century. Springer, Cham-Dordrecht
9. Cariani P (2001) Temporal codes, timing nets, and music perception. Journal of New Music Research 30(2):107–137
10. Caselli G, Massetti G, Cecchi G (2018) Tartini, the third tone and the cochlea. pH 1:2–24
11. Cernuschi A (1996) D'Alembert pris au jeu de la musique. Ses interventions musicographiques dans l'Encyclopédie. Recherches sur Diderot et sur l'Encyclopédie 21:145–161
12. Chandler BG (2017) Jean Philippe-Rameau and the corps sonore. Athens Journal of Humanities & Arts 4(1):7–24
13. Christensen T (1989) Music theory as scientific propaganda: The case of d'Alembert's Élémens de musique. J Hist Ideas 50(3):409–427
14. Christensen T (1993) Rameau and musical thought in the Enlightenment. Cambridge University Press, Cambridge
15. Christensen T (2019) Rameaus langer Schatten: Studien zur deutschen Musiktheorie des 18. Jahrhunderts. By Ludwig Holtmeier. Music Theory Spectrum 41(2):369–376
16. d'Alembert J (1743a) Essai d'une nouvelle théorie de la résistance des fluides. David, Paris
17. d'Alembert J (1743b) Traité de dynamique. David, Paris
18. d'Alembert J (1759) Essai sur les elemens de la philosophie. In: d'Alembert J (ed) Melanges de littérature, d'histoire et de philosophie (vols 4), Catelain, Amsterdam
19. d'Alembert J (1762) Élémens de musique, théorique et pratique suivant les principes de M. Rameau, éclaircis, développées et simplifiés. Second edition. Bruyset, Lyon
20. d'Alembert J (1894) Discourse preliminaire de l'Encyclopédie. Edited by Picavet F. Colin, Paris
21. Descartes R (1650) Musicae compendium. Ackersdijck and Zijll, Utrecht
22. Di Stefano N, Bertolaso M (2014) Understanding musical consonance and dissonance: epistemological considerations from a systemic perspective. Systems 2(4):566–575
23. Di Stefano N, Vuust P, Brattico E (2022) Consonance and dissonance perception. a critical review of the historical sources, multidisciplinary findings, and main hypotheses. Physics of Life Review 20:273–304
24. Euler L (1739a) Explicatio phaenomenorum quae a motu successivo lucis oriuntur. Commentarii Academiae Scientiarum Petropolitanae 11:150–193
25. Euler L (1739b) Tentamen novae theoriae musicae ex certissismis harmoniae principiis dilucide expositae. Academiae Scientiarum, Saint Petersburg
26. Euler L (1748) De propagatione pulsuum per medium elasticum. Novi Commentarii Academiae Scientiarum Petropolitanae 1:67–105
27. Euler L (1766a) Conjecture sur la raison de quelques dissonances généralement reçues dans la musique. Mémoires de l'Académie des Sciences et Belles Lettres de Berlin 20:165–173
28. Euler L (1766b) Du véritable caractère de la musique moderne. Mémoires de l'Académie des Sciences et Belles Lettres de Berlin 20:174–179
29. Euler L (1770-1774) Lettres à une princesse d'Allemagne sur divers sujets de physique & de philosophie (3 vols). Steidel & Compagne, Mietau, Leipzig
30. Euler L (2010) Tentamen novae theoriae musicae ex certissismis harmoniae principiis dilucide expositae. Italian translation and introduction by Del Piero A. Memorie della Accademia delle Scienze di Torino 34:2, 244
31. Forgia A (1999) Fisica nella musica. Zanichelli, Bologna
32. Fourier JBJ (1822) Théorie analytique de la caleur. Didot, Paris
33. Girdlestone C (2014) Jean-Philippe Rameau: His life and work. Dover, New York
34. Glarean H (1547) Dodekachordon. Petri, Basel
35. Grant RM (2013) Leonhard Euler's unfinished theory of rhythm. Journal of Music Theory 57(2):245–286

36. Guanti G (2003) Chi ha paura della Scienza platonica fondata nel cerchio' di Tartini? Società Italiana di Musicologia 1(38):41–73
37. Hällström GG (1832) Von den Combinationstönen. Annalen der Physik und Chemie 100(3):438–466
38. Harrison D (1994) Harmonic function in chromatic music. University of Chicago, Chicago
39. Haynes B (2002) A history of performing pitch. A story of A, Scarecrow, Lanham, Maryland
40. Helmholtz H (1877) Die Lehre von den Tonempfindungen als physiologische Grundlage für die Theorie der Musik, 4th edn. Vieweg und Sohn, Braunschweig
41. Helmholtz H (1882) Ueber combinationstöne. In: Helmholtz H (ed) Wissenschaftliche Abhandlungen (3 vols), vol 1. Barth, Leipzig, pp 256–302
42. Helmholtz H (1885) On the sensations of tone as a physiological basis for the theory of music. Longmans, London
43. Helmholtz H (1896) Handbuch der physiologischen Optik, 2nd edn. Voss, Leipzig
44. Hiebert E (2014) The Helmholtz legacy in physiological acoustics. Springer, Dordrecht
45. Kameoka A, Kuriyagawa M (1969) Consonance theory part I: Consonance of dyads. The Journal of the Acoustical Society of America 45(6):1451–1459
46. Kameoka A, Kuriyagawa M (1969) Consonance theory part II: Consonance of complex tones and its calculation method. The Journal of the Acoustical Society of America 45(6):1460–1469
47. Kromhout MJ (2020) George Simon Ohm and the mathematical analysis of sound. Isis 111(3):471–492
48. Kursell J (2018) Epistemologie des Hörens: Helmholtz' physiologische Grundlegung der Musiktheorie. Wilhelm Fink, Leiden
49. Lagrange JL (1759) Recherches sur la nature et la propagation du son. In: Serret JA, [Darboux G] (1867-1892) (ed) Oeuvres de Lagrange (14 vols), vol 1, Gauthier-Villars, Paris, pp 39–148
50. Lalo C (1908) Esquisse d'une esthétique musicale scietifique. Alcan, Paris
51. Lhori A, Carral S, Chatziioannou V (2011) Combination tones in violins. Archives of Acoustics 36(4):727–740
52. Lindsay BR (ed) (1973) Acoustics: Historical and philosophical development. Dowden, Hutchinson & Ross, Stroudsburg, Pennsylvania
53. Lippius J (1612) Synopsis musicae novae omnino verae atque methodicae universae. Ledertz, Strasburg
54. Mersenne M (1636-1637) Harmonie universelle (2 vols). Cramoisy, Paris
55. Meyer MF (1957) Tartini and beat-tone pitches. Am J Psychol 70(4):646–650
56. Montucla JE (1758) Histoire des mathématiques (2vols). Jombert, Paris
57. Nayfeh AH, Mook DT (1979) Nonlinear oscillations. Wiley, New York
58. Ohm GS (1843) Ueber die Definition des Tones, nebst daran geknüpfter Theorie der Sirene und ähnlicher tonbildender Vorrichtungen. Annalen der Physik und Chemie 59:513–565
59. Ohm GS (1844) Noch ein Paar Worte über die Definition des Tones. Annalen der Physik und Chemie 62:1–18
60. Olson D (1969) Musical combination tones and oscillations of the ear mechanism. Am J Phys 37:730–733
61. Pesic P (2013) Thomas Young's musical optics: Translating sound into light. Osiris 28(1):15–39
62. Pischner H (1963) Die Harmonielehre Jean-Philippe Rameaus, ihre historischen Voraussetzungen und ihre Auswirkungen im französischen, italienischen und deutschen musiktheoretischen Schrifttum des 18. Jahrhunderts: ein Beitrag zur Geschichte des musikalischen Denkens. Breitkopf & Hartel, Leipzig
63. Polzonetti P (2001) Tartini e la musica secondo natura. LIM, Lucca
64. Rameau JP (1722) Traité de l'harmonie réduite à ses principes naturels. Ballard, Paris
65. Rameau JP (1726) Nouveau système de musique theorique. Ballard, Paris
66. Rameau JP (1737) Generation harmonique, ou traité de musique theorique et pratique. Praut, Paris
67. Rameau JP (1750) Démonstration du principe de l'harmonie, servant de base à tout l'art musical théorique & pratique. Durand et Pissot, Paris

68. Rameau JP (1752) Nouvelle réflections de M. Rameau sur sa démonstration du principe de l'harmonie. Servant de base à tout l'art musical téorique et pratique. Durand et Pissot, Paris
69. Rameau JP (1760) Nouvelle réflections sur le principe sonore. In: Rameau (ed) Code de musique pratique et Nouvelle réflections sur le principe sonore, Imprimerie Royale, Paris, pp 189–237
70. Rameau JP (1762a) Conclusions sur l'origine des sciences, URL http://www.rameau2014.fr/RESSOURCES/Catalogue/Theorie/Conclusions-sur-l-origine-des-sciences, accessed 20 September 2024
71. Rameau JP (1762b) Origine des sciences, suivie d'une controverse sur le même sujet. Bibliothèque nationale de France, département Arsenal, 8-S-18131
72. Rameau JP (1971) Treatise on harmony. Translation and introduction by Gossett P. Dover, New York
73. Rehding A (2003) Hugo Riemann and the birth of modern musical thought. Cambridge University Press, Cambridge
74. Riemann H (1882) MusikLexikon. Bibliographischen Instituts, Leipzig
75. Riemann H (1896) Harmony simplified. Augener, London
76. Roederer J (2008) The physics and psychophysics of music: An introduction, 4th edn. Springer, Dordrecht
77. Romieu JB (1879) Nouvelle découverte des sons harmoniques graves. In: Roche E (ed) Notice sur les travaux de J.B. Romieu, Typographie et Lithographie de Boehm et Fils, Montpellier, pp 21–26
78. Rousseau JJ (1768) Dictionnaire de musique. Duchesne, Paris
79. Schneider H (1986) Jean-Philippe Rameaus Letzter Musiktraktat. Vérités également ignorées et intéressantes tirées du sein de la nature (1764). Steiner, Stuttgart
80. Schoenberg A (1983) Theory of harmony.Translated into English by Carter RE. University of California Press, Berkeley
81. Serravezza A (1996) Musica e scienza nell'età del positivismo. Il Mulino, Bologna
82. Sethares WA (2005) Tuning, timber, spectrum, scale, 2nd edn. Springer, Dordrecht
83. Shirlaw M (1917) The theory of harmony. Novello & Company, London
84. Sorge GA (1744) Anweisung zur Stimmung und Temperatur. Piscators Scrhriften, Hamburg
85. Strutt JW (1877-1878) The theory of sound (2 vols). Macmillan, London
86. Stumpf C (1883-1890) Tonpsychologie (2 vols). Hirzel, Leipzig
87. Tartini G (1754) Trattato di musica secondo la vera scienza dell'armonia. Stamperia del Seminario, Padua
88. Tartini G (1762) Regola del terzo suono. Original manuscript conserved in the British Museum, London
89. Tartini G (1767) De' principi dell'armonia musicale. Stamperia del Seminario, Padua
90. Tartini G (1767-1770) Scienza platonica fondata sul cerchio, URL https://www.discovertartini.eu/Scienza_platonica.pdf, archivio regionale di Capodistria. ms. 233: antica 128, inventario: 4-233; cc. 95. Accessed 18 September 2024
91. Tartini G (1977) Scienza platonica fondata sul, cerchio. Todeschini CA, CEDAM, Padua
92. Tenney J (1988) A history of consonances and dissonances. Excelsior Music Publishing Company, New York
93. Terhardt E (1984) The concept of musical consonance: A link between music and psychoacoustic. Music Percept 1(3):276–295
94. Turner SR (1977) The Ohm-Seebeck dispute, Hermann von Helmholtz, and the origins of physiological acoustics. The British Journal for the History of Science 10:1–24
95. Tymoczko D (2012) The generalized tonnetz. Journal of Music Theory 56(1):1–52
96. Various (2025) New Grove Dictionary of Music and Musicians. URL https://doi.org/10.1093/gmo/9781561592630.article.27643, accessed 1 January 2025
97. Verba C (2017) Music and the French Enlightenment. Rameau and the philosophes in dialogue. Oxford University Press, Oxford
98. Vieth G (1805) Ueber Combinationstöne, in Beziehung auf einige Streitschriften über sie zweier englischer Physiker, Th. Young und Jo. Gough. Annalen der Physik und Chemie 21(11):265–314
99. Weber W (1892) Über die Tartini'schen Töne. In: Königlichen Gesellschaft der Wissenschaften (ed) Wilhelm Weber's Werke (6 vols), vol 1, Springer, Berlin, pp 360–364

Index

D. Capecchi and G. Capecchi, *Joseph Sauveur and His Time: Music and Mathematicians*, https://doi.org/10.1007/978-3-032-04707-6

MIX
Papier aus verantwortungsvollen Quellen
Paper from responsible sources
FSC® C105338

If you have any concerns about our products,
you can contact us on
ProductSafety@springernature.com

In case Publisher is established outside the EU,
the EU authorized representative is:
Springer Nature Customer Service Center GmbH
Europaplatz 3, 69115 Heidelberg, Germany

Printed by Libri Plureos GmbH
in Hamburg, Germany